ELECTROMAGNETIC PHENOMENA IN COSMICAL PHYSICS

INTERNATIONAL ASTRONOMICAL UNION
SYMPOSIUM No. 6

HELD IN STOCKHOLM, AUGUST 1956

ELECTROMAGNETIC PHENOMENA
IN COSMICAL PHYSICS

EDITED BY

B. LEHNERT

Royal Institute of Technology, Stockholm

*Printed with
financial assistance from
U.N.E.S.C.O.*

CAMBRIDGE
AT THE UNIVERSITY PRESS
1958

CAMBRIDGE
UNIVERSITY PRESS

University Printing House, Cambridge CB2 8BS, United Kingdom

Cambridge University Press is part of the University of Cambridge.

It furthers the University's mission by disseminating knowledge in the pursuit of education, learning and research at the highest international levels of excellence.

www.cambridge.org
Information on this title: www.cambridge.org/9781316612859

First published 1958
First paperback edition 2016

A catalogue record for this publication is available from the British Library

ISBN 978-1-316-61285-9 Paperback

CONTENTS

vii

viii

PREFACE

An increasing interest in electromagnetic phenomena has characterized research in cosmical physics during the last decade. This development also reflects itself in an enlargement of the space devoted to electrodynamics in some recently held meetings. Two symposia on cosmical gas dynamics, at Paris in 1949 and at Cambridge in 1953, were organized by the International Astronomical Union (I.A.U.) in co-operation with the International Union of Theoretical and Applied Mechanics (I.U.T.A.M.). The programmes of these symposia included discussions on magneto-hydrodynamics, but the first meeting to be concentrated entirely on this subject was held at The Royal Society in London on the initiative of Sir Edward Bullard.

This forms the background of a subsequent symposium 'Electromagnetic Phenomena in Cosmical Physics' which was organized by the International Astronomical Union. It was held on 27–8 and 30–1 August 1956 at the Royal Institute of Technology in Stockholm, and on 29 August at the Stockholm Observatory in Saltsjöbaden. Financial support was given by I.A.U. ($3000), I.U.P.A.P. (Union of Pure and Applied Physics; $1500) and U.G.G.I. (Union of Geodesy and Geophysics; $750) from U.N.E.S.C.O. funds. The Swedish Government contributed 10,000 Sw. Crs. (about $2000) and further support was given by Telefonaktiebolaget L. M. Ericsson, Stockholm.

The organizing committee consisted of H. Alfvén (Stockholm), chairman; L. Block (Stockholm) and B. Lehnert (Stockholm), secretaries; H. W. Babcock (Pasadena), L. Biermann (Göttingen) and T. G. Cowling (Leeds). Together with the invitations a preliminary programme was sent out in advance by the organizing committee. A shortened version of this programme has been published in a circular of the I.A.U. in April 1956. Some weeks before the start of the symposium summaries of more than three-quarters of the contributions were distributed among the participants.

The following persons were present at the meetings:

Australia: J. H. Piddington (Sydney).

Belgium: P. Ledoux (Liège).

Finland: J. Tuominen (Helsinki).

France: Alice Daudin (Paris), J. F. Denisse (Paris), J. Heidmann (Paris), L. Leprince-Ringuet (Paris), E. E. Schatzmann (Paris).

Germany: L. Biermann (Göttingen), A. Ehmert (Weissenau), W. Fricke (Heidelberg), A. Schlüter (Göttingen), S. Temesváry (Göttingen).

Great Britain: P. M. S. Blackett (London), R. Hanbury Brown (Jodrell Bank, Manchester), O. Buneman (Cambridge), T. G. Cowling (Leeds), J. W. Dungey (Cambridge), H. Elliot (London), A. von Engel (Oxford), V. C. A. Ferraro (London), T. Gold (Greenwich Observatory), R. Hide (Harwell), F. D. Kahn (Manchester), R. Latham (London), A. C. B. Lovell (Jodrell Bank, Manchester), F. J. Lowes (S.M.R.E., Harpur Hill), D. Mestel (Leeds), R. S. Pease (Harwell), P. H. Roberts (A.W.R.E., Aldermaston), Pamela Rothwell (London), P. A. Sweet (London), R. J. Tayler (Harwell), W. B. Thompson (Harwell), P. C. Thonemann (Harwell).

India: V. Sarabhai (Ahmedabad), D. Venkatesan (Ahmedabad; temporarily in Stockholm).

Italy: Francesca Bachelet (Roma), G. Righini (Firenze).

Japan: Y. Fujita (Tokyo).

Netherlands: H. C. van de Hulst (Leiden).

Norway: Guro Gjellestad (Bergen), E. Jensen (Oslo), H. Trefall (Bergen).

Poland: K. Serkowski (Warsaw), W. Zonn (Warsaw).

Sweden: H. Alfvén (Stockholm), E. Åström (Stockholm), L. Block (Stockholm), E. Å. Brunberg (Stockholm), A. Dattner (Stockholm), D. Eckhartt (Stockholm), Aina Elvius (Saltsjöbaden), T. Elvius (Saltsjöbaden), N. Herlofson (Stockholm), G. Larsson-Leander (Saltsjöbaden), B. Lehnert (Stockholm), B. Lindblad (Saltsjöbaden), S. Lundquist (Stockholm), E. Lyttkens (Uppsala), Y. Öhman (Saltsjöbaden), A. Reiz (Lund), A. E. Sandström (Uppsala), W. Stoffregen (Uppsala).

U.S.A.: H. W. Babcock (Pasadena, Calif.), A. Baños (Los Angeles, Calif.), W. H. Bostick (Hoboken, N.J.), G. R. Burbidge (Pasadena, Calif.), L. Davis (Pasadena, Calif.), A. Deutsch (Pasadena, Calif.), S. E. Forbush (Washington, D.C.), E. O. Hulburt (Washington, D.C.), A. Kantrowitz (Everett, Mass.), G. F. W. Mulders (U.S. Office of Naval Research, London), E. Parker (Chicago, Ill.), K. Prendergast (Yerkes Observatory, Wis.), J. A. Simpson (Chicago, Ill.), S. F. Singer (College Park, Md.), L. Spitzer (Princeton, N.J.), W. F. G. Swann (Swarthmore, Penn.).

U.S.S.R.: L. A. Artsimovich (Moscow), I. N. Golovin (Moscow), A. J. Kipper (Tartu), E. R. Mustel (Moscow), A. B. Severny (Simeis), J. P. Terletzsky (Moscow).

The discussions were confined to such questions as could be investigated by theory, experiments and observations with a reasonable hope of success. Further considerations on the purpose and formation of the symposium programme are given by H. Alfvén in the 'Opening Address' of this volume. The main subjects of the sessions as well as the titles of the presented papers are given in the table of contents.

A chairman and a secretary were elected for each session. The arrangements were as follows:

Monday, 27 August
  Morning. Chairman: W. F. G. SWANN, Secretary: R. HIDE. Opening Address and Papers 1–4.
  Afternoon. Chairman: J. H. PIDDINGTON; Secretary, R. HIDE. Papers 5–7.

Tuesday, 28 August
  Morning. Chairman: P. M. S. BLACKETT; Secretary, B. LEHNERT. Papers 8–11.
  Afternoon. Chairman, G. RIGHINI; Secretary, O. BUNEMAN. Papers 12–17.

Wednesday, 29 August
  Chairman: P. LEDOUX; Secretary, E. PARKER. Papers 18–25.

Thursday, 30 August
  Morning. Chairman: V. C. A. FERRARO; Secretary: A. SCHLÜTER. Papers 26–30.
  Afternoon. Chairman: L. SPITZER; Secretary: E. JENSEN. Papers 31–35.

Friday, 31 August
  Morning. Chairman: L. LEPRINCE-RINGUET; Secretary: R. TAYLER. Papers 36–42.
  Afternoon. Chairman: A. B. SEVERNY; Secretary: J. HEIDMANN. Papers 43–46.

In addition, notes were taken, a tape-recorder was used and the comments were written down on special forms by the speakers taking part in the discussions.

On 27 August a tour was arranged around the Department of Electronics. The present work on cosmic ray intensity variations, cosmic ray orbits, electron orbits, plasma-resonance, model experiments on the aurorae and magnetic storms and magneto-hydrodynamic experiments with mercury was shown. Further, the Stockholm Observatory in Saltsjöbaden was visited on Wednesday, 29 August. Professor B. Lindblad gave a survey of the history of the observatory, and the instruments and some observational results were demonstrated to the visitors.

On Saturday, 1 September, and Monday, 3 September, some papers on high current discharges were added to the original programme. Many of the participants were still in Stockholm and took the opportunity to listen to these reports. They are included in Part VI. Part VII contains papers connected with subjects discussed at the symposium. They were not read at the conference, partly because of lack of time, and partly because some of the authors were not able to join the meetings.

Further research in the field of cosmical electrodynamics was certainly stimulated by a great number of interesting discussions. The motion of magnetic-field lines in an electric conductor was considered in connexion with a hydromagnetic dynamo and the behaviour of a magnetic field at a neutral point. Final conclusions were not reached, but the discussions clearly showed the many changing aspects of magneto-hydrodynamics and the danger of making generalizations. The importance of pressure-balanced magnetic fields was stressed in connexion with the magneto-hydrostatic equilibrium of cosmic gas masses.

A rigorous theory on magneto-turbulence has not yet been established. This makes it somewhat difficult to assess the importance of turbulence for phenomena in solar physics, interplanetary space and cosmic radiation. There has earlier been some confusion about the electrical conductivity in a magnetic field. The difficulties now seem to have been overcome and the 'friction coefficient' between the ion and electron fluids in a fully ionized gas is accepted as a fruitful approach.

Experiments are valuable tools in magneto-hydrodynamic research, and there are even interesting experimental results which are not predicted by theory or observations, e.g. the plasmoids. Care is necessary, however, when an extrapolation is made of experimental results to cosmical physics.

In the session on solar electrodynamics interesting attempts were made to explain the solar flares as a pinch effect, as the form of a neutral point discharge or, finally, as the result of collisions between an oscillating plasma and a neutral gas. Some new aspects were presented on the origin and structure of sunspots.

The problem of magnetic variable stars is still open for discussion. The observed field variations may be due to magneto-hydrodynamic oscillations, to the motion of the star as a rigid rotator, or to both effects.

Too little is still known of the physics of interplanetary space. Therefore the connexion between solar phenomena and associated terrestrial phenomena such as magnetic storms and aurorae is rather speculative, and entirely different theories can exist side by side in this field.

Finally, cosmic radiation and its time variations can be explained in many ways. One important question is how the obtained data are affected by observational methods.

The Unions express their gratitude to U.N.E.S.C.O. for the financial help to this symposium. A publication grant for this volume is gratefully acknowledged. I want to express my sincere thanks to Mrs B. Törnell for invaluable help during the symposium and for skilled assistance with the manuscripts and discussions of this volume. I am also indebted to my father Prof. E. Lehnert for valuable help with the manuscripts.

B. LEHNERT

ROYAL INSTITUTE OF TECHNOLOGY

STOCKHOLM
April 1957

OPENING ADDRESS

BY H. ALFVÉN

Royal Institute of Technology, Stockholm

On behalf of the organizing committee and of the Royal Institute of Technology I wish you welcome to this symposium. I need not stress the importance of the subject which we are going to discuss—as participants in this symposium all of us are aware of it. You also know very well how this field has been opened by Birkeland and Störmer, by Chapman, Cowling and Ferraro, by Hale, Swann, and many others. However, it is not until the last decade that the interest in this subject has become more general. Many different research groups are active but their ideas differ very much, which gives a good reason for meeting and exchanging ideas.

The first meeting in this field was called by Professor Bullard in London last year. The present one is the second meeting, and considering the rapid development I expect that there will be many similar meetings in the near future. Moreover, our field now tends to become important even for the peaceful use of thermonuclear energy, which adds technological interest to the purely scientific interest.

Concerning the programme it is obvious that a full coverage of the title 'Electromagnetic phenomena in cosmical physics' should also have included ionosphere physics, radioastronomy, the problems concerning the origin of cosmic radiation, and perhaps also the origin of the solar system. However, these branches have been excluded from the programme for different reasons. For example, radioastronomy was discussed only a year ago in Manchester. Even so there remain three important fields for the symposium:

(1) *Fundamental magneto-hydrodynamics* including currents in gases in the presence of a magnetic field. Although the emphasis has been and still is lying on the theoretical investigations, important experimental investigations will also be reported.

(2) *Stellar magnetism* is a rapidly developing branch of astronomy, which directs the interest to the importance of electromagnetism to stellar phenomena.

(3) *Electromagnetic phenomena within the solar system* may be an appropriate subheading including *solar electrodynamics, magnetic storms and aurorae,* and

cosmic ray intensity variations. It is an important purpose of the symposium to stimulate the fusion of these different fields into a common field of research, which at the same time gives us an example of the application of fundamental magneto-hydrodynamics to astrophysics.

The symposium will also include a visit to the Stockholm Observatory, Saltsjöbaden, and a demonstration of laboratory experiments with cosmical applications at the Royal Institute of Technology.

PART I

MAGNETO-HYDRODYNAMICS

A. THEORY
B. EXPERIMENTS
C. IONIZED GAS IN A MAGNETIC FIELD

A. THEORY

PAPER I

MAGNETIC FIELDS IN ASTROPHYSICS

H. C. van de HULST

University Observatory, Leiden, Netherlands

This symposium is held under the auspices of the International Astronomical Union at the department of electronics of a technical university. This combination of astronomy and technical electronics has already become so familiar, that it hardly strikes us as peculiar. This shows how strongly the case for electric and magnetic phenomena in a wide variety of astrophysical problems has been proven. However, all of this is comparatively recent history. One glance at an older epoch may illustrate this point and, perhaps, may help us to take fewer things for granted during our discussions.

Just fifty years ago, in 1906, Agnes Clerke[1] wrote a modern textbook, *Problems in Astrophysics*. In it the term magnetic field occurs only in one context. She devotes three pages to a problem that has an important place also in this symposium, namely the role of the sunspots in causing the terrestrial magnetic storms. Her discussion is concluded by the words: 'The machinery by which electromagnetic impulses are propagated from the sun to the earth, completely evades scrutiny. Sundry conjectures on the subject have been hazarded, but none of them rests on any sure basis. What we know about modes of communication is chiefly negative.' The discussions at this symposium will show to what extent the situation has improved.

I. SOLAR PHENOMENA

Two years after Miss Clerke's book, Hale discovered the magnetic fields of sunspots and again five years later, he believed he had found the general magnetic field of the sun. Knowledge about the sunspot fields has rapidly increased since that time, but the general field of the sun has been under constant debate. Only recent techniques allow the measurement of fields as low as a few gauss over the entire disc. Even then, caution is required in the interpretation.

Both types of investigations, of the spot-fields and of the general solar field, were prompted by superficial arguments: the apparent vortex structure in the Hα photographs in a spot-region and the striking pattern of the streamers on corona photographs, which look very much like magnetic lines of force. These arguments are suggestive but they ought to be justified by a later, more thorough investigation. If I am right, the precise interpretation of both phenomena is still discussed; we hope to hear more about them during this symposium. However, they form only a few of a multitude of phenomena known by the collective term *solar activity*.

Solar activity includes everything that is changing on the sun (with the traditional exception of the moving elements in the convective zone, the (consequent) granulation in the photosphere, and the (consequent) existence of a chromosphere with spicules). The observed changes included in the term solar activity have very different time scales. One extreme is the deep-seated cause of the 22-year cycle, which shows beautifully in the migration of the sunspot and prominence zones and in the reversal of dominant magnetic polarity at each half-cycle. The other extreme is formed by the storm bursts in the radio emission of the sun, which last for a fraction of a second. Between these extremes lie, for instance, the sunspots, which appear, develop and vanish in a month or so, and also the flares with all their associated effects that take about fifteen minutes.

It is not difficult to assign a rough order of cause and effect to these events. Roughly, the deep-seated and long-lived phenomena are cause, the others effect. If we wish to compare with the waves of the sea, the 22-year solar cycle corresponds to the fundamental tidal wave, and the solar radio bursts to the splashes and foam which we enjoy at the beach. It is, of course, an enormous task to track the chain of cause and effect in detail, but this is one of the jobs we are here for.

2. HYDROMAGNETICS

I have not started, as Dr Bullard did on an earlier occasion [2] by giving a definition of magneto-hydrodynamics (which I shall call hydromagnetics): '*the study of the motion of fluids in the presence of magnetic fields*'. The reason for mentioning this topic in the second place, in spite of the fact that the symposium programme starts with it, lies in the relation between observations and theory.

Most of us would tend to call this a mainly theoretical symposium, but we should do well to consider what situation would exist if, for some unfortunate reason, all astrophysical and geophysical observations had been impossible. We then would have calculated by theory the possible existence of gas spheres in equilibrium like the sun; we might have a hunch of the solar corona but not of its beautiful streamers. In all probability, we would not even have guessed the existence of sunspots, nor of flares, nor of cosmic rays. Consequently this symposium would have been a good exercise of applied mathematics, illustrated by some laboratory experiments. It is only fair to say that, conversely, the absence of theoretical notions would have been even more disastrous: the observations then would have been an incoherent, nonsensical set of records.

This fiction story demonstrates that we do not really hope to grasp completely the complexity of the actual events in our theories. What we do hope is that our theories will help us to make sense of the observations, that is, to see which phenomena have a common cause, to distinguish certain chains of predictable events, and also to conceive of new observations, which may solve a crucial question in the interpretation of the data.

The theoretical investigations that are most fruitful deal with *models*. A model is a fictitious situation or experiment in which (unlike the real situation) all conditions are known. In discussing the real events, there may be many good reasons for giving a tentative or uncertain interpretation and for defending conflicting theories. In discussing model situations, however, there is no excuse for conflicting theories or obscure answers, for a precise question should (in the long run) receive a precise answer.

Just what are the model situations relevant to the topics of our symposium? In looking over the huge collection of papers that have appeared in the last seven years or so, I find it impossible to give even a brief summary of what has been accomplished [3, 4]. But a brief classification of the problems, with some comments, may be useful (Table 1).

Table 1. *Schematic classification of the theoretical problems*

I Fluid 'Hydromagnetics'	II Ionized gas 'Plasma dynamics'
A. Basic equations	A. Basic equations
B. Problems with external field	B. Problems with external field
C. Problems with self-field	C. Problems with self-field

The fertile field of problems in column I has been opened by Alfvén with his studies of magneto-hydrodynamics, or, by a shorter term that has become quite popular, hydromagnetics. His main applications [5] were to astrophysical problems, where the conductor is an ionized gas, so that the term hydromagnetics might be assumed to include both columns I and II. However, the problems in column II have a longer standing. Appleton's magneto-ionic theory, if extended to lower frequencies to include the motion of the positive ions, is one of the central topics of IIB.

About five years ago the question whether the theories of I were applicable to the gases in II seemed a quite difficult one, especially as it was hard to tell what conductivity σ to use: the full one along the magnetic lines of force, or the reduced one across the lines of force. By the work of Schlüter, Cowling, Piddington and others, this situation has now been cleared up. At the same time most authors have become more conscious of the various pitfalls. A theory is usually developed now either for I (fluids) or for II (ionized gas). This has caused a shift in the terminology: the term hydromagnetics (or magneto-hydrodynamics) is now very often reserved for column I.

Another important distinction is between the problems in lines B and C. In B an external magnetic field is given, the magnitude of which does not have to be questioned. If, moreover, it is supposed that the additional magnetic field caused by the induction currents in the investigated fluid or gas is smaller by an order of magnitude than the given field, the dynamic equations can be linearized. This was not needed in Alfvén's original presentation of the hydromagnetic waves, because of the particular orientation of the fields. But in all later extensions about waves with arbitrary polarization running through a medium in an arbitrary direction, the linearization is essential.

The problems in C, in which both the character and the magnitude of the field are directly linked with the state of motion, are far more difficult. Hydromagnetic turbulence and the famous dynamo problem [6] belong to this class. They are basically non-linear and I shall not try a review. Certainly the problems in IIC are the hardest and have hardly been approached, in spite of their evident importance for the interstellar gas, stellar magnetic fields, etc. On the other hand the problems in IB are simplest; it is Dr Alfvén's merit to have found the right place to start the explorations.

3. STELLAR AND INTERSTELLAR FIELDS

It should not be inferred from the introduction that the sun presents more exciting, or more important problems in the field of our symposium than astrophysics at large. In one respect, solar studies are unique: they show us so much detail that we lose at once our belief in simplified theories. But the sun is in all respects an average star, so that by sheer logic we may assume that anything we observe on the sun will appear in a more pronounced fashion in some other type of star.

Stellar magnetic fields have indeed been observed and offer a number of spectacular problems. Dr Babcock, the discoverer and almost exclusive author of the observations on this topic is present here, so there is no need for me to anticipate his lecture.

Interstellar fields have not so simple a history; perhaps I may relate a personal recollection. Oort and Burgers had before 1945 studied some problems in the aerodynamics of the interstellar gas. It seemed worthwhile to pursue these problems. In 1949 an international symposium on these matters was organized by I.A.U. and I.U.T.A.M. in Paris[7]. At that time magnetic fields had hardly been mentioned in this connexion and Oort was worried that only one or two participants would be able to judge their possible importance. It turned out, however, that almost everybody picked up this point and hydromagnetic turbulence formed one of the central topics of the discussions. At that time we were even unaware of work in the same directions that had started in other parts of the world.

The interstellar problems have continued to attract attention in connexion with the observable details of interstellar clouds, with the expansion of nova shells, and with the origin and acceleration of cosmic rays. Several conferences[8, 9] since that time have been devoted exclusively to one of these objects. The organizers of the present symposium have decided to place the emphasis on problems of the sun and the solar system and on stellar magnetism. Only a few papers remind us of the vast and interesting field outside these main topics.

4. INTERPLANETARY SPACE

Seven papers in the programme deal with problems presented by interplanetary space. Interest in this subject has sprung up recently from many sides. Before that time, we knew that the earth was at one astronomical unit from the sun with above it the ionosphere, which in practice could be studied only to the reflexion level in the F2 layer, say, at 250 km. Theorists had ventured to about 1000 or 2000 km and there were also some ideas about a ring current at the order of an earth radius away from the surface, i.e. at a distance still less than 0·0001 a.u. At the other end of the line we knew that the normal solar corona extended to roughly five solar radii, or 3,500,000 km = 0·02 a.u. Occasional coronal streamers extended slightly further. For the rest, i.e. for 98 % of the distance, interplanetary space seemed sufficiently empty not to worry about it. This has now changed. A list of the new lines from which evidence has come, or may still come, may be arranged as follows:

Near the earth

(*a*) Whistlers, by the current theory, present evidence of a density of the order of 10^3 electrons/cm^3 about one earth radius above the surface.

(*b*) Observation of radio sources or moon echoes through the ionosphere at frequencies of the order of 15–20 Mc/s has made it possible to obtain, in principle, some information about the entire ionosphere, including the top half.

(*c*) The suggestion has been made that the counterglow ('Gegenschein') of the zodiacal light should be explained by radiation from a gas tail of the earth, extending to about ten earth radii.

(*d*) Artificial satellites are now scheduled to go up to 500 km but may in a later stage of development give valuable information about the interplanetary medium.

Near the sun

(*e*) Red and infra-red eclipse photometry from an aircraft has shown the corona to extend to at least eighteen solar radii.

(*f*) Radio observations of the Crab nebula at the time when the sun passes it have consistently shown that the radio waves are scattered in the outer corona. This effect becomes noticeable at twenty solar radii.

(*g*) New attention has been given to theories of the escape of electrons and protons (evaporation) from the outer parts of the corona.

Between sun and earth

(*h*) Observations of comet tails indicate the action of corpuscular streams with densities of the order of 10^3 particles/cm^3.

(*i*) Older measurements of the polarization of the zodiacal light have been confirmed. Unless extreme assumptions about the polarization of light scattered by interplanetary dust are made, these measurements prove the existence of 200 to 800 free electrons/cm^3 at one a.u. from the sun.

(*j*) Observations of cosmic rays strongly indicate that the interplanetary medium affects the intensity and time of arrival both of the galactic cosmic rays and of the cosmic rays from solar flares.

It would be inadequate, in the present context, to attempt a more complete discussion. Reference may be made to various review articles [10, 11] and to the reports of the I.A.U. General Assembly at Dublin [12] for details. It would seem that the data provided by points (*h*) and (*i*) in the list are sufficiently certain as a basis of the discussion of the effects (*j*), which will form a topic of discussion at this symposium.

REFERENCES

[1] Clerke, A. M. *Problems in Astrophysics* (London: A. and C. Black, 1903).
[2] Bullard, E. and others. *Proc. Roy. Soc.* A, **233**, 289, 1955.
[3] Lundquist, S. *Arkiv för Fysik*, **5**, 297, 1952.
[4] Spitzer, L. *Physics of Fully Ionized Gases* (Interscience Publishers, Inc., New York, 1956).
[5] Alfvén, H. *Cosmical Electrodynamics* (Oxford University Press, 1950).
[6] Elsasser, W. M. *Rev. Mod. Phys.* **28**, 135, 1956.
[7] *Problems of Cosmical Aerodynamics* (ed. J. M. Burgers and H. C. van de Hulst), Proceedings Paris Symposium 1949 (Central Air Documents Office, Dayton, Ohio, 1951).
[8] *Gas Dynamics of Cosmic Clouds* (ed. H. C. van de Hulst and J. M. Burgers), Cambridge Symposium 1953, I.A.U. Symposium Report Number 2 (North Holland Publishing Company, Amsterdam, 1955).
[9] Guanajuato conference on the origin of cosmic rays, September 1955; no report published.
[10] van de Hulst, H. C. *The Solar System* (ed. G. P. Kuiper), vol. I, *The Sun*, Univ. Chicago Press, ch. 5, 'The Chromosphere and the Corona', 1953.
[11] van de Hulst, H. C. *The Solar System* (ed. G. P. Kuiper), vol. IV, *Minor Planets, Meteorites, Comets and Origin of Solar System*, ch. II, 'The zodiacal light' (in preparation).
[12] *I.A.U. Transactions*, vol. 9 (Dublin), (see reports of commissions 13, 15 and 22 *a*), (Cambridge University Press, 1957).

Discussion

Biermann: I would like to make a remark concerning one special point of Dr van de Hulst's survey. The long time scale of the cycle of solar activity (twenty-two years) does not necessarily indicate that it should be regarded as directly connected with the cause of all the phenomena of shorter time scale. Since the subject will come up again on Thursday, I only wish to point out that there are reasons to believe that the turbulence of the hydrogen convection zone—which is to be regarded as deep for this purpose—is the main cause of most of the phenomena such as the solar activity. This was discussed at Dublin at the conference on turbulence in stellar atmospheres at some length (*I.A.U. Transactions*, vol. 9 (1957)). As I am going to explain in more detail on Thursday, meridional circulations should be present in the convective zone, the properties of which are connected with the structure of the turbulence. Its period is necessarily long compared with the characteristic time scale of turbulence elements even at larger depths.

van de Hulst: I look forward to the discussion on those points on Thursday.

Swann: One can think of several phenomena associated with a body of the size of a star and which have periods associated with that size. Thus, there are the periods of mechanical vibration, the period of electrical oscillation depending upon equivalent self-inductance and capacity and so on. I should like to ask Dr van de Hulst whether he can cite to us phenomena of these or analogous kinds which can have periods of the order of twenty-two years. Does my question make sense?

van de Hulst: The question makes sense and I hope someone will answer.

Gold: None of the phenomena the chairman mentioned has periods of the right order. The discussions I am aware of seemed to show that none other than some form of torsional magnetically coupled oscillations could be invoked. For such oscillations, however, the quantities involved can be rather arbitrarily chosen to fit the period.

Alfvén: Van de Hulst has remarked that a very important question is the density of interplanetary space. This topic will not be the subject of any particular paper during this symposium, but it will certainly be of importance to many of the problems to be considered here. I think it would be interesting to discuss it a little now although we shall probably come back to it many times later. Van de Hulst said that the value of the density is derived under the assumption that only half of the polarization is due to grains and the other half of the polarization is due to the electrons. Of course there are very seldom two independent effects just about equal so one may hesitate a little to accept this value.

Further, the value of the density from the whistlers seems not to be very definite. The whistlers, namely, are measured in a very low-frequency range and the theory which gives about the same density as that from polarization measurements is based on the assumption that only the motion of electrons is taken into account. But the whistlers occur at such low frequencies that it is possible that even ions interfere, and that would modify the values obtained. In this case there seems to be a possibility that we have a much lower value of density.

The emission of beams in the interplanetary space and quite a few other phenomena which we will discuss later indicate a much lower density. Different estimates point in the direction that a value of one, or even a lower value, would be in better agreement with these phenomena.

Dungey: I agree with Professor Alfvén on the importance of determining the interplanetary electron density from the zodiacal light. The whistler value refers to particles which are trapped by the earth's field and need not be the same as the interplanetary value. There should be a permanent geomagnetic effect due to the interplanetary ionized gas.

van de Hulst: I wish to answer this question about the zodiacal light. The ranges given in the paper were obtained with estimates of the dust polarization that I consider extreme. About 500–600 electrons/cm^3 are obtained if the dust polarization is zero. If it is positive, then the electron density gets lower; if it is negative the electron density gets higher. We can at most assume 15 % dust polarization under these angles on the basis of theoretical computations. So I feel fairly sure that the range of estimates is correct. But if somebody assumes 25 % instead of 15 % dust polarization, then of course the lower limit may go down from 200 electrons/cm³ to zero.*

Swann: If it should be trapping of electrons by an external magnetic field that is responsible for the somewhat higher values, we have a situation where there will be a rate of disappearance in that trapped region, by recombination, and a rate of supply, in which the balance determines the number present.

Ferraro: Like Professor Alfvén I should be happier if the electron density of the interplanetary gas were lower than the value of 200–800 electrons/cm³ suggested by van de Hulst. Corpuscular theories of magnetic storms indicate that the density of solar streams of corpuscles is of the order of 1–100 electrons/cm³. It is difficult to see how the streams could push their way through this denser gas.

van de Hulst: The data I have presented do not indicate that the gas should be stationary. It may be moving away from the sun continuously.

Ferraro: If the interplanetary gas were streaming towards the earth we should expect some sort of geomagnetic effect to be made manifest at the earth's surface.

Dungey: I think that you should certainly get the Chapman-Ferraro situation all the time; this is one possible happening. In such a situation the field is confined by the surrounding gas stream and is formed into a cavity surrounding the earth. Its dimensions can be estimated and depend on the electron density. The dimension of the cavity is proportional to the inverse one-sixth power of the density and with Siedentopf's analysis the radius of the cavity is about ten earth radii. If you put the density down by a factor of 1000 you come up to thirty radii which is still quite a small thing.

Alfvén: If you eject a piece of matter from the sun and suppose that it moves radially outwards with constant velocity, then the density will decrease as $1/r^2$. If you assume that this emission is just below the visibility in the corona, i.e. if you assume that the density at a distance of five solar radii just equals the electron density in the corona, then the value you get for the density in the beam at the earth's orbit is less than 10 electrons/cm³. For any assumption about the radial emission of beams which actually pass through the corona we cannot suggest a density inside the beam at the earth's orbit which is above this figure.

Simpson: The question of temperatures for the ionized gas in interplanetary space is also important for our understanding of the possible range of ion densities, which could exist in space. Chapman has recently calculated that, even at the distance of one a.u., the equivalent temperature may be remarkably high (about 10^5 °K). Although we know very little experimentally about these temperatures, perhaps you would be willing to discuss this question and its relation to the expected range of gas densities in the interplanetary medium.

van de Hulst: In the outer corona we may speak of a temperature of the order

* 'Les particules solides dans les astres', Symposium Report, *Mém. Soc. Roy. Sci. Liège*, 15 (1955), H. C. van de Hulst, 'On the polarization of the zodiacal light', p. 89.

of 10^6 °K. But I do not agree with Chapman's assumptions of a stationary gas with a well defined temperature. At one a.u. from the sun the mean free path is not small and it is difficult to talk about any temperature at all.

Parker: Chapman (Smithsonian Contr. to *Astrophysics*, **2**, 1, 1957), using the thermal conductivity of a tenuous ionized gas, and assuming that the interplanetary medium is in hydrostatic equilibrium in the solar gravitational field, has calculated that $n_e = 500/cm^3$ and $T = 2 \times 10^5$ °K at the orbit of the earth. To attack his results one must probably argue that there are sufficiently strong interplanetary fields to reduce the thermal conductivity.

MAGNETO-HYDRODYNAMIC WAVES IN COMPRESSIBLE FLUIDS WITH FINITE VISCOSITY AND HEAT CONDUCTIVITY*

ALFREDO BAÑOS, JR.

Department of Physics, University of California, Los Angeles, U.S.A.

ABSTRACT

The general theory of magneto-hydrodynamic waves in an ideal conducting fluid embedded in a uniform field of magnetic induction, and the application of the theory to the systematic analysis of the various modes of propagation in incompressible and compressible fluids have been presented by the author in two earlier papers [1,2]. In these papers, however, no effort was made to include the thermodynamics of the situation, which amounts to the tacit assumption that the fluid is of zero heat conductivity. In this case the resulting modes are of two kinds: isothermal (*v*-modes) and adiabatic (*p*-modes).

In this paper we first establish the conservation laws of momentum and energy for a (macroscopic) compressible fluid with finite viscosity and finite thermal and electrical conductivities, which is embedded in a uniform field of magnetic induction, and we then derive quite generally the exact (non-linearized) equation governing the distribution of temperature in such a fluid. Next, making use of the linearized magneto-hydrodynamic wave equation in the fluid velocity, combined with the resulting heat diffusion equation and with the equation of state of the fluid, and applying the mathematical techniques developed earlier, we obtain a higher order partial differential equation in the fluid temperature from which ensue all the temperature modes.

In particular, we examine in detail the behavior of plane homogeneous waves, and it is shown that a compressible fluid with the indicated properties sustains altogether six different modes, two of which are pure shear modes, devoid of density, pressure, and hence temperature fluctuations (*v*-modes), while the remaining four are shear-compression waves accompanied necessarily by density, pressure, and temperature fluctuations (*p*-modes). The two shear modes, which are isothermal, comprise a slightly attenuated Alfvén wave, and a highly attenuated viscous mode, sometimes referred to as a vorticity mode. The four shear-compression modes have in general very complex properties, but in the low frequency and low heat conductivity case they are easily identified as (1) a modified (adiabatic) sound wave slightly attenuated;

* This research was supported by the United States Air Force, through the Office of Scientific Research of the Air Research and Development Command.

(2) a slightly attenuated modified Alfvén p-wave; (3) a highly attenuated viscous wave; and (4) a highly attenuated thermal wave governed in the main by the thermal properties of the medium.

I. INTRODUCTION

The underlying fundamental notions in the theory of magneto-hydro-dynamic waves in incompressible fluids were originally due to Alfvén and his co-workers [3], but there had been lacking for some time a systematic analysis of the linearized, unbounded media, and boundary value problems in the field of magneto-hydrodynamic waves in incompressible and compressible fluids. To this end we undertook to give, in two earlier papers [1, 2], hereinafter to be referred to as I and II respectively, such a systematic study. The first paper deals mainly with the general theory of plane homogeneous waves and of time harmonic cylindrical waves propagating in a homogeneous and isotropic conducting fluid of infinite extent embedded in a uniform field of magnetic induction. The medium is assumed to consist of an ideal fluid devoid of viscosity and expansive friction, which is characterized (in rationalized mks units) by the rigorously constant macroscopic parameters μ, ϵ, and σ, where $\mu\epsilon = c^{-2}$ and σ is the (ohmic) conductivity. The second paper deals with the application of the general theory to the determination of the modes of propagation and to the computation of the corresponding propagation constants in incompressible and compressible fluids.

However, in these two papers no effort was made to include the thermodynamics of the situation, which amounts to the tacit assumption that the fluid is of zero heat conductivity. In this case the resulting modes are of two kinds: isothermal (v-modes) and adiabatic (p-modes). In this paper we continue with the purely macroscopic approach, for we believe that the results obtained are of considerable value and may serve as a guide to the more complicated problems in which the macroscopic approach is no longer tenable. The medium is now assumed to be a conducting compressible fluid endowed with finite viscosity and heat conductivity, embedded in a uniform field of magnetic induction.

First, we examine anew the conservation laws of momentum and energy and we then derive quite generally the exact (non-linearized) equation governing the distribution of temperature in such a fluid. Next, making use of the linearized magneto-hydrodynamic wave equation in the fluid velocity, combined with the linearized form of the heat diffusion equation and with the equation of state of the fluid, and applying the mathematical

16

techniques developed in II, we obtain a higher order partial differential equation in the fluid temperature from which ensue all the temperature modes.

We examine, in particular, the structure of plane homogeneous waves in a conducting compressible fluid with finite viscosity and heat conductivity, and it is shown that a fluid with the indicated properties sustains altogether six different modes, two of which are pure *shear* waves, devoid of density, pressure, and temperature fluctuations (*v*-modes), while the remaining four are *shear-compression* modes which of necessity are accompanied by density, pressure, and temperature fluctuations (*p*-modes).

The two shear modes, which are isothermal, consist of a slightly attenuated Alfvén wave and a highly attenuated viscous wave, sometimes referred to as a vorticity mode. The four shear-compression modes have in general very·complex properties, but in the low frequency and low heat conductivity case they are readily identified as: (1) a slightly attenuated modified (adiabatic) sound wave; (2) a slightly attenuated modified Alfvén *p*-wave; (3) a highly attenuated viscous wave, and (4) a highly attenuated temperature wave governed in the main by the thermal properties of the medium.

2. CONSERVATION LAWS

In order to establish the equation governing the distribution of temperature in an unbounded magneto-hydrodynamic field we need to examine first the conservation laws of momentum and energy as they apply to a rigid volume V within a bounding surface S rigorously fixed in the observer's inertial frame of reference. The heat diffusion equation then ensues quite generally by combining the two conservation laws as indicated below.

Conservation of momentum

The law of conservation of momentum states simply that the time rate of change of the total mechanical plus electromagnetic momentum contained within the fixed volume V is equal to the mechanical force acting across the bounding surface on the fluid contained within the volume, plus the *influx* of both electromagnetic and mechanical momentum across the surface S. Expressed in tensor notation the law becomes

$$(d/dt) \int_V (\rho v_i + g_i) \, d\tau = \int_S P_{in} \, da + \int_S T_{in} \, da - \int_S (\rho v_i) \, v_n \, da, \qquad (1)$$

in which the subscript n refers to the outward normal. In the volume integral ρv_i denotes the mechanical momentum density and $g_i = \mu \epsilon S_i$ the electromagnetic momentum density.

The first surface integral on the right of (1) denotes the total mechanical force acting on the fluid contained within the bounding surface as deduced from the mechanical stress tensor [4]

$$P_{ik} = -\left(p + \tfrac{2}{3}\bar{\mu}\frac{\partial v_\alpha}{\partial x_\alpha}\right)\delta_{ik} + \bar{\mu}\left(\frac{\partial v_k}{\partial x_i} + \frac{\partial v_i}{\partial x_k}\right), \tag{2}$$

wherein p is the pressure and $\bar{\mu}$ the coefficient of viscosity, and whose (tensor) divergence leads to the familiar Stokes–Navier equation. The second surface integral represents the rate at which electromagnetic momentum is flowing into the volume and is computed in terms of the *total* Maxwell's electromagnetic stress tensor, Eq. (I-8), which in the present notation becomes

$$T_{ik} = \epsilon(e_i e_k - \tfrac{1}{2}e^2\delta_{ik}) + \mu(H_i H_k - \tfrac{1}{2}H^2\delta_{ik}), \tag{3}$$

and in which it is recalled $\mu\epsilon = c^{-2}$. Finally, the third surface integral on the right of (1) represents merely the influx of mechanical momentum transported across the surface S by the moving fluid.

To obtain from (1) the differential form of the law of conservation of momentum we first convert all four integrals into simple volume integrals by transposing under the sign of integration the time derivative acting on the volume integral and by applying the (tensor) divergence theorem to the remaining surface integrals. Then, making use of the law of conservation of mass (equation of continuity)

$$\partial\rho/\partial t + \partial(\rho v_\alpha)/\partial x_\alpha = 0, \tag{4}$$

we obtain from (1) the differential form

$$\rho(dv_i/dt) + \partial g_i/\partial t = \partial P_{i\alpha}/\partial x_\alpha + \partial T_{i\alpha}/\partial x_\alpha. \tag{5}$$

Finally, introducing into (5) the Lorentz force density of electromagnetic origin which, according to Eq. (I-10), can be written in the form

$$f_i = \partial T_{i\alpha}/\partial x_\alpha - \partial g_i/\partial t, \tag{6}$$

we obtain the Eulerian equations of motion,

$$\rho(dv_i/dt) = f_i + \partial P_{i\alpha}/\partial x_\alpha, \tag{7}$$

for a compressible fluid with finite viscosity.

Conservation of energy

The law of conservation of energy states in the present instance that the time rate of change of the *total* (kinetic plus internal plus electromagnetic) energy stored within the fixed volume V is equal to the sum of three terms: the rate of doing work of the mechanical forces acting on the fluid within the surface S, the influx of kinetic plus internal energy transported across the bounding surface by the moving fluid, and the influx of heat plus electromagnetic energy across the surface S. Expressed in tensor notation the law becomes

$$(d/dt) \int_V [\tfrac{1}{2}\rho v^2 + \rho U + (\tfrac{1}{2}\epsilon e^2 + \mu H^2)] \, d\tau = \int_S v_\alpha P_{\alpha n} \, da$$
$$- \int_S (\tfrac{1}{2}\rho v^2 + \rho U) \, v_n \, da - \int_S (q_n + S_n) \, da, \qquad (8)$$

in which U denotes the intrinsic internal energy of the fluid, q_i the heat flow vector, and S_i the familiar Poynting's vector. Once again the subscript n refers to the outward normal.

To obtain the differential form of the law of conservation of energy we proceed as before by transforming every integral in (8) into a simple volume integral. Thus, applying the divergence theorem to the surface integrals and again making use of the equation of continuity (4), we obtain the law in the form

$$\rho v_\alpha \frac{dv_\alpha}{dt} + \rho \frac{dU}{dt} + \frac{\partial}{\partial t} (\tfrac{1}{2}\epsilon e^2 + \tfrac{1}{2}\mu H^2) = \frac{\partial(v_\alpha P_{\alpha\beta})}{\partial x_\beta} - \frac{\partial q_\alpha}{\partial x_\alpha} - \frac{\partial S_\alpha}{\partial x_\alpha}. \qquad (9)$$

To reduce this equation further we note from (7) that

$$\rho v_\alpha (dv_\alpha/dt) = f_\alpha v_\alpha + v_\alpha (\partial P_{\alpha\beta}/\partial x_\beta) \qquad (10)$$

and we recall that, according to Eq. (I-7), we have in the present notation

$$f_\alpha v_\alpha = - J^2/\sigma - (\partial/\partial t) (\tfrac{1}{2}\epsilon e^2 + \tfrac{1}{2}\mu H^2) - \partial S_\alpha/\partial x_\alpha. \qquad (11)$$

Hence, replacing $f_\alpha v_\alpha$ in (10) by (11) and making use of the resulting expression in (9), we obtain the much simpler expression

$$\rho \frac{dU}{dt} = - \frac{\partial q_\alpha}{\partial x_\alpha} + \frac{\partial v_\alpha}{\partial x_\beta} P_{\alpha\beta} + \frac{J^2}{\sigma} \qquad (12)$$

which expresses in differential form the principle of conservation of energy for a conducting compressible fluid with finite viscosity.

Heat diffusion equation

To deduce from (12) the equation governing the distribution of temperature in a magneto-hydrodynamic field we assume first that the fluid is endowed with a *constant* heat conductivity K. Thus, the heat flow vector q_i may be written as

$$q_i = - K(\partial T/\partial x_i), \qquad (13)$$

2-2

where T is the absolute temperature, whence the divergence of the heat flow vector becomes

$$\partial q_\alpha / \partial x_\alpha = -K(\partial^2 T / \partial x_\alpha^2) = -K\nabla^2 T. \qquad (14)$$

Next, we observe that the second term on the right of (12) may be resolved into two terms

$$\frac{\partial v_\alpha}{\partial x_\beta} P_{\alpha\beta} = -p\,\frac{\partial v_\alpha}{\partial x_\alpha} + \Phi, \qquad (15)$$

where the first term denotes the rate at which work is done by the pressure p in compressing the fluid inside the surface S, and where

$$\Phi = \bar{\mu}\left[\frac{\partial v_\alpha}{\partial x_\beta}\left(\frac{\partial v_\beta}{\partial x_\alpha} + \frac{\partial v_\alpha}{\partial x_\beta}\right) - \frac{2}{3}\left(\frac{\partial v_\alpha}{\partial x_\alpha}\right)^2\right] \qquad (16)$$

is the viscous dissipation function (Goldstein, 1943[5]), a quadratic function in the velocity components which is always positive definite[6].

Substituting (14) and (15) into (12) and reverting to Gibbsian vector notation we obtain

$$\rho(dU/dt) + p(\nabla \cdot \mathbf{v}) = K\nabla^2 T + \Phi + J^2/\sigma, \qquad (17)$$

which is the equation governing the distribution of temperature in a conducting compressible fluid with finite viscosity and heat conductivity embedded in a uniform field of magnetic induction. We note that the equation contains two quadratic source terms: the dissipation function Φ due to finite viscosity and the electromagnetic dissipation function J^2/σ due to finite electrical conductivity. From a purely macroscopic point of view Eq. (17) is exact, having assumed that the fluid is endowed with a constant thermal conductivity K and a constant ohmic conductivity σ. To apply (17) to a specific case it is of course necessary to invoke an equation of state linking the intrinsic energy U to the temperature and to other pertinent thermodynamic variables.

3. LINEARIZED EQUATIONS

The foregoing discussion is quite general and in order to apply the theory to the determination of the plane wave modes in a compressible fluid with finite viscosity and heat conductivity we must of course linearize (17) and relate it to the magneto-hydrodynamic wave equation applicable to the present case.

Assuming at the outset that we can neglect the electric displacement current ($\epsilon = 0$), and confining our attention exclusively to time harmonic waves, we obtain from Eq. (I-66) the linearized magneto-hydrodynamic wave equation

$$\nabla^2(\mathbf{F}/\sigma + B_0^2 \mathbf{v}_t) = -i\omega\mu\mathbf{F} + B_0^2(\nabla_t^2 \mathbf{v}_t - \nabla_t\nabla_t \cdot \mathbf{v}_t), \qquad (18)$$

in which B_0 denotes the externally applied uniform field of magnetic induction and in which the vector $\mathbf{F}$, as deduced from Eqs. (I-21) and (7), becomes in the present instance

$$\mathbf{F} = -i\omega\rho_0\mathbf{v} + \nabla p - \tfrac{1}{3}\rho_0\nu\nabla\nabla \cdot \mathbf{v} - \rho_0\nu\nabla^2\mathbf{v}, \qquad (19)$$

where ρ_0 denotes the equilibrium density and ν the kinematic viscosity.

Next, we take up the linearization of the heat diffusion equation (17). Although it is possible to proceed quite generally with an arbitrary equation of state for the fluid, we find it convenient to assume initially that the fluid obeys the law of perfect gases,

$$p = (\gamma - 1)\, c_v\rho T, \qquad (20)$$

where c_v is the specific heat at constant volume and γ the ratio of specific heats, $\gamma = c_p/c_v$. In this case the internal energy depends only on the absolute temperature, $dU = c_v dT$.

Hence, letting ρ_0, p_0, and T_0 denote the constant values of the chosen thermodynamic variables corresponding to the equilibrium state, and letting ρ, p, and T denote from now on the small departures from the equilibrium state, we obtain from (17), upon dropping all quadratic terms, the linearized form

$$(K\nabla^2 + i\omega\rho_0 c_v)\, T = p_0\nabla \cdot \mathbf{v}, \qquad (21)$$

which yields $\nabla \cdot \mathbf{v}$ once we know the temperature distribution. Finally, to make the system determinate in the three dependent variables p, T, and $\nabla \cdot \mathbf{v}$ we need, in addition to (18) and (21), the expression

$$i\omega p = p_0\nabla \cdot \mathbf{v} + (\gamma - 1)\, i\omega\rho_0 c_v T, \qquad (22)$$

which is readily deduced by eliminating the (excess) density ρ with the aid of the linearized forms of (4) and (20).

It now remains to make use of the foregoing equations to determine the structure of the plane wave modes which can exist in the presence of finite viscosity and heat conductivity. For the purpose, we adopt here the elementary plane wave solutions illustrated in Figs. I-1 and II-12 and described in detail in §§ II-2 and II-4·1. It is shown that the solutions generated by the velocity vector $\mathbf{v}_1$, Eq. (II-11), lead in the present instance to two distinct pure shear *velocity* modes, which are devoid of density, pressure, and temperature fluctuations, and which are therefore *isothermal*. Similarly, the solutions generated by the linear combination $\mathbf{v}$, Eq. (II-33), lead in this case to four distinct shear-compression *pressure* modes, which are necessarily accompanied by density, pressure, and temperature fluctuations, and which therefore will henceforth be referred to as *temperature* modes.

4. ISOTHERMAL MODES

Following the techniques outlined in §II-1, we first insert (19) into (18), and we then proceed to the reduction of the resulting vector equation (18) to three scalar equations by seeking the z-component, the divergence, and the z-component of the curl. This last procedure yields the equation

$$\{[1 - ia(1 - iq\nabla^2)]\,\nabla^2 + k_a^2 - \nabla_t^2\}\,(\ell_z \cdot \nabla \times \mathbf{v}) = 0 \qquad (23)$$

in which the unit vector ℓ_z denotes the direction of the externally applied magnetic field. Here, k_a is the wave number associated with Alfvén's phase velocity

$$k_a = \omega/V_a = \omega(\mu\rho_0)^{1/2}/B_0 \qquad (24)$$

and a and q are two convenient parameters,

$$a = \omega\rho_0/\sigma B_0^2 \quad \text{and} \quad q = \nu/\omega, \qquad (25)$$

which measure respectively the hydromagnetic coupling and the viscous damping. The parameter a, which vanishes in the limit of infinite conductivity, is dimensionless, whereas q has the dimensions of a cross-section and vanishes when the kinematic viscosity goes to zero. As a check, it is observed that (23), after putting $q = 0$, becomes identical to Eq. (II-4) upon placing $\epsilon = 0$.

Next, we observe that for plane waves the velocity vector $\mathbf{v}_1$, Eq. (II-11), is perpendicular to the plane defined by the direction of the magnetic field and the vector propagation constant $\mathbf{k}$. Therefore, as shown in Eq. (II-12), this vector is divergenceless (pure shear) and has no z component; furthermore, assuming that the vector $\mathbf{k}$ does not coincide with the direction of the magnetic field, we have in (23) that $\ell_z \cdot \nabla \times \mathbf{v}$ is non-zero. Hence, to satisfy (23) we need merely replace ∇ by $i\mathbf{k}$ and, equating the bracket to zero, we obtain the quadratic in k^2

$$aqk^4 + (\cos^2\theta - ia)\,k^2 - k_a^2 = 0, \qquad (26)$$

in which θ denotes the angle between the vector $\mathbf{k}$ and the direction of the magnetic field (Fig. I-1). As a check we note that, putting $q = 0$ in (26), yields immediately the limiting ($\epsilon \to 0$) form of Eq. (II-13).

Equation (26) has two distinct roots in k^2 and therefore leads to two shear modes. The exact roots of (26) can be readily written down, but we prefer

22

to examine the limiting form of the roots when $aqk_a^2 \ll 1$, which corresponds to the case of high electrical conductivity, low viscosity, and moderately low frequencies. In this case, the roots of (26) are given approximately, to first order of small quantities, by

$$k_+^2 \approx \frac{k_a^2}{\cos^2\theta - ia}\left\{1 - \frac{aqk_a^2}{(\cos^2\theta - ia)^2}\right\} \qquad (27)$$

and

$$k_-^2 \approx -\frac{\cos^2\theta - ia}{aq}\left\{1 + \frac{aqk_a^2}{(\cos^2\theta - ia)^2}\right\}, \qquad (28)$$

from which we can readily deduce the corresponding phase velocities and attenuation factors. The first mode, characterized by the wave number k_+, is an ordinary Alfvén wave slightly attenuated by the presence of finite conductivity and finite viscosity. The second mode, governed by k_-, is a highly attenuated pure shear or vorticity mode characteristic of viscous layer phenomena. Both modes are solenoidal and, according to (21), isothermal. Therefore, as pointed out before, these modes are entirely devoid of density, pressure, and temperature fluctuations. Finally, we note in passing that, if $\cos^2\theta \ll a$, then the leading terms of (27) and (28) become respectively

$$k_+^2 \approx i\omega\mu\sigma \quad \text{and} \quad k_-^2 \approx i/q = i\omega/\nu, \qquad (29)$$

indicating that in this limit the Alfvén wave degenerates into a 'skin' wave governed in the main by the electromagnetic properties of the medium, whereas the viscous mode becomes a true vorticity mode characterized mainly by the kinematic viscosity.

5. TEMPERATURE MODES

Continuing with the method of attack outlined in the preceding section, we first insert (19) into (18) and then proceed to compute the z-component and the divergence of the resulting vector equation. In this manner we obtain two scalar equations involving the variables v_z, $\nabla \cdot \mathbf{v}$, and the (excess) pressure p. Combining these two equations we first eliminate v_z, obtaining a single equation in $\nabla \cdot \mathbf{v}$ and p. Next, making use of (22) we eliminate p in terms of $\nabla \cdot \mathbf{v}$ and T to obtain finally, instead of Eqs. (II-7) and (II-8), the more involved expressions

$$\gamma k_s^2(1 - iq\nabla^2)\,v_z = -(1 - \tfrac{1}{3}iq\gamma k_s^2)\,(\partial/\partial z)\,(\nabla \cdot \mathbf{v}) - (i\omega/T_0)\,(\partial T/\partial z), \qquad (30)$$

$$\{[\nabla^2 + (k_a^2 - ia\nabla^2)\,(1 - iq\nabla^2)]\,[\nabla^2 + \gamma k_s^2(1 - \tfrac{4}{3}iq\nabla^2)] - (1 - \tfrac{1}{3}iq\gamma k_s^2)\,\nabla^2\nabla_t^2\}$$
$$\times (\nabla \cdot \mathbf{v}) + [\partial^2/\partial z^2 + (k_a^2 - ia\nabla^2)\,(1 - iq\nabla^2)]\,(i\omega/T_0)\,\nabla^2 T = 0, \qquad (31)$$

23

in which we now have the additional variable T. In these equations we have introduced, in addition to the wave number k_a and the parameters a and q, as defined by (24) and (25), the wave number associated with the (adiabatic) velocity of sound in the medium $k_s = \omega/V_s$, $V_s^2 = \gamma p_0/\rho_0$. As a check we observe that, in the absence of viscosity ($q = 0$) and in the adiabatic limit of vanishing heat conductivity ($K \to 0$), making use of (21) to eliminate T, equations (30) and (31) reduce respectively to the limiting ($\epsilon \to 0$) form of Eqs. (II-7) and (II-8). Finally, to obtain the higher order partial differential equation governing the distribution of temperature we need only substitute (21) into (31) to eliminate $\nabla \cdot \mathbf{v}$; however, no purpose is served by writing down this more complicated equation, since we wish to examine plane waves at once.

For the purpose, we choose a velocity vector $\mathbf{v}$ which lies in the plane of the wave normal and the direction of the externally applied field in accordance with Eqs. (II-33) and (II-34), as illustrated in Fig. II-12. We observe that these shear-compression modes, Eqs. (II-33) and (II-34), are such that $\boldsymbol{\ell}_z \cdot \nabla \times \mathbf{v} = 0$; hence, (23) is identically satisfied, and we must now make use of (31) and (21) to determine the various temperature modes. To this end we substitute (21) into (31), and replacing ∇ by ik, we obtain finally a fourth-order algebraic equation in k^2,

$$\{[k^2 - (k_a^2 + iak^2)\,(1 + iqk^2)]\,[k^2 - \gamma k_s^2(1 + \tfrac{4}{3}iqk^2)]$$

$$- (1 - \tfrac{1}{3}iq\gamma k_s^2)\,k^2 k_x^2\}\,(k^2 - i\omega\rho_0 c_v/K)$$

$$- \{[k^2 - (k_a^2 + iak^2)\,(1 + iqk^2)]\,c - k_x^2\}\,(\gamma - 1)\,i\omega\rho_0 c_v k^2/K = 0, \qquad (32)$$

which has four distinct roots and which, therefore, yields four shear-compression modes accompanied by density, pressure, and temperature fluctuations; that is, the counterpart of the *pressure* modes discussed in § II-4.

No attempt will be made here to examine in detail the exact roots of (32), which probably can only be handled numerically, but we can apply to (32) various tests of its validity and we can examine the limiting form of the roots in various cases of practical interest. As a first test, let us make the externally applied field vanish; i.e. let us remove all hydromagnetic coupling ($B_0 = 0$). In this case, both k_a^2 and a become infinite as B_0^{-2}, which reduces (32) to the simpler equation

$$[k^2 - \gamma k_s^2(1 + \tfrac{4}{3}iqk^2)]\,(k^2 - i\omega\rho_0 c_v/K) - (\gamma - 1)\,i\omega\rho_0 c_v k^2/K = 0, \qquad (33)$$

from which ensue the acoustic, vorticity, and thermal modes characteristic of an acoustic field with finite thermal conductivity. As an example, let us examine (33) in the limit of small frequencies ($\omega \to 0$); in this case we obtain from (33), as long as K remains finite, a mode with the wave number

$$k^2 \approx k_s^2 (1 - \tfrac{4}{3}iqk_s^2)^{-1}, \tag{34}$$

which corresponds to an *adiabatic* sound wave slightly attenuated by the presence of finite viscosity. On the other hand, in the limit of very large frequencies ($\omega \to \infty$), we deduce from (33) a mode with the wave number

$$k^2 \approx \gamma k_s^2 (1 - \tfrac{4}{3}iq\gamma k_s^2)^{-1}, \tag{35}$$

which represents, for small viscosities, a slightly attenuated sound wave propagating with the *isothermal* phase velocity.

As a second test, let us examine the limiting form of (32) in the case of infinite electrical conductivity ($a = 0$), zero viscosity ($q = 0$), and zero heat conductivity ($K = 0$). In this case (32) reduces to

$$(k^2 - k_a^2)\,(k^2 - k_s^2) = k^2 k_x^2, \tag{36}$$

which fully confirms the limiting ($\epsilon \to 0$) form of Eq. (II-36) and from which we deduced in § II-4 the properties of the ideal magneto-acoustic modes.

Finally, to illustrate with one example the application of (32) to special cases of practical interest, we propose to examine the limiting form of (32) in the case of vanishingly small heat conductivity. Thus, letting $K \to 0$ in (32), we obtain the cubic in k^2

$$[k^2 - (k_a^2 + iak^2)\,(1 + iqk^2)]\,[k^2 - k_s^2(1 + \tfrac{4}{3}iqk^2)] = (1 - \tfrac{1}{3}iqk_s^2)\,k^2 k_x^2, \tag{37}$$

which now supersedes our earlier Eqs. (II-41) and (II-43), and from which ensue three shear-compression temperature modes: a modified (adiabatic) sound wave, a modified Alfvén pressure wave, and a modified vorticity mode. The fourth temperature mode, which has disappeared from (32) by putting $K = 0$, is seen to be governed mainly by the thermal properties of the medium and is, therefore, a highly attenuated wave.

Other cases of interest that can be examined profitably include infinitely high heat conductivity, which according to (21) leads to an isothermal temperature distribution, and the cases of both high and low frequencies. In all cases the computations can be greatly simplified if we can assume that the fluid possesses extremely high electrical conductivity ($a \ll 1$) and very low viscosity ($qk_a^2 \ll 1$), for then familiar perturbation methods such as were employed in II are available to us. Finally, to complete the discussion we observe that, once the wave number k has been determined from (32) or from any of its limiting forms for a particular shear-compression mode, then the corresponding angular parameter ϕ which defines the chosen linear combination (II-33) can be readily obtained by applying the technique outlined in § 4·1.

REFERENCES

[1] Baños, Alfredo, Jr. *Phys. Rev.* **97**, 1435–43, 1955.
[2] Baños, Alfredo, Jr. *Proc. Roy. Soc.* A, **233**, 350–67, 1955.
[3] Alfvén, H. *Cosmical Electrodynamics* (Oxford, Clarendon Press, 1950).
[4] Lamb, Horace. *Hydrodynamics* (New York, Dover Publications, 1945), p. 574.
[5] Goldstein, S. *Modern Developments in Fluid Dynamics* (Oxford, Clarendon Press), vol. II, 1943, p. 603, eq. (8).
[6] Eckart, Carl. *Phys. Rev.* **58**, 267–9, 1940.

Discussion

Spitzer: How many modes vanish if the coefficient of viscosity goes to zero?

Baños: With finite viscosity I get six modes, two shear modes and four temperature modes. With vanishing viscosity one shear mode and three temperature modes remain.

Swann: You have referred to the thermodynamics involved in the derivations. When the mechanism is expressed explicitly in terms of viscosity one needs no thermodynamics in the ordinary sense of the word except when you use the equation of a perfect gas. Am I right in saying that you do not use thermodynamics except in that case?

Baños: Yes.

Schatzman: What would come out of the equations if the conductivity depends on the temperature and the density?

Baños: This is a difficult question which I cannot answer immediately.

Swann: A perturbation method could perhaps give the answer.

Spitzer: Professor Baños has given a very complete and elegant analysis of infinitesimal waves in a fluid—van de Hulst's category I B. An interesting result on finite hydromagnetic waves in a plasma has been obtained by Kruskal, Rosenbluth and others in the U.S.A.; this provides at least one result under category II C. The analysis considers a solitary hydromagnetic disturbance, traveling perpendicular to the magnetic field, in a plasma in which no collisions occur. The orbits of the charged particles in the time variable magnetic field are taken into account; the gas temperature is assumed zero. The analysis goes through without difficulty for a velocity up to twice the Alfvén velocity for infinitesimal disturbances. At this critical velocity the magnetic field rises to three times its value in front of and behind the pulse.

A FLUID SELF-EXCITED DYNAMO

LEVERETT DAVIS, Jr.

California Institute of Technology, Pasadena, California, U.S.A.

ABSTRACT

The possibility that a simply connected perfectly conducting fluid body could generate an increasing external magnetic field by acting as a self-excited dynamo is demonstrated by exhibiting a cycle of motions that doubles the external field each cycle. The essential feature of the motion is that interior points become surface points. This requires points which were originally adjacent to become widely separated. The possibility of such motions is demonstrated and some of the conditions that might lead to their development are considered.

Ever since 1919 when Larmor[1] suggested that the magnetic fields of sunspots, the earth, and the sun might be maintained by self-excited dynamo action, there have been discussions as to whether such a dynamo was really possible in a simply connected body of fluid in which there is nothing that resembles a commutator. Cowling[2] showed that a self-maintaining dynamo is impossible in a medium of finite conductivity if one requires axial symmetry. Elsasser[3] and Bullard[4] have treated much more complicated motions described by series of harmonics with results that suggest, although they are not always regarded as demonstrating, that such a dynamo is possible. Parker[5] finds that in a sphere of fluid which rotates non-uniformly and has a convective zone the eddies provide an important contribution to the dynamo action. Since these models are quite complicated, it seems worthwhile to investigate some of the essential features of a self-excited dynamo with the aid of an over-simplified model.

One simplification is to regard the conductivity as being infinite. In this case any field that is once established is maintained if the fluid remains stationary: here the problem of interest is to find a motion that increases the field external to the body. Bondi and Gold[6] have shown the great importance in this connexion of the theorem[7] that in a perfectly conducting fluid the material inside a magnetic tube of force at one instant will always lie in a tube which may be regarded as the same tube of force.

They considered only motions in which, because of a continuity condition, the surface is always made up of the same particles. Hence the same tubes of force always end on the surface. Since their number cannot increase, one can get nothing resembling a self-excited dynamo in a simply connected body. One can also get nothing like a pair of sunspots that suddenly appear when a loop of magnetic field lines is pushed up through the surface. It therefore appears likely that Bondi and Gold's continuity condition need not always hold.

Let us consider more general fluid motions for which the continuity conditions are not quite so stringent. First consider a simple cycle[8] of fluid flow that doubles the dipole field of a sphere of perfectly conducting fluid. Increase by any power of two can be obtained by repeating the cycle the required number of times. Although the model is so over-simplified that the cycle is not closely related to the processes that maintain the earth's magnetic fields or the processes that occur in stars, it does seem to cast light on the necessary properties of more realistic models. Fig. 1 shows the motion schematically in cross-section, any parallel cross-section appearing the same except for scale. One starts with a uniform magnetic field inside, and hence a dipole field outside, all produced by surface currents. A counter-clockwise eddy in the upper hemisphere and a clock-wise eddy in the lower carry the configuration from (a) through (b) to (c), where the fields in the two hemispheres are parallel but oppositely directed. Thus the field is now largely quadrupole. A rigid body rotation of the lower hemisphere through 180° then gives an internal field which has the same direction everywhere and twice the flux of the original field. If one were dealing with a cube the internal field would be uniform and the cycle would be complete. With a sphere the internal field is not uniform but can be made so by further motion in which the tubes of force are shifted while remaining parallel to themselves. The simplest such motion lies in the planes shown in Fig. 1, but it changes the fluid densities. If the fluid is incompressible, a motion involving four eddies in the plane normal to the diameter through 1 and 1' can lead to a uniform field, although some further separation of adjacent points is required. The ultimate location of each tube of force must be such that its length is the same as that of the original tube of force of which it formed one half. The ultimate result is a uniform internal and dipole external field of twice the original strength and a different direction as shown in (d).

It is at once apparent that the essential reason for the success of this mechanism is the crowding of all of the original surface points into a fraction of the final surface and the appearance on the remainder of the

28

surface of the ends of tubes of force that were originally joined in the interior. In the model considered, all the surface points of Fig. 1 (a) are carried to the left half of (c), and the right half of the surface of (c) is made

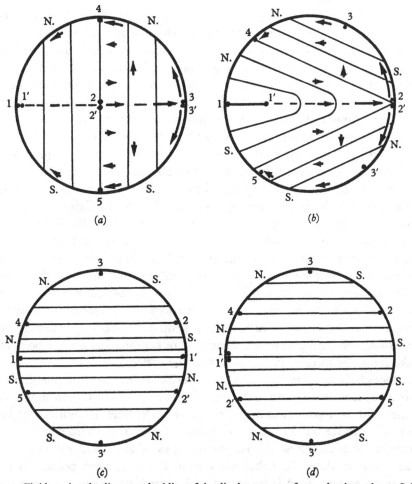

Fig. 1. Fluid motions leading to a doubling of the dipole moment of a conducting sphere. Solid lines represent tubes of force, the dotted line a plane along which the fluid will separate, numbers label certain particles of fluid, and arrows give the fluid velocities that will carry each configuration into the succeeding one. N. and S. give the nature of the effective surface poles. To go from c to d the bottom half of the system is rotated through 180° about the axis from 3' to 3 and the tubes are shifted parallel to themselves to give a uniform internal field.

up of points which were on the equatorial plane of (a). Thus points such as 2 and 2' must move through the stagnation point at the original location of 3 in a finite time; and points which were originally adjacent must become widely separated. Such motions do not occur at the usual stagnation points of a perfect fluid when there is no magnetic field because very close to these points the velocity is at most of the order of the distance from the stagnation point and a logarithmically infinite time is required for the fluid to pass through it.

It is now clear that if new tubes of force are to pass through the surface, as in a growing sunspot or a star with an increasing total magnetic field, there must be loci on the surface at each point of which the flow lines split, as at the extreme right of Fig. 1 (a). It is also clear that if r is the distance from such a point the velocity in its neighborhood must be of order r^p with $p < 1$ in order that a fluid particle may pass through the point in a finite time and allow room for interior points to spread out on the surface. It will be noted that the velocity and, if $\frac{1}{2} < p < 1$, the acceleration go to zero, not infinity, at the singularity. Thus the essential question is whether fluid motion of this character can take place. Such fluid motions can be described in terms of spherical harmonics following Elsasser[3] and Bullard[4], but the series converge very slowly (the coefficient of the nth harmonic in general being of order n^{-1}) and are hence difficult to use. If only a finite number of terms in the series are retained, p becomes unity and the motion is no longer of the required character. If there is a sharp groove in the surface at the splitting point so that the flow line turns through less than 90° at a single point, then the ordinary motion of an ideal fluid is all that is required. This provides a counter-example for the conjecture of Bondi and Gold[6] that the simple-connectivity of the volume containing the fluid prevents a separation of neighboring surface points to make room for interior points to rise to the surface.

Although it is relatively easy to describe a flow having the desired character, it is less easy to find forces that will produce it. An upwelling of the fluid at a point where the magnetic field lines are horizontal just below the surface is a necessary condition. If the surface is fixed, continued upwelling will compress the magnetic tubes of force and displace their fluid contents along the tubes away from the stagnation point. In this way the magnetic field is increased and ultimately the forces associated with it will become important. In an earlier paper[8] it was thought plausible that these forces would produce a fluid motion of the desired kind and that new tubes of force would break through the surface. But further analysis suggests that this is unlikely and that the magnetic forces are more likely to lead to a stoppage of the fluid motion. However, if the surface is not fixed by a rigid non-conducting barrier but is instead a region of continuously decreasing density as on a star, there seems likely to be no difficulty of this kind.

There is an alternative way to get the new tubes of force to intersect the surface. It has been assumed throughout that the conductivity is infinite. This really means that it is so large that the current density $\mathbf{i}$, necessary to produce the magnetic fields, can flow without significant dissipation for times long compared to those considered. But an upwelling will increase the internal field strength without affecting the external field strength and this will lead to a large curl $\mathbf{H} = 4\pi\mathbf{i}/c$ on the surface. The infinite conductivity condition cannot be regarded as holding in the thin surface layer and the tubes of force move upward through and out of the fluid there.

Although the model described by Fig. 1 is so simplified that it would be expected to have little resemblance to any model that would explain the earth's magnetic field, it may be worth noting that in both there are pronounced secular variations and external quadrupole fields. However the main value of this model is in the light it can cast on the nature of the fluid motion required for the existence of a self-excited dynamo in a simply connected body, and in the proof that such dynamos can exist in a perfect conductor.

REFERENCES

[1] Larmor, J. *Brit. Assoc. Reports* (1919), 159; *Engineering*, **108**, 461, 1919.
[2] Cowling, T. G. *Mon. Not. R. Astr. Soc.* **94**, 39, 1934.
[3] Elsasser, W. M. *Phys. Rev.* **69**, 106, 1946; **70**, 202, 1946; **72**, 821, 1947.
[4] Bullard, E. C. *Proc. Roy. Soc.* A, **197**, 433, 1949; **199**, 413, 1949.
[5] Parker, E. N. *Astrophys. J.* **122**, 293, 1955.
[6] Bondi, H. and Gold, T. *Mon. Not. R. Astr. Soc.* **110**, 607, 1950.
[7] Truesdell, C. *Phys. Rev.* **78**, 823, 1950.
[8] Davis, L. *Phys. Rev.* **102**, 939, 1956.

Discussion

Buneman: The solution presented does not appear to be 'simply connected'. The dip, taken to its logical conclusion, together with a corresponding dip at the antipodes, i.e. a volcano together with an anti-volcano at the antipodes, are connected by a funnel right through the center of the earth. We now have a toroidal topology and therefore no violation of the Bondi–Gold theorem.

Davis: In the model there is no anti-volcano at the antipodes. This is necessary in order to avoid swallowing up there as many tube endings as are put to the surface by the upwelling. It does not seem to me that such a flow gives the sphere the connectivity of a torus unless one produces the volcano not by convection but by the insertion of a pump with fixed rigid pipes.

Gold: The Bondi–Gold theorem referred to must not be thought of as saying more than it did. But with its limitations—whether they approach physical reality or not—it may not go very far, but it is right. The external field of a simply connected body of a perfect conductor cannot increase as a result of a hydrodynamic motion in which neighboring fluid particles remain neighbors forever. If any of these limitations are dropped we no longer see any reason why the external field should not be caused to increase. Consequently, if the conductivity is finite, or if the connectivity of the body is changed by allowing the material to be cut at a dip as in Dr Davis' case, a dynamo could certainly be made.

Davis: I wish to raise no questions concerning the basic theory of Bondi and Gold or concerning the conditions for its validity, which they have carefully stated; and I think that we are in essential agreement as to the way in which the theory applies to the model under discussion here. But am I not right in feeling that you have now retreated slightly from the statement in the original paper that the absence of a 'tearing' of the fluid, i.e. a separation of originally adjacent points, 'is in any case implied if the motion is not to affect the topological connectivity of the body'? And do we not attach different meanings to 'connectivity'? I say that a pool of water remains simply connected as a knife is thrust into it until the knife touches the bottom; while you feel that the connectivity changes as soon as the knife enters the water.

I should also like to emphasize that a sharp edge of the groove gives no stagnation point of the flow even if the groove has a finite angle. Thus, the fluid 'turns around the corner' in a finite time.

Gold: I think that in the statement of the theorem neighbours must remain neighbours for two reasons, namely: (a) Without this the Bondi–Gold theorem would not be true. (b) We cannot even define the topological connectivity of the body if a splitting is allowed. This is normal formal hydrodynamics.

Swann: Does not your model have an infinite time constant for the decay of currents and fields because of the infinite conductivity?

Davis: The attempt here was to produce a mechanism which could defeat this infinite time constant by suitable motions. The time constant would depend on that of the hydrodynamic motions.

Ferraro: How does the external field fit the internal field?

Davis: The answer is given by the process described in Fig. 1. We start in Fig. 1 (a) with a uniform internal field and an external dipole field. The configuration is produced by suitable surface currents. At the last stage in Fig. 1 (d) the internal field strength is doubled. The flux out of the sphere should be conserved and the external field can be fitted to the internal by means of a suitably chosen distribution of surface currents.

Ferraro: Does that mean that the doubled field can also be doubled?

Davis: Yes.

Ferraro: I think this is unbelievable.

Davis: I do not see why this should be impossible. A doubled external field just implies doubled surface currents produced by the motion.

NON-STABLE MAGNETO-HYDRODYNAMICAL PROCESSES IN STARS

A. J. KIPPER

Tartu Observatory, Tartu, Estonia, U.S.S.R.

ABSTRACT

1. The behaviour of the magnetic field of a star was investigated by Cowling, Lamb and Wrubel on the assumption of the star as a solid body and on the assumption that the field variations caused by the extinction do not create any motions of the stellar matter. The important results obtained make possible the interpretation of magneto-hydrodynamical processes in stars only quite roughly. When motions of stellar matter caused by the electromagnetic forces are taken into account new properties may be revealed and the non-stability of the magneto-hydrodynamical processes in stars established. Entangled magnetic fields are the general expression of non-stability.

2. The mathematical criterion of a confused stellar magnetic field is given in the preceding paper from the magneto-hydrodynamical equations for an ideal incompressible fluid. The conception of a confused field as a superposition of various fields of different scales is introduced.

3. It is known that the presence of a magnetic field causes an increase of the stability of hydrodynamical processes, as compared with processes in the absence of a magnetic field. It may most probably be concluded that the extent of confusion of a magnetic field is not the same, as it happens as a result of ordinary turbulent motions in a liquid. However, owing to magneto-hydrodynamical formulae, it may be stated in the present paper that if the field confusion is considered as a superposition of fields of different dimensions and not as a field of entangled magnetic lines, no stability of the above kind can be expected.

4. The theory of a totally entangled magnetic field, as a field in a state of maximum confusion is examined. A very close similarity between a totally entangled magnetic field and the turbulent motion in a viscous liquid is established. The question of the time of extinction of a magnetic field of a star is discussed.

NOTE. Sections 1, 2, 4 of the paper are based upon the results obtained by the author, reported by him at the 4th Cosmological Conference. Section 3 contains unpublished results.

I. EINLEITUNG

Nach der im Jahre 1942 veröffentlichten bekannten Untersuchung von H. Alfvén über magneto-hydrodynamische Wellen begann sich das neue

Gebiet der Astrophysik—die kosmische Elektrodynamik—schnell zu entwickeln. Heutzutage kann man nicht mehr daran zweifeln, dass elektromagnetische Erscheinungen eine entscheidende Rolle in vielen Erscheinungen der kosmischen Physik spielen. Insbesondere können sich Fragen über den inneren Aufbau der Sterne wie auch über ihre Atmosphären nicht weiter entwickeln, ohne die Gesetzmässigkeiten zu berücksichtigen, die sich aus den Gleichungen Maxwells für ein gasartiges und durch hohe Leitfähigkeit gekennzeichnetes Medium kosmischer Dimensionen ergeben.

Die weitere Entwicklung der auf Sterne angewandten kosmischen Elektrodynamik ist mit den Namen von Cowling[1], Lamb[2], Wrubel[3], Chandrasekhar[4], Syrowatskij[5] u. a. verbunden. Es ist erwiesen worden, dass das ausschliesslich durch Selbstinduktion aufrechterhaltene Magnetfeld äusserst stabil ist und erst nach Verlauf von 10^{10} Jahren erlischt. In vielen theoretischen und experimentellen Arbeiten ist auch die erhöhte Stabilität der magnetò-hydrodynamischen Prozesse in Vergleich zu den gewöhnlichen hydrodynamischen, ohne Magnetfeld verlaufenden Prozessen festgestellt worden. Hieraus ergibt sich die in der wissenschaftlichen Literatur vertretene Ansicht, dass das Vorhandensein des Magnetfeldes die Entwicklung der Instabilität der hydrodynamischen Bewegung hindert und die turbulenten Strömungen in laminare umwandelt.

Es muss jedoch darauf hingewiesen werden, dass die Frage der Stabilität der magneto-hydrodynamischen Prozesse kosmischer Dimensionen noch nicht als endgültig geklärt betrachtet werden kann. In einem flüssigen Medium ruft die Dämpfung des langlebigen Magnetfeldes immer Wirbel hervor, die ihrerseits die Kraftlinien verwirren. Die Gesetzmässigkeiten des verwirrten Feldes unterscheiden sich wesentlich von denjenigen der ordnungsmässigen Felder. In Erwägung auch des Umstandes, dass im Laboratorium mit dem Magnetfeld angestellte Experimente keine vollständige Vorstellung vom Wesen des Magnetfeldes von Sternendimension geben können, ergibt sich die Notwendigkeit einer weiteren und allseitigen Erforschung der Frage der Stabilität der magneto-hydrodynamischen Prozesse kosmischer Dimensionen.

Zu Beginn dieses Vortrags wird der Versuch gemacht, zu beweisen, dass das Magnetfeld eines Sternes nur in verwirrtem Zustand existieren kann. Sodann wird die anschauliche Darstellung des verwirrten Feldes mit Hilfe von Bildern der Kraftlinien behandelt und werden einige die Gesetzmässigkeiten dieses Feldes ausdrückende Formeln gegeben. Zum Schluss wird die Frage der Zeit der Dämpfung des Magnetfeldes berührt. In bezug auf einige Einzelheiten der Berechnung ist der Verfasser dieses

Vortrages genötigt, auf seine früheren, in den Veröffentlichungen der vierten Konferenz über Fragen der Kosmogonie, Moskau, 1955, erschienenen Forschungen zu verweisen [6].

2. ÜBER DIE VERWIRRUNG DES MAGNETFELDES IN FLÜSSIGEM LEITFÄHIGEM MEDIUM

Betrachten wir die Veränderungen eines Magnetfeldes in einem aus reibungsfreier inkompressibler Flüssigkeit bestehenden Medium. Diese Veränderungen sind in den wohlbekannten Gleichungen der Magneto-Hydrodynamik niedergelegt. Als gesuchte Grössen figurieren in den erwähnten Gleichungen die Stärke des Magnetfeldes und die Geschwindigkeit des Mediums v. Die Stärke des Magnetfeldes kann im allgemeinen als unendliche Reihe dargestellt werden:

$$H = \Sigma_\lambda b_\lambda(t) \exp[-t/(4\pi\sigma\lambda^2)]h_\lambda(x), \qquad (2.1)$$

wo die Vektoren $h_\lambda(x)$ nicht von der Zeit t abhängig sind und die folgende Gleichung befriedigen:

$$\nabla^2 h_\lambda + \frac{1}{\lambda^2} h_\lambda = 0. \qquad (2.2)$$

In (2.1) bezeichnen: σ—die Leitfähigkeit in elektromagnetischen Einheiten, $b_\lambda(t)$—die skalare Funktion der Zeit t. Bei der Lösung der Gleichung (2.2) müssen die Randbedingungen der Aufgabe berücksichtigt werden. Folglich bilden die Vektoren $h_\lambda(x)$ ein vollständiges Orthogonalsystem der Eigenvektoren mit den entsprechenden Eigenwerten λ [2].

Die Entwicklung (2.1) ist eine Entwicklung des Vektors der Stärke des Magnetfeldes H nach den Eigenvektoren der Gleichung (2.2).

Es ist leicht festzustellen, dass (2.1) die folgende Gleichung befriedigt:

$$\frac{\partial H}{\partial t} - \frac{1}{4\pi\sigma}\nabla^2 H = 0, \qquad (2.3)$$

wenn b_λ nicht von der Zeit abhängig sind, d. h. wenn $b = 0$. Andererseits nehmen die Gleichungen der Magneto-Hydrodynamik bei einem Medium vom Charakter eines starren Körpers ($v = 0$) gerade die Form (2.3) an. Folglich, mit konstanten b_λ, ergibt die Formel (2.1) die Stärke des Feldes in einem strömungsunfähigen Medium.

Um die Stärke des Magnetfeldes H in einem flüssigen Medium auszudrücken, ist es zweckmässig, sich derselben Entwicklung (2.1) zu bedienen. Doch müssen dann b_λ als gewisse Funktionen der Zeit betrachtet werden. Diese werden aus dem System

$$\dot{b}_\lambda = \sum_{\lambda'} A_{\lambda'\lambda} b_\lambda \qquad (2.4)$$

bestimmt, wobei die von der Zeit t abhängigen $A_{\lambda\lambda'}$ vom Verfasser dieses Vortrags auf Grund der magneto-hydrodynamischen Gleichungen in der oben angeführten Arbeit berechnet worden sind.

Schreiben wir (2.1) auf folgende Weise um:

$$H = \sum_\lambda H_\lambda, \qquad (2.5)$$

$$H_\lambda = b_\lambda \exp^{[-t/T_\lambda]} . h_\lambda(x), \quad T_\lambda = 4\pi\sigma\lambda^2 \qquad (2.6)$$

und bezeichnen wir mit H_λ die Teilfelder des allgemeinen Feldes H. Die Grösse λ mit einer Dimension von cm^{+1} nennen wir den Maszstab des Teilfeldes H_λ und die Grösse T_λ mit einer Dimension von sek^{+1}—die Zeit der exponentiellen Dämpfung dieses Teilfeldes. Die Formel 2.5 weist darauf hin, dass im allgemeinen das Magnetfeld eines Sternes als Superposition von Feldern verschiedenen Maszstabes dargestellt werden kann. Ferner muss erwähnt werden, dass der Maszstab λ eines gewissen Teilfeldes H_λ eine solche Entfernung bedeutet, auf welche hin sich H_λ wesentlich verändert. Wir werden das Magnetfeld H als verwirrt bezeichnen, wenn es aus Teilfeldern aller möglichen Maszstabe besteht, vom grössten $\lambda = l$ bis zum kleinsten $\lambda = \lambda_0$. l werden wir den Hauptmaszstab und λ_0 den inneren Maszstab des allgemeinen Feldes nennen. In einem verwirrten Felde sind alle b_λ aus der Formel (2.1) von Null verschieden, in einem ordnungsmässigen Feld dagegen verschwinden fast alle b_λ, und nur einzelne von ihnen haben einen endlichen Wert.

Nehmen wir an, dass in einem gewissen Zeitpunkt $t = 0$ das Feld ein ordnungsmässiges war, d. h.

$$b_l \neq 0, \quad b_\lambda = 0, \quad \lambda \neq 1.$$

Dann kann man auf Grund von (2.4) ohne jegliche Schwierigkeiten schliessen, dass in den folgenden Zeitpunkten $t > 0$ alle b_λ von Null abweichen werden. Mit anderen Worten—das Feld verwirrt sich.

Präziser lässt sich die Aufgabe von der Verwirrung des Feldes auf folgende Weise formulieren. Es gilt, die Bedingungen zu finden, unter denen in einem Zeitraum, der klein ist im Vergleich zu der Zeit T_l der exponentiellen Dämpfung des Teilfeldes H_l mit dem Hauptmaszstab l, das Verhältnis

$$b_\lambda / b_l$$

sich bei allen bis zur Grösse der Grössenordnung Eins vergrössert. Die Lösung dieser Aufgabe ist vom Verfasser in der oben angeführten Arbeit gegeben. Wenn die dimensionslose Grösse

$$K = \frac{\bar{H} . \sigma . l}{\rho^{1/2}} \qquad (2.7)$$

viel grösser ist als eins, verwirrt sich das Magnetfeld immer; wenn dagegen $K < 1$, findet keine Verwirrung statt; das Feld erlischt nach dem exponentiellen Gesetz und bleibt dabei ordnungsmässig. In der Formel (2.7) bezeichnen: $\bar{H}$—den Mittelwert der Stärke des allgemeinen Magnetfeldes, σ—die Leitfähigkeit in elektromagnetischen Einheiten, l—den Hauptmaszstab des Feldes, der nach der Grössenordnung dem Radius des Sternes gleich ist, ρ—die Dichte des Mediums. Für Sterne, die ein Magnetfeld von mehr als 0.1 Oerstedt besitzen, mit $\rho = 1$ g/cm, $\sigma = 10^{-5}$ sec/cm², $l = 10^{11}$ cm ergibt sich

$$K > 10^5 \gg 1.$$

Folglich, in allen Fällen, in denen das Magnetfeld eines Sternes beobachtet werden kann, muss es verwirrt sein, und dem Verfasser scheint es, dass sich ein solcher Schluss aus Gleichungen der Magneto-Hydrodynamik ohne irgendwelche ergänzende Hypothesen ziehen lässt.

3. ÜBER DIE ANSCHAULICHE DARSTELLUNG DES VERWIRRTEN MAGNETFELDES

Zu Beginn der vorliegenden Abhandlung wurde erwähnt, dass in der wissenschaftlichen Literatur die Ansicht über die erhöhte Stabilität der magneto-hydrodynamischen Prozesse im Vergleich zu den gewöhnlichen hydrodynamischen ohne Magnetfeld verlaufenden Prozessen vertreten wird. Andrerseits haben wir den Versuch gemacht, zu beweisen, dass das Magnetfeld eines Sternes immer verwirrt ist. In der Verwirrung äussert sich aber gerade die Instabilität der entsprechenden Prozesse. Somit ergibt sich ein Widerspruch zwischen den Angaben der einschlägigen Literatur und den Ergebnissen der vorliegenden Untersuchung. Es ist jedoch möglich, diesen Widerspruch durch ein anschauliches Bild des verwirrten Feldes mit graphischer Darstellung der Kraftlinien aufzuklären.

In der Literatur wird das verwirrte Feld als ein aus unregelmässig gekrümmten Kraftlinien bestehendes betrachtet. Die letzteren schliessen sich in überwältigender Mehrzahl ausserhalb des Sternes, da der Stern ein allgemeines Feld besitzt (Abbildung 1). Auf Grund eines solchen Bildes wird die beruhigende Wirkung des Magnetfeldes auf die Bewegung des Mediums folgendermassen dargestellt.

In einem Medium von hoher Leitfähigkeit kleben die Kraftlinien des Magnetfeldes an der sich bewegenden Flüssigkeit. Die quer über die Kraftlinien verlaufende Bewegung verlängert die Linien und vergrössert das Feld. Jedoch bringt die Vergrössung des Feldes endlich die Wirbel und das Entstehen von Feldern kleineren Maszstabes zum Stehen. In Anbetracht dessen, dass auch die Dämpfung der Felder kleineren Masz-

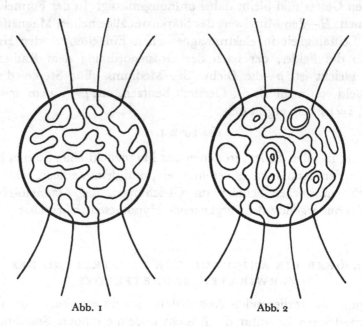

Abb. 1   Abb. 2

stabes sich schnell vollzieht, wenn sie nicht fortwährend erneuert werden, scheint es, dass die Verwirrung sich nicht entwickeln kann (Batchelor[7]).

Andrerseits würde die beschriebene Erscheinung die Verwandlung der ursprünglichen Unordnung in Ordnung bedeuten, was einem Grundsatz der Natur widerspricht (Heisenberg[8]). Um die mit der Verwandlung der Unordnung in Ordnung verbundenen Schwierigkeiten zu überwinden, hat eine Reihe von Forschern die Ansicht geäussert, dass die gesamte Energie des Magnetfeldes in Teilfeldern kleineren Maszstabes konzentriert ist. Jedoch widerspricht auch diese Hypothese den Ergebnissen der vorliegenden Untersuchung. Wenn in irgendeinem Zeitpunkt die Teilfelder grossen Maszstabes fehlen, so könnte man auf Grund der Gleichung (2.4) folgern, dass sie in den folgenden Augenblicken erscheinen. Ausserdem würde die Konzentration der Magnetenergie in

Feldern ausschliesslich kleinen Maszstabes bedeuten, dass der Stern als Ganzes kein messbares Magnetfeld besitzt.

In der vorliegenden Abhandlung wird das verwirrte Feld als eine Superposition von Feldern verschiedenen Maszstabes dargestellt. Eine solche Darstellung ist formell und dem Wesen nach gehaltvoller. Sie umfasst die unregelmässig gekrümmten Kraftlinien, die sich ausserhalb des Sternes schliessen, aber auch die Mehrzahl der sich innerhalb des Sternes schliessenden Linien. Die durch die Entwicklung des Feldes hervorgerufene Verkürzung der Kraftlinien bewirkt die Ausglättung derselben, wobei aber die Geschlossenheit der Linien nicht zerstört und ihre Zahl nicht vermindert wird. Eine Darstellung des verwirrten Sternfeldes ist auf der Abbildung 2 gegeben. Das allgemeine System der Kraftlinien eines Sternes besteht aus einer Menge von abgeschlossenen Systemen, zwischen denen sich die sich ausserhalb des Sternes schliessenden Linien hinziehen (Abbildung 2). Jedes abgeschlossene System der Kraftlinien ist in gewissem Sinne selbstständig. Bewegungen kleineren Maszstabes als die Grösse des betrachteten Subsystems bilden die inneren Bewegungen dieses Systems. Strömungen jedoch, die ihrem Maszstab nach grösser als die geschlossenen Kraftlinien sind, tragen das System als Ganzes von einem Ort zum andern.

Die beschriebene Darstellung des verwirrten Magnetfeldes eines Sternes dürfte nicht die zu Beginn des Abschnitts angeführten Widersprüche hervorrufen. Die geäusserten Grundgedanken weiter verfolgend, kann man voraussetzen, dass die einzelnen Systeme der geschlossenen Kraftlinien durch Flächen von den übrigen Teilen abgetrennt sind, die fast Unstetigkeitsflächen darstellen. Syrobatskij hat darauf hingewiesen, dass tangentiale Unstetigkeiten im Magnetfeld stabil sind, wobei betont werden muss, dass in der gewöhnlichen Hydrodynamik gerade die umgekehrte Sachlage festgestellt worden ist, d. h. dass die tangentialen Unstetigkeiten absolut instabil sind. Nach der aufgestellten Hypothese besteht die beruhigende Wirkung des Magnetfeldes nicht im Zerstören der Verwirrung der Bewegung, sondern im Erhalten der Unstetigkeitsflächen, die als das Hauptmerkmal des verwirrten Magnetfeldes zu betrachten sind. Die Materie des Sternes nimmt unter dem Einfluss des Magnetfeldes eine gewissermassen tropfenartige Struktur an. Bei der Entwicklung des Magnetfeldes versetzt sich ein Teil dieser Tropfen oder Granulae, aber an ihrer Stelle entstehen neue. Weitere Untersuchungen müssen den Beweis für die Glaubwürdigkeit dieser aufgestellten Hypothese erbringen.

4. DAS VOLLSTÄNDIG VERWIRRTE MAGNETFELD

Aus den Gleichungen der Magneto-Hydrodynamik geht hervor, dass Veränderungen des Magnetfeldes stets Veränderungen in der Bewegung der Flüssigkeit hervorrufen und umgekehrt. Die Verwirrung des Magnetfeldes ist von einer Verwirrung der Bewegung des Mediums begleitet, die von einem wirbelartigen Charakter ist. Auf diese Weise besteht ein enger Zusammenhang zwischen den Problemen des verwirrten Magnetfeldes und der turbulenten Bewegung der Flüssigkeit. Daher ist es möglich, bei der Untersuchung des verwirrten Feldes sich der schon ausgearbeiteten Methoden der Turbulenztheorie zu bedienen.

Chandrasekhar, Kaplan u. a. sind diesen Weg gegangen, doch hat es sich alsbald herausgestellt, dass die Lösung dieser Aufgabe auf viele Schwierigkeiten stösst, da in der Magneto-Hydrodynamik die Zahl der die Prozesse bestimmenden Parameter viel grösser ist als in der gewöhnlichen Hydrodynamik. Um trotz alledem einige einfache Gesetze für das verwirrte Magnetfeld zu erhalten, ist es notwendig, das Problem durch einige ergänzende Hypothesen zu vereinfachen. Nehmen wir an, es gebe einen gewissen Grenzzustand des verwirrten Magnetfeldes, dem die Felder im Prozesse der Verwirrung zustreben, und dass dieser Zustand von den Anfangsbedingungen und der Vorgeschichte des Systems nicht abhängig ist. Indem wir diesen Zustand mit 'vollständig verwirrt' bezeichnen, nehmen wir an, dass das Feld eines Sternes im Lauf seiner Entwicklung diesen Zustand während eines bedeutend kürzeren Zeitraumes erreicht, als zur Dämpfung des Feldes notwendig ist, unter der Bedingung, dass die dimensionslose Grösse K genügend gross ist.

Die Hypothese von dem Vorhandensein des vollständig verwirrten Magnetfeldes eines Sternes ist eine Übernahme der Ideen Kolmogorows, Onsagers, Weizsäckers und Obuchows aus der Theorie der Turbulenz in die Magneto-Hydrodynamik. Die von besagten Forschern ausgearbeitete Theorie des universalen Gleichgewichts stützt sich auf die Voraussetzung von der Existenz eines gewissen statistischen Gleichgewichts, dem alle Komponenten kleineren Maszstabs der turbulenten Bewegung im Laufe ihrer Entwicklung zustreben. Je grösser die Reynolds'sche Zahl ist, einen desto grösseren Intervall der Maszstabe umfasst das Gleichgewicht und erstreckt sich im Grenzfall auf das ganze Spektrum. In analogem Sinn wird auch in der vorliegenden Abhandlung das vollständig verwirrte Magnetfeld eines Sternes betrachtet.

Sich auf die Hypothese des statistischen Gleichgewichts stützend, ist es dem Verfasser dieser Abhandlung gelungen, in der oben ausgeführten

Arbeit einige Wechselbeziehungen für das vollständig verwirrte Magnetfeld zu ermitteln, die für eine weitere Untersuchung des Problems von Wichtigkeit sein dürften. Wegen Platzmangels können in der vorliegenden Arbeit nur einige davon angeführt werden.

(a) In dem vollständig verwirrten Magnetfeld bis zu einer Genauigkeit der dimensionslosen Koeffizienten der Grössenordnung Eins ist die magnetische Energie des Feldes gleich der kinetischen Energie der Wirbel

$$E^M \approx E^K. \tag{3.1}$$

Die Beziehung (3.1) ist eine der wesentlichsten Beziehungen der Magnetfelder galaktischen Ursprungs, die zuallererst von Batchelor ermittelt wurde. Doch gelangte Batchelor zu (3.1) auf anderen Wegen und von anderen Voraussetzungen ausgehend. Der entscheidende Prozess, der zu einer Gleichwerdung der magnetischen und der kinetischen Energie führt, ist nach Batchelor das Unterbrechen der Bewegung des Mediums durch das sich vergrössernde Magnetfeld. In der vorliegenden Abhandlung ergibt sich die Gleichheit beider Energieformen als Folge der Hypothese vom Vorhandensein eines Grenzzustandes, des Zustandes der völligen Verwirrung des Magnetfeldes. Falls die Theorie des vollständig verwirrten Magnetfeldes eine statistische ist, muss sich die Beziehung (3.1) schon aus allgemeinen Vorstellungen von der Gleichheit der Energie der verschiedenen Freiheitsgrade ergeben.

(b) Unter der Voraussetzung, dass das Magnetfeld durch die Ausströmung der Jouleschen Wärme sich quasistationär verändert, kann die Abnahme der gesamten Energie E mit der Zeit t mittels der Differentialgleichung

$$\frac{dE}{dt} = -AE^{3/2}l^{-5/2}\rho^{-1/2}, \tag{3.2}$$

erfasst werden, wo A die dimensionslose Konstante von der Grössenordnung Eins darstellt und die übrigen Benennungen die oben angegebenen Bedeutungen haben. Die Lösung der Gleichung ist

$$E = \frac{E_0}{\left(1 + \dfrac{t}{T'}\right)^2}, \tag{3.3}$$

wobei

$$T' = \frac{Tl}{K}. \tag{3.4}$$

In den zuletzt gegebenen Formeln (3.3) und (3.4) bezeichnen: E_0 – die Energie des Systems im Anfangszeitpunkt $t = 0$, T' – die Zeit der exponentiellen Dämpfung des Feldes vom Typ Cowling-Wrubel.

41

Die Formel (3.3) gibt das Gesetz der Dämpfung eines vollständig verwirrten Magnetfeldes wieder, wobei T' als Zeit der Dämpfung dieses Feldes gedeutet werden kann. Falls bei den Sternen $K < 10^5$ ist, ist nach (3.4) die Zeit der Dämpfung T' um 10^5 oder mehr Male kleiner als im Falle von ordnungsmässigen Feldern, als deren Beispiel ein Feld vom Typ Cowling-Wrubel dienen mag. Eine grosse Stabilität des Feldes kommt bei realen Sternen anscheinend nicht vor.

(c) Da die Gesetze des vollständig verwirrten Magnetfeldes die allbekannten Gesetze der im Gleichgewicht bleibenden Turbulenz in Formeln der Magnetodynamik wiederholen, liegt kein Grund vor, alle aufzuzählen. Es sollen hier nur einige von ihnen angeführt werden.

Das bekannte Gesetz Kolmogorow-Obuchows in der Magneto-Dynamik des flüssigen Mediums lautet: Der Mittelwert der Differenz der Feldstärken an zwei verschiedenen Punkten von der Entfernung λ ist der Kubikwurzel dieser Entfernung proportional, unter der Bedingung dass

$$l \gg \lambda \gg \lambda_0$$

ist.

Der innere Maszstab nach der Grössenordnung wird aus der Formel

$$\lambda_0 \approx \frac{1}{K^{3/4}} \tag{3.5}$$

berechnet.

Die Feldstärke der Komponente des Feldes mit innerem Maszstab ergibt sich aus der Formel

$$H_{\lambda_0} \quad \frac{H}{K^{1/4}} \tag{3.6}$$

usw. Es sei noch bemerkt, dass das Gesetz Kolmogorow-Obuchows wie auch die Formeln (3.4) und (3.5) die Möglichkeit offen lassen, die Hypothese von dem völligen verwirrten Magnetfeld auf ihre Richtigkeit zu prüfen, wenn z. B. ausführliche Messungen des Magnetfeldes der Sonne vorliegen sollten.

5. SCHLUSSBEMERKUNGEN

Die Theorie des Magnetfeldes in einem gasartigen Medium von hoher Leitungsfähigkeit hat, auf kosmische Nebel angewandt, eine besondere Entwicklung durchgemacht. Es ist darauf hingewiesen worden, dass die Kraftlinien des galaktischen Magnetfeldes sich in Gaswolken kosmischer Dimensionen verwirren. Die Theorie des Magnetfeldes von Nebeln ist die Theorie des verwirrten Feldes.

Die Erforschung der Magnetfelder von Sternen ist einen etwas anderen Weg gegangen. Ausser verschiedenen Fragen über die Stabilität eines

Sternes mit einem Magnetfeld wurden Probleme über die Struktur des durch Selbstinduktion aufrechterhaltenen Feldes in einem Medium vom Charakter eines starren Körpers gelöst. Die Möglichkeiten der Verwirrung des Magnetfeldes eines Sternes sind fast gar nicht untersucht worden.

Die Darlegungen der vorliegenden Untersuchung machen den Versuch zu beweisen, dass die Verwirrung des Magnetfeldes eines durch Selbstinduktion aufrechterhaltenen Sternes seine Grundeigenschaft bildet. Die Verwirrung des Feldes ergibt sich aus den Gleichungen der Magneto-Hydrodynamik ohne irgendwelche ergänzende Hypothesen. Letztere erweisen sich nur bei der Erforschung der Eigenschaften des verwirrten Feldes als notwendig.

Die Grundhypothese der vorliegenden Untersuchung besteht in der Voraussetzung des Vorhandenseins eines gewissen Grenzzustandes des Feldes, der einen Zustand der maximalen Unordnung darstellt, und dem die realen Felder im Prozess ihrer Entwicklung zustreben. Die Entwicklung des Feldes ist vor allem durch seine Dämpfung gekennzeichnet. Diese Darstellung der Veränderung der Felder ist in der ersten Näherung den Sternen angemessen, doch kann sie anscheinend bei kosmischen Nebeln, wo das Feld durch das allgemeine Feld der Milchstrasse aufrechterhalten wird, nicht angewandt werden. Der Prozess der Verwirrung verläuft anscheinend in beiden betrachteten Fällen auf verschiedene Weise. Somit sind die Schlussfolgerungen der vorliegenden Untersuchung als erste Näherung den magneto-hydrodynamischen Prozessen eines Sternes bei deren langsamem, quasistationärem Verlauf anwendbar.

Fragen über die Entwicklung der Sterne sind mit Fragen über ihren Aufbau verbunden. Beim gegenwärtigen Stand der Wissenschaft ist es unmöglich, beim Erforschen des Aufbaus der Sterne, die Rolle der Elektrodynamik unbeachtet zu lassen. Zu einer erfolgreichen Anwendung der Gesetzmässigkeiten der magneto-hydrodynamischen Prozesse in einem Medium von hoher Leitfähigkeit beim Lösen astrophysischer Probleme muss jedoch die statistische Theorie des Feldes weiter entwickelt werden. Die Errungenschaften der Theorie der turbulenten Bewegung der Flüssigkeit können dabei von grossem Nutzen sein.

43

LITERATURVERZEICHNIS

[1] Cowling, T. G. *M.N.* **105**, 166, 1945.

[2] Lamb, H. *Phil. Trans.* **174**, 519, 1883.

[3] Wrubel, M. N. *Astrophys. J.* **116**, 291, 1952.

[4] Chandrasekhar, S. *Proc. Roy. Soc.* A, **204**, 435, 1951; **207**, 301, 1951; **216**, 293, 1953.

[5] Syrovatskij, S. I. *Журнал экспериментальной и теоретической физики*, **24**, 622, 1953. (*Journal der Theor. und Exp. Physik*, **24**, 622, 1953.)

[6] Kipper, A. *Труды четвертого совещания по вопросам космогонии* (М. 1955). (Veröffentlichungen der vierten Konferenz über Fragen der Kosmogonie, Moskau, 1955.)

[7] Batchelor, G. K. *Problems of Cosmical Aerodynamics* (I.A.U. Symposium, Paris), Central Air Documents Office, Dayton, Ohio, 1951.

[8] Heisenberg, W. *Problems of Cosmical Aerodynamics* (I.A.U. Symposium, Paris), Central Air Documents Office, Dayton, Ohio, 1951.

Discussion

Lehnert: In Ihrem Modell in Abb. 2 enthält das Gebiet des Sternes abgeschlossene Systeme, die nicht mit dem äusseren Magnetfeld verbunden sind. Die Grenzflächen dieser abgeschlossenen Systeme können dann als Unstetigkeitsflächen auftreten. Ist es möglich, dass Instabilitäten an diesen Flächen entstehen können? Ich denke dann besonders an einen gewöhnlichen Prozess in der Hydrodynamik, das heisst, an eine Prandtlsche Trennungsschicht. Bei genügend hohem Geschwindigkeitsgradienten entsteht hier eine Instabilität. Im magneto-hydrodynamischen Falle hat man anderseits auch die stabilisierende Einwirkung eines Magnetfeldes zu beachten. Es entsteht dann die Frage, ob die destabilisierende Einwirkung des Geschwindigkeitsgradienten, oder ob die stabilisierende Einwirkung des Magnetfeldes an so einer Grenzfläche eine dominierende Rolle spielen wird.

Kipper: Ich kann in diesem kurzen Vortrag leider keine vollständige Beschreibung der Resultate der vorliegenden Arbeit geben. Es kann aber bewiesen werden, dass das Magnetfeld die Unstetigkeitsflächen stabil macht.

Lehnert: Ich bin nicht ganz überzeugt, dass eine solche Unstetigkeitsfläche unter allen Umständen stabil ist. Bei einem magneto-hydrodynamischen Experiment mit Quecksilber (*Proc. Roy. Soc.* A. **233**, 299, 1955) ist nämlich in einem Spezialfall gezeigt worden, dass das Magnetfeld den Geschwindigkeitsgradienten derartig vergrössern kann, dass eine Instabilität entsteht. Man könnte hier natürlich den Einwand machen, dass die Bedingungen des Experiments nicht denjenigen der Astrophysik entsprechen. Es ist deshalb nicht sicher, dass das Experiment als ein Argument gegen die Stabilität im astrophysikalischen Falle von stark 'eingefrorenen' magnetischen Feldlinien verwendet werden kann.

Dann habe ich noch eine andere Frage. Falls der Parameter K in Ihrer Gleichung (2.7) viel grösser als eins ist, verwirrt sich das Magnetfeld. Anderseits findet bei $K > 1$ keine Verwirrung statt. Die Bedingung $K > 1$ ist identisch mit der Bedingung, dass magneto-hydrodynamische Wellen auftreten können (siehe Lehnert, Gleichung (19)) dieses Bandes). Wenn $K > 1$, entstehen anderseits keine Wellen, sondern aperiodische Bewegungen, die nach einem exponentiellen Gesetz abklingen. Kann man aus Ihrer Bedingung $K > 1$ und aus meiner entsprechenden Bedingung schliessen, dass die zunehmende Verwirrung des Magnetfeldes von magneto-hydrodynamischen Wellen verursacht wird, die fortlaufend das Innere des Sternes in allen Richtungen durchqueren?

Kipper: Diese Gleichung ist eine Differentialgleichung nach der Zeit und enthält in sich auch die periodischen Lösungen; aber in dieser Arbeit habe ich diese Lösungen nicht speziell betrachtet. Ich habe nur ein Kriterium gefunden, das ermöglicht, die Bedingung zu stellen, dass diese b_λ ungefähr gleich werden, also dass alle Komponenten vorhanden sind.

THE AXISYMMETRIC CASE IN HYDROMAGNETICS

S. CHANDRASEKHAR AND KEVIN H. PRENDERGAST

Yerkes Observatory, University of Chicago, U.S.A.

ABSTRACT

Recent work at the Yerkes Observatory has been concerned with the study of configurations in which the magnetic and velocity fields possess a common axis of symmetry. In those cases where the density ρ may be assumed constant, it has proved advantageous to employ a representation suggested by Lüst and Schlüter [1]: in cylindrical co-ordinates (ϖ, ϕ, z) let

$$\mathbf{h} = \frac{\mathbf{H}}{\sqrt{4\pi\rho}} = \varpi T \hat{\phi} + \text{curl}\,(\varpi P \hat{\phi}),$$

and
$$\mathbf{v} = \varpi V \hat{\phi} + \text{curl}\,(\varpi U \hat{\phi}),$$

where $\hat{\phi}$ is a unit vector and T, P, V, and U are independent of the azimuthal angle ϕ. The hydrodynamic equation

$$\rho \frac{\partial \mathbf{v}}{dt} + \rho \mathbf{v} . \text{grad } \mathbf{v} = -\text{grad } p + \rho \text{ grad } V + \frac{1}{4\pi}\,(\text{curl } \mathbf{H}) \times \mathbf{H},$$

may then be replaced by the pair of equations (cf. Chandrasekhar [2])

$$\varpi^3 \frac{\partial V}{\partial t} = \frac{\partial(\varpi^2 T, \varpi^2 P)}{\partial(z, \varpi)} - \frac{\partial(\varpi^2 V, \varpi^2 U)}{\partial(z, \varpi)}, \tag{1}$$

and
$$\varpi\,\Delta_5 \frac{\partial U}{\partial t} - \frac{\partial(\Delta_5 P, \varpi^2 P)}{\partial(z, \varpi)} + \frac{\partial(\Delta_5 U, \varpi^2 U)}{\partial(z, \varpi)} = \varpi \frac{\partial T^2}{\partial z} - \varpi \frac{\partial V^2}{\partial z}, \tag{2}$$

where Δ_5 is the Laplacian operator in 5 dimensions. The equation for the magnetic field,
$$\frac{\partial \mathbf{H}}{\partial t} = \text{curl}\,(\mathbf{v} \times \mathbf{H}) - \frac{1}{4\pi\sigma}\,\text{curl curl } \mathbf{H},$$

may similarly be replaced by the pair of equations

$$\varpi^3\,\Delta_5 P - \varpi^3 \frac{\partial P}{\partial t} = \frac{\partial(\varpi^2 P,\,\varpi^2 U)}{\partial(z, \varpi)}, \tag{3}$$

and
$$\varpi\,\Delta_5 T - \varpi \frac{\partial T}{\partial t} = \frac{\partial(T, \varpi^2 U)}{\partial(z, \varpi)} - \frac{\partial(V, \varpi^2 P)}{\partial(z, \varpi)}. \tag{4}$$

Three classes of problems have been studied with the aid of these equations: first, the equilibrium of a mass of fluid of infinite electrical conductivity; second, the decay time of a magnetic field in the presence of fluid motions; third, the stability of static equilibrium configurations. Problems in the first category are governed by equations (1) and (2), with $\partial/\partial t \equiv 0$. If in addition the velocity is zero, these equations possess the two general integrals (cf. Chandrasekhar and Prendergast [3])

$$\varpi^2 T = F(\varpi^2 P),$$

and
$$\Delta_5 P = -T \frac{d}{d(\varpi^2 P)} (\varpi^2 T) + \Phi(\varpi^2 P), \tag{5}$$

where F and Φ are arbitrary. Eq. (5) with $\Phi \equiv 0$, and $F(\varpi^2 P) = \alpha \varpi^2 P$, $\alpha =$ constant gives rise to a class of force-free fields, one example of which was studied by Lüst and Schlüter. The general solution for this case has been given by Chandrasekhar [4]. It has also been shown by Prendergast [5] that the case $F(\varpi^2 P) = \alpha \varpi^2 P$, $\Phi = \kappa$ determines a spherical equilibrium configuration in which the magnetic forces do not vanish. More recently an equilibrium solution embodying a toroidal velocity field has been obtained by Sykes [6].

For an assigned velocity field (V, U) Eqs. (3) and (4) govern the decay of a magnetic field in a fluid conductor. Cowling's [7] theorem on the impossibility of a self-excited axisymmetric dynamo with a purely poloidal field follows from Eq. (3); the analogous result for a toroidal dynamo has been proved from Eq. (4) by Backus and Chandrasekhar [8]. The problem of the prolongation of decay times by means of suitable patterns of fluid motion has been investigated by Chandrasekhar [9], who has obtained results which are valid for small velocities. An upper limit to the decay time of the external field of a sphere in which arbitrarily strong fluid motions are present has recently been derived by Backus [10]. He finds that in no case can the decay time be greater than four times that which would prevail in the absence of internal motions.

Stability problems have been treated using the linearized versions of Eqs. (1) to (4). The stability of Prendergast's equilibrium sphere has been investigated by this method and a variational principle has been obtained for the frequencies of oscillation. The stability of a force-free field in a compressible medium has been studied by S. K. Trehan [11], using an extension of the same technique.

REFERENCES

[1] Lüst, R. and Schlüter, A. Zs. f. Ap. **34**, 263, 1954.
[2] Chandrasekhar, S. Astrophys. J. **124**, 232, 1956.
[3] Chandrasekhar, S. and Prendergast, K. H. Proc. Nat. Acad. Sci., Wash., **42**, 5, 1956.
[4] Chandrasekhar, S. Proc. Nat. Acad. Sci. **42**, 1, 1956.
[5] Prendergast, K. H. Astrophys. J. **123**, 498, 1956.
[6] Sykes, J. B. Astrophys. J. **125**, 615, 1957.
[7] Cowling, T. G. Mon. Not. R. Astr. Soc. **94**, 39, 1934.
[8] Backus, G. and Chandrasekhar, S. Proc. Nat. Acad. Sci., Wash., **42**, 105, 1956.
[9] Chandrasekhar, S. Astrophys. J. **124**, 244, 1956.
[10] Backus, G. Astrophys. J. **125**, 500, 1957.
[11] Trehan, S. K. Astrophys. J. **126**, 429, 1957.

Discussion

Spitzer: As I shall report on Wednesday, I have obtained some exact results for the axisymmetric case which seem to suggest that fluid motions have relatively minor effect on the decay rate of a dipole field. In view of this result I would like to question how certain you are of the convergence of the series which you obtain in your deductions. Is it not possible that the higher modes would seriously modify the results?

Prendergast: Yes, this is quite true. The convergence of this series is a rather difficult question. It seems to be established that one can prolong the decay time in the axisymmetric case by a factor of 4 at least. Whether one can prolong it by a factor of 15 is extremely doubtful.

Blackett: I want to make two remarks about the use of the results of the rock magnetic data. If the reversely magnetized rocks do indicate a reverse field then the fluid dynamo producing the field must have properties like a flip-flop circuit with two positions of stability characterized by equal currents in opposite directions. The currents remain constant for a time of the order of 1 million years and then suddenly change over a period of some 10^4 years. However, one must not be too certain as yet that the earth's field has reversed. The possibilities of physical-chemical reversal are so numerous and complex that it is difficult to rule them out for certain. The crucial test of tracing a given reversal at a given moment of geological time over the major continents has not yet been achieved. Till this has been done, I think it would be wise to assume that reversals of the earth's field, though probable, have not finally been proved.

Prendergast: It is, of course, realized that there may be some resemblance with a flip-flop circuit. The discussion of the decay time here leans on considerations which are valid also when one has a flip-flop. The decay may go on for some time until the field gets weak, and then something happens, i.e. in the form of a reversal. On the other hand, one has not as yet found any mechanism which describes the reversal.

Schlüter: A special case of the static equilibrium models considered in the paper presented here is that where the magnetic field does not act at all on the matter ('force-free' magnetic fields). The corresponding equations have been generally solved by R. Lüst and A. Schlüter, if axial symmetry and proportionality between electrical current and magnetic field are supposed. If one restricts the geometry even more, namely to the symmetry of an infinite cylinder (all lines of force and of electric current form co-axial helices), then also the non-linear case can be solved generally. This model of a field can serve to demonstrate that force-free fields exist, which can transport along an axis of symmetry angular momentum around this axis.

Mestel: I wish to emphasize the seriousness of the assumption made here and in other recent papers that the material is incompressible. The equations of hydrostatic equilibrium then restrict the magnetic field to a certain shape; otherwise mass motions are set up. A similar result holds if the density is not constant but a function of pressure only. However, in a real non-degenerate star, an arbitrary magnetic field can be balanced by non-spherical variations in temperature as well as in pressure. In a convectively stable region the pressure field reacts on the meridional magnetic field not through the equation of support, but through the energy equation. As long as the magnetic field is not too large, this reaction is small, and so the star can support an arbitrary imposed field. Hence the results for barotropic and perfect-gas stars are *qualitatively* quite different, and so the barotropic or incompressible approximation is misleading. Whether the temperature field can adjust itself to support an arbitrary magnetic field depends on the rate at which the field is imposed. A proto-star condensing in the Kelvin–Helmholtz time scale, for example, will automatically adjust its density and temperature fields to balance the magnetic force. On the other hand a rapidly imposed field will lead to density variations accompanied by adiabatic temperature changes, so that the resulting pressure field will not in general balance the magnetic force. The resulting mass-motions and the distortion of its field are difficult to compute; but certainly assumptions of incompressibility or barotropy are again misleading.

B. EXPERIMENTS

PAPER 6

MAGNETO-HYDRODYNAMIC EXPERIMENTS

B. LEHNERT

Royal Institute of Technology, Stockholm, Sweden

ABSTRACT

Comparisons are made between magneto-hydrodynamics on cosmical and on laboratory scale. Magneto-hydrodynamic waves, turbulence, the generation of magnetic fields and thermal convection are discussed and a review is given of earlier experimental investigations. The possibilities are examined of realizing cosmical phenomena of this type in the laboratory.

I. INTRODUCTION

It often happens that astrophysical problems have a degree of complication high enough to hamper a rigorous theoretical approach. An imaginable way to obtain a solution is provided by the method of 'scaling down' the astrophysical configuration to laboratory dimensions in proportions such as to conserve its main properties. Thus, besides forming valuable tests of theory, magneto-hydrodynamic experiments may serve the purpose of solving astrophysical problems directly in the laboratory.

The significance of such model experiments depends upon the possibility of realizing cosmical conditions in the laboratory. This question can be answered by means of similarity laws stated for sets of configurations of different linear dimensions and equal physical character. It should be stressed that the form of the laws depends upon the physical properties which are stated to be conserved in a transformation from cosmical to laboratory dimensions. As a matter of fact, similarity laws for gaseous discharges, where the potential difference per mean free path is kept constant, are inapplicable to magneto-hydrodynamic waves (Alfvén[1]). In an accompanying paper Dr Block will discuss the conditions for model experiments on gaseous discharges in a magnetic field and apply the results to the auroral phenomenon.

2. CONDITIONS ON COSMICAL AND ON LABORATORY SCALE

The similarity laws of magneto-hydrodynamics are now formulated without inclusion of the Coriolis force. Relativistic effects and displacement- and convection currents are assumed to be negligible and the permeability to be nearly equal to its value *in vacuo*. The basic equations are

$$\frac{\partial \mathbf{B}}{\partial t} + \lambda \operatorname{curl}^2 \mathbf{B} = \operatorname{curl} (\mathbf{v} \times \mathbf{B}); \quad \lambda = 1/\mu\sigma, \tag{1}*$$

$$\operatorname{div}(\rho\mathbf{v}) + \frac{\partial \rho}{\partial t} = 0, \tag{2}$$

$$\rho \frac{d\mathbf{v}}{dt} = \frac{1}{\mu}(\operatorname{curl} \mathbf{B}) \times \mathbf{B} - \nabla p - \rho\nabla\phi + \eta\nabla^2\mathbf{v} + \tfrac{1}{3}\eta\nabla(\operatorname{div} \mathbf{v}), \tag{3}$$

$$\rho \frac{d}{dt}(e + \tfrac{1}{2}\mathbf{v}^2) = -p \operatorname{div} \mathbf{v} + \mathbf{v}.[-\nabla p + \eta\nabla^2\mathbf{v} + \tfrac{1}{3}\eta\nabla(\operatorname{div} \mathbf{v})]$$
$$+ \eta(\operatorname{curl} \mathbf{v})^2 + \tfrac{4}{3}\eta(\operatorname{div} \mathbf{v})^2$$
$$+ 4\eta\left[\frac{\partial w}{\partial y}\frac{\partial v}{\partial z} + \frac{\partial u}{\partial z}\frac{\partial w}{\partial x} + \frac{\partial v}{\partial x}\frac{\partial u}{\partial y} - \frac{\partial u}{\partial x}\frac{\partial v}{\partial y} - \frac{\partial v}{\partial y}\frac{\partial w}{\partial z} - \frac{\partial w}{\partial z}\frac{\partial u}{\partial x}\right] + W \tag{4}$$

and

$$f(p, \rho, T) = 0, \tag{5}$$

where $\mathbf{B}$ is the magnetic field strength, σ the electric conductivity, ρ the density, p the pressure, ϕ the gravitation potential, η the viscosity, $e = e(\rho, T)$ the internal (thermal) energy per unit mass, $\mathbf{v} = (u, v, w)$ the velocity, and T the temperature. Equation (4) expresses the conservation of energy. The left-hand member is the change of internal and kinetic energy per unit time in a co-ordinate system which follows the motion of a fluid element. This is supplied by the right-hand member which is the sum of the rate at which the hydrodynamic stresses are doing work on an element of the fluid (Lamb[2]) and the rate W at which energy is supplied from other sources than these stresses. Eq. (5) is the equation of state. We have

$$W = i^2/\sigma + \mathbf{v}.(\mathbf{i} \times \mathbf{B}) - \rho\mathbf{v}.\nabla\phi + \theta\nabla^2 T, \tag{6}$$

where $\mathbf{i}$ represents the current density and θ the thermal conductivity. The two first terms in Eq. (6) are the power $\mathbf{E}.\mathbf{i} = (\mathbf{i}/\sigma - \mathbf{v} \times \mathbf{B}).\mathbf{i}$ supplied by the electromagnetic field, the third is the work done per unit time by the gravitation field and the fourth is the heat supplied through thermal conduction.

We assume that the internal energy has the form

$$e = c_v T, \tag{7}$$

---

* In this paper Ohm's law is assumed to have the simple form $\mathbf{i} = \sigma(\mathbf{E} + \mathbf{v} \times \mathbf{B})$.

with c_v as the specific heat at constant volume. Eq. (7) holds for polytropic gases and with good approximation for experiments with liquid metals. We also introduce dimensionless variables according to the following transformations:

$$x_\alpha = L_c x'_\alpha; \quad t = t_c t'; \quad L_c/t_c = V_c, \tag{8}$$

$$B_\alpha = B_c B'_\alpha; \quad B_c(\mu\rho_c)^{-\frac{1}{2}} = V; \quad v_\alpha = V_c v'_\alpha \tag{9}$$

and $$p = p_c p'; \quad \phi = \phi_c \phi'; \quad T = T_c T', \tag{10}$$

the indices (α), (c) and $(')$ implying an arbitrary component, characteristic quantities and dimensionless quantities, respectively. Further, expressions (6) and (7) and the conservation theorems, (2) and (3), of mass and momentum are introduced into the energy theorem (4) (compare Lehnert[3]). After introduction of transformations (8), (9) and (10) the basic equations (1), (2), (3) and (4) are easily shown to become

$$\frac{\partial \mathbf{B}'}{\partial t'} + \frac{\lambda}{V_c L_c} \operatorname{curl}' \mathbf{B}' = \operatorname{curl}' (\mathbf{v}' \times \mathbf{B}'), \tag{11}$$

$$\operatorname{div}' (\rho' \mathbf{v}') + \frac{\partial \rho'}{\partial t'} = 0, \tag{12}$$

$$\frac{d\mathbf{v}'}{dt'} = -\frac{p_c}{\rho_c V_c^2} \frac{1}{\rho'} \nabla' p' + \frac{V^2}{V_c^2} \frac{1}{\rho'} (\operatorname{curl}' \mathbf{B}') \times \mathbf{B}'$$

$$- \frac{\phi_c}{V_c^2} \nabla' \phi' + \frac{\nu}{V_c L_c} \frac{1}{\rho'} [\nabla'^2 \mathbf{v}' + \tfrac{1}{3} \nabla'(\operatorname{div}' \mathbf{v}')] \tag{13}$$

and $$\frac{dT'}{dt'} = \frac{p_c}{\rho_c V_c^2} \frac{V_c^2}{c_v T_c} \frac{p'}{\rho'^2} \frac{d\rho'}{dt'} + \frac{\lambda}{V_c L_c} \frac{V^2}{V_c^2} \frac{V_c^2}{c_v T_c} \frac{1}{\rho'} (\operatorname{curl}' \mathbf{B}')^2$$

$$+ \frac{\nu}{V_c L_c} \frac{V_c^2}{c_v T_c} \frac{1}{\rho'} \left\{ (\operatorname{curl}' \mathbf{v}')^2 + \tfrac{4}{3}(\operatorname{div}' \mathbf{v}')^2 \right.$$

$$+ 4\left[\frac{\partial w'}{\partial y'} \frac{\partial v'}{\partial z'} + \frac{\partial u'}{\partial z'} \frac{\partial w'}{\partial x'} + \frac{\partial v'}{\partial x'} \frac{\partial u'}{\partial y'} - \frac{\partial u'}{\partial x'} \frac{\partial v'}{\partial y'} - \frac{\partial v'}{\partial y'} \frac{\partial w'}{\partial z'} - \frac{\partial w'}{\partial z'} \frac{\partial u'}{\partial x'} \right]\right\}$$

$$+ \frac{\kappa}{V_c L_c} \frac{1}{\rho'} \nabla'^2 T'. \tag{14}$$

$\nu = \eta/\rho_c$ is the kinematic viscosity and $\kappa = \theta/\rho_c c_v$ the thermometric conductivity.* The basic equations contain the characteristic parameters V_c/V, $V_c L_c/\lambda$, $V_c L_c/\nu$, $V_c L_c/\kappa$, $V_c^2/c_v T_c$, $p_c/\rho_c V_c^2$ and ϕ_c/V_c^2. If the dimensionless variables $x'_\alpha, t', \ldots, T'$ are given and these parameters are kept constant,

* After the completion of this manuscript the author has been informed that the generalization of the energy theorem to a form including viscous dissipation and heat diffusion has already been given by Dr Baños in the paper presented at this symposium.

a set of similar configurations of different linear dimensions can be constructed. The ratio between forces of different types is invariant at corresponding points within such a set.

Consider a strongly pronounced magneto-hydrodynamic phenomenon where the magnetic energy is of the same order of magnitude as the kinetic energy, i.e. $V^2 \approx V_c^2$. Dissipation of energy due to Joule heat, viscosity and heat conduction will be negligible if

$$B_c L_c / \lambda (\mu \rho_c)^{\frac{1}{2}} \gg 1, \tag{15}$$

$$B_c L_c / \nu (\mu \rho_c)^{\frac{1}{2}} \gg 1 \tag{16}$$

and $$B_c L_c / \kappa (\mu \rho_c)^{\frac{1}{2}} \gg 1, \tag{17}$$

which relations are easily deduced from the forms of the characteristic parameters. Condition (15) implies that the magnetic flux is almost 'frozen' in the fluid (Lundquist [4]). Condition (16) corresponds to the same statement for vortex lines and condition (17) corresponds to a negligible 'slip' between matter and isothermal surfaces being caused by thermal conduction. When the compression work is negligible the isothermal surfaces will also be 'frozen' in the fluid.

Some examples are given in Table 1, where parts of the data have been taken from works by Lyon [5], Jackson [6], Elsasser [7, 8], Alfvén [1], Kuiper [9, 10] Chandrasekhar [11] and Allen [12]. The considerable increase of the effective thermometric conductivity caused by radiation at high temperatures has been taken into account. From Eq. (15) and Table 1 is seen that Joule dissipation plays an important role in magneto-hydrodynamic experiments. An exact similarity transformation of cosmical configurations to the laboratory is rendered impossible, even with the best available conductors such as liquid sodium, sodium-potassium alloy and ionized gases. For the characteristic linear dimensions given in Table 1 viscous dissipation is seen to be of minor importance in experiments and in most cosmical applications. This does not mean that viscous dissipation is negligible under all circumstances, which is readily seen from experiments on flow in ducts in transverse magnetic fields. In such cases a flow exists in narrow boundary layers with considerable viscous forces and the magnetic and kinetic energies are not necessarily of the same order of magnitude.

3. MAGNETO-HYDRODYNAMIC WAVES

The question now arises how significant is the departure of laboratory conditions from the real astrophysical situation. This may be discussed by means of a concrete, simple example. Consider a conducting incompressible

Table 1. *Some characteristic data for magneto-hydrodynamic experiments and for cosmical applications of magneto-hydrodynamics*

	Temperature T_0 (°K)	Magnetic field B_0 (Vs/m²)	Linear dimension L_0 (m)	Density ρ_0 (kg/m³)	Electromagnetic viscosity† λ (m²/s)	Kinematic viscosity† ν (m²/s)	Thermometric conductivity† κ (m²/s)	$\dfrac{B_0 L_0}{\lambda \sqrt{(\mu\rho_0)}}$‡	$\dfrac{B_0 L_0}{\nu \sqrt{(\mu\rho_0)}}$‡	$\dfrac{B_0 L_0}{\kappa \sqrt{(\mu\rho_0)}}$‡
Experiments										
Mercury	293	1	0·1	1350	0·780	$1·14 \times 10^{-7}$	$4·43 \times 10^{-6}$	1·0	$6·8 \times 10^{6}$	$1·6 \times 10^{5}$
Sodium	373	1	0·1	928	0·0765	$6·31 \times 10^{-7}$	$67·3 \times 10^{-6}$	38	$4·6 \times 10^{6}$	$4·2 \times 10^{4}$
Gallium	308	1	0·1	6093	0·216	$3·11 \times 10^{-7}$	$14·0 \times 10^{-6}$	5·3	$3·7 \times 10^{6}$	$7·4 \times 10^{4}$
Tin	513	1	0·1	6910	0·378	$2·76 \times 10^{-7}$	$20·0 \times 10^{-6}$	2·8	$3·8 \times 10^{6}$	$5·3 \times 10^{4}$
Sodium-potassium (22 % Na, 78 % K)*	293	1	0·1	905	0·270	$7·86 \times 10^{-7}$	$24·2 \times 10^{-6}$	11·0	$3·8 \times 10^{6}$	$1·2 \times 10^{5}$
Ionized gas hydrogen	10^{5}	0·1	0·1	10^{-7}	15	10	10	2000	4000	4000
Cosmical applications										
Earth's interior	10^{4}	10^{-3}?	2×10^{6}	10^{4}	1	10^{-6}?	2×10^{-5}	2×10^{4}	2×10^{10}	10^{9}
Sunspots	4×10^{3}	0·2	10^{7}	0·1	20	10^{-2}‖	10^{9}¶‖	10^{9}	2×10^{11}	10
Solar granulation	6×10^{3}	10^{-2}	10^{6}	10^{-4}	100§	10‖	10^{9}¶‖	10^{7}	10^{8}	1
Magnetic variable stars	10^{8}	1	10^{10}	10^{8}?	1	3×10^{-7}‖	10^{10}?¶‖	3×10^{11}	10^{18}	30
Interstellar space; more condensed regions	10^{4}	10^{-9}?	10^{20}	10^{-21}?	10^{3}?§	10^{17}‖	?	3×10^{21}	3×10^{7}	?
Interplanetary space	10^{5}	10^{-9}?	10^{13}	10^{-20}	10^{2}?§	10^{16}‖	?	10^{16}	10	?
Corona	10^{6}	10^{-4}?	10^{9}	10^{-15}	1?§	10^{13}‖	?	3×10^{15}	300	?

* Experiments with sodium-potassium alloy are being planned by Lochte-Holtgreven [13].

† The electromagnetic viscosity is defined by $\lambda = 1/\mu\sigma$, where σ is the electrical conductivity, the kinematic viscosity by $\nu = \eta/\rho_0$, where η is the coefficient of viscosity, and the thermometric conductivity by $\kappa = \theta/\rho_0 c_v$, where θ is the coefficient of thermal conductivity and c_v the specific heat at constant volume.

‡ If the parameters of the third and second columns from the right are much greater than unity conditions become favourable for a strongly developed magneto-hydrodynamic phenomenon. In such a case there is almost no 'slip' between matter and magnetic field lines', as well as between matter and vorticity lines, respectively. If, in addition, the parameter of the first column from the right is much greater than unity there is almost no 'slip' between matter and isothermal surfaces, provided that the compression work can be neglected.

§ Values given by the resistivity (the inverted 'parallel conductivity') have been chosen.

‖ Radiative viscosity is not included but may increase ν by a factor of about 100 (Cowling [9], p. 555). Turbulent viscosity is not included in ν.

¶ Effective value of the thermometric conductivity with inclusion of radiative exchange of heat. Figures give the order of magnitude. The conduction part is affected by the magnetic field.

liquid between two parallel, infinitely conducting planes in a perpendicular, homogeneous external magnetic field $\mathbf{B_0}$ (Fig. 1). We assume a perturbation field $\mathbf{b} = (b, 0, 0)$ to exist as given by Fig. 1. From the basic equations is easily deduced the special solution

$$b(z, t) = \{b_1 \exp [j(\lambda - \nu) k^2 (\zeta^2 - 1)^{\frac{1}{2}} t]$$

$$+ b_2 \exp [-j(\lambda - \nu) k^2 (\zeta^2 - 1)^{\frac{1}{2}} t]\} \exp [-(\lambda + \nu) k^2 t] \sin kz, \quad (18)$$

with $\quad k = 2\pi/L, \quad \lambda > \nu, \quad \zeta = 2B_0/(\mu\rho)^{\frac{1}{2}} k^2 (\lambda - \nu) \quad$ and $\quad j = (-1)^{\frac{1}{2}}.$

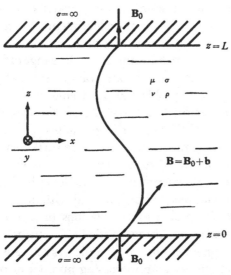

Fig. 1. Plane state of motion of an electrically conducting, viscous liquid between two infinitely conducting planes in a homogeneous, perpendicular magnetic field, $\mathbf{B_0}$.

The initial perturbation field is $b(z, 0) = (b_1 + b_2) \sin kz$. The magnetic field lines will perform damped oscillations if $\zeta > 1$, whereas the motion will be aperiodic if $\zeta \leqslant 1$ (Lehnert[14]). The critical aperiodic case, $\zeta = 1$, is given by

$$B_0 L/\lambda(\mu\rho)^{\frac{1}{2}} = \pi, \quad (19)$$

where we have put $\lambda \gg \nu$, which is a good approximation in most cases according to Table 1. The third column from the right in the same table shows that magneto-hydrodynamic waves exist in the laboratory only in the strongest available magnetic fields. A rough approach to the astrophysical situation of virtually undamped waves, i.e. waves where the damping length at least exceeds 10 wave-lengths, say, requires magnetic field strengths of 2 Vs/m² (= 20,000 gauss) within dimensions of 0·25 m for liquid sodium and of 1 m for sodium-potassium alloy.

55

Investigations on forced magneto-hydrodynamic oscillations in mercury and in liquid sodium have already been carried out by Lundquist [15, 16] and Lehnert [17]. Torsional oscillations were fed into the lower end of a cylindrical column of liquid with a magnetic field directed along the axis of symmetry. The amplitude ratio and phase difference between the lower end and the upper free surface were studied as functions of frequency and magnetic field strength. In the case of high conductivity a strong resonance peak and a phase shift of 180° would be expected to occur at the free surface. The peak occurs when the length of the column corresponds to a quarter of a wave-length of transverse magneto-hydrodynamic waves. However, only a very flat maximum arises in the experiment with mercury; this is not a resonance phenomenon but merely a dispersion effect. The data of the mercury experiment correspond to the aperiodic range of free motion. Thus, the forced oscillations are 'degenerate waves', somewhat similar to the temperature 'wave' in a problem of heat conduction with periodically varying temperature at the boundary. In the liquid sodium experiment there occurs only a slightly developed resonance phenomenon, showing that Joule dissipation is of major importance even in this case. Consequently, both these experiments have to be regarded merely as verifications of theory and not as model experiments on cosmical physics.

An improvement can possibly be obtained with an ionized gas. Bostick and Levine [18, 19, 20] have observed oscillations in a toroidal plasma embedded in an annular stationary magnetic field. The oscillations are interpreted as standing magneto-hydrodynamic waves. An increase of the magnetic field strength gives an increasing number of oscillating modes. This could be explained by the passage of an increasing number of higher modes from the aperiodic range into the range of wave motion, provided that the damping is small enough. However, it should be pointed out that the configuration of this experiment is bounded by rigid walls; the oscillations do not necessarily consist purely of transverse Alfvén waves and the dissipation is mainly determined by the narrow cross section of the tube.

4. TURBULENCE

Turbulence is likely to be a common phenomenon in masses of cosmic dimensions. It is seen from Table 1 that a large part of the turbulent spectrum of wave-lengths will be situated in the periodic range of motion in most cosmical applications. In the simple case of Fig. 1 a turbulent field of cosmical dimensions will consist of a spectrum of waves, the decay time of which is not affected by an external magnetic field B_0 (Eq. (18)). In the

56

laboratory, on the other hand, the spectrum falls entirely within the aperiodic range of motion and a strong suppression of turbulence is caused by the external field, as found experimentally by Hartmann and Lazarus [21, 22].

In spite of this difference between cosmical and laboratory conditions experimental investigations may serve the purpose of deepening the understanding of such a complicated phenomenon as magneto-turbulence. The experiments by Hartmann and Lazarus show that the onset of turbulence in a channel is delayed by a transverse magnetic field. A plot of the critical velocity against the corresponding field strength gives a linear relationship (Lehnert [23]). This is interpreted theoretically in terms of the characteristic parameters mentioned earlier (Lehnert [23], Lundquist [4], Lock [24]). Murgatroyd [25, 26] has previously extended Hartmann's measurements to a wider range of data.

Experiments on magneto-hydrodynamic channel flow of liquid metals are of great technical importance. Further contributions in this field have been given by Kolin [27, 28], Arnold [29], Murgatroyd [30], Shercliff [31, 32, 33, 34, 35], Greenhill [36, 37], Robin [38], and Barnes [39].

5. GENERATION OF MAGNETIC FIELDS

The velocity field is capable of stretching and twisting the magnetic field lines in a fluid of high conductivity. The process, which may take place in the form of turbulence or as a regular macroscopic phenomenon, provides a mechanism for converting kinetic energy into magnetic. For the generation of the magnetic fields of the stars and of the earth a feed-back mechanism has been suggested in the form of a magneto-hydrodynamic dynamo. The possibility of making an experiment on such a dynamo may be discussed with the help of the simple configuration of Fig. 2 (Bullard and Gellman [40]). The condition for the dynamo to be self-excited becomes $v \approx 40\lambda/a$, where $v = \omega a$ is the velocity at the periphery of the rotating disc, a is the radius of the dynamo and the resistance has been put equal to about $10a/\sigma a^2$. For a volume of 100 litres of liquid sodium $a = 0.3$ m and $v = 10$ m/sec. Besides being rigid and multiply connected this dynamo has the important feature of being asymmetric. It has been shown that a complete symmetry gives no feed-back of a singly connected fluid dynamo and that the feed-back is caused by asymmetric modes. Consequently, the fluid dynamo will be much less efficient than the simple dynamo in Fig. 2. Velocities considerably higher than 10 m/sec will probably be needed for liquid sodium in the example discussed and the experiment will be difficult to carry out.

57

A strong twisting of a magnetic field gives rise to an instability which may serve as a mechanism for an increase of magnetic energy (Alfvén[41], Lundquist[42]). The onset of the instability may possibly be studied in an experiment with liquid sodium. The field will be twisted to a high degree within a vessel of the size of 0·3 m if the differential velocity highly exceeds the diffusion velocity of the field lines, which is about 0·5 m/sec.

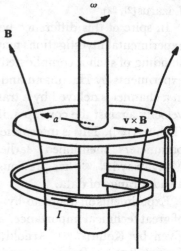

Fig. 2. A simple dynamo (Bullard and Gellman [40]).

Another mechanism for the generation of magnetic fields is provided by the diffusion of charged particles in a turbulent field. This causes a separation of charge and gives rise to electric currents and magnetic fields (Biermann[43], Schlüter [44]). An experimental confirmation of the theory has been given with rotating flames and with mercury by Lochte-Holtgreven and Schilling[45, 46, 47] and by Burhorn, Griem and Lochte-Holtgreven[48, 49, 50].

6. INSTABILITY AND THERMAL CONVECTION

A complete description of the great variety of different types of magneto-hydrodynamic instability is not within the scope of the present paper. We may only mention an interesting type of torus-shaped disturbance in a plasma, which will be described in an accompanying paper by Dr Bostick (see also Bostick[51]). Some additional experiments on instability are given in the references of this paper (Colgate[52], Lehnert[53, 54]).

The following discussion will be limited to the onset of thermal convection in a magnetic field. In the laboratory cellular convection will be set up in a layer of fluid which is heated from below with a sufficiently large vertical temperature gradient. Table 1 shows that a cell size of 1 cm in mercury will correspond to the aperiodic range of magneto-hydrodynamic motion. Since $\kappa < \lambda$ the diffusion of isothermal surfaces occurs slowly as compared to the diffusion of magnetic field lines. The dissipation is increased considerably by the Joule heat and the onset of thermal convection is a sensitive function of the magnetic field strength as shown theoretically by Thompson[55] and by Chandrasekhar[11]. The theory agrees with experiments in mercury performed by Nakagawa[56], Jirlow[57] and Lehnert and Little[58].

However, thermal convection under astrophysical conditions differs fundamentally from that of the experiments. Radiative exchange of heat increases the effective thermometric conductivity and $\kappa \gg \lambda \gg \nu$. Dissipation due to Joule heat and gas viscosity is relatively small. A large range of wavelengths of convective motion will correspond to the periodic case of virtually undamped magneto-hydrodynamic waves. For these wave-lengths the magnetic field lines are 'frozen' in matter, whereas the isothermal surfaces diffuse rapidly through it. Variations of the strength of an external magnetic field do not influence dissipation in such a case and the onset of convection is not critical with respect to the field strength, as shown by Chandrasekhar[11]. For wave-lengths in the periodic range the onset of convection occurs as over-stability, i.e. in the form of oscillations. This may be understood by the fact that the field lines are 'frozen' in matter and cannot be twisted indefinitely by a stationary cellular pattern.

7. CONCLUSIONS

The examples given in this paper clearly show the difficulties involved in making a reproduction of cosmical phenomena in the laboratory. Further progress in the problem of magneto-hydrodynamic model experiments can only be made with the largest achievable magnets or with ionized gases.

REFERENCES

[1] Alfvén, H. *Cosmical Electrodynamics* (Oxford University Press, 1950).
[2] Lamb, H. *Hydrodynamics*, Dover Publ., New York, sixth ed., 1932, p. 580.
[3] Lehnert, B. *Tellus*, 8, no. 2, 241, 1956.
[4] Lundquist, S. *Ark. Fys.* 5, no. 15, 297, 1952.
[5] Lyon, N. *Liquid metals handbook*, Atomic Energy Commission, Dept. of the Navy, Washington D.C., NAVEXOS P-733 (Rev.) 2nd ed., June 1952.
[6] Jackson, B. *Liquid metals handbook, sodium-NaK supplement*, Atomic Energy Commission, Dept. of the Navy, Washington D.C., NAVEXOS P-733 (Rev.) 3rd ed., July 1955.
[7] Elsasser, W. M. *Rev. Mod. Phys.* 22, no. 1, 28, 1950.
[8] Elsasser, W. M. *Phys. Rev.* 95, no. 1, 1, 1954.
[9] *The Sun*, ed. by Kuiper, G. P. Chicago University Press, 1953.
[10] *The Earth as a Planet*, ed. by Kuiper, G. P. Chicago University Press, 1954.
[11] Chandrasekhar, S. *Phil. Mag.* (7), 43, no. 340, 501, 1952 and 45, 1177, 1954.
[12] Allen, C. W. *Astrophysical Quantities* (University of London, The Athlone Press, 1955).
[13] Lochte-Holtgreven, W. Private communication, 1956.
[14] Lehnert, B. *Quart. Appl. Math.* 12, no. 4, 321, 1955.
[15] Lundquist, S. *Nature, Lond.* 164, 145, 1949.
[16] Lundquist, S. *Phys. Rev.* 76, 1805, 1949.
[17] Lehnert, B. *Phys. Rev.* 94, 815, 1954.

[18] Bostick, W. H. and Levine, M. A. *Phys. Rev.* **87**, 671, 1952.
[19] Bostick, W. H. and Levine, M. A. *Magneto-hydrodynamic waves generated in an ionized gas in a toroidal tube having an annular D–C magnetic field*, Sci. Rep. no. 3, 1952, Dept. of Physics, Tufts College, Mass.
[20] Bostick, W. H. and Levine, M. A. *Phys. Rev.* **97**, 13, 1955.
[21] Hartmann, J. *Math.-fys. Medd.* **15**, 1937, no. 6.
[22] Hartmann, J. and Lazarus, F. *Math.-fys. Medd.* **15**, 1937, no. 7.
[23] Lehnert, B. *Ark. Fys.* **5**, no. 5, 69, 1952.
[24] Lock, R. C. *Proc. Roy. Soc.* A, **233**, 105, 1955.
[25] Murgatroyd, W. *Nature, Lond.* **171**, 217, 1953.
[26] Murgatroyd, W. *Phil. Mag.* (7), **44**, 1348, 1953.
[27] Kolin, A. *Rev. Sci. Instrum.* **16**, 109, 1945.
[28] Kolin, A. *Rev. Sci. Instrum.* **16**, 209, 1945.
[29] Arnold, J. S. *Rev. Sci. Instrum.* **22**, 43, 1951.
[30] Murgatroyd, W. *The model testing of electromagnetic flow meters*, Atomic Energy Res. Est., Harwell, Rep. X/R, 1053, 6+iii pp., 1953.
[31] Shercliff, J. A. *Proc. Camb. Phil. Soc.* **49**, 136, 1953.
[32] Shercliff, J. A. *The theory of the d.c. electromagnetic flow meter for liquid metals*, Atomic Energy Res. Est., Harwell, Rep. X/R, 1052, 23+iv pp., 1953.
[33] Shercliff, J. A. *J. Appl. Phys.* **25**, 817, 1954.
[34] Shercliff, J. A. *J. Sci. Instrum.* **32**, 441, 1955.
[35] Shercliff, J. A. *Proc. Roy. Soc.* A, **233**, 396, 1955.
[36] Greenhill, M. *Electromagnetic pumps and flowmeters*, Atomic Energy Res. Est., Harwell, Inf./Bib. 93, 3, 1954.
[37] Greenhill, M. *Electromagnetic pumps and flowmeters*, Atomic Energy Res. Est., Harwell, Inf./Bib. 93, 4 (second edition), 1955.
[38] Robin, M. *J. Rech. Cent. Nat. Rech. Sci.* **5**, 187, 1953.
[39] Barnes, A. H. *Nucleonics*, **11**, 16, 1953.
[40] Bullard, E. C. and Gellman, H. *Phil. Trans.* **247**, 213, 1954.
[41] Alfvén, H. *Tellus*, **2**, 74, 1950.
[42] Lundquist, S. *Phys. Rev.* **83**, 307, 1951.
[43] Biermann, L. *Z. Naturf.* **5a**, Heft 2, 65, 1950.
[44] Schlüter, A. *Z. Naturf.* **5a**, Heft 2, 72, 1950.
[45] Lochte-Holtgreven, W. and Schilling, P. O. *Naturwissenschaften*, **40**, 387, 1953.
[46] Schilling, P. and Lochte-Holtgreven, W. *Nature, Lond.* **172**, 1054, 1953.
[47] Schilling, P. and Lochte-Holtgreven, W. *Z. Naturf.* **9a**, Heft 6, 520, 1954.
[48] Burhorn, F., Griem, H. and Lochte-Holtgreven, W. *Naturwissenschaften*, **40**, 387, 1953.
[49] Burhorn, F., Griem, H. and Lochte-Holtgreven, W. *Nature, Lond.* **172**, 1053, 1954.
[50] Burhorn, F., Griem, H. and Lochte-Holtgreven, W. *Z. Phys.* **137**, 175, 1954.
[51] Bostick, W. H. *Experimental study of ionized matter projected across a magnetic field*, Univ. of California, Radiation Laboratory, Livermore, California, Contract no. W-7405-eng-48, 1956.
[52] Colgate, S. A. *Liquid sodium instability experiment, Part I, 1955*, Univ. of California, Radiation Laboratory, Livermore Site, Livermore, California, Contract no. W-7405-eng-48, UCRL-4560, Cy 6.
[53] Lehnert, B. *Tellus*, **4**, no. 1, 63, 1952.
[54] Lehnert, B. *Proc. Roy. Soc.* A, **233**, 299, 1955.
[55] Thompson, W. B. *Phil. Mag.* (7), **42**, 1417, 1951.
[56] Nakagawa, Y. *Nature, Lond.* **175**, 417, 1955.
[57] Jirlow, K. *Tellus*, **8**, no. 2, 252, 1956.
[58] Lehnert, B. and Little, N. C. *Tellus*, **9**, 97, 1957

EXPERIMENTAL WORK AT THE UNIVERSITY OF CHICAGO ON THE ONSET OF THERMAL INSTABILITY IN A LAYER OF FLUID HEATED FROM BELOW

YOSHINARI NAKAGAWA

Enrico Fermi Institute of Nuclear Studies, University of Chicago, U.S.A.

AND

KEVIN H. PRENDERGAST

Yerkes Observatory, University of Chicago, U.S.A.

This paper will summarize the experimental work at the University of Chicago on the problem of the onset of thermal instability in a layer of fluid heated from below. The purpose of this work has been to test certain theoretical predictions of the Rayleigh number at which instability sets in, and to determine the type of instability which appears at the critical point. The earlier experiments of this series were done at the hydrodynamics laboratory of the University of Chicago in connexion with a program of meteorological reseach [1, 2, 3, 4]. The current work is being done at the newly organized hydromagnetics laboratory of the Enrico Fermi Institute of Nuclear Studies. This laboratory utilizes the magnet of the old Chicago cyclotron, with pole pieces 92·7 cm in diameter and a gap of 22·1 cm. The magnet was reconstructed to allow the field strength to be varied from o to 13,000 gauss; the field is uniform to better than 1 % over the experimental area. The new laboratory is under the administrative supervision of Professors S. K. Allison and S. Chandrasekhar; the experiments are being done by Y. Nakagawa. The theoretical investigations are primarily the work of Chandrasekhar [5, 6, 7, 8, 9, 10] and it will be convenient to review some of his results before discussing the experiments.

The general situation which is envisaged is the following: suppose that a horizontal layer of fluid is heated from below in such a way as to establish a mean adverse temperature gradient $\beta = -dT/dz$. In addition to the force of gravity $\mathbf{g}$ acting on the fluid there may be a prevailing magnetic field $\mathbf{H}$, or the layer may be in rotation about some axis with an angular velocity $\mathbf{\Omega}$. The fluid may be confined between rigid boundaries, or one or both of the bounding surfaces may be free. For simplicity we will consider only the case where $\mathbf{g}$, Ω and $\mathbf{H}$ are parallel and both surfaces are free. For very small rates of heating (that is, β very small) the transport of heat will take place by conduction. If the heating rate is gradually increased, however, the conductive regime eventually breaks down and the transport of heat thereafter takes place by convection. At the transition point a regular pattern of vertical convective cells appears whose horizontal wave number may be denoted by a. For a cell of hexagonal section with sides of length L,

$$a^2 = \left(\frac{4\pi}{3L}\right)^2 d^2,$$

where d is the depth of the layer.

The onset of instability is treated as a problem in perturbation theory. The solutions of the first-order equations for the velocity, the temperature and the magnetic field are assumed to be of the form

$$f(x, y)\, g(z)\, e^{pt},$$

where $f(x, y)$ is common to all the physical variables of the problem. One then eliminates all of these variables except the velocity of vertical motion $w(z)$; this process leads to an ordinary linear differential equation of high order for $w(z)$ which, together with the boundary conditions, constitutes an eigenvalue problem for p. The adverse temperature gradient β enters this equation only through the Rayleigh number

$$R = \frac{\alpha \beta g}{\kappa \nu}\, d^4,$$

where α is the coefficient of volume expansion, κ is the thermometric conductivity and ν is the kinematic viscosity. Similarly, the magnetic field $\mathbf{H}$ enters only through the dimensionless parameter

$$Q = \frac{\sigma \mu^2 H^2}{\rho \nu}\, d^2,$$

where ρ is the density, σ the electrical conductivity, and μ is the permeability. The dependence of the solution on the angular velocity Ω is determined by the value of the Taylor number

$$T = \frac{4\Omega^2}{\nu^2}\, d^4.$$

For a solution $w(z)$ satisfying the boundary conditions, the characteristic equation of the problem is of the form

$$F(R, a^2, p;\quad Q, T) = 0.$$

The meaning of this equation is the following: for an adverse temperature gradient given by R, a perturbation whose horizontal wave number is a will behave as e^{pt}. If the real part of p is greater than zero the perturbation will grow with time—if the real part is less than zero, it will decay. The fluid will first become unstable *in the mode characterized by* a for R such that the real part of p is zero. In particular, the lowest value of R (regardless of a^2) for which the real part of p is equal to zero gives the value of the adverse temperature gradient for which *any* sort of instability can occur. This value will be called R_c. The type of instability which appears at the critical point is said to be *convective* if both the real and imaginary parts of p are zero, or *overstable* if the imaginary part of p is not zero. In the case of convection the perturbation increases monotonically with time—in the case of over-stability the perturbation grows as an oscillation of increasing amplitude.

With these preliminary remarks we may draw up a table listing the experiments which have already been performed, and those which are now in progress.

In each of these experiments the problem is to determine the temperature gradient β_c (and hence the critical Rayleigh number R_c) at which the fluid first becomes unstable. One must further ascertain whether the instability is of the convective or the overstable type, and if the latter is the case the period of oscillation must also be determined. In the latest version of the apparatus the fluid is contained in a shallow Pyrex dish which is electrically heated from below by means of a non-inductive resistance coil. The upper surface of the fluid is cooled by nitrogen to facilitate the establishment of a linear temperature gradient. The actual gradient is measured by means of a non-inductive 9-element copper-constantan thermopile immersed in the liquid. The output of the thermopile goes through a D.C. amplifier and is displayed on an Easterline-Angus pen recorder. The accuracy with which β can be determined with this equipment is about $\pm 0.05°$ C/cm; one can, however, see fluctuations in β whose amplitude is only $0.01°$ C/cm. In the present set-up (experiment IV of Table 1) the entire apparatus is placed between the poles of the cyclotron magnet and rotated about a vertical axis. The circuits for the heating coil and thermopile are brought out of the experimental area through mercury-filled troughs.

In order to determine the critical Rayleigh number one measures a series of temperature gradients corresponding to various rates of heating. A typical set of recordings from the thermopile is shown in Plate I. (These data were taken during the experiment on convection in a rotating layer of water [2].) The three tracings give β as a function of time for $\beta < \beta_c$ (top), $\beta \simeq \beta_c$ and $\beta > \beta_c$. The times at which the heating coil was turned on and off are indicated. In each case the value of β is estimated from the highest portion of the trace. Plate II is taken from the experimental data on overstability in a rotating layer of mercury, [2] again with $\beta < \beta_c$ (top), $\beta \simeq \beta_c$ and $\beta > \beta_c$. Notice particularly the wiggles which appear in the middle and bottom traces—these establish the presence of overstability and enable the period of oscillation to be determined. An enlargement of the last trace appears in Plate III.

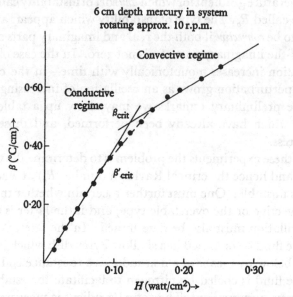

Fig. 1. Determination of β_c. (From D. Fultz and Y. Nakagawa[2].)

The final determination of β_c is facilitated by plotting β against the heating rate, as shown in Fig. 1. For the case illustrated, β is a linear function of the heating rate H for $H < 0.08$ watts/cm² (corresponding to a definitely conductive regime), and again for $H > 0.16$ watts/cm² (corresponding to convection). Between these limits, however, there is a transition region in which it is possible that the convective cells have started to form near the bottom of the fluid but do not yet extend to the top. The adopted value of β_c is obtained by extrapolating the linear portions of the curve.

The procedure outlined above is sufficient to determine one value of the critical Rayleigh number. The experiment on thermal instability in a rotating liquid involved the determination of R_c as a function of the Taylor number for both the convective and overstable cases. The results are presented in Figs. 2 and 3, and it may be seen that the theoretical

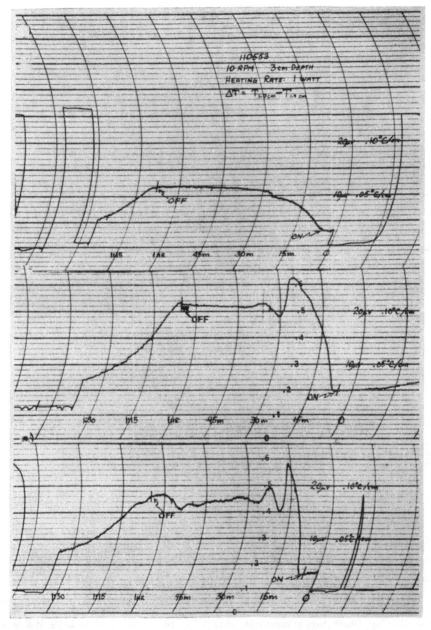

Plate I. Convection in a rotating layer of water: adverse temperature gradient β as a function of time for $\beta < \beta_c$ (top), $\beta \simeq \beta_c$ (middle) and $\beta > \beta_c$ (bottom). The time increases to the left.

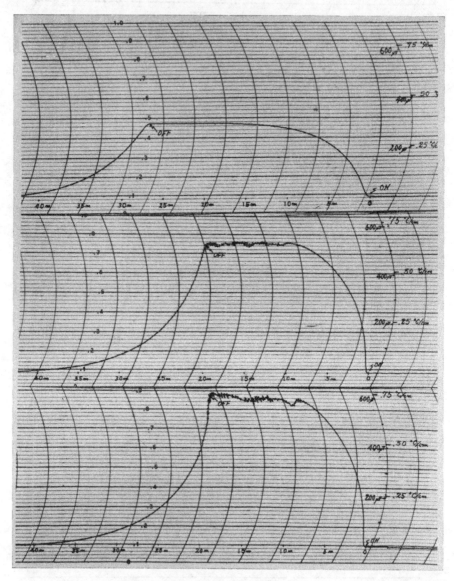

Plate II. Overstability in a rotating layer of mercury: adverse temperature gradient β as a function of time for $\beta < \beta_o$ (top), $\beta \simeq \beta_o$ (middle) and $\beta > \beta_o$ (bottom). The time increases to the left.

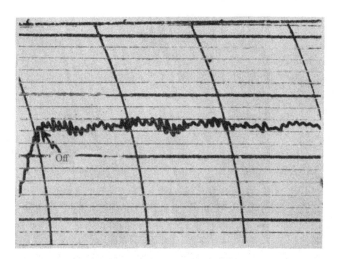

Plate III. Enlargement of the bottom trace shown in Pl. II.

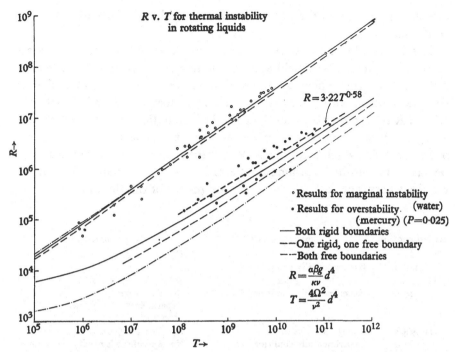

Fig. 2. Comparison of the predicted and observed Rayleigh numbers for rotating liquids as a function of the Taylor number T. (From D. Fultz and Y. Nakagawa. [2].)

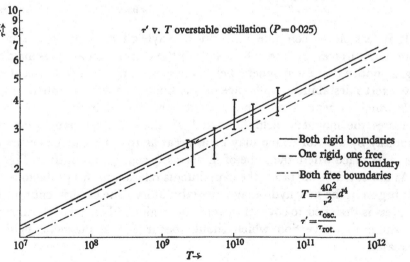

Fig. 3. Comparison of the predicted and observed periods of overstable oscillation for rotating layer of mercury. (From D. Fultz and Y. Nakagawa. [2].)

predictions of R_c and the periods of the overstable oscillations have been confirmed.

The last study of the series summarized in Table 1—the case where there is both a magnetic field and an angular velocity—is by far the most difficult. The theoretical treatment[9] involves the solution of a double eigenvalue problem of the twelfth order: for assigned Q and T both the critical Rayleigh number and the period of oscillation must be determined as eigenvalues. The experiment requires one to rotate the apparatus in a magnetic field which may reach 8000 gauss. Extreme precautions must be taken to eliminate inductive effects if β is to be measured with the required accuracy (a few microvolts) by means of the thermopile.

Table 1

Experiment no.		Convective case	Overstable case
I	$H=0$ $\Omega=0$	R_c is constant Experimentally confirmed	Cannot occur
II	$H=0$ $\Omega\neq0$	$R_c \propto T^{\frac{4}{3}}$ as $T\to\infty$ Experimentally confirmed for water	$R_c^* \propto T^{\frac{4}{3}}$ as $T\to\infty$ Exists for $\nu \ll \kappa$ Experimentally confirmed for mercury
III	$H\neq0$ $\Omega=0$	$R_c \propto Q$ as $Q\to\infty$ Experimentally confirmed for mercury	$R_c^* \propto Q$ as $Q\to\infty$ No experiment is possible, as η must be less than κ for overstability to occur. Occurs under astrophysical conditions
IV	$H\neq0$ $\Omega\neq0$	$R_c \propto Q$ as $Q\to\infty$ for fixed T	$R_c^* \propto Q$ as $Q\to\infty$ for fixed T. Experiment in progress[†]

It is feasible to compute the critical Rayleigh number only if both bounding surfaces are free. The results of the calculation are presented in Fig. 4; notice that the magnetic field (in a certain range of Q, for fixed T) may exert a destabilizing influence on the fluid. The wave numbers of the cells which appear at marginal stability are plotted in Fig. 5. As one increases the magnetic field (for fixed T) the cell size changes discontinuously by a factor which may be as great as 10. This change occurs at the same place where the slope of R_c versus Q changes abruptly in Fig. 4.

As indicated in the table, the experiments relating to this problem have just begun in the new hydromagnetics laboratory. The experiment now in progress is designed to cover the overstable region in Fig. 4 and the abrupt transition to convection which should occur as the magnetic field is increased.

† Note added in proof: These experiments have now been completed by Y. Nakagawa and are reported in *Proc. Roy. Soc.* A, **240**, 108, 1957; *Proc. Roy. Soc.* A, **242**, 81, 1957; and *Rev. Sci. Instrum.* **21**, 603, 1957.

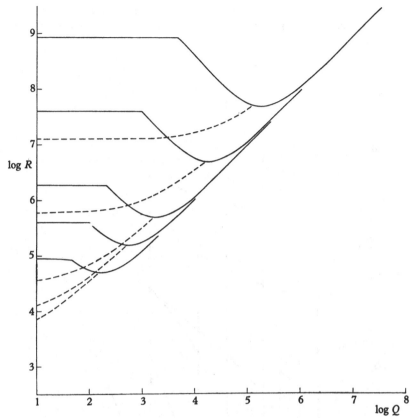

Fig. 4. Predicted critical Rayleigh number R as a function of magnetic number Q and Taylor number T. The Taylor numbers corresponding to each pair of curves are, from the top down, $10^{10}, 10^8, 10^6, 10^5, 10^4$. The dashed curves correspond to overstability, the full curves to convection. (From S. Chandrasekhar [9].)

REFERENCES

[1] Fultz, D., Nakagawa, Y. and Frenzen, P. *Phys. Rev.* **94**, 1471, 1954.
[2] Fultz, D. and Nakagawa, Y. *Proc. Roy. Soc.* A, **231**, 211, 1955.
[3] Nakagawa, Y. and Frenzen, P. *Tellus*, **7**, 1, 1955.
[4] Nakagawa, Y. *Nature, Lond.* **175**, 417, 1955.
[5] Chandrasekhar, S. *Phil. Mag.* (7), **43**, 501, 1952.
[6] Chandrasekhar, S. *Proc. Roy. Soc.* A, **217**, 306, 1933.
[7] Chandrasekhar, S. *Phil. Mag.* (7), **45**, 1177, 1954.
[8] Chandrasekhar, S. *Proc. Roy. Soc.* A, **225**, 173, 1954.
[9] Chandrasekhar, S. *Proc. Roy. Soc.* A, **237**, 476, 1956.
[10] Chandrasekhar, S. and Elbert, D. *Proc. Roy. Soc.* A, **231**, 198, 1955.

Discussion

Biermann: When I discussed the subject of the cooling of sunspots by magnetic inhibition of convection, in agreement with Chandrasekhar, the results of this

5·2

work, in a general way, seemed to support the line of reasoning underlying the theoretical conceptions. Are there any more recent results which might throw a light on this problem?

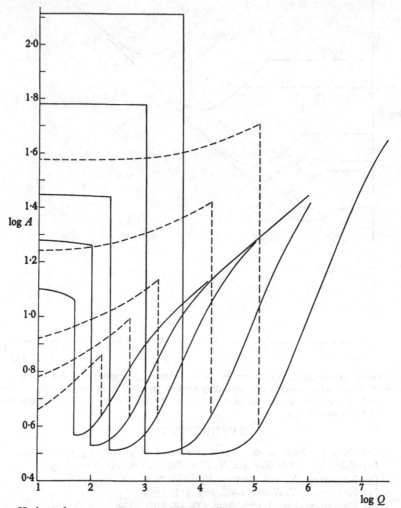

Fig. 5. Horizontal wave number at marginal stability as a function of Q and T. The Taylor numbers are the same as those in Fig. 4. The dashed curves correspond to overstability, the full curves to convection. (From S. Chandrasekhar [9].)

Prendergast: I can only answer that the experiment has been extended over a much wider range of values than in earlier investigations by Nakagawa, and that it agrees with theory. The only case which has been examined is that without the influence of the Coriolis force. An experiment is being planned where a magnetic field and a rotation are acting simultaneously. This is, of course, the most interesting case.

Lehnert: I should like to point out that convection on laboratory scale corresponds to aperiodic motions (as I have called them in my paper), whereas convective motions in astrophysics mainly consist of periodic modes. The former type of convection is critically suppressed by an external magnetic field, but not necessarily the latter. Consequently, the types of convection which can be realized in the laboratory are quite different from those to be considered in astrophysical problems, such as in sunspots. This has also been pointed out by Chandrasekhar.

Schatzman: Recently Rösch at Pic du Midi observed granulation inside sunspots. I want to ask Professor Biermann about his opinion of these observations.

Biermann: I talked with Rösch about his observations recently. It seemed then that more complete details must become known, e.g. the time scale of the spot granulae, before conclusions regarding the postulated magnetic inhibition of convection in the spots can be drawn.

Gold: Also one ought to mention the case of the penumbra. One would like to find a reason for a sudden change in the type of convection before its complete inhibition in order to account for the sharp distinction in the appearance of granulation, penumbra and umbra.

Lehnert: One might feel tempted to identify the regions of granulation, penumbra and umbra with states of turbulent convection, regular cellular convection and thermal conduction, respectively. The boundaries should then be given by the critical magnetic field strengths for the suppression of turbulent convection and of regular convection (*Tellus*, **9**, 102, 1957). However, it is not at all certain that the onsets of regular convection and of turbulent convection are critical in the astrophysical situation.

Thompson: Are any attempts being made to observe cell shapes in a magnetic field?

Prendergast: No, there is no room for a mirror between the pole faces.

Thompson: Are there any experiments planned in which the magnetic field and the temperature gradient are not parallel; for in such a case the cell shape should be considerably changed?

Prendergast: An experiment is planned with vertical temperature gradient and a horizontal field.

Lehnert: A simple experiment with an oblique magnetic field has recently been performed in Stockholm (*Tellus*, **9**, 101, 1957).

Hide: As Dr Prendergast has explained, when hydromagnetic forces or Coriolis forces act separately, the critical Rayleigh number, R_c, increases monotonically with the parameters Q or T, as the case may be. Thus, these agencies, acting separately, tend to *inhibit* instability. However, when rotation and a magnetic field act simultaneously there are certain values of T and Q for which R_c *decreases* with T and Q. Is there any simple physical interpretation of this extremely interesting result that it is possible to choose values for the magnetic field and the rate of rotation such that any increase in them will *encourage* instability?

Prendergast: This point is clarified in a recent paper by S. Chandrasekhar, *Dædalus*, **86**, 323, 1957.

Lehnert: I think that Dr Hide's question has some connexion with an earlier derived result (*Quart. Appl. Math.* **12**, 335, 1955). Consider a plane hydromagnetic mode in an external magnetic field. We restrict the discussion to the aperiodic case, where the dissipation is too strong (or the field too weak) for eigen-oscillations to exist. Then, an increase in magnetic field strength causes an increase in the induced currents and in the dissipation. If a rotation is introduced the Coriolis force may be shown to *counteract* this effect; the increase of the dissipation caused by the magnetic field is *reduced*.

A possible physical explanation of this is that the Coriolis force provides a contribution to the effective 'tension' of the magnetic field lines, which may be regarded as elastic 'strings'. Thus, for a given momentum balance between the forces acting on a fluid element, the electrodynamical force can be reduced, part of its role being taken over by the Coriolis force. A reduction of the electrodynamic force gives a corresponding reduction in induced currents and in Joule heat dissipation. If this interpretation is right one could perhaps also state that, for a given value of Q, an increase of T would destabilize the motion in some cases.

Lowes: Does overstability occur over the whole vessel simultaneously or at random in one or more convection cells? By using a thermopile you are presumably averaging the temperature fluctuations. Might you therefore miss some cases?

Prendergast: You certainly get a kind of averaging by using a thermopile. The cell dimension is, let us say, of the order of a centimeter and the thermopile causes an average over several cells. One gets an average of the fluctuations in temperature in cells. The order of fluctuation is $0 \cdot 01 °/cm$ and the amplitude in an individual cell is probably considerably higher than that. From one cell to the next, under experimental realizable conditions and certainly not over the whole vessel, there need not be a whole region where the cells move like that. They move in one direction at one time and in another at another time.

C. IONIZED GAS IN A MAGNETIC FIELD

PAPER 8

IONIZED GAS IN A MAGNETIC FIELD

A. SCHLÜTER

Max Planck Institut für Physik, Göttingen, Germany

ABSTRACT

The ionized gas is described as a mixture of several fluids; each obeying a quasi-hydrodynamic equation of motion with additional terms describing the mechanical interaction. Particularly, two- and three-fluid models are considered. The nature of the approximations ('quasi-neutrality', 'creeping diffusion') is discussed. Conservation-laws are formulated for the case of negligible effect of mutual encounters and of pressure diffusion. These models lead to a generalization of Ohm's law; it is shown that the additional terms are of practical importance if one has three components, of which one may be neutral.

I. INTRODUCTION AND GENERAL FORMULATION OF THE MODEL

The dynamics of an ionized gas in an electromagnetic field can be treated by different methods. An exact theory would consist of the solution of the Boltzmann equation of the kinetic theory of gases together with Maxwell's equations, taking all interactions into account. Since this is impracticable one has to rely upon approximate methods. One such approximation consists in using a hydrodynamic equation of motion for the ionized gas supplemented by a term representing the Lorentz force, together with Maxwell's equations (usually neglecting the displacement current) and a form of Ohm's law appropriate to moving conductors. So one arrives at a set of simultaneous equations, known as the 'hydromagnetic' equations

$$\rho \frac{d\mathbf{v}}{dt} = \mathbf{k} + \frac{1}{c}\mathbf{j} \times \mathbf{H}; \quad \text{div}\,(\rho\mathbf{v}) = -\frac{\partial\rho}{\partial t}, \qquad (1a)$$

$$\mathbf{j} = \sigma \mathbf{E}^c; \quad \mathbf{E}^c = \mathbf{E} + \frac{1}{c}\mathbf{v} \times \mathbf{H}, \qquad (1b)$$

$$c\,\text{curl}\,\mathbf{H} = 4\pi\mathbf{j}; \quad \text{div}\,\mathbf{H} = 0, \qquad (1c)$$

and

$$c \text{ curl } \mathbf{E} = -\frac{\partial \mathbf{E}}{\partial t} \qquad (1\,d)$$

($\mathbf{k}$ = density of all non-electromagnetic forces, for instance $\mathbf{k} = -\text{grad}\, p + \rho\mathbf{g}$; ρ = mass density; $\mathbf{g}$ = gravitational acceleration; σ = ohmic conductivity; all other symbols have their usual meanings).

A different method can be applied, particularly when the electromagnetic field can be considered known and when the effect of pressure is small. It consists of solving the equation of motion of some representative kind of the differently charged particles—either strictly or by some approximate method, as the one first used by Alfvén[1]—and thence one infers the behaviour of the ionized gas more or less intuitively. The difficulties lie here in knowing the electromagnetic field and in selecting the representative particles; because of this some of the results gained by this method are demonstratively spurious.

I shall discuss a third method, which is in some respects intermediate between the two approaches I have mentioned. Here all particles of one kind (characterized by their charge per mass ratio) are taken to constitute a fluid, and the ionized gas is described as a mixture of (at least two) such fluids penetrating each other. The equation of motion of each fluid component is the conventional hydrodynamic one apart from additional terms describing the interaction with the Maxwell field due to the electric charge and the electric current carried by the component considered and further terms representing the frictional interaction between all components. We assume that the frictional force between the jth and the kth component is proportional to the relative velocity $\mathbf{v}_j - \mathbf{v}_k$ and proportional to the density of either component. We therefore write for it

$$\beta_{jk}\rho_j\rho_k(\mathbf{v}_j - \mathbf{v}_k); \qquad (2)$$

ρ_j = mass density of jth component.

In a free path theory the parameter β_{jk} depends on the collisional cross-section q_{ik}, on the root mean square relative speed v_{jk} and on the masses approximately

$$\beta_{jk} \approx q_{jk}v_{jk}/(m_j + m_k). \qquad (3)$$

One would therefore expect it to be essentially independent of a possible magnetic field. This is borne out by an exact kinetic treatment, the total change of β_{jk} between zero magnetic field and infinite field strength being by about a factor 2.

The equation of motion of the jth component is then

$$\rho_j \frac{d_j \mathbf{v}_j}{dt} = \rho_j \sum_k \rho_k \beta_{kj}(\mathbf{v}_k - \mathbf{v}_j)$$
$$+ \rho_j z_j \left\{ \mathbf{E} + \frac{1}{c} \, \mathbf{v}_j \times \mathbf{H} \right\}$$
$$+ \mathbf{k}_j \qquad (4)$$

(d_j/dt = time derivative following the motion of the jth component, $\beta_{kj} = \beta_{jk} > 0$ is the friction parameter, z_j = charge to mass ratio, k_j = density of all non-electromagnetic forces including the gradient of the partial pressure p_j).

For later convenience we give here a list of abbreviations:

$\rho = \sum_k \rho_k$ (total mass density),

$\mathbf{k} = \sum_k \mathbf{k}_k$ (total non-electromagnetic force),

$\rho \mathbf{v} = \sum_k \rho_k \mathbf{v}_k$ (mean mass velocity),

$\mathbf{j} = \sum_k \rho_k z_k \mathbf{v}_k$ (electric current density),

$\mathbf{E}^c = \mathbf{E} + (1/c) \, \mathbf{v} \times \mathbf{H}$ (electric field in a system of reference moving with the mean mass velocity).

In addition to Eq. (4) we have Maxwell's equations including the relation between the electric field and the total charge density

$$\mathrm{div}\ \mathbf{E} = 4\pi \sum_k \rho_k z_k. \qquad (5)$$

Together with the approximate equations determining the non-electromagnetic force (which we shall however not consider in detail) the system of equations is complete. To simplify it, we use two approximations.

1. We assume quasi-neutrality. That is, we drop Eq. (5) and replace it by

$$\sum_k \rho_k z_k = 0. \qquad (6)$$

With this relation the system is again complete and suffices to determine the electric field, including its curl-free part. The approximation is consistent when it turns out that

$$|\,\mathrm{div}\ \mathbf{E}\,| \ll 4\pi \sum_k \rho_k |\,z_k\,|, \quad \text{say}, \qquad (7)$$

so that the relative magnitude of the two sides of this inequality is of such an order that the error in the densities ρ_k induced by the approximation does not appreciably influence the solution of the equation of motion.

2. We assume the diffusion to be 'creeping' by neglecting the difference of acceleration of the components:

$$\frac{d_j \mathbf{v}_j}{dt} = \frac{d\mathbf{v}}{dt}. \tag{8}$$

Since usually the diffusion velocities are small compared with the mean velocity, Eq. (8) holds in most practical cases too. The physical meaning of this assumption is that the diffusion equilibrium sets in instantaneously whether or not the gas as a whole is being accelerated.

The two assumptions are of quite a different nature. Assumption 1 will be good in almost all cases of astrophysical interest, except for problems like a non-linear theory of plasma oscillations of great amplitude. The simplification gained is relatively minor, however, since we have replaced one instantaneous differential equation (namely Eq. (5)) by an algebraic relation. There are however many cases where diffusion is certainly not at all 'creeping', but the inertia due to the relative velocities is of decisive importance. Particularly, plasma oscillations are excluded. The approximation 2 will be good in all cases where a 'magnetohydrodynamic'approach is reasonable. While the restriction imposed is severe, the simplification is considerable since we are now left with just as many time derivations as in ordinary hydromagnetics.

We shall first discuss the Eqs. (4) together with our assumptions in the special case of a two-component plasma and only briefly deal with the more general case of a three-component plasma.*

2. THE TWO-FLUIDS MODEL

If only two constituents are present our equations are readily solved in terms of those variables which describe the behaviour of the plasma as a whole:

$$\rho \left(\frac{\partial \mathbf{v}}{\partial t} + \mathbf{v} \cdot \operatorname{grad} \mathbf{v} \right) = \frac{1}{c} \mathbf{j} \times \mathbf{H} - \operatorname{grad} p, \tag{9a}$$

$$c\mathbf{E} + \mathbf{v} \times \mathbf{H} = \frac{c}{\sigma} \mathbf{j} + \frac{\alpha}{\rho} \{ \mathbf{j} \times \mathbf{H} - c \operatorname{grad} \gamma p \}, \tag{9b}$$

* For the case of a two-component field, equations corresponding to those given here were first considered by A. Schlüter[2] and independently by M. H. Johnson and E. O. Hulburt [3] (the latter without the inertial terms). The three-component case was considered by A. Schlüter[4] and applied to interstellar magnetic fields by A. Schlüter and L. Biermann[5] and to the ionosphere by I. Lucas and A. Schlüter[6]. The relation to the kinetic theory of gases was established in the case of creeping diffusion by M. H. Johnson[7].

where $\sigma = \beta_{12}/z_1 . z_2$ (ohmic conductivity),

$$\alpha = 2(\rho_1 - \rho_2)/(\rho_1 z_1 - \rho_2 z_2)$$

and $$\gamma = (\rho_1 p_2 - \rho_2 p_1)/(\rho_1 - \rho_2) p.$$

We have here taken pressure as the only non-electromagnetic force acting.

α is a constant which only depends on the nature of the plasma, not on its actual state. For a mixture of ions (of mass m_i and charge $+e$) and electrons it is practically $\alpha = m_i/e$. γ depends on temperature only as far as the ratio of the partial pressures does, so in a simple plasma where the electron temperature equals that of the ions, $\gamma = \frac{1}{2}$.

The difference between these equations and the hydromagnetic equations (1) lies only in the two terms multiplied by α. These are the Hall term $(\alpha/\rho c)$ $(\mathbf{j} \times \mathbf{H})$ which produces an electric field when a current flows across the lines of force, and the pressure diffusion term (α/ρ) grad γp. Both terms are the more important the smaller the density becomes. The Hall term is more important than the Ohm term $\mathbf{j}/\sigma$ if $\alpha H/c > \rho/\sigma$, that is if the mean of the gyro-frequencies is larger than the collision frequency, and this is the case for practically all cosmical magnetic fields, except for the interior of stars, planets, and the like. We are particularly interested in the deviations from ordinary hydromagnetics and shall therefore discuss the extreme case of vanishing ohmic resistivity.

If $\sigma \to \infty$, the Eqs. (9) allow a number of transformations. One of these is gained by eliminating in Eq. (9a) the Lorentz term by means of the Hall term of Eq. (9b). We then obtain:

$$\rho \frac{d\mathbf{v}}{dt} = \frac{\rho}{\alpha} \left(\mathbf{E} + \frac{\mathbf{v}}{c} \times \mathbf{H} \right) - \text{grad} \ (1 - \gamma) \ p. \tag{10}$$

This is the equation of motion of a charged fluid with a mass-to-charge ratio α and a pressure $(1 - \gamma) \ p \approx p/2$. So, if the electromagnetic field is known, it is essentially the same problem to solve the equation of motion for a quasi-neutral plasma as for a gas consisting of charged particles of one kind only. This is one of the many possible transformations in this field which are correct in a formal sense but completely useless in practice. The point is, in this case, that we never know the electromagnetic field beforehand. There are situations where we may neglect the influence on the magnetic field by the currents flowing in the plasma, or where we may treat it as a perturbation. But the electric field is always determined by the space charges in the fluid, whenever the motion of a quasi-neutral plasma is applicable. There is no other way but to solve both the equation of motion

75

($9a$) (or equivalently (10)) *and* the diffusion equation ($9b$) simultaneously. Because of the relative unimportance of **E** it is advisable to eliminate **E** in Eq. ($9b$) by taking its curl:

$$\frac{\partial \mathbf{H}}{\partial t} = -\operatorname{curl}\left(\frac{c^2}{4\pi}\operatorname{curl}\mathbf{H}\right)$$

$$+ \operatorname{curl}\left\{\left(\mathbf{v}-\frac{\alpha}{\rho}\mathbf{j}\right)\times\mathbf{H}\right\}$$

$$-\frac{\alpha c}{\rho^2}(\operatorname{grad}\rho \times \operatorname{grad}\beta p). \tag{11}$$

We consider again the case of vanishing resistivity $1/\sigma$. If furthermore $\alpha/\rho \to 0$, we have the well-known hydromagnetic relation

$$\frac{\partial \mathbf{H}}{\partial t} = \operatorname{curl}(\mathbf{v}\times\mathbf{H}). \tag{12}$$

This equation has a simple meaning: the magnetic flux through every closed line which is moving with the fluid is constant—the magnetic lines of force are frozen in. Returning to the case $\alpha \neq 0$, it is tempting to introduce instead of the mean mass velocity a slightly different velocity **v'** by

$$\mathbf{v}' = \mathbf{v} - \frac{\alpha}{\rho}\mathbf{j}$$

or

$$\rho\mathbf{v}' = \rho_2\mathbf{v}_1 + \rho_1\mathbf{v}_2. \tag{13}$$

If we use this mean velocity Eq. (11) reads (with $\sigma \to \infty$)

$$\frac{\partial \mathbf{H}}{\partial t} = \operatorname{curl}(\mathbf{v}'\times\mathbf{H}) - \frac{\alpha c}{\rho^2}[\operatorname{grad}\rho \times \operatorname{grad}p]. \tag{14}$$

The only effect of the Hall term is therefore that the lines of force are not moving with the mean mass velocity (**v**) but with a velocity which differs from this by a quantity of the order of the relative diffusion velocity. This result is not surprising, since the concept of the mean mass velocity has been introduced for its obvious importance for the equation of motion, but it is certainly not appropriate to the diffusion problem when the forces are not proportional to the masses. Besides this motional-induction term we have the pressure term; it describes real creation and annihilation of the magnetic lines of force and is of particular importance if one wants to treat the first origin of magnetic fields in fluid conductors. It disappears if p and ρ are uniquely related to each other as they are in many simple cases of interest.

76

Another useful relation is obtained by taking the curl of Eq. (10). Introducing the vorticity $\boldsymbol{\omega} = \text{curl } \mathbf{v}$, we have:

$$\frac{\partial (\mathbf{H} + \alpha c \boldsymbol{\omega})}{\partial t} = \text{curl } \{\mathbf{v} \times (\mathbf{H} + \alpha c \boldsymbol{\omega})\} + \frac{\alpha c}{\rho^2} \{\text{grad } \rho \times \text{grad } (1 - \gamma) \, p\}. \tag{15}$$

We have seen that the magnetic field moves with a velocity which is different from the mean mass velocity, now we see which quantity it is that is transported by $\mathbf{v}$; in the case of a mixture of ions and electrons the conserved quantity is

$$\frac{e\mathbf{H}}{(m_i - m_e) \, c} + \text{curl } \mathbf{v}. \tag{16}$$

For the influence of the pressure term the same remarks apply as above. Eq. (15) has two interesting limiting cases. If $\alpha \to 0$, we return to the hydromagnetic case previously discussed. For $\mathbf{H} \to 0$ we obtain, however, the well-known vorticity theorem of hydrodynamics, but with a small deviation:

$$\frac{\partial \, \text{curl } \mathbf{v}}{\partial t} = \text{curl } \{\mathbf{v} \times \text{curl } \mathbf{v}\} + \frac{1}{\rho^2} \{\text{grad } \rho \times \text{grad } (1 - \gamma) \, p\}, \tag{17}$$

the deviation consisting in the term with grad γp, which we have already found to be responsible for the creation of magnetic flux. The two quantities, the sum of which is conserved in the considered sense, are comparable to one another when curl $\mathbf{v}$ has the order of the smaller gyrofrequency of either component. So again, the modification to hydromagnetics due to the Hall terms is very small indeed for all cases where the application of hydromagnetics is at all reasonable.

The fact that the real importance of the Hall effect is very small is contrary to what one would expect, if one describes its effect as a reduction of conductivity across the lines of force. In our case, where we have considered the case of vanishing ohmic resistivity ($\sigma \to \infty$), the cross-conductivity would indeed be zero.

3. THE THREE-FLUIDS MODEL

From the treatment of the two-fluids model we have learned how to handle the diffusion equations: we have to solve them with respect to the electric field $\mathbf{E}$ or $\mathbf{E}^c$ in terms of the magnetic field (and thereby the electric current) and the partial pressures only. So we arrive at a modification of Ohm's law, which is then used to determine curl $\mathbf{E}$ and thereby $\partial \mathbf{H}/\partial t$.

If we carry through this programme for the case where we have three

77

different constituents, and solve the Eqs. (4), remembering our assumptions (6) and (8), we arrive at the following somewhat lengthy formula:

$$(\beta_{23}\rho_1 z_1^2 + \beta_{13}\rho_2 z_2^2 + \beta_{12}\rho_3 z_3^2)\ \mathbf{E}^c + z_1 z_2 z_3 (\mathbf{E}^c \times \mathbf{H}/c)$$

$$= (\beta_{12}\beta_{13}\rho_1 + \beta_{12}\beta_{23}\rho_2 + \beta_{13}\beta_{32}\rho_3)\ \mathbf{j}$$

$$- (1/\rho)\ (\beta_{23} z_1(\rho - 2\rho_1) + \beta_{13} z_2(\rho - 2\rho_2) + \beta_{12} z_3(\rho - 3\rho_3))\ (\mathbf{j} \times \mathbf{H}/c)$$

$$+ (1/\rho^2)\ (\rho_1 z_2 z_3 + \rho_2 z_3 z_1 + \rho_3 z_1 z_2)\ (\mathbf{j} \times \mathbf{H}/c) \times \mathbf{H}/c$$

$$+ (1/\rho)\ \{\beta_{23}\rho_1 z_1 + \beta_{13}\rho_2 z_2 + \beta_{12}\rho_3^3 z_3\}\ \mathbf{k}$$

$$- \{\beta_{23}\rho_1 z_1 \mathbf{k}_1 + \beta_{13}\rho_2 z_2 \mathbf{k}_2 + \beta_{12}\rho_3 z_3 \mathbf{k}_3\}$$

$$+ (1/\rho^2)\ \{\rho_1 z_2 z_3 + \rho_2 z_1 z_3 + \rho_3 z_1 z_2\}\ \mathbf{k} \times \mathbf{H}/c$$

$$- (1/\rho)\ \{\rho_1 z_2 z_3 \mathbf{k}_1 + \rho_2 z_1 z_3 \mathbf{k}_2 + \rho_3 z_1 z_2 \mathbf{k}_3\}\ \times \mathbf{H}/c.$$

We have by this formulation not completely fulfilled our aim, the term $\mathbf{E}^c \times \mathbf{H}$ not being removed. This could easily be done, but in the case of greatest interest—namely if one component is not charged—its coefficient disappears. The essential novel features appearing here are the terms which contain the square of H/ρ. They are larger than the Hall term in the ratio given by a certain average value of the gyro-frequencies relative to the collision frequencies. Their occurrence is most easily explained in the case of one neutral component. Then, the Lorentz force $\mathbf{j} \times \mathbf{H}/c$ acts on the charged components only, hence these move relative to the neutral component ('ambipolar diffusion'). It is then the mean velocity of the charged components which determines the motional induction $\mathbf{v} \times \mathbf{H}$ and this velocity differs from the mean mass velocity by a term proportional to the Lorentz force. By this effect the dissipation of energy is really increased if an electric current flows perpendicular to the lines of force and—as found by A. Schlüter and L. Biermann [5]—it might well be that this sink of energy is of importance in the case of magnetic fields in interstellar H II-regions. It also seems that by this mechanism the tidal currents in the ionosphere are effectively limited to the lower layers (I. Lucas [8]). A further effect of this term is a modification of shock conditions compared to the hydromagnetic case, while the Hall term does not contribute.

REFERENCES

[1] Alfvén, H. *Cosmical Electrodynamics* (Oxford University Press, 1950).
[2] Schlüter, A. *Z. Naturf.* **5**a, 72, 1950.
[3] Johnson, M. H. and Hulburt, E. O. *Phys. Rev.* **79**, 802, 1950.
[4] Schlüter, A. *Z. Naturf.* **69**, 73, 1951.
[5] Schlüter, A. and Biermann, L. *Z. Naturf.* **5**a, 237, 1950.
[6] Lucas, I. and Schlüter, A. *Arch. Elektr. Übertr.* **8**, 27, 1954.
[7] Johnson, M. H. *Phys. Rev.* **84**, 566, 1951.
[8] Lucas, I. *Arch. Elektr. Übertr.* **8**, 123, 1954.

Discussion

Cowling: I have during the last year derived results essentially equivalent to those given by Schlüter. It appears that the question normally posed as to how the conductivity is affected by a magnetic field is too imprecise for the answer to have any value. The more important question is how collision processes affect the dissipation of a magnetic field; an answer can be given to this but only if the physical circumstances are clearly defined.

Buneman: Conservation of vortices (in the electrodynamic sense, i.e. vortices of momentum plus vector potential) is a very fundamental property. It applies to each species separately when there are no collisions, even under extreme relativistic conditions. When there are collisions the vortices of the total momentum are conserved—hence Dr Schlüter's result. Conservation of vortices is an extremely useful fact for resolution of problems and has been employed successfully by myself in calculations for conditions where no collisions take place, such as in interplanetary space.

Piddington: Dr Schlüter has considered each component of the gas as having a separate motion. This is undoubtedly necessary to obtain a complete solution but it may lead to great complexity in some astrophysical problems. These problems are usually so complicated in any case that some simplifying assumptions are necessary. One simplification which is often permissible is to consider two or perhaps more of the different gas components as a single gas with a single mass motion. An example is a hydromagnetic disturbance in a gas containing heavy ions and electrons and perhaps neutral particles. There is no doubt that, because of their greater mobility, the electrons move to some degree separate from the heavy ions and so cause space-charge electric fields within the ion plasma. This results in 'an electron plasma wave' or space-charge wave as an integral part of the whole hydromagnetic wave. However, the electric current which flows to cause this wave is small; in fact it is equal to the displacement current which, as Dr Schlüter has pointed out, is negligible, except when relativistic effects are significant.

Perhaps it is desirable to examine each particular astrophysical problem with a view to reducing as far as possible the total number of gas components considered. This may avoid the development of equations which cannot be solved.

von Engel: What is the relative importance of the production of charges (e.g. in the ionosphere) which has not been considered in the theory?

Schlüter: The equations which I have given describe only the balance of momentum, so they hold irrespective of the presence of ionization and recombination processes. These may, however, influence the coefficients of friction. In the case of three co-existing fluids one has to introduce a condition on the transmutations between the constituents. In the work on the tidal currents in the ionosphere the approximation was made that the degree of ionization is controlled by the instantaneous equilibrium between radiative ionization and recombination, independent of the state of motion.

Terletzsky: Do you agree with me that for extremely rarefied gases it is better to solve your first equations—the equations of mutually penetrating ideal gases?

Schlüter: As far as the assumptions (quasi-neutrality, creeping diffusion) hold, both approaches are mathematically equivalent. Otherwise, one has either to solve the original equations directly or to use transformations which do not imply the correctness of these approximations.

WAVES IN A HOT IONIZED GAS IN A MAGNETIC FIELD

ERNST ÅSTRÖM

Royal Institute of Technology, Stockholm, Sweden

The electromagnetic state of vacuum is characterized by two vector quantities, namely **E** and **B**. They are related to current and charge density by the equations

$$\operatorname{curl} \mathbf{B}/\mu_0 - \epsilon_0 \frac{\partial \mathbf{E}}{\partial t} = \mathbf{i}, \tag{1}$$

$$\operatorname{div} \mathbf{E} = \rho/\epsilon_0, \tag{2}$$

$$\operatorname{curl} \mathbf{E} + \frac{\partial \mathbf{B}}{\partial t} = 0, \tag{3}$$

$$\operatorname{div} \mathbf{B} = 0. \tag{4}$$

Sometimes it is suitable and possible to introduce two new quantities, **H** and **D**, so defined that the equations (1) and (2) appear in the new form (Stratton[1])

$$\operatorname{curl} \mathbf{H} - \frac{\partial \mathbf{D}}{\partial t} = 0, \tag{5}$$

$$\operatorname{div} \mathbf{D} = \rho \tag{6}$$

and with a linear relation between **E** and **D** and also between **B** and **H**. This formalism is common when dealing with fluid and solid media, and has also been introduced to ionized media of zero temperature (Nichols and Schelleng[2], Alfvén[3], p. 85, Åström[4]). Here we shall say a few words about this matter for ionized media of non-zero temperature. In this connexion we also get an opportunity to discuss the meaning of the conception of diamagnetism.

Let us assume a medium which initially is homogeneous in a homogeneous magnetic field. Let us assume that we can neglect collisions. In this case it seems suitable to introduce fictitious particles situated at the guiding centres. Their motion shall be equal to the drift velocity. The magnetic moment due to the spiralling motion of the actual particle becomes an intrinsic property of the fictitious particle. By introducing these particles we have established that the motion of these particles is a single valued function of the space co-ordinates. The random motion is taken into

6

account by introducing a temperature, and the corresponding electric current by introducing the magnetic moment (compare the alternative view discussed by Cowling[5] and Spitzer[6], p. 25). It may be necessary to keep the frequencies well below the gyro-frequency for the treatment to be valid, but we get simple relations for this case.

In the first approximation the motion of charged particles is the drift in crossed electric and magnetic fields. The drift is independent of the sign of the charge and hence, in a neutral plasma, the corresponding electric current vanishes. Therefore we ought to use the second approximation (Alfvén[3], p. 18),

$$v_{\text{drift}} \approx \frac{\mathbf{E} \times \mathbf{B}}{B^2} + \omega^{-2}(e/m)\,\frac{\partial \mathbf{E}}{\partial t}, \tag{7}$$

where $\mathbf{E}$ is the perturbing electric field and ω is the gyro-angular frequency. If n is the density then the current is

$$\mathbf{i}_1 = \Sigma n e \mathbf{v} \approx \frac{\mathbf{E} \times \mathbf{B}}{B^2}\,\Sigma n e + \frac{\Sigma n m}{B^2}\,\frac{\partial \mathbf{E}}{\partial t}, \tag{8}$$

where the sum is to be extended over all types of particle present. Since we treat a macroscopically neutral medium the first term on the right-hand side vanishes.

In a region where the magnetic field is inhomogeneous we also get a drift of charged particles. In contrast to the drift in crossed fields the direction does depend on the sign of the charge of the particles, and hence we do not get any cancellation of the corresponding current in a neutral plasma. This current is

$$\mathbf{i}_2 \approx \frac{\Sigma n e v_\perp^2}{4B^4}\,\mathbf{B} \times \nabla B^2. \tag{9}$$

The current due to the spiralling motion is a multivalued function of the space co-ordinates and therefore ought to be introduced in another way. Let us compute the field at a point P in a homogeneous plasma. Assume a cylinder generated by the magnetic-field lines through a circle with its centre at P and the radius equal to the Larmor radius R. The contribution from particles with their centres of circular motion outside is easily seen to be zero and the contribution from those inside is

$$(1/2)\,\mu_0 n e v_\perp R.$$

This contribution has opposite direction to the magnetic field created by other sources. If therefore $\mathbf{B}$ is the field we actually have and $\mathbf{B}_0$ is the field from all sources but with the present one excluded we get

$$\mathbf{B} = \mathbf{B}_0 - a\mathbf{B}; \tag{10}$$

$$a = \mu_0 B^{-2} \Sigma \tfrac{1}{2} n m v_\perp^2. \tag{11}$$

From what is said we find that $\mathbf{B_0}$ and not $\mathbf{B}$ shall appear in Eq. (1). Then Eqs. (1), (7)–(11) give

$$\operatorname{curl} \mathbf{H} = \frac{\partial \mathbf{D}}{\partial t} + (\mathbf{grad}\ a) \times \mathbf{B}, \tag{12}$$

$$\mu^{-1} = 1 + 2a, \tag{13}$$

$$\epsilon = 1 + \mu_0 B^{-2} \Sigma n m c^2, \tag{14}$$

$$\mathbf{H} = \mu^{-1} \mu_0^{-1} \mathbf{B} \quad \text{and} \quad \mathbf{D} = \epsilon \epsilon_0 \mathbf{E}. \tag{15}$$

These equations are of well-known form when the last term in (12) vanishes, i.e. when the ratio of the thermal to the magnetic-field energy is independent of the co-ordinates perpendicular to the magnetic-field lines. Let us for a moment keep to the case when this condition is fulfilled. We have thus defined the permeability of the medium. Since $0 < \mu < 1$ the medium is diamagnetic.

That the magnetic moment $\frac{1}{2} m v_\perp^2 / B$ is a constant for motion along a flux tube and for time variations in the magnetic field is well known for slow variations, but it seems to be true also for fast variations. If we accept the Minkowski notation this quantity is also constant for relativistic velocities (cp. Leverett Davis, Jr.[7]).

If we discuss only small disturbances from homogeneity along the magnetic-field lines n/B, also, where n is the particle density, is constant. From what is said follows that the permeability is also a constant, even for large disturbances. If we had kept the whole discussion relativistically correct we should have found that the permeability is not constant when account is taken of the relativistic effects.

The dielectric constant is known before but we may add that it cannot be treated as a constant for large disturbances.

By our procedure we have eliminated the mechanical quantities but we find that the magnetic-field energy of the medium, $(1/2)\ \mathbf{BH}$, now is the sum of $\frac{1}{2} B^2 / \mu_0$ and $2/3$ of the thermal-energy density if the velocity distribution is isotropic and $(1/2)\ \mathbf{ED}$ includes both $\frac{1}{2} \epsilon_0 E^2$ and the kinetic-energy density due to the drift in crossed fields.

Let us discuss waves travelling perpendicular to the magnetic-field lines. The phase velocity is $c(\mu \epsilon)^{-\frac{1}{2}}$. After introducing the values of μ and ϵ we find three cases of special interest.

1. $\frac{1}{2} B^2 / \mu_0 \gg n m c^2$. The magnetic field is so strong that the motion of the charged particles is greatly hindered. In the limit we get electromagnetic waves *in vacuo*.

2. $n m c^2 \gg \frac{1}{2} B^2 / \mu_0 \gg n k T$. The density of thermal energy can still be neglected compared with the density of magnetic energy but the kinetic

6-2

energy due to the drift in crossed fields exceeds the electric-field energy. The phase velocity is equal to the Alfvén velocity. Looked upon in the present way they are transverse electromagnetic waves, but if we concentrate on the mechanical properties they are longitudinal compression waves.

3. $nkT \gg \frac{1}{2}B^2/\mu_0$. The electromagnetic-energy densities can be neglected besides the corresponding mechanical energies. The electric and magnetic fields are nevertheless necessary for the transfer of momentum in the medium. In this case we get diamagnetic sound waves. They have a phase velocity which is almost the sound velocity in the same gas if it were not ionized. Again it has to be looked upon as a transverse wave viewed from the present treatment but it can as well be accepted as a longitudinal wave (cp. N. Herlofson [8] and van de Hulst [9]).

Under the present assumptions the validity of the deduction is restricted to frequencies which are small compared to the gyro-frequency, but the collision frequency does not enter into consideration.

When we keep to regions where μ is constant we encounter no trouble, but when this is no longer the case the extra term in Maxwell's equations becomes important. This for instance means that $H_{\parallel}$ is no longer continuous at a boundary.

We have introduced **H** and μ in order to be able to use the results conventional electrodynamics offer. Since there exist other methods for solving the present problem it is perhaps suitable to restrict our present method to the case when μ is space-independent.

(1/2) **BH** is actually a pressure. When we have a solid boundary this quantity is no longer treated as an entity since the wall discriminates between two components. The wall can take up the mechanical part but does not react with the magnetic part. In equilibrium then the magnetic field has the same value on both sides of the boundary (Bohr [10]). To take such an experiment as a definition of permeability, i.e. to say that the gas is not diamagnetic, seems to be unrealistic since permeability ought to be a property of a homogeneous medium (here plasma) and not something characterizing a specific boundary value problem (cp. H. Alfvén [3], p. 57, Spitzer [11], p. 27).

REFERENCES

[1] Stratton, J. A. *Electromagnetic Theory* (New York, 1941), p. 2.
[2] Nichols, H. W. and Schelleng, J. C. *Bell Syst. Tech. J.* **4**, 215, 1925.
[3] Alfvén, H. *Cosmical Electrodynamics* (Oxford University Press, 1950).
[4] Åström, E. *Ark. Fys.* **2**, 443, 1950.
[5] Cowling, T. G. *Mon. Not. R. Astr. Soc.* **90**, 140, 1929 and **92**, 407, 1932.
[6] Spitzer, L. *Physics of Fully Ionized Gases*, Interscience Publishers, Inc. (New York, 1956).
[7] Davis, Leverett, Jr. *Phys. Rev.* **101**, 351, 1956.
[8] Herlofson, N. *Nature, Lond.* **165**, 1020, 1950.
[9] van de Hulst, H. C. *Problems of cosmical aerodynamics*, Central Air Documents Office, Dayton, Ohio, 1951.
[10] Bohr, N. *Studier over metallernes elektronteori*, Dissertation, Copenhagen, 1911.
[11] Spitzer, L., Jr. *Astrophys. J.* **116**, 299, 1952.

Discussion

Cowling: I would like to ask about the term $nmv_\perp^2/B^2$; where does it come from? Is it a term arising from the diamagnetism?

Åström: This term comes from two sources. One is the drift velocity perpendicular to the magnetic field caused by its inhomogeneity. The velocity is proportional to $mv_\perp^2$. Further, the momentum of the rotational motion in the magnetic field also gives a term of this kind.

Cowling: If the transverse pressure effect arises as I expect from diamagnetism, care is necessary, or the macroscopic approach may prove misleading. Care is, of course, necessary in a microscopic approach in defining the relation between **B** and **H**, as is evidenced by the lengthy discussions on Lorentz' polarization in the ionosphere. But I am not sure that the difficulties of a macroscopic approach are less than for a microscopic approach.

Spitzer: Dr Åström has made the point that in a hot ionized gas, where collisions are infrequent, one must go to the microscopic picture for a detailed solution. I should like to agree entirely with this result, subject to the proviso that this approach involves difficulties of its own. In particular, the velocity distribution in any particular situation is no longer Maxwellian and must be determined directly from the Boltzmann equation. Problems of this sort are sufficiently complicated to keep many theorists busy for a long time.

There is one result in this area which I should like to report at this time. One may ask how constant is the diamagnetic moment of a gyrating particle. Professor Alfvén showed that the quantity is constant to the first order in an expansion parameter, t, which is essentially the ratio of the Larmor radius to the distance over which the magnetic field changes substantially. Hellwig demonstrated that the diamagnetic moment is constant to the second order in t. Recently Kruskal and Kulsrud at Princeton demonstrated that this quantity is constant to all orders in t. This does not mean that the diamagnetic moment is rigorously constant, but rather that in the asymptotic expansion of the magnetic moment in powers of t, all the coefficients of t are zero. Thus we may conclude that the diamagnetic moment is very constant, indeed!

Swann: Starting from the basic equations in the microscopic form one can derive macroscopic relations by taking averages. It appears that the averaging of the ρu-terms gives rise to a conduction current and a polarization term and a complicated term which has to be subtracted from **B** to give **H**. In order to realize the macroscopic equations one has to average over a macroscopic element of volume of space. One then hopes to be able to formulate simple relations between j and **E**, etc., to complete the equations in usable form. Only if this is the case is there any sense in introducing the macroscopic equations. If this is not possible it is better to proceed directly from the microscopic relations.

Åström: It is true that one has to be careful when treating these problems, but I also think that if one divides the motion of the individual particle into a circular motion and a drift motion one avoids trouble. This does not mean that I want to state that the solution I have presented here is necessarily the best one. This paper has been presented mainly to stress that before one uses some terms one has to define them. In this special case it is about diamagnetism and related quantities.

EXPERIMENTAL STUDY OF PLASMOIDS

WINSTON H. BOSTICK*

University of California Radiation Laboratory, Livermore, California, U.S.A.

ABSTRACT

A plasma source can be used to project ionized matter across a magnetic field. The configuration of plasma observed when an electromagnetic braking action is produced by the presence of low-pressure gas (about $1\,\mu$) in the vacuum chamber provides an insight into the manner in which magnetic-field lines can be dragged and twisted. By firing several sources simultaneously, it is possible to simulate the production of spiral galaxies and barred spirals. The paper presented here forms an extension of earlier experiments performed by the author on plasmoids [1].

I. PRODUCTION OF A PLASMOID IN
FIELD-FREE SPACE

(a) The π_1 plasmoid

It has not only been demonstrated [2,3] that plasma can be projected from a plasma gun or plasma source with speeds up to 2×10^7 cm/sec, but the experimental observations suggest that the plasma travels (even in field-free space) not as an amorphous blob, but as a structure (called a plasmoid) whose form is determined by the magnetic field it carries along with itself. The mechanism whereby the plasma is propelled from the source has already [2] been outlined, and the hypothesis that the plasma travels in field-free space in the form of a torus has been supported with a Kerr-cell picture of the plasma in a toroidal form leaving the source.

The probe traces shown in Plate I (*a*) and (*b*) are further evidence that the projected plasma has structure. The exact structure of the plasma cannot be accurately delineated from these probe traces, but at least the probe traces are consistent with the hypothesis that the plasma is in the form of a torus. It is unfortunate, from the observational point of view, that these plasmoids move so rapidly. If they could be produced so that their center of mass was stationary in the laboratory, they could be photographed much more easily. (Later in this paper there will be described stationary

* Now at Stevens Institute of Technology, Hoboken, N. J., U.S.A.

plasmoids formed in a magnetic field which definitely exhibit toroidal structure.)

It has been possible with a magnetic coupling loop to pick up signals which are believed to be associated with the magnetic fields trapped by the plasmoid of the type shown in Figs. 2, 3, and 4 in Reference [2]. (Note that no external D.C. magnetic field is employed here.) Examples of such signals are shown in Plate II. Although the structure of these signals is too complex for analysis, these signals are nevertheless experimentally identified with the magnetic fields carried by the plasmoid.

In the magnetic coupling loop-signals, as with the probe signals, the outstanding feature is the steep leading edge, which can in no way be associated with an ordinary shock wave. The steepness of this leading edge of the plasmoid may possibly explain the abrupt onset of magnetic storms approximately 24 hr after a solar disturbance. It is entirely possible that ions and electrons ejected from the sun come to the earth in the form of a plasmoid.

In fact if a probe is immersed in a local magnetic field (e.g. due to a bar magnet) and a plasma source is fired at the probe at 1 m distance, the signal from the probe not only has a steep leading edge, but it exhibits large irregular oscillations which indicate rapidly varying ion densities and electric fields produced by the plasma encountering the stationary magnetic field.

The type of plasmoid (to be designated the π_1 plasmoid, because it is unstable) diagrammed in Fig. 4 of Reference [2] is expected to expand in directions of both increasing R and r, which are respectively the large and small radii of the torus. If m_T is the total mass of the plasmoid in grams and I is the total circulating current in amperes, an approximate differential equation (with the neglect of the logarithmic term) is

$$m_T d^2R/dt^2 = 8I^2/50\pi. \tag{1}$$

If at $t=0$ the original flux $\phi_0 \approx 2\pi^2 I_0 R_0/5$, where I_0 and R_0 are the initial current and radius, and $dR/dt=0$, the solution is given by

$$R_0 \sqrt{(R(1+R/R_0))} - 1/2R_0^{3/2} \ln\left[(\sqrt{(1+R/R_0)} + \sqrt{(R/R_0)})/\sqrt{R_0}\right]$$

$$- R_0^{3/2}[\sqrt{(2)} - 1/2 \ln\{(\sqrt{(2)}+1)/\sqrt{R_0}\}] \cong \sqrt{(2\gamma t)}, \tag{2}$$

where $$\gamma \equiv 8I_0^2 R_0^2/50\pi m_T.$$

From this solution it can be seen that when $R/R_0 \gg 1$, $R \sim t$. The expected expansion of the π_1 plasmoid may thus be thought of as a magnetic explosion where most of the outward velocity is picked up when the plasmoid is small.

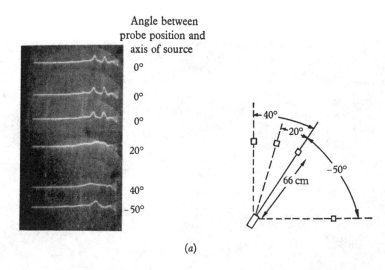

Angle between
probe position and
axis of source

0°

0°

0°

20°

40°

−50°

(a)

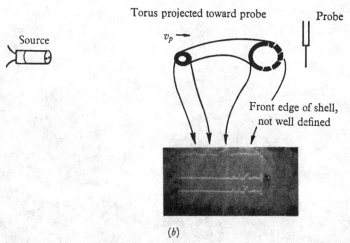

Source

Torus projected toward probe

Probe

v_p

Front edge of shell,
not well defined

(b)

Plate I (*a*). Probe tracks taken with alnico bar-magnet probes at 66 cm from the source of plasma at the various angles shown. The peak current in the source is about 3000 amperes. The sweep speed is 5 μsec/cm. Time goes from right to left and the source is fired at the beginning of the trace. The alnico bar-magnet is 1 × 1 × 2 cm.

(*b*) Probe traces taken at a distance of 10 cm from the source at sweep speeds of 0·5 μsec/cm. Sensitivity is 15 v/cm. The signal is developed across a 50-ohm resistance to ground. The probe is 1 mm in diameter and 2 cm long. Time goes from right to left and the trace starts at the firing of the source. An attempt has been made to identify the various parts of the hypothesized torus with the portions of the 0·5 μsec/cm trace.

(*facing p.* 88)

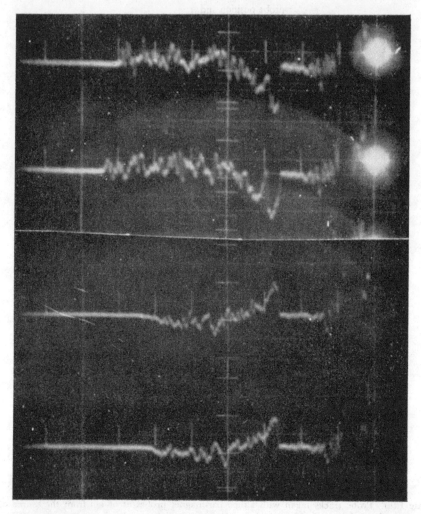

Plate II. Signals from a one-turn, 0·64 cm diameter coupling loop terminated with 50 ohms, where there is no external magnetic field applied. The coil is oriented in the plane of the paper, with respect to the source shown in Fig. 2., Reference [1], and located 10 cm directly in front of the source. The sensitivity is 0·5 volt/cm. The sweep speed is 0·5 μsec/cm, with the time going from right to left. The first two traces represent the current in one direction of the source, the second two traces represent the current in the opposite direction. The true plasma signals arrive at the loop at a time of about 1·3 μsec after the firing of the source which triggers the sweep. These signals are an indication of trapped magnetic fields within the plasma.

(b) The S-plasmoid

It is possible to conceive of a plasmoid which at first sight seems to be more stable than the π_1 plasmoid. This plasmoid, which we shall designate the S-plasmoid, is diagrammed in Fig. 1. Conceivably this plasmoid could be produced by winding a thin metallic ribbon into a helix and then bending the helix into a torus. If the metallic ribbon is then suddenly energized with a current so that it is vaporized and ionized, there presumably would be formed the S-plasmoid. Though this method of producing the S-plasmoid in field-free space would be tedious, it might nevertheless be fruitful and should eventually be tried.

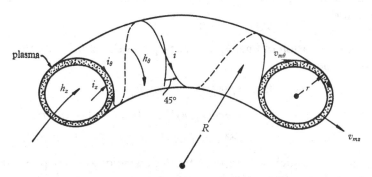

Fig. 1. The S-plasmoid. This configuration of magnetic field and plasma is essentially the same as that of the H-centered pinch in a toroidal tube which was first proposed and investigated experimentally in the United States in 1955 by M. A. Levine and L. S. Combes [Tufts University, Dept. of Physics Report, January 30, 1956 (unpublished)]. The configuration has since received theoretical development by M. N. Rosenbluth and C. L. Longmire [*Annals of Physics*, **1**, 120, 1957]. Papers 47–49 of this volume are reports on experimental and theoretical work on the same configuration in a straight tube (see also I. N. Golovin *et al.*, *J. Atomic Energy U.S.S.R.*, **5**, 26, 1956; V. D. Shafranov, *J. Atomic Energy U.S.S.R.*, **5**, 38, 1956).

From the theoretical point of view let us briefly examine the S-plasmoid in field-free space (i.e. no external D.C. magnetic field) and see if it is stable.

If we neglect the effects of nkT and centripetal acceleration due to any rotary motions of the mass of plasma, we note that equilibrium about r (see Fig. 1) requires that

$$h_z = h_\theta \quad \text{or that} \quad i_z = i_\theta.$$

Then since $\qquad i_z = 2\pi r i_z \quad \text{and} \quad I_\theta = 2\pi R i_\theta,$

$$I_\theta / I_z = R/r.$$

Let us assume that at time $t = 0$, the initial current $I_z = I_0$, and that the plasma is such a good conductor so that for all subsequent time the initial magnetic fluxes ϕ_z and ϕ_θ will be preserved. For purposes of simplification, electromagnetic units are used here. Then

$$\phi_z = L_z I_z = 4\pi R[\ln (8R/r) - 2]\, I_z$$

$$= L_{z0} I_0 = 4\pi R[\ln (8R_0/r_0) - 2]\, I_0 \tag{3}$$

and

$$W_z = (1/2)\, L_z I_z^2 = 2\pi R[\ln (8R/r) - 2]\, I_z^2. \tag{4}$$

Also,

$$\phi_\theta = L_\theta I_\theta - 2\pi r^2 I_0/R = 2\pi r^2 R I_z/Rr = 2\pi r I_z \tag{5}$$

$$= L_{\theta 0} I_{\theta 0} = 2\pi r_0 I_0 \tag{6}$$

and

$$W_\theta = 1/2 L_\theta I_\theta^2 = 1/2\, 2\pi r^2 I_\theta^2/R = \pi R I_z^2. \tag{7}$$

Now let us note that if ϕ_z and ϕ_θ are constant, $\alpha \equiv \phi_z/\phi_\theta$ must remain constant, and

$$\phi_z/\phi_\theta = 2R/r[\ln (8R/r) - 2].$$

Hence, preservation of flux ϕ_z and ϕ_θ requires that R/r be a constant. The total energy

$$W_T = W_z + W_\theta = 1/2\phi_z^2/L_z + 1/2\phi_\theta^2/L_\theta$$

$$= 1/2\phi_z^2\{1/4\pi R[\ln (8R/r) - 2] + R/2\pi r^2 \alpha^2\}$$

$$= (\phi_z^2/4\pi R)\, \{r/2R[\ln (8R/r) - 2] + R/r\alpha^2\}. \tag{8}$$

Now the force which will expand the S-plasmoid, and yet preserve the fluxes ϕ_z and ϕ_θ and hence R/r, is

$$-\left(\frac{\partial W_T}{\partial r}\right)_{R/r} = (\phi_z^2/4\pi r^2)\, \{r/2R[\ln (8R/r) - 2] + R/r\alpha^2\}. \tag{9}$$

This force is to be compared with the force which expands only the π_1-plasmoid:

$$-\left(\frac{\partial W_z}{\partial R}\right)_{\theta_z} = \phi_z^2/8\pi R^2[\ln (8R/r) - 2]. \tag{10}$$

Eqs. (9) and (10) give the somewhat unexpected result that for all values of $R/r \geqslant 1$ we have $\left|\dfrac{\partial W_T}{\partial r}\right| > \left|\dfrac{\partial W_z}{\partial R}\right|$, and hence the S-plasmoid should actually expand faster than the π_1-plasmoid.

It is interesting to note that for $W_z = W_\theta$, $\phi_z = \phi_\theta$ (i.e. where stability might possibly occur) the corresponding value of R/r is $1 \cdot 0$. Unfortunately, the formulae used for L_z, ϕ_z, and W_z do not hold below $R/r \geqslant 2 \cdot 5$. It is further interesting to note that for $R/r = n$, where n is an integer, a current stream-line will coincide cyclically with itself after one complete revolution of the plasma. And since R/r is a constant, the preservation of $R/r = n$ amounts to a kind of macroscopic quantum condition.

It is possible to plot W_T as a function of R for various values of r. The minima of these curves represent the situation where $W_\theta = W_z$, $\phi_\theta = \phi_z$, $\alpha = 1\cdot0$, and $R/r = 1\cdot0$. The expansion of the S-plasmoid will presumably occur along the minima of the curves and in the direction of increasing r. However, the portions of the curves to the left of the minima have no meaning because here $R/r < 1$. Hence, any realizable conditions involve a trajectory considerably to the right of the minima. It will be necessary to make the computation with expressions for L_z and L_θ which hold for values of R/r which are close to 1.

Furthermore, the stability relationships should be examined for configurations (like a muff) where the cross-section of the S-plasmoid is not circular, but oblong.

This brief analysis suggests that the S-plasmoid is unstable, but it cannot be stated definitely that all shapes of S-plasmoids in field-free space are unstable. There is a very real possibility that the S-plasmoid immersed in an external D.C. magnetic field would be stable.

2. PRODUCTION OF PLASMOIDS IN A MAGNETIC FIELD

(a) Introduction

It has already been demonstrated[2,3] that plasmoids can be projected across a magnetic field. It is quite possible that ionized material ejected from the surface of the sun proceeds and escapes across the magnetic field of the sun in the same manner that laboratory-produced plasmoids cross a magnetic field.

Plasmoids produced in a good vacuum are elongated cylinders[2,3] which travel as a wave front of constant velocity across the magnetic field. These plasmoids bear very little resemblance to the S-plasmoid and move so rapidly as to make instantaneous photography very difficult if not impossible. However, these plasmoids leave a wake or track[2] which enables us to photograph their path quite easily.

It is further observed[2] that these plasmoids experience an electromagnetic braking action which decelerates and deflects them when they encounter one another or when they travel through gas at a pressure of about $1\,\mu$. Indeed, several of these plasmoids can be made to spiral[2] in consort to produce a ring of plasma. The organic relation between this laboratory-observed process and the evolution of spiral galaxies and stars has already been suggested. More recent measurements[4] with a Kerr cell portray a time-sequence in the formation of this torus or ring, and show that not only is the torus produced automatically, but also that it is stationary, i.e. with regard to translation, and stable over a period of at least 30 μsec. During this time the torus appears to retain a circular form of about the same large and small radii.

There is some temptation to identify this observed [2] torus with the S-plasmoid immersed in a magnetic field. Before succumbing to this temptation, let us examine the results of some more recent measurements which teach us the prudent lesson that somehow we must learn to understand the way in which magnetic-field lines are dragged, spun, and interwoven if we are to give an adequate description of the resultant plasma configurations.

(b) Barred spirals

One of the simplest (relatively, that is) results which must be understood is the 'barred spiral' [4] which is produced by firing two sources at one another across a magnetic field. Over a wide range of variation of parameters the two plasmoids seem to seek each other out unerringly. The sequential stereo photographs of Plate III show the process of production of these 'barred spiral' structures at a pressure of 2 μ.

It can be seen from Plate III that the leading edges of the two plasmoids seem to seek each other and latch on to one another. The same process can be observed in Plate IV, where the pressure is 1 μ, where the bond between the two plasmoids apparently does not hold as well as in Plate III. The photographs of Plate IV also show the interesting feature that the tails of the spiral arms become forked.

Plate V shows a sequence of photographs where the pressure is 4 μ and where the leading edges of the plasmoids are positioned by the twisting of the plasmoids so that they are in no position to attach to one another. Under these circumstances the two plasmoids press tightly against one another but remain separated for at least 6 μsec. In Plate VI (a) where the pressure is 6 μ, the plasmoids remain separated for at least 10 μsec. Furthermore, the stereo photographs of Plates III, IV, V and VI (a) show that the plasmoid in proceeding across the magnetic field at these fairly high pressures assumes the form of a helix of progressively increasing diameter.

The photographs of Plates III, IV, V and VI (a) give us enough information to suggest that the configuration of plasma and magnetic field when one plasma source is fired across a magnetic field is that shown in Fig. 2. Eventually it may be possible to analyze this process quantitatively. For the moment, a description by a drawing will have to suffice.

It is now possible to see how two plasmoids fired at one another across a magnetic field can latch on to one another, as shown in Fig. 3. Such a

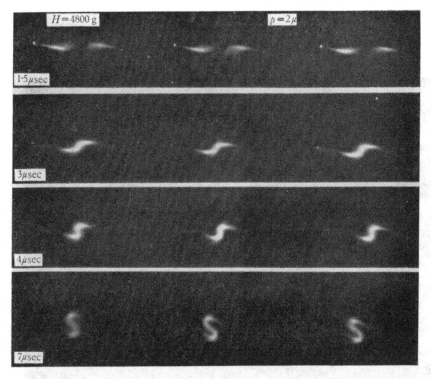

Plate III. A sequential study of barred spirals which are produced by firing two plasmoids from sources 10 cm apart at one another simultaneously across a magnetic field of 4800 gauss. The Kerr cell exposure times are 2 μsec and the various delay times of the sequence are indicated in μsec. The pressure in the chamber is 2 μ. The plasmoids and their trajectories are rendered luminous primarily by the recombination light of the titanium and deuterium ions which come from the plasma source. The photographs on the left are the left stereo photos and those on the right, the right stereo photos, with an angle between the two views of 10°. The middle photograph is taken straight ahead along the direction of the magnetic field. The current (3000 amperes for 0·4 μsec) through the source produces a magnetic field which opposes the D.C. field and diminishes the velocity of projection of the plasmoid across the magnetic field.

(*facing p.* 92)

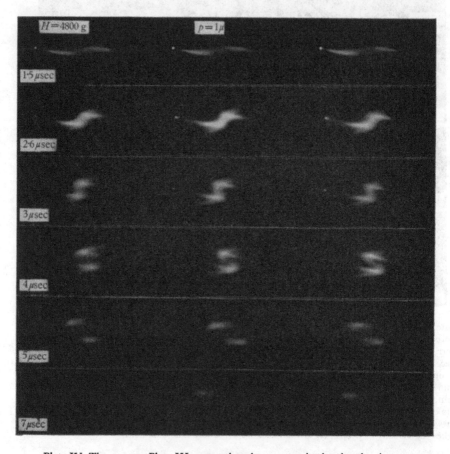

Plate IV. The same as Plate III except that the pressure in the chamber is 1 μ.

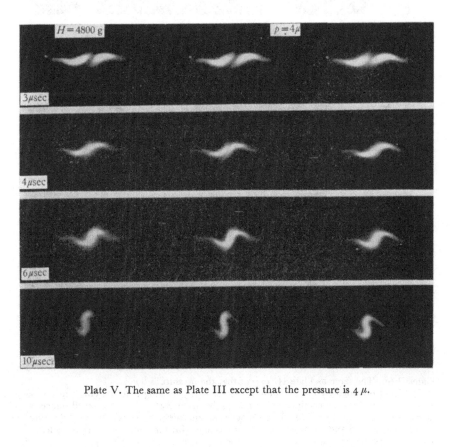

Plate V. The same as Plate III except that the pressure is 4 μ.

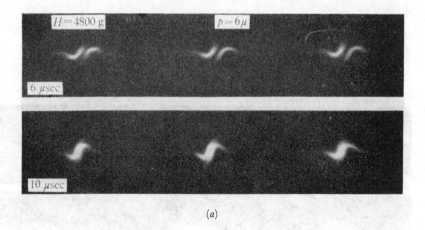

(a)

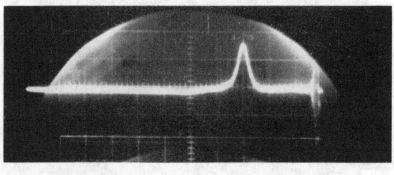

(b)

Plate VI (a). The same as Plate III except that the pressure is 6 μ.

(b) Signal obtained by probe 1 mm in diameter, 0·5 cm long placed 1 cm from a grounded probe with 50 ohms connecting them. The probe assembly is placed 30 cm distance, axially (away from the camera) down the solenoid from the position of formation of the ring. The sensitivity is 2 v/cm and the sweep speed is 2 μsec/cm with time going from right to left. Probe assembly is placed laterally off the axis at the approximate radius of the plasma ring which is formed. The solenoid is 44 cm long and 13 cm in diameter. The pressure in the chamber was 2 μ. Much more work is necessary in the study of this type of signal; we tentatively identify these signals with a whirl-ring moving away from the camera.

plasma-magnetic field configuration can also explain the forked tail on each plasmoid seen very clearly in Plate IV. Apparently if the leading loops of the two plasmoids have twisted into such a position that no stagnation

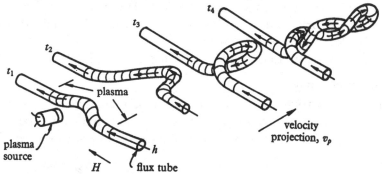

Fig. 2. Suggested configuration of plasma and magnetic field of a single plasmoid projected across a magnetic field when an electromagnetic braking action occurs because of a pressure of about 1 μ in the vacuum chamber. The tightly twisted configuration will itself assume a general helical configuration (try twisting two strands of wire, rope, or rubber tightly). This helical configuration is seen especially clearly in Plates V and VI (*a*).

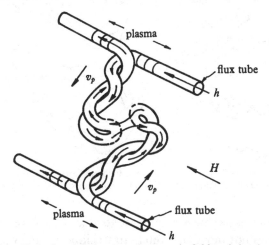

Fig. 3. Suggested configurations of plasma and magnetic fields existing in the formation of a barred spiral. The stagnation point produced when the two leading loops of the plasmoids approach one another permits the lines of force to leap from one plasmoid to another carrying plasma across and tying the two plasmoids together.

point is reached, the plasmoids studiously avoid one another, as shown in Plates V and VI (*a*).

After the union of the two plasmoids has been accomplished, as in Plates III and IV, the angular momentum will wind them up into a spiral, to a certain extent, until the angular momentum has been brought to zero by the stretching of the field lines. The resultant plasma and magnetic configuration then seems to be stable. The barred spirals have been followed in time as long as 15 μsec and still preserve their shapes with well-defined boundaries. Furthermore the plasma does not seem to migrate in the direction of the original D.C. magnetic field. It is rather astonishing that such a bizarre configuration of plasma and magnetic field should appear to be stable. No theoretician known to the author has *a priori* dreamed of such a configuration, to say nothing of contemplating its stability.

(c) Production of rings

We must now try to understand, at least in a qualitative way, how rings or toruses can be produced by plasmoids. Measurements already reported [4] show that with four plasma sources a ring can be produced which

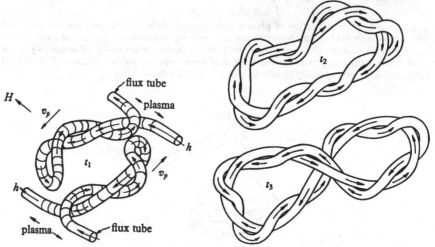

Fig. 4. Suggested sequence of plasma and magnetic-field configurations to explain the flattened ring and 'figure 8' observed in Plate VII. The velocity of projection v_p of the original plasmoids is indicated.

apparently maintains its shape for at least 30 μsec. Moreover, it is observed that this ring does not move or stretch appreciably in the direction of the D.C. magnetic field during this time interval. The magnetic-field configuration in the ring must, therefore, be such as to confine the plasma in this fairly stable ring.

It has been possible to produce a ring with only two sources (see Plate VII), but the ring produced is flattened. In fact, as time goes on the ring configuration develops a constriction and flips over into a 'figure 8'. Before attempting to understand the rings which are produced with four or more plasma sources, let us first try to understand this flattened ring which is produced by two sources. Fig. 4 suggests a plasma-magnetic field

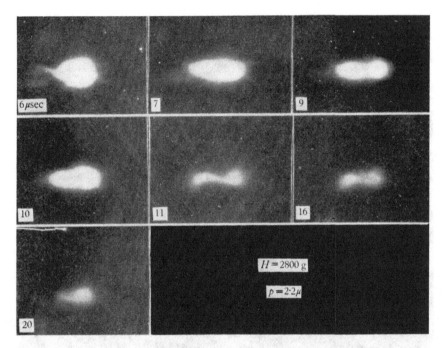

Plate VII. Sequence of photographs of formation of a flattened ring which flips into a 'figure 8'. The ring is formed by firing two sources across a magnetic field (into the paper) of 2800 gauss with a pressure of 2·2 in the chamber. The exposure time is 2 μsec and the various delay times are indicated in μsec. The current in the source is in such a direction as to diminish the velocity of propagation in the magnetic field.

(*facing p.* 94)

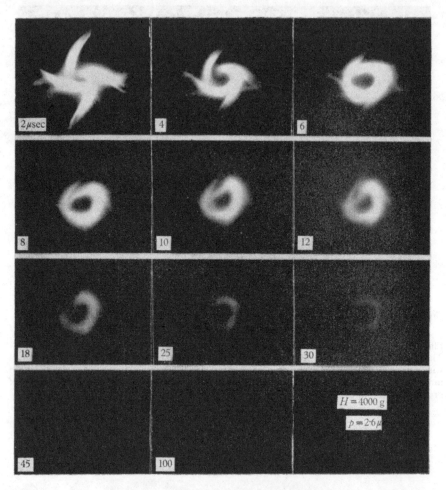

Plate VIII. An example of the formation of a ring by firing four sources across a magnetic field of 4000 gauss (into the paper) at a pressure of 2·6 μ. The exposure time is 2 μsec and the various delay times are indicated in μsec. The current in the sources is in such a direction that the velocity of projection of the plasma is retarded by the D.C. magnetic field. The 100 μsec delay photograph actually shows a faint ring on the original.

configuration to explain the ring shown in Plate VII. It is readily understandable that a flattened ring formed by tightly twisted strands will constrict in the center and form a 'figure 8'. It can be seen from the hypothesis of Fig. 4 that we might actually expect two rings, one formed from each strand, but that they are topologically intertwined.

As has already been reported [4], it is possible to form rings by firing four sources. It is believed that these rings have essentially the same structure as the ring shown in Plate VII and Fig. 4, except that they are initially circular instead of flattened, and they have more angular momentum. Therefore, we may expect the rings formed by four sources to have a tendency to preserve their circular shape instead of flipping into a 'figure 8'. An example of a photographic sequence of the formation of a ring from four sources is given in Plate VIII.

It is possible to produce rings somewhat similar to those of Plate VIII by firing eight sources. Various examples of photographic sequences with eight sources are given in Plates IX, X, XI and XII. Very likely the rings formed by eight sources have higher peripheral velocity and hence more angular momentum than those formed from four sources. This higher peripheral velocity may account for the deformations of the ring which are observed in Plates IX, X and XI at the later times.

It is to be emphasized that these rings which have been produced by having the source current in the direction so as to diminish the velocity of propagation of the original plasmoids across the D.C. magnetic field, remain in focus up to at least 30μsec, and in some cases up to 100μsec. Therefore we can say that the ring does not move or stretch appreciably in the direction of the magnetic field.

The situation is quite different when the source current is in such a direction as to increase the velocity of propagation across the D.C. magnetic field. Under these circumstances, it has been observed [2, 3] that the initial velocity of propagation across the field is greater and the initial diameter of the plasmoid is greater. A sequence of photographs taken under these circumstances with eight sources firing simultaneously is shown in Plate XIII, where the ring which is produced now apparently moves along the D.C. lines of force toward the camera. Probe traces (see Plate VI (b)) taken at a position 30 cm behind the source suggest that there is also plasma' (presumably in the form of a ring) traveling along the lines of force in the opposite direction with a velocity $\leqslant 3 \times 10^6$ cm/sec. Here now is a situation which appears to be the simultaneous production of two whirl-rings which move away from one another along the D.C. magnetic-field lines. Plate XIV shows stereo photos taken of the ring which moves toward the camera. The ring

95

appears to have a helical twist which probably represents the direction of magnetic field and motion of the plasma within the ring. A suggested description of the process of formation of these two rings is given in Fig. 5, where it can be seen that the two rings which are now produced are not topologically entangled but are free to move away from one another. Each of these rings now is similar to the S-plasmoid of Fig. 1, except that they exist in a D.C. magnetic field, and h_z and h_θ fields have time to penetrate

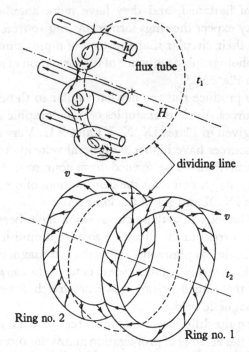

Fig. 5. Suggested description of formation of two whirl-rings which move away from one another, as observed in Plates VI (b) and XIV.

the plasma and add vectorially to produce a helical magnetic-field configuration within the ring. The measurements as yet do not yield quantitative information on whether the rings maintain their diameter as they move along the D.C. magnetic-field lines. It will be necessary to construct a longer solenoid for the magnetic field than the 44-cm solenoid which was used in these experiments in order to examine the radial stability of these rings and to measure their velocity along the field. It will also be necessary to devise a suitable technique for exploring the magnetic fields which are trapped in these rings.

96

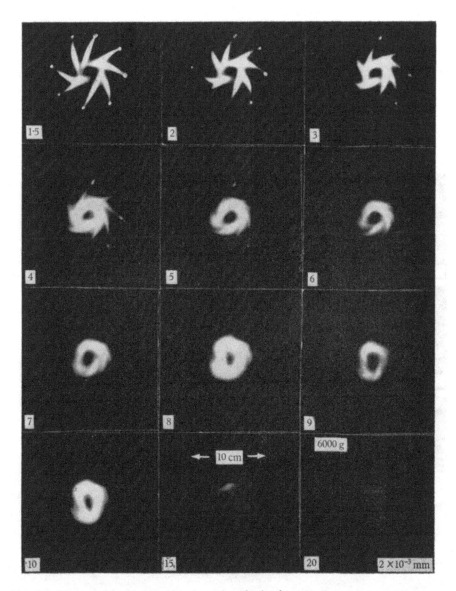

Plate IX. Photographic sequence of the formation of a ring from the plasmoids from eight sources fired across a magnetic field of 6000 gauss, into the paper. The pressure is 2 μ. The exposure time is 2 μsec, and the delay times are indicated in μsec. The current in the sources is in the direction to diminish the velocity of projection of the plasmoids across the D.C. magnetic field. Note that at 15 μsec delay, the ring has grown an ear.

(*facing p.* 96)

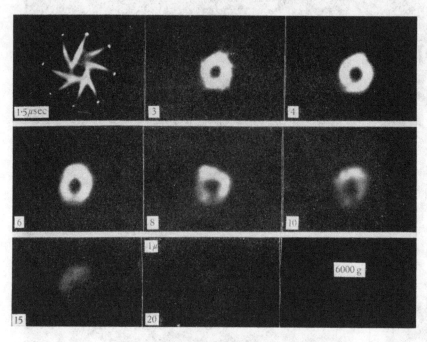

Plate X. Same as Plate IX except that the sources are oriented more symmetrically, and the pressure is 1 μ.

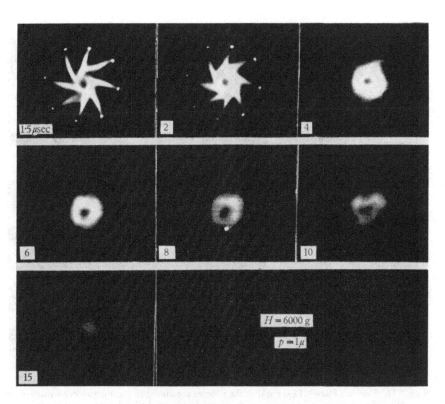

Plate XI. Same as Plate IX except that the sources are aimed so as to produce a smaller diameter ring, and the pressure is 1 μ.

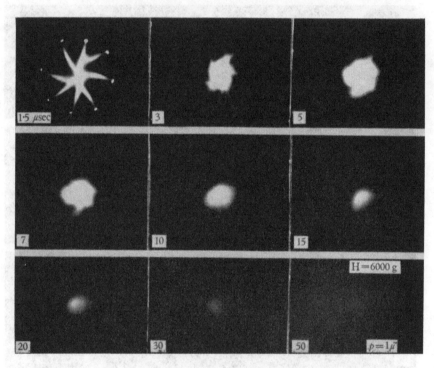

Plate XII. Same as Plate XI except that the sources are aimed to produce an even smaller ring. The original photograph shows an illuminated blob of plasma lasting out to 50 μsec.

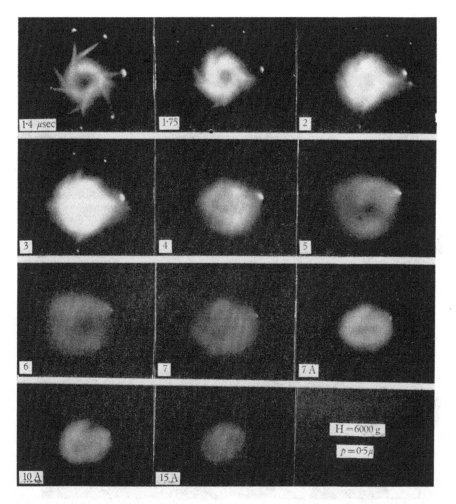

Plate XIII. Sequence of photographs taken in a manner similar to those of Pl. IX except that the source current is in the direction which *aids* the velocity of propagation of the plasmoid across the magnetic field, and the pressure is 0·5 μ. The whirl-ring which is formed gets progressively out of focus as it apparently moves toward the camera, and at 7 μsec delay, it is badly blurred. The delay times '7A, 10A, and 15A' correspond to moving the camera back 15 cm which, at least for 7A sharpens up the picture. It is believed that for delays 10A and 15A, and perhaps for 7A, the ring has travelled until it has encountered the lucite window of the vacuum system which is 20 cm from the position where the ring is formed.

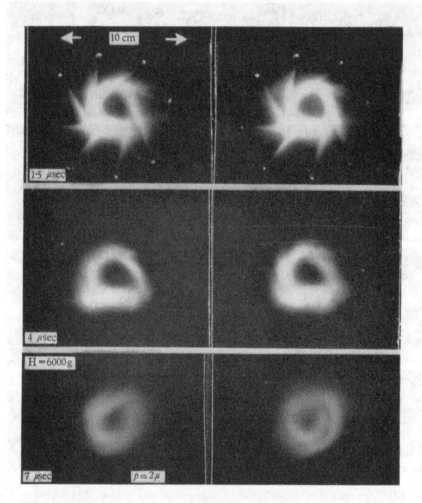

Plate XIV. A sequence of stereo photos (20° between the two) of the rings produced, as in Pl. XIII, with the pressure equal to 2 μ. For the 7 μsec delay the camera was refocused for an object distance closer by about 15 cm, to compensate for the fact that the ring apparently moves toward the camera. The 4 μsec delay photo, especially, suggests that the ring is constructed of material which has a helical twist.

3. CONCLUSION

By firing simultaneously two or more plasmoids across a magnetic field it has been possible to produce co-operative phenomena which not only simulate the production of spiral galaxies and astonomical barred spirals but which permit us to study these processes in the laboratory. Hypotheses to explain the experimental effects have been advanced in outline. Accurate quantitative work and detailed theoretical analysis should now begin.

4. ACKNOWLEDGMENTS

The author wishes to acknowledge the fine experimental work of Orrin A. Twite who assisted in the work and who took many of the photographs, and to V. G. McIntosh who constructed the plasma sources. Gratitude is due to W. R. Baker, O. A. Anderson, Jack Reidel, and N. J. Norris for the loan of the Kerr-cell equipment. The author wishes to thank S. A. Colgate and C. M. Van Atta for their support, encouragement, and advice in this work.

REFERENCES

[1] Bostick, W. H. *Experimental study of ionized matter projected across a magnetic field*, Univ. of California, Radiation Laboratory, Livermore Site, Rep. no. UCRL-4695, 1956.
[2] Bostick, W. H. *Phys. Rev.* **104**, 292, 1956.
[3] Harris, E., Theus, R. and Bostick, W. H. *Phys. Rev.* **105**, 46, 1957.
[4] Bostick, W. H. and Twite, O. A. *Nature, Lond.* **179**, 214, 1957.

Discussion

Blackett: I must say that I appreciate very much the new kind of experiments that Dr Bostick has done. I think it is an extraordinary exciting new field. Perhaps the theoretical solution of those phenomena lies somewhere in Dr Schlüter's equations.

Ferraro: This is an interesting and important paper. I was particularly interested in two of the phenomena described by Dr Bostick. The first concerns the fact that a plasmoid was able to move freely in a magnetic field perpendicular to its direction of motion. I believe the explanation that Dr Bostick suggested is in fact likely to be the true one, that is, that the polarization electric field generated by the motion of the plasmoid in the magnetic field balances the electromagnetic deflecting force. The two tracks left behind by the plasmoid are in fact the polarization surface charges which are trapped by the magnetic field. This is the solution Chapman and I first gave in our papers on magnetic storms and it is gratifying to see these experiments bearing on the theory.

The second comment I wish to make refers to Dr Bostick's remark that the magnetic impulse registered by a magnetic coupling loop as a plasmoid passed over it may have some bearing on the theory of sudden commencements of magnetic storms. It seems to me that a 'solar plasmoid' is unlikely to imitate even qualitatively the variation of the horizontal force observed during a sudden commencement. This remains fairly constant for an hour or so after the sudden commencement, unlike the variations of the magnetic field which are associated with a passage of the plasmoid over the loop.

Bostick: I must say that there is in no way an exact analogy between conditions in cosmical physics and in the measurements we made. But the plasma enters the magnetic field of the earth and—this may be pure speculation—it may enter the region of the earth in the form of a ring (plasmoid), similar to what is observed in the laboratory. In that way it might give a fast-rising but sustained impulse to the magnetic field by forming a type of ring-current. Also, our laboratory plasmoids are the result of the emission of plasma in one short impulse, whereas the solar emission of plasma may be a relatively sustained emission.

I should like to say that if we can shoot plasmoids across a magnetic field, in the laboratory, it may be possible for a plasma to be shot across the solar magnetic field into interplanetary space. It is probable that when such a plasmoid leaves the sun it will take a part of the sun's magnetic field with itself; I do not know what the shape of the plasma configuration will be.

Alfvén: These are extremely fascinating results. They stress again the importance of experimental approach to astrophysical problems. But before applying the results to cosmical physics it would be very important to find a relevant criterium for the existence of plasmoids in astrophysics. Could you by some similarity transformation give such a criterium?

Bostick: Of course. The speeds we have are comparable with the astronomical speeds; one can say that we have the same order of magnitude in the speed-situation. Concerning the densities we are way out by a factor of 10^{15}. All these things have to be worked out more in detail.

von Engel: How did you measure the magnetic moment of a free plasmoid and what numerical results have been obtained?

Bostick: A signal is picked up from the plasmoid by a magnetic coupling loop. The field measured was of the order of 100 gauss.

von Engel: What was the nature of the light emitted and did it show lines from the metal and the gas?

Bostick: We have made some measurements lately and practically all the light is contained in the lines of titanium I and titanium II and in the Balmer series. Presumably this light is recombination light. If there are other lines they are relatively very dim.

Swann: Concerning the photographs which showed a kind of elastic scattering of one ring by another, the orbits and impact parameters seem sufficiently definite to enable one to calculate the forces concerned in the collisions if one knows the masses of gas in the rings and the velocities. Has such a calculation been made and if so, what is the implication as regards the nature of the force?

Bostick: No, we have not made such calculations. But the amount of momentum which is carried by the plasma coming out of the source can be measured by mechanical macroscopic means quite easily.

Ferraro: Have you measured the densities?

Bostick: No, we have not, but we have inferred a value of $10^{11}/cm^3$ from probe measurements 100 cm from the source. The measurement is difficult to carry out since it is transient and the ion density is high.

A CHARACTERISTIC LINE CURRENT IN A FULLY IONIZED GAS

R. S. PEASE

Atomic Energy Research Establishment, Harwell, England

ABSTRACT

Standard formulae for the electrical resistance and for the radiating properties of a fully ionized gas have been combined with pinch effect relations to obtain the stationary state radial distribution functions and current—voltage characteristics of a filamentary current. The calculations suggest that the radiation cooling permits a pinched discharge to exist, with a maximum current of about one or two million amperes.

Alfvén[1] has drawn attention to the prevalence of line currents in cosmic physics and has pointed out the role which might be played by the pinch effect in producing them. Tonks[2] and Alfvén[3] have both discussed a possible disrupture of a high current discharge due to the pinch effect, when the current exceeds a certain value, such as is observed for instance in mercury arc rectifiers. Both these treatments appear to ignore radiation cooling, which is likely to be an important source of energy dissipation in cosmic physics. The present calculations of the characteristics of a filamentary current in a fully ionized gas assume that all the power input is dissipated in 'Bremsstrahlung' radiation.

We suppose the current filament extends over a cylinder of radius b containing N atoms/unit length, and is actuated by a uniform applied axial electric field of strength E. We suppose the gas is completely ionized hydrogen, and radiates power according to Cillié's[4] 'Bremsstrahlung' formula. The energy balance is then expressed by

$$EI = \int_0^b 2\pi r B n^2 T^{1/2} \, dr, \qquad (1)$$

where I is the total current, n is the number of electrons (and of protons) per unit volume, T is the electron temperature, and B is a constant equal to $1 \cdot 4 \times 10^{-40}$ M.K.S. units; r is the distance from the axis of the current filament. In the axial direction, the momentum imparted to the electrons by the field is equal to that lost in collisions with protons; and from the work of Gvosdover[5] we obtain

$$EeN = \int_0^b \pi r n^2 e^2 w T^{-3/2} G \log X \, dr, \tag{2}$$

where e is the electronic charge, w is the electron drift velocity, G is a constant equal to 65 M.K.S. units, and $X = (kT/e^2 n^{1/3} . 137\beta)^2$ where β is the ratio of the electron speed to that of light; discussions of the factor 137β are given by Williams[6] and Spitzer[7]. Finally, in the radial direction, if the filament is neither expanding nor contracting there is a balance between the hydrostatic pressure and the pinch forces; and so:

$$\mu_0 i \frac{di}{dr} + 2\pi r^2 k \frac{d}{dr} (nT) = 0, \tag{3}$$

where μ_0 is the permeability of free space, i is the current enclosed within a radius r, and k is Boltzmann's constant. i is related to w by the formula

$$\frac{di}{dr} = 2\pi r n e w. \tag{4}$$

In (3) we have assumed that the ion temperature is equal to the electron temperature; if the temperature is zero, the factor of 2 is omitted from the second term of (3).

These equations cannot be solved fully without information about transport processes in the gas. We may solve however for certain simple cases, notably: zero viscosity and thermal conductivity; infinite viscosity and thermal conductivity; and zero viscosity and infinite thermal conductivity. When this is done, it is found that the solutions in each of these cases are very similar, and that the last case is the best of these simple approximations. The radial distribution functions for the case are:

$$n = \frac{2N}{b^2} (1 - r^2/b^2) \tag{5a}$$

and

$$\omega = \bar{\omega}/2 (1 - r^2/b^2). \tag{5b}$$

T is, of course, a constant. So also is the current density (new). This configuration is the same as that discussed by Schlüter[8]. The radius of the filament is given by

$$b = (2B/3\pi E)^{1/2} (\mu_0 N^3/k)^{1/4}. \tag{6}$$

The current is found to be a constant, independent of b, N and E provided the filament is supported entirely by the self-magnetic field, and is equal to

$$I = \frac{k}{\mu_0} (12G (\log X)/B)^{1/2}. \tag{7}$$

The effect of increasing the electric field is thus solely to increase the constriction of the current channel. The extra power fed in is dissipated solely by the consequent increase of the factor n^2 in equation (1). The term $\log X$ contains n, and the increased constriction lowers this term slightly: however, as has often been pointed out, $\log X$ is large and very insensitive to large changes of X, and so the current is virtually independent of E. If the electric field becomes very low, it must be expected that in practice some of the pressure is supported by means other than the self-magnetic field. It is easily shown that in such circumstances the current drops below the above value. With $\log X$ equal to 30, the current is about 2×10^6 amperes. In the other two cases mentioned, the numerical factors of Eqs. (6) and (7) are slightly different. When the constants G and B are expressed in terms of fundamental atomic constants, it is found that

$$I \sim me^{-2}\hbar^{1/2}c^{7/2} \quad \text{(e.s.u.)}.$$

This particular combination of constants can be written $(mc^3e^{-1})\,(e^2/\hbar c)^{-1/2}$. With an ionized gas containing positive ions of charge Z, the current is found to be altered by the factor $2Z/(Z+1)$.

These results depend on a number of assumptions implicit in Eqs. (1)–(3). Two particularly important ones are as follows. First, it is assumed that the gas has a Maxwellian velocity distribution superimposed on a slow drift velocity. Thus the results might well be inapplicable when a beam of fast electrons with predominantly ordered motion passes through a cloud of ions. Secondly, Eq. (2) implies that the electric current heats the gas entirely by collisions with the electrons and protons. In cases where there is a large electrodynamic voltage, the gas might be heated by collisions arising from the bulk motion of the gas. Such motions are known to arise from inherent instability of a current carrying plasma (Kruskal and Schwarzschild[9]).

REFERENCES

[1] Alfvén, H. *Proc. Roy. Soc. A,* **233**, 296, 1955.
[2] Tonks, L. *Trans. Electrochem. Soc.* **72**, 167, 1937.
[3] Alfvén, H. *Cosmical Electrodynamics* (Oxford University Press, 1950).
[4] Cillié, G. *Mon. Not. R. Astr. Soc.* **92**, 820, 1931.
[5] Gvosdover, S. D. *Phys. Z. Sowjet,* **12**, 164, 1937.
[6] Williams, E. J. *Rev. Mod. Phys.* **17**, 217, 1945.
[7] Spitzer, L. *Physics of Fully Ionized Gases,* Interscience Publ. (New York, 1956).
[8] Schlüter, A. *Z. Naturf.* **5a**, 72, 1950.
[9] Kruskal, M. and Schwarzschild, M. *Proc. Roy. Soc. A,* **223**, 348, 1954.

Discussion

Spitzer: Have you taken into account the reduction of thermal conductivity by the magnetic field? How does the temperature come out from that?

Pease: The mean temperature comes out from the pinch relation solely. You have $\mu_0\, I^2 = 4Nk\overline{T}$ where N is the ion density per unit length of the column. For $N = 10^{20}$ m^{-1} the temperature comes out about 5×10^7 °K. Regarding the thermal conductivity question, I have taken the reduction of thermal conductivity by the magnetic field into account when estimating the effect of conductivity on the radial temperature gradients. The gradients obtained when the conductivity is assumed zero, in fact require heat flow from each plasma element which is comparable with $Bn^2 T^{\frac{1}{2}}$, the radiation loss. This is calculated for the thermal conductivity of the electrons. But the reduction of the thermal conductivity of the positive ions is not so great. Hence I conclude that radial temperature gradients are largely eliminated by radial thermal conduction.

Artsimovich: Have you also investigated other forms when the radius varies with time? If it varies with time, and the contraction is strong enough, the temperature will increase considerably and 'Bremsstrahlung' can probably not be neglected.

Pease: I have not studied the non-stationary case, which I believe you have treated, i.e. where the radius collapses. There is no inertial term in my equations.

Artsimovich: If you neglect the 'Bremsstrahlung' effects, have you then investigated the time variation rates?

Pease: No, but your colleagues have done it, surely.

Artsimovich: I should think that these investigations are too specialized and have only historical interest. In reality the plasma column is not stable.

A great number of solutions can be obtained by means of relations which express a current-compression. If the temperature is low enough for 'Bremsstrahlung' not to be of importance, then the radius of the pinch can be determined and the current becomes proportional to $T^{1/7}$.

Pease: I have investigated cases where the energy dissipation is by other mechanisms, for instance when the 'Bremsstrahlung' radiation is not produced by proton-electron collisions but is produced by magnetic spiralling.

Artsimovich. If this investigation is carried on still further then cases can be found in which the limiting current is larger than the values you have found.

Pease: Is this the situation when one is working with transient conditions?

Artsimovich: Yes. Such investigations are based upon simple conditions expressing thermal equilibrium between ions and electrons.

Spitzer: What is the time dependence of the transient solutions which you have obtained?

Artsimovich: We have got many different solutions the forms of which depend on the boundary conditions; I do not think that I can give a simple answer to Dr Spitzer's question.

Thonemann: Pease talked about the steady state limiting current. Instabilities do in fact develop rapidly as the following figures illustrate. Plate I a shows a straight discharge in a tube with only slight perturbations near the cathode end. As time proceeds and the current increases, striations appear (Plate I b). Finally at a still later time the line current breaks up or develops into a badly defined helix with a pitch of about 45° (Plate I c). Plate II shows another example of an unstable discharge in a toroidal glass tube, again of helical form.

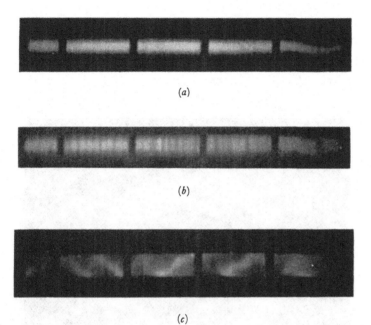

(a)

(b)

(c)

Plate I. Gaseous discharge in a cylindrical tube. (a) The current is relatively weak. A slight perturbation is seen near the cathode in the right-hand corner of the picture. (b) Striations occur when the current is raised. (c) If the current is raised further it breaks up and a helix is formed.

(facing p. 102)

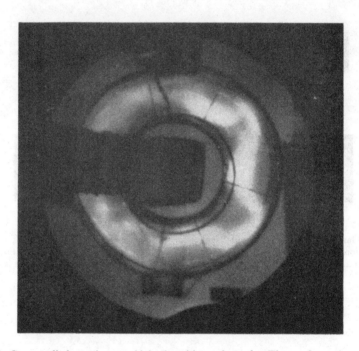

Plate II. Gaseous discharge in a toroidal tube without electrodes. The gas forms the secondary of a pulse transformer. For strong currents a helical configuration arises as seen from the picture.

PART II
SOLAR ELECTRODYNAMICS

SOLAR ELECTRODYNAMICS

T. G. COWLING

The University, Leeds, England

ABSTRACT

A historical account of the subject's development is attempted. Prior to 1940, the most significant papers were those by Larmor (1919) and Cowling (1934) on dynamo theories of solar fields: by Kiepenheuer (1935) on the corona; and by Ferraro (1937) on isorotation. These indicated the importance of electromagnetic forces and were groping towards the idea of frozen-in fields. The latter idea was, however, not clearly stated before Alfvén's 1941–2 papers.

Theory since then is divided into sections concerned with mechanical effects of magnetic fields, theories of sunspots, and the nature and origin of solar magnetic fields. The first includes theories of magnetic control of support of coronal filaments and prominences (van de Hulst, Alfvén, Dungey) and theories of magnetic influence on sunspot equilibrium. The second includes Alfvén's and Walén's theories of the solar cycle, and Biermann's explanation of sunspot coolness in terms of magnetic inhibition of convection. Sunspot theories, being discussed more fully by Biermann, are considered only briefly.

Electromagnetic heating covers theories of coronal heating and flares, discharge phenomena, particle acceleration and radio emission. Many of the older theories (Alfvén's on coronal heating, Giovanelli's on flares, that of Bagge and Biermann on cosmic rays) are set aside because of their neglect of self-induction effects and inadequacy of the mechanism of conversion. The relative motion of charged particles and neutral atoms (Piddington, Cowling) is described as supplying a powerful heating effect.

As regards the magnitude of the general solar magnetic field, it is suggested that the observed value can be discarded only if decisive reasons are given. Other theories having so far proved inadequate, dynamo theories of the origin of solar fields are regarded as the most promising. These can be partial, as when a toroidal field capable of explaining spot fields is supposed to be generated from the general field (Walén and others), or when a turbulent field is supposed to be generated from a smaller regular field (Alfvén and others): or total, when a simultaneous explanation of all fields is attempted (e.g. Parker). A general appraisal is made of the different theories.

In what follows, a historical account of the development of solar electrodynamics will be attempted. Emphasis will be rather on theories advanced to explain the different phenomena than on the observations; the phenomena are well known, their explanation is difficult.

Solar electrodynamics may be said to have begun in 1889, when Bigelow inferred a solar magnetic field from the form of coronal streamers near the poles. However, real interest in the subject began only after Hale's measurements of sunspot magnetic fields and his announcement of the existence of a general solar magnetic field. His discoveries stimulated Larmor, in 1919, to suggest a dynamo theory of the origin of such fields. Even though the suggestion was shown by Cowling in 1934 to be untenable in the form in which he had advanced it, the general idea of dynamo maintenance of cosmic fields is one that is still fruitful.

The Mount Wilson workers originally suggested that the sun's general field, and probably sunspot fields, are limited to low layers in the sun's atmosphere. In 1928 Chapman advanced a theory of the radial limitation of the general field, based on a study of the drifts of charged particles in crossed electric, magnetic and gravitational fields; a second theory, based on the diamagnetic properties of free electrons, was advanced by Ross Gunn. Cowling showed in 1929 that such theories were untenable, and that any correct theory must be based rather on a study of the electric currents flowing. Later the phenomenon of rapid radial limitation was recognized as non-existent. Indeed, the very existence of a general solar field was for a time regarded as extremely doubtful; only in the last few years has the work of H. W. and H. D. Babcock established the existence of such a field, though they have found that its properties are far different from those earlier ascribed to it. Before 1945, most theoretical work was based on the assumption of a dipole-like field, with maximum strength about 50 gauss near the poles.

During the decade 1930–40 some progress was made in understanding the basic laws of motion of an ionized gas in a magnetic field. The developments were parallel to those made in theories of the ionosphere, put forward about 1930. The importance of mechanical interaction between material and field in a sunspot was clearly recognized by Cowling in his 1934 paper, and there were glimmerings of the idea that lines of force are frozen into the material. Cowling did not, however, fully realize the extent to which lines of force can be frozen into the material, because of a mistaken belief that a polarization electric field permits material to slip freely through a magnetic field.

Kiepenheuer, in a paper on the corona in 1935, made use of the idea of frozen-in fields; he supposed that masses of gas ejected into the corona break off, and carry with them, pieces of the photospheric field. The lines of force are probably too firmly frozen into the material to permit pieces of the field to be broken off like this, but even this partial recognition of frozen-in fields is interesting. The same principle provides the simplest explanation of Ferraro's law of isorotation (1937) that in a star in steady non-uniform rotation the angular velocity must be the same at all points of a line of force; but this explanation was not, in fact, given.

Just before the last war, moving pictures of prominence motions were first seen. These offered clear evidence of the influence of magnetic fields. Jets of material were seen to be continually thrown up along curved paths, returning along the same paths. Since charged particles are known to follow lines of force, the inference was obvious that these curved paths are along lines of force. It was also speculated, both then and later, that quiescent prominences might in part be supported by magnetic forces.

The subject developed more rapidly after 1941, stimulated by Alfvén's work. Alfvén clearly stated the principle of frozen-in fields; he showed that in consequence magneto-hydrodynamic waves could be propagated along the lines of force in a conducting fluid, and re-derived the law of isorotation; and he indicated again the importance of magnetic forces to sunspot equilibrium. The importance of his work was in his emphasis on the two-way interaction between magnetic fields and material motions. True, his results were all implicit in the theory of Maxwell's stresses; others had earlier attained some of his ideas; and sometimes magneto-hydrodynamic waves were invoked when clearly the actual interaction between field and motion was far more complicated than a wave motion. But these facts should not obscure the importance of the impact of Alfvén's ideas.

From this point on, the ramifications of the subject become too great for it to be treated as a whole. Developments in the theory of sunspots will first be considered. Alfvén advanced a theory of sunspots, based on the assumed production of whirl rings in the sun's interior, and their progression along lines of force of the general field to the sun's surface. This theory was developed and made more precise by Walén. Alfvén never wholly accepted Walén's account, and Walén has since repudiated it; but it still remains the most complete account of Alfvén's theory. It is unfortunate that Alfvén has never given more details, in particular of the way in which the surface magnetic field is created from a whirl ring; at present his theory is a theory of the solar cycle without any proof that it can explain the existence of individual spots. However, it would not be appropriate for me here to add to the criticisms of Alfvén's work which I have made elsewhere in more than one place.

Walén, after abandoning Alfvén's ideas, has himself put forward a theory of the solar cycle. This is based on the idea of torsional oscillations round the sun's axis, due to the stiffness introduced by the general magnetic field. The oscillations are excited by periodic convulsions in the sun's interior. Once again, since I have criticized Walén's work elsewhere I shall not go into details here.

In 1939, Biermann suggested to me in a letter that sunspots might be due to the inhibition of convection by a magnetic field, and the consequent reduction of heat transported to the surface. This suggestion appeared, almost as a chance remark, in a paper by him in 1941. The inhibition of small-scale convection by a magnetic field was inferred, on general physical arguments, by Walén in 1949; detailed mathematical work by Thompson and Chandrasekhar, while confirming that a magnetic field interferes with convection, showed that it does not make it impossible with a sunspot. I have tried myself to develop the theory that sunspot darkening is due to the reduction of convection in a spot; Hoyle has suggested a theory on rather different lines, convection being not so much reduced as restricted to motion along the lines of force.

Theories of electromagnetic heating will next be considered. In 1940, Alfvén suggested that some prominences might be the visible signs of electromagnetic discharges. Such a belief was possible when a prominence could be regarded as shining because hotter than the corona, but it is untenable now that the coronal temperature is known to be $10^6 \,^\circ\mathrm{C}$. All that survives of it is the suggestion, which Dungey has elaborated, that electric currents flowing along prominence arches may lead to increased densities through a pinch effect. Even this is subject to strong objections, both because of the difficulty in making currents flow along such an arch, and because of the instability of such currents. Prominence arches appear rather to consist of material moving along lines of force; however, a real difficulty, not so far discussed, is to explain how coronal material condenses into prominences in spite of the resistance provided by a magnetic field.

In 1948, Giovanelli suggested that solar flares are due to electromagnetic discharges along magnetic lines of force. He supposed the discharges to arise because the electrical conductivity of an ionized gas increases rapidly with the temperature, so that the beginnings of a discharge produce a highly conducting channel along which the further discharge can readily proceed. This idea is attractive, but Piddington and Cowling have shown that it encounters overwhelming difficulties. The electric field available to drive the discharge was only guessed, and in view of the extent to which lines of force are frozen into the material it is doubtful whether fields of the magnitude required can exist. Moreover—and this is the fundamental objection—self-induction ensures that an increase in conductivity along a channel of the size of a flare does not lead to any appreciable increase in the current flowing, within a time of the order of a day. Thus increases in conductivity cannot lead to a sudden increase in radiation within a few minutes, as actually observed; a decreased resistance should rather be expected to lead to decreased heating.

In 1947, Alfvén suggested that the high temperature of the corona might be due to Joule heating by magneto-hydrodynamic waves. His suggestion was based on a formula due to Cowling, which indicated a reduction in conductivity transverse to the lines of force, this leading to enhanced heating. Schlüter later showed that in a fully ionized gas the reduction in conductivity is of such a nature as to produce no increase in the heating effect of electric currents; this is reasonable, since the magnetic field does not increase the number of collisions between ions and electrons, from which Joule heating arises. This disposes of the suggestion that Joule heating is important in the corona itself. However, Piddington and Cowling have recently shown that in a partially ionized gas the reduction in conductivity does correspond to a real increase in the Joule heating. Thus the production of fast coronal particles may be possible in the upper layers of the chromosphere, where a small number of neutral particles are still present.

Piddington suggests that a similar mechanism may be responsible for solar flares. Magneto-hydrodynamic waves can be supposed to be generated in sub-surface layers of a sunspot by convection which, though held in check by a magnetic field, is none the less present there on a reduced scale. Such waves in certain circumstances travel upward with relatively little loss in energy, provided that they are associated with mainly horizontal motions; when they reach a sufficient heat, the conversion of energy into Joule heat becomes important, and a flare is observed. On this theory, a flare is simply an enhanced form of an activity present all the time, presumably that responsible for plages. A different theory of flares has been put forward by Dungey, who suggested that an instability near a neutral point of the magnetic field might lead to a progressive increase in the electric currents flowing near such a point. This provides a discharge theory which is an advance on Giovanelli's but which nevertheless still appears a little artificial.

Since 1945, a large flare has on a number of occasions been observed to be accompanied by the emission of numerous soft cosmic-ray particles from the sun. Their mechanism of acceleration is almost certainly electromagnetic. Bagge and Biermann suggested that they might arise near a magnetic neutral point, accelerated by an electric field due to the relative motion of two magnetic fields. Such a suggestion appears untenable, in view of the closeness which with lines of force are frozen into the material; the necessary electric fields can hardly exist. One possibility appears to be that particles may on occasion travel with the phase-velocity of magnetohydrodynamic waves, speeding up with these waves as they travel into regions of less density. This is a more ambitious version of an idea advanced by the Babcocks, according to which constrictions in a bundle of lines of force may accelerate particles from lower levels into the corona. The mechanism has certain affinities with a Fermi mechanism of acceleration, and also with the mechanism suggested by Menzel and Salisbury, according to which particles are accelerated by riding on the crest of low-frequency electromagnetic waves.

Solar radio emission is in a different category from other phenomena considered above since, apart from certain outbursts, it seems to have little connexion with magnetic fields. The origin of the outbursts is generally believed to be some form of plasma oscillations. Any theoretical discussion of such oscillations is normally based on assumptions which preclude the possibility of escape of the radiation generated, but there is no obvious reason why such assumptions are essential.

Finally, one may comment on theories of the nature and origin of solar magnetic fields. Alfvén's theory of sunspots posited a general solar field of order 25 gauss, whereas the latest observations indicate a field of order 1 gauss. Alfvén has suggested that a field of order 25 gauss may in fact exist, overlaid by a turbulent field several times greater, and that the lower observed figure is due to lower observability of the turbulent + general fields when they reinforce each other than when they are opposed. A theorist has a reluctance, sometimes misplaced, to trying to explain away the observations; and in any case, whereas Alfvén's ideas may be reasonable, one cannot be satisfied with less than an argument which shows that the observed field must inevitably be smaller than the real one.

Attempts to explain solar fields, either the general field or that of sunspots, have so far met with limited success. Explanations in terms of thermo-electric effects have proved inadequate. Biermann has indeed shown that in the presence of non-uniform rotation such effects may produce toroidal fields some hundreds of gauss strong, but since such fields cannot be reversed from one sunspot cycle to the next, they can hardly explain the observed phenomena. The time of decay calculated by

Cowling for the general field leaves open the possibility that this may be a relic of an interstellar field existing before the sun's formation, but only if turbulence does not materially reduce the time of decay. Apart from this, the most promising possibility of explaining the fields seems to be in terms of a dynamo theory.

Some partial theories attempt to explain one of the solar fields in terms of another. For example, Walén and others have attempted, by invoking torsional oscillations or other periodic changes in the angular velocity, to derive from the general field a toroidal field capable of explaining spot fields. This suggestion involves difficulties about orders of magnitude, unless one supposes the general field to be stronger below than at the surface. Some workers have preferred to reverse the argument, seeing in the sun's general field the survival of fields of previous sunspot cycles. Again, turbulence in the sub-surface layers is sometimes supposed to twist the lines of force of the general field to give a stronger but irregular turbulent field. On the other hand, Walén has suggested that turbulence may actually operate to prevent a magnetic field from penetrating the turbulent layer, and arguments by Sweet and Elsasser indicate that a field which does penetrate the turbulent layer will, at least, decay rapidly.

A complete dynamo theory must explain all solar fields, general, spot and the rest, by induction due to a motion which, though not necessarily completely regular, should at least possess certain regular characteristics. The most ambitious theory of this kind to date is that of Parker, who invokes as a regular characteristic the effect of Coriolis forces in twisting rising and falling convection currents. He succeeds in deducing a field largely confined to the sun's outer layers, which is largely poloidal at high latitudes, toroidal at low. The surface field steadily travels down towards the equator, and has a zonal structure, a belt with a toroidal field of one sign being followed at higher latitudes by one with a field of opposite sign. The theory is difficult to express in precise mathematical form, but some deeper investigation of it is really required.

This concludes my historical account. I have unfortunately had to omit any account of Russian work, and hope that others will repair the omission. I have tried to avoid undue bias; at the same time, one cannot present an account like this without voicing one's own personal opinions. I do not expect anyone to agree with everything that I have said.

Our subject is a young one, even though its beginnings were earlier than is sometimes realized. Workers too often think of the subject as having begun with their own work; indeed, because it is so young and because connected accounts of it are not numerous, the rediscovery of earlier results has not been rare. The law of isorotation, first found by Ferraro, was certainly found independently by Alfvén and Sweet. The fact that inequalities of angular velocity might produce from the general field a toroidal field of importance in the theory of sunspots was certainly realized by me before 1945; I have since seen it appear under the names of at least four authors. Certain basic mathematical equations have similarly been derived independently by more than one author, and interpreted differently by them. Such repetitions are inevitable in so young a subject as our own; so to speak, ideas are in the air, and different workers pluck them out of the air at different times and in diverse manners. For this reason, though I have tried to attribute priority where due, questions of priority do not seem to me to be those of prime importance. The main triumphs of our subject are still in the future; the past is relevant only as it guides our future work.

Discussion

Gold: A new situation arises in the study of flares and their correlation with high-energy particles. We see these particles coming from flares but we do not know how. The mechanism is not understood. One conjecture, and to my mind a very useful one, is that the flare represents the sudden instability of a volume current in its own magnetic field (pinch effect), as suggested earlier by Alfvén.

If this is right it may be of interest to refer to what may be the most interesting laboratory analogy. Dr Kurchatov reported in April about the Russian experiments on high intensity discharges, and one result was the production of particles with energies in excess of the energies available from the applied field. This may be a laboratory analogy to the cosmic ray production in flares.

Biermann: The limitations of the simple model suggested that Bagge and myself in 1949 were only in part considered in that paper. A discussion based on less limitations was carried further by Schlüter in 1952. In all my own later discussions of the subject of the electromagnetic acceleration to cosmic ray energies (e.g. in *Kosmische Strahlung*, ed. by Heisenberg (Springer, 1953), and in *Amer. Rev. Nuclear Sci.* vol. 2) care was taken to emphasize the limitations as well as the positive aspects of the theory.

Bostick: To return to the subject of solar prominences and a laboratory analogue of these, a pair of sources at the poles of a magnet produce a picture as given by Fig. 1. The streamers occur only when current flows from A to B or from B to A, but not otherwise.

Another experiment consists of firing plasmoids at a screen along a magnetic field with no current flowing between the screen and the plasma source. This produces on the screen a cross instead of a 'spot', that is, no streamers (see Fig. 2). Finally, a source placed near the end of a coil as shown in Fig. 3 produces several spots on a screen which is placed perpendicularly to the axis of the coil when a current of several hundred amperes flows from the source to the screen. The number of spots observed at the screen indicates that the current from the source is shredded into streamers directed along the magnetic-field lines. Along each streamer there presumably results a helical magnetic field, and this may possibly act as a plasma guide along which the projected plasma may travel. There exists a hypothesis to account for the shredding of current along a magnetic field into streamers, but it is too complicated to be developed here in a brief discussion.

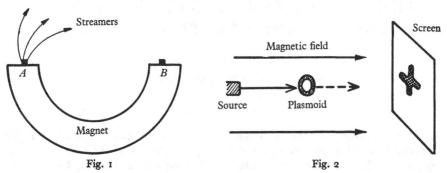

Fig. 1   Fig. 2

Fig. 1. A pair of plasmoid sources (A, B) placed at the poles of a magnet. The streamers occur only when current flows from A to B or from B to A, but not otherwise.

Fig. 2. Firing of plasmoids at a screen, along a magnetic field. A 'cross' is produced at the screen.

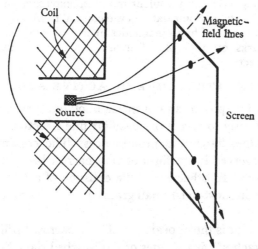

Fig. 3. Plasmoid source placed near the end of a coil. Spots are observed at a screen which indicates that the beam is shredded into streamers along the magnetic field.

Cowling: I should like to emphasize the difficulties involved in applying the results of experiments to cosmic problems. In the laboratory discharges are limited by tubes; there are electrodes to introduce potential differences; and gravity, which is certainly important in prominences, has very little influence. A careful examination of the effects of these and of all the scale effects involved is necessary before cosmic phenomena can be interpreted in the light of laboratory results.

THE PROCESSES IN ACTIVE REGIONS ON THE SUN AND ELECTROMAGNETICS

A. B. SEVERNY

Astrophysical Observatory, Crimea, U.S.S.R.

ABSTRACT

1. Brief summary of the dynamics of flare development, based on the analysis of moving picture and spectroscopic data. Some new spectroscopic data on the fine structure of emission in active regions, evidencing the peculiar character of motions in them: outbursts of corpuscules, ascending grains of continuous emission, explosive processes and shock-waves.

2. Summary of the observational data on the motions in prominences. Regular electromagnetic and turbulent motions, 'explosive' motions and their admissible interpretation. Some data on the coronal forms above active regions and on the possible role of hydromagnetics in their formation.

3. Some remarks on the role of hydromagnetics, admissible, from the observational aspect.

I. FINE STRUCTURE OF ACTIVE REGIONS

The high resolving power and dispersion of our new solar tower has permitted us recently to reveal the peculiar fine structure in the emission spectra of facculae, flares and prominences. The continuous and line emission were observed in the form of very narrow bright threads cutting the solar spectrum. In other words the emission of active regions is concentrated mostly in short-lived small grains or nuclei of some hundred kms in size [1].

The most peculiar is the appearance of very narrow brilliant wings (the so-called 'moustaches') at the sides of undisturbed dark Fraunhofer lines spreading sometimes up to 15 Å from the center of the lines.

These moustaches appear always on the background of the continuous emission. But the most important feature of these phenomena is that the blue wing of these moustaches is *brighter* and *broader* than the red one; sometimes only *one* wing is observed at the blue side of the line. Example of these phenomena are given in Plates I and II.

The dark lines of the solar spectrum are shifted to the violet at the very laces where these grains of continuous and linear emission are observed.

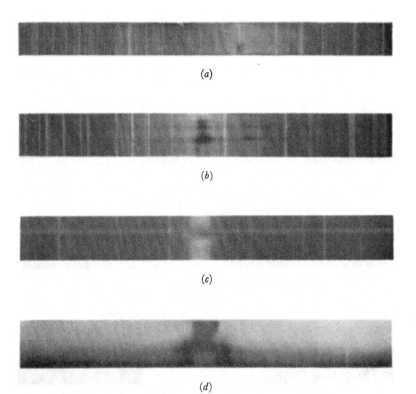

(a)

(b)

(c)

(d)

Plate I. Examples of fine structure of emission of active regions. (a) The continuous emission near H-line; (b) the 'moustaches' in H-line; (c) the typical 'moustaches' in Hα-line; (d) the 'moustaches' in Hα above the limb.

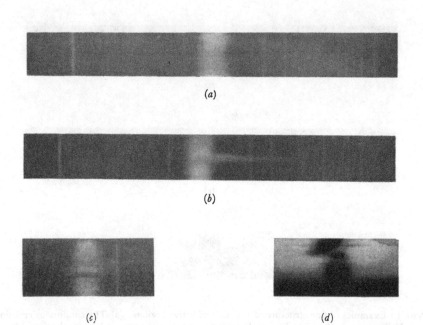

Plate II. The examples of 'moustaches': (a) in Hα with one blue wing; (b) in Hα in absorption; (c) in Hα cutting the line; (d) at the bottom of eruptive prominence.

Sometimes the Fraunhofer lines are slightly filled up with this emission. These nuclei of emission were observed in several cases above the limb. This shows that the grains of emission are emerging out of the deep layer and emitted at different levels in the solar atmosphere. The velocity of the ascending motions does not exceed 5 km/sec.

Photometric investigation showed that the continuous emission originates in a semi-transparent mass and possesses an energy distribution which looks like that of Ao–B5-type stars (Fig. 1). The rate of energy generation in these nuclei does not exceed 20 % of the solar flux of radiation. Owing to the small size of these nuclei the temporary liberation of radiative energy appears to be very large and of the order of the

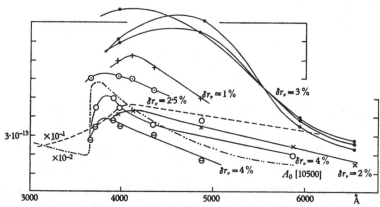

Fig. 1. The distributions of intensity in the grains of continuous emission (full lines); the dotted curve represents the same in undisturbed regions of the sun reduced by 10^{-1}; the dotted-point line the same in a Ao-type star reduced by 10^{-2}, and δr_v is the mean value of filling up of Fraunhofer lines with the continuous emission.

energy liberated in the deep interior of the sun owing to thermonuclear reactions.

The nature of this continuous emission is not clear as yet. It can be definitely said that this emission is not connected with the recombinations in the optically thin mass of hydrogen, as in this later case it should increase with the wave-length. It should not be produced by the process of scattering by free electrons, as this process could not secure the observed distribution of intensity along the spectrum. This emission is possibly connected with the relativistic electrons, although the effect of collisions makes the injection of these electrons somewhat improbable. Fig. 2 shows that the mean observed distribution of intensity of continuous emission might be explained by the emission of relativistic electrons possessing a differential spectrum of energy $dN(E) \propto E^{-1}$.

The preliminary results of measurements of polarization showed that this continuous emission is slightly polarized by 10 % in the plane perpendicular to the line of sight for grains near the limb of the sun. This fact also favours the hypothesis of relativistic electrons.

The results of spectrophotometric investigations of moustaches, as shown on Fig. 3 reveal clearly that the blue wing is brighter and broader than the red one in almost every case, which serves as an evidence of the

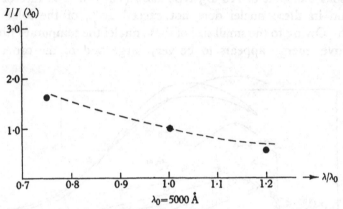

Fig. 2. The mean intensity distributions in grains (dots) compared with the theory for relativistic electrons with $dN(E) \propto E^{-1}$ (dashed line).

process of ejection of atoms out of the grain. The asymmetry of position indicates the velocity of emitting atoms up to 1000 km/sec. This effect of asymmetry *does not depend* on the position on the disk, showing that the process of corpuscular ejection is similar to a process of rapid symmetrical expansion or explosion of previously small mass but not to the process of pure radial ejection. (Particles moving from the observer are always going into the layers of larger optical depth and decelerating more rapidly than the approaching ones.)

Practically the same width of hydrogen and metallic moustaches shows that neither the Stark effect nor the natural damping are responsible for the broadening. The macroscopic motions (similar to turbulence) are most probably the cause of this broadening, which is insensitive to molecular weight. There is some evidence that collisions with neutral hydrogen also play an important role in the process of broadening going on in these moustaches.

Furthermore, the moving picture records show that the growth of line intensity emission is accompanied by the growth of area, that also evidences a process similar to explosion (Fig. 4). The duration of this growth of

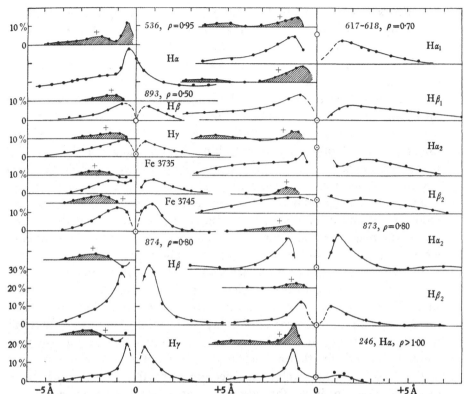

Fig. 3. The examples of profiles of emission in moustaches for the grains at different distances from the center of the sun's disc (ρ). The dashed areas are the asymmetry of these profiles showing the excess of emission in the blue wings. (All intensities in units of continuous spectrum of sun.)

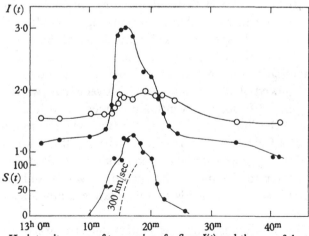

Fig. 4. The Hα-intensity curve of two grains of a flare $I(t)$ and the run of the area of this grain $S(t)$. (The dotted curve is the run of linear dimension of the brightest grain.)

intensity and expansion in flares and in knots of eruptive prominences is several minutes. The velocity of expansion of area or, probably, of the front of a shock-wave is of the order of several tens of km/sec. The process is highly dissipative: the lifetime increases with the area. The appearance of surges moving with supersonic velocities and subjected to extra-gravitational accelerations cannot be connected with any process, except an explosion-like process commencing with a shock-wave. These surges are accompanied by bursts of radio emission, i.e. by the process of non-linear oscillations of plasma. When falling at a right-angle on the system of magnetic lines of force of the nearest spot the shock-wave is reflected, as if it were reflected from a flexible wall. This can explain the observed appearance of a bright patch at the very border of spots' umbrae and the following lifting up of gases (the outburst of filament) along the curved path—the magnetic line of force of the spot's field. In several moving films we measured the explosive motions directly. The appearance of moustaches at the bottom of eruptive prominences should also be mentioned.

All these data lead to the conclusion that some unstable formations of small dimensions are lifted up on the surface of the sun. These formations are disintegrating. The process of decay, or instability of these formations manifests itself: (1) in the liberation of continuous emission in amounts comparable with the thermonuclear generation of energy in the interiors of the sun, (2) in the explosional ejection of particles with velocities up to 1000 km/sec, (3) in the formation of a shock-wave and macroscopic (or turbulent) motions, and (4) in the collisional excitation and ionization of atoms.

The probable source of these phenomena might be a sort of nuclear processes. Some general evidences concerning anomalous abundances of some elements, and especially the investigation of deuterium on the sun recently completed by us, lead to this conclusion. The spectrophoto-electric investigation by means of instruments of very high resolving power (600,000) showed that the depression between two water vapour lines λ 6561,105 and λ 6560,570, eventually connected with D_α increases from the center to the limb for 1·5–1·7 times (Fig. 5). The marked change of the contour λ 6561,105 (as a result of blending of the water vapour line with one of the D_α fine structure components) is also observed when passing from the center to the limb. The measured width and contour of this depression ($W = 1·8$ mÅ) leads to the relative abundance of D/H from 3 to 5×10^{-5} [1].

The most striking is that this depression is markedly increasing in moustaches. It was recorded in several cases. Burbidge, Burbidge and

Fowler[2] have recently proposed a mechanism similar to bethatron, explaining the abundance of Li, Be, D, providing the existence of magnetic fields up to 10^6 gauss. This process secures $D/H > 10^{-5}$ in a hot spot, the energy of particles being $kT \approx 0.5$ MeV. It is interesting that the amount of energy found above is of the same order of magnitude, when being related to the uppermost layers of photosphere. However, further investigations are highly desirable, especially regarding the magnetic properties of these fine-structure formations.

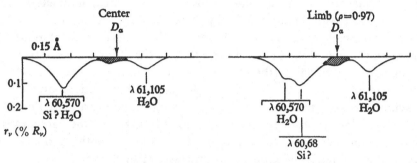

Fig. 5. The profiles of depression presumably connected with the D_α-line at the center and at the limb of the disc (50 photo-electric records).

2. ON THE ROLE OF ELECTROMAGNETICS IN ACTIVE REGIONS

The role of magnetic fields is more or less definite, when we are dealing with the space limitations of motions of solar plasma in the vicinity of a spot. The predominant motions along the magnetic lines of force in the case of electromagnetic prominences and filaments can be established from a comparison of curved paths of their knots and streams with the probable topography of the magnetic field of neighboring spots. The measured uniform motions along these curved paths can not be explained by means of field of forces of radiation pressure and gravity. The marked accelerations appear, as a rule, at the beginning and at the end of the path. The knots and streams do not penetrate, as a rule, into the umbrae of a spot. Its magnetic field prevents this process. Moreover the magnetic field of a spot restricts the possibilities of turbulent motions in the nearest prominences.

At the upper parts of the corona the curvature of the paths is less than in the lower ones and the knots of eruptive prominences are moving along spirals with radially increasing step as if it were a motion of an isolated charge in the field of a single magnetic pole[3].

The difference between the free paths of ions and electrons (and consequently the difference of retardation) in a knot moving rapidly through the corona may induce a current in the direction of motion and load to spiralling of the knot (Pikelner, [4]). The Evershed effect, the comparatively low temperature of a spot, the lowering of chromosphere above a spot—all these data show, that predominantly horizontal pressure gradients may appear in the vicinity of a spot. These gradients, as well as the above mentioned explosional processes, might be the cause of motions of solar plasma along the lines of force. Theoretical magneto-hydro-dynamics of plasma, taking the gravity and compressibility into account, shows that small motions of the plasma should be accomplished along the lines of force even if the excess of pressure is of the order of several per cent of the equilibrium value (the theoretical accelerations were found to be of the order of the observed ones [5]). But supersonic motions and extra-gravitational accelerations (eruptions and surges) require a more powerful agent—similar to that described above. Moustaches, the large radial velocities and splashes of brightness in the bottom of eruptive prominences are further evidences of explosional processes of such a kind.

The sunspot's magnetic field, as well as a general magnetic field, can, of course, disturb the almost symmetrical outburst of particles from the grain of emission. But the ejection of particles can disturb somewhat, in its turn, the topography of the general magnetic field of the sun. In the case of the simple dipole field and the pure radial ejection of particles the Hall current and the induced magnetic field connected with it can appear in the stream of corpuscles. According to Ponomareff the topography of the resulting magnetic field should be very similar to the observed coronal forms. The stability of these forms evidences the possibility of such a magnetic field in the corona, which excludes the transfer of matter across the lines of force.

Far enough from the active regions and the sunspot's field well pro-nounced turbulent motions are observed (quiescent prominences). Here we have metamorphoses, which are similar to that observed in the earth's clouds and smokes. Trajectories of knots are similar to that of Brown's particles, the distribution of knots according to velocities being different from the simple law of random values. The Reynolds numbers are large ($\geqslant 10^5$). The observed correlative function between the difference in the brightness of knots and their distances agrees fairly good with the theoretical one for the case of local isotropic turbulence (Dubov [6]). This permits us to consider the knots as turbulent pulsations of density in the continuous plasma of prominences (the mean characteristic dimension $\approx 2 \times 10^9$, the characteristic time is 2×10^3 sec, the mean turbulent velocities are ≈ 10 km/sec). Owing to the process of entangling of the lines of force the turbulence should lead to an increase of the magnetic field up to a certain limiting value (about 1 gauss). It means that all motions in such a prominence should be transmitted by the hydromagnetic waves along the lines of force from one part of the mass to another. This process favours the development of the isotropic turbulence.

REFERENCES

[1] Severny, A. B. *Astr. J., Moscow*, no. 3, 1956. *I.A.U. Transactions*, vol. 9 (Dublin meeting), (Cambridge University Press, 1957).
[2] Burbidge, G., Burbidge, E. M. and Fowler, W. *Astrophys. J.* **122**, 271, 1955.
[3] Severny, A. B. *C.R. Acad. Sci. U.R.S.S.* **82**, no. 1, 1952; *Publ. Crim. Astroph. Obs.* **10**, 3, 1952.
[4] Pikelner, S. B. *Russian Astron. Journ.* **38**, 641, 1956.
[5] Severny, A. B. *Publ. Crim. Astroph. Obs.* **11**, 129, 1954.
[6] Dubov, E. E. *Publ. Crim. Astroph. Obs.* **15**, 1955.

Discussion

Burbidge: What are the average energies of the relativistic electrons which you suggest may be responsible for continuous emission in the grains?

Severny: This has not been computed. All we know is the total photometric energy which is about 10^5 erg/cm^3.

Burbidge: What is the magnetic field to account for the energy? I think that a very high electron density, some 10^8 relativistic electrons, might be needed if $H \approx 1$ gauss.

Severny: It is shown by Gordon that not many electrons are needed—it depends on their individual energies and, of course, on the magnetic field.

Spitzer: What is the angular diameter and the linear size of these bright grains?

Severny: Perhaps 2 sec of arc. But only in perfect weather can this be observed, the limitation being set by the circle of scattering. One calculates perhaps 300 km as the linear extent.

Spitzer: How frequent are the grains? How many are present on the solar disk at one time?

Severny: Sometimes three to five at a time; they occur especially in growing spots. They are very short-lived (10–15 min.) and often disappear while preparing for observation.

Alfvén: Are these phenomena associated with the emission of magnetic storms producing beams?

Severny: I do not know, but one could measure the depression in the wings of the spectral line emission (H and K or Hα). In this way one can try to predict magnetic storms with 80 % certainty.

Alfvén: The displacement corresponds to a velocity of 300 to 1000 km/sec. Is there a possibility that this velocity continues outwards so that there is a radial emission?

Severny: In the case of pure radial ejection out of these grains the asymmetry of the moustaches should disappear near the limb, but the effect of asymmetry is even more pronounced near the border of the disk.

Alfvén: Can you estimate the total mass of gas which moves here?

Severny: There should be about 10^4 particles/cm^3 above the surface to secure the observed extra emission in the wings of H- and K-moustaches.

Alfvén: But what is the total mass integrated over the whole phenomenon?

Severny: The linear extent is 300 km squared so that one can estimate the total number.

Biermann: Did you want to suggest a relative abundance of deuterium of several times 10^{-5}, or else did you want to indicate an upper limit?

Severny: There was a change of the depression in the region of the D_α-line by a factor 1·5 when going from the center to the border of the sun. Comparing theoretical results with my observations, the abundance comes out to be $3-5 \times 10^{-5}$.

Öhman: Is the mean life time of these 2 min. elements as long as 40 min.?

Severny: Moustaches have 1–10 min. life time. On the other hand, there is sometimes continuous emission from half an hour to one hour.

Tuominen: Could the 'moustaches' represent the continuous spectrum of the prominence superimposed on the solar spectrum?

Severny: No, in the case of scattering by an electron condensation we should not observe the dependence of intensity on the wave-length.

Alfvén: The density 10^4/cm^3 is less than mean density in the corona. How did you get such a value?

Severny: The density was measured from the widths of additional depression in the wings of the H and K-lines.

Alfvén: But this density is much lower than in the photosphere.

Spitzer: Is this perhaps the number of excited atoms?

Severny: The value I have given for the density refers to the particle density of the additional flux which is able to produce an additional excitation of the chromosphere by means of collisions.

Alfvén: Yes, but I mean what total density in the moving piece of matter do you expect?

Severny: I am not able to answer that question. I want to keep to the observations as much as possible.

Cowling: The density 10^4/cm^3 must correspond to an appropriate height.

Severny: Yes, this height will not be more than that of the chromosphere, i.e. 10^3 km to 10^4 km.

PAPER 14

THE NEUTRAL POINT THEORY
OF SOLAR FLARES

P. A. SWEET

University of London Observatory, England

ABSTRACT

It is shown that under certain conditions hydrostatic equilibrium becomes unstable in a conducting medium in the presence of a magnetic field containing a neutral point if the gas pressure is less than a limiting value.

The motion resulting from breakdown of hydrostatic equilibrium in the solar chromosphere above complex sunspot groups could produce solar flares and cosmic ray particles.

I. INTRODUCTION

Cowling[1] has pointed out that, if solar flares were due to Joule heating and cosmic ray particles due to electric fields in the chromosphere, electric current sheets of no more than a few metres thick would have to be set up. Dungey[2] indicated that the field near a neutral point is unstable and would constrict itself to produce current sheets of the narrowness required. This effect provided the germ of the ideas in the present paper although it is shown that the gas pressure and conditions far from the neutral point, neglected by Dungey, play essential parts in the development of the high currents.

2. THE MECHANICAL FORCES DUE TO A DISTORTED
FIELD WITH A NEUTRAL POINT

Consider the fields in Fig. 1, the lines of force being directed in the xy-plane and are independent of the z-coordinate. Introduce a vector potential $(0, 0, A)$, so that $A =$ constant on a line of force. It is supposed that the electric currents producing the field are situated below PQ. The two regions bounded by the lines $A = A_2$ in Fig. 1 may be regarded as the original fields due to the currents in regions A and B, and the field outside as the common field resulting from the partial interpenetration of these fields.

Suppose the current systems A and B are displaced towards each other by an appreciable fraction of their initial separation. The medium below PQ is taken to be a perfect conductor, and the normal flux, and hence the value of A, at every point of PQ in each region A and B remain unaltered.

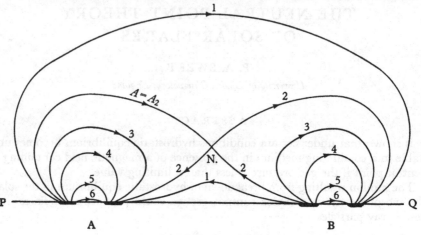

Fig. 1. Potential field of two bipolar systems.

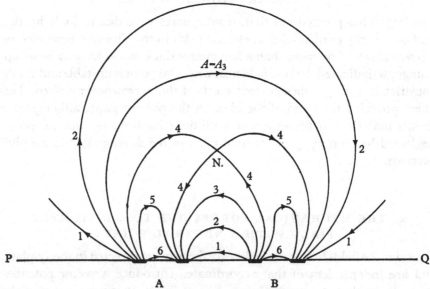

Fig. 2. Potential field of displaced systems.

If the medium above PQ were a non-conductor the two systems would suffer a further interpenetration and the value of A at the neutral point would increase to values A_3, A_4, etc., according to the extent of the

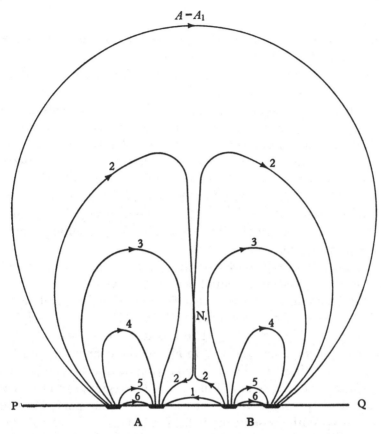

Fig. 3. Field of displaced systems in perfectly conducting medium.

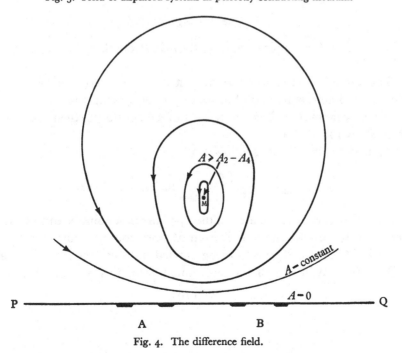

Fig. 4. The difference field.

displacement, as in Fig. 2. No currents are induced above PQ and the field here is called the equivalent potential field.

If, on the other hand, the medium is a perfect electrical conductor the lines of force through the neutral point move with the medium and the two fields A and B do not interpenetrate further. The fields may be said to collide. The value of A at the neutral point remains unaltered and the field is of the general form represented in Fig. 3. It is not unique as it depends on how the medium moves during the displacement. It has, however, the same normal components on PQ as the equivalent potential field. The latter is uniquely determined by these components, hence the colliding field, which has a different value of A at the neutral point, can under no circumstances be potential. The magnitude of the resulting currents will now be investigated.

Let $(0, 0, A')$ and $(0, 0, A'')$ be the vector potentials for the colliding field and the equivalent potential field, respectively, above PQ, and consider the function $A = A' - A''$. Therefore $A = 0$ on PQ, and $\Delta^2 A' = \Delta^2 A$. The lines through the neutral point of the colliding field, $A' = A_2$ in Fig. 3, connect regions A and B on PQ. The lines through the neutral point of the equivalent potential field, $A'' = A_4$ in Fig. 2, also connect regions A and B. The latter lines intersect PQ at points lying within the pairs of points of intersection of $A' = A_2$ with PQ in each of the regions A and B. The lines $A'' = A_4$, $A' = A_2$ must therefore intersect at least one point M. Therefore $A_M = A_2 - A_4$. With the sense of the field as in Figs. 1, 2 and 3 A_r decreases with increasing r, hence $A_2 > A_4$ and we may write

$$A_{max} \geqslant A_2 - A_4 = o(H_0 L), \tag{1}$$

where H_0 and L are typical values of the field strength and linear dimensions of the field.

The analysis that follows was suggested by some of the classical work of Picard[3]. There is no loss of generality in supposing that there is only a single maximum in A. The system of lines $A = $ constant is then of the general form shown in Fig. 4.

A_M may be expressed as follows:

$$A_M = -\frac{1}{2\pi} \iint_{1/2\text{-plane}} \ln (r_1/r)\, \Delta^2 A_X\, dS, \tag{2}$$

where dS is an element of area of the xy-plane at a point X and r and r_1 are, respectively, the distances of X from M and from the mirror image, M_1, of M in PQ, the integration being carried out over the half-plane above PQ. $\ln (r_1/r)$ is in fact the Green's function for this region. There are by definition no currents above PQ other than those induced by the displacement. The principal contribution to A_M therefore arises in some region D whose dimensions, and in particular whose farthest point from M_1, do not exceed a value of order L. From (2) it is then seen that

$$A_M = -\frac{\lambda}{2\pi} \iint_D \ln{(r_1/r)}\, \Delta^2 A_X\, dS, \qquad (3)$$

where $\lambda = 0\,(1)$. A is a maximum at M, therefore $\Delta^2 A < 0$ in a neighbourhood of M; let D_1 be the region within D formed from all points at which $\Delta^2 A < 0$. Hence

$$A_M \leqslant \frac{\lambda}{2\pi} \iint_{D_1} \ln{(r_1/r)}\, |\,\Delta^2 A_X\,|\, dS, \qquad (4)$$

therefore

$$|\,\Delta^2 A\,|_{max} \geqslant \frac{2\pi A_M}{\lambda \iint_{D_1} \ln{(r_1/r)}\, dS}. \qquad (5)$$

But the denominator on the right-hand side of (5) cannot exceed a value of the order of magnitude L^2, hence, using (1), (5) shows that

$$|\,\Delta^2 A\,|_{max} > \lambda_1 H_0/L, \qquad (6)$$

where $\lambda_1 = 0\,(1)$. The current density in the z-direction is given, in gaussian units, by $j = -c\Delta^2 A/4\pi$, therefore

$$j_{max} > c\lambda_1 H_0/4\pi L. \qquad (7)$$

The mechanical force F_{em} exerted by the current is jH'/c where H' is the field strength. If the currents exceed the limit in (7) over a region with dimensions comparable with L, there must be points in the region where $H' = 0(H_0)$. Hence

$$F_{max} > \lambda_2 H_0^2/4\pi L, \qquad (8)$$

where $\lambda_2 = 0\,(1)$. This inequality holds *a fortiori* when the currents are concentrated into a smaller region.

Consider now the hydrostatic pressure P required to maintain equilibrium with the force F_{em}. In hydrostatic equilibrium

$$\text{grad}\, P = -\Delta^2 A'\, \text{grad}\, A'/4\pi. \qquad (9)$$

P and $\Delta^2 A'$ are therefore functions of A', and (9) may be integrated to give

$$P_M - P_X = -\frac{1}{4\pi} \int_{A'_X}^{A'_M} \Delta^2 A'\, dA'. \qquad (10)$$

This shows that at all points X in a region D_2 where the current exceeds the limit in (7),

$$P_M - P_X > \frac{\lambda_1 H_0 (A'_M - A'_X)}{4\pi L}. \qquad (11)$$

127

If D_2 has dimensions of order L, X can be chosen far enough from M so that $A'_M - A'_X = o(H_0 L)$. (11) then shows that

$$P_M > \lambda_3 H_0^2 / 4\pi, \tag{12}$$

where $\lambda_3 = o\,(1)$. Alternatively, suppose D_2 is narrow in one of its dimensions, corresponding to a current sheet of thickness $h \ll L$ and of width $l = o(L)$. Then taking $D = D_2$ in (3),

$$A_M > \lambda \mid \Delta^2 A \mid_{max} \iint_{D_2} \ln\ (r_1/r)\ dS/2\pi = \lambda_4 \mid \Delta^2 A \mid_{max} lh, \tag{13}$$

where $\lambda_4 = o\,(1)$. Hence

$$\mid \Delta^2 A \mid_{max} = o(A_M/lh). \tag{14}$$

Let X be a point on the boundary of D_2 near M, then

$$A_M - A_X = o(h^2 \mid \Delta^2 A \mid_{max}) = o(A_M h/l). \tag{15}$$

Therefore

$$A'_M - A'_X = A''_M - A''_X + o(A_M h/l). \tag{16}$$

If M does not coincide with the neutral point of the equivalent potential field the sign of $A''_M - A''_X$ can be changed by taking X on the boundary of D_2 opposite to the original point selected. A point can therefore be chosen such that

$$A'_M - A'_X > \lambda_5 A_M h/l, \tag{17}$$

where $\lambda_5 = o\,(1)$. If M coincides with this neutral point then

$$A''_M - A''_X = o(h^2 A_M/L^2) \tag{18}$$

and (17) still holds. Finally, on applying (17) and (14) to (10),

$$P_M = o(A_M^2/l^2) = o(H_0^2), \tag{19}$$

the same result as in (12). (12) holds *a fortiori* when both dimensions of the current region are small.

In all cases, therefore, P_M must exceed a value of order H_0^2 if hydrostatic equilibrium is to be possible without additional external forces. In applying the above criteria it must be remembered that the forces immediately below PQ required to force the current systems A and B together are themselves of order H_0^2/L. In a continuous system, therefore, the forces immediately above PQ may also be of this order of magnitude. The criteria would apply, however, whenever the scale height of the forces below PQ is small compared with L. This certainly obtains in the case of sunspot fields. For spot fields the gravitational forces and the gas pressures in the chromosphere fall far short of the limits in (8) and (12). Hydrostatic equilibrium would therefore break down over a complex spot group in which, along a line of minimum resultant field strength, the field was much smaller than the typical value,* and in which the components of the group were moving relative to each other.

3. PHYSICAL INTERPRETATION OF THE EQUILIBRIUM CRITERIA

As the displacements in the photosphere proceed the magnetic field transmits the sub-photospheric force to the chromosphere. The resulting distortion in the chromospheric field will be greatest where the field is weakest, i.e. near the neutral line. The field there will tend to flatten, the effect

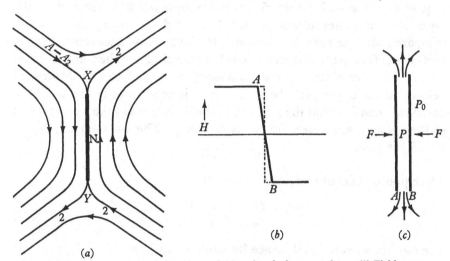

Fig. 5. The collision layer. (a) Field in neighbourhood of current sheet. (b) Field across current sheet. (c) Idealized hydrodynamic model.

being analogous to the flattening of a motor tyre when loaded. A limit is reached when hydrostatic equilibrium breaks down. A thin collision layer of gas is then formed as in Figs. 5 (a) and (b). There is an inward magnetic pressure at the boundaries approximately equal to $H_1^2/8\pi$ where H_1 is the field strength there. This is uncompensated along the middle line of the layer where the field strength is small. Hydrostatic equilibrium then requires a gas pressure of approximately $H_1^2/8\pi$ in the middle of the layer, and since the layer is effectively open at its ends this means that the gas pressure surrounding the layer must be of the same order of magnitude. For large sub-photospheric displacements $H_1 = \mathrm{o}(H_0)$, and the criterion in (12) follows.

* Such lines are generalizations of the neutral lines in the present two-dimensional analysis.

4. THE DYNAMICS OF THE COLLISION LAYER

An exact treatment of the motion is not considered in the present paper. Instead, an analogy in pure hydrodynamics is considered. Consider two parallel rigid plates, as illustrated in Fig. 5 (c), with infinite extension in the z-direction. Let the distance h between the plates be small compared with their length 2l. The region between the plates is occupied by gas and the surrounding gas pressure is taken to be negligible. Suppose the plates are being forced together by a constant force F per unit z-length. The gas offers little resistance to the plates until sufficiently compressed. After this a quasi-steady state is set up in which gas flows out from the ends of the layer and the plates approach each other with a velocity that is small compared with the gas velocities along the layer. Take axes centred at the mid-point of one plate with the x-axis along the plate and the y-axis directed toward the other plate. Let the gas velocity be (u, v, o) and the downward velocity of the upper plate be V. The flow is nearly steady and is almost one-dimensional so that the pressure P, the density ρ and the x-component of the velocity are nearly functions of x only. The y-component of the velocity is given by

$$v = -Vy/h. \tag{20}$$

The equation of continuity div $\rho\mathbf{v} = 0$ reduces to

$$u\frac{d\ln\rho}{dx} + \frac{du}{dx} = \frac{V}{h}. \tag{21}$$

The motion is irrotational, hence Bernoulli's equation

$$u^2 = 2\int_P^{P_0} \frac{dP}{\rho} \tag{22}$$

holds, where the suffix o denotes values at the origin. For simplicity in the analysis isothermal flow is assumed. The results for isentropic flow do not differ radically from the isothermal case, so that in the circumstances the extra analysis involved in the isentropic case is not justified. A fuller analysis would in any case have to take into account the effects of radiative cooling and Joule heating. In flare conditions the rate of radiative cooling is sensitive to the temperature. This would keep the temperature low in spite of heavy Joule heating. Thus take

$$\frac{P}{\rho} = \frac{P_0}{\rho_0} = \frac{\Re T}{\mu}, \tag{23}$$

where T is the temperature, $\Re$ is the gas constant and μ is the mean molecular weight of the gas in units of the mass of the hydrogen atom. On solving (20)–(23) u is given by $u = q\xi$, where $q^2 = \Re T/\mu$ and

$$\xi(1 - \tfrac{1}{3}\xi^2) = Vx/qh. \tag{24}$$

P and ρ are given by $\qquad \rho/\rho_0 = P/P_0 = e^{-\frac{1}{2}\xi^2}. \tag{25}$

The boundary condition at the ends of the layer, $x = \pm 1$, is determined by the external pressure. As this is assumed to be negligible compared with P_0 it is therefore below the so-called critical pressure for outflow from a tube (see for example Prandtl[4]). In the isothermal case this pressure is $P_0/\sqrt{e}$; the pressure at the ends is equal to this, therefore (25) shows that $\xi = 1$ at $x = 1$. On integrating along the plates,

$$F = 2 \int_0^1 P \, dx. \tag{26}$$

When reduced by means of the previous equations, this shows that

$$F = \frac{2q}{\sqrt{e}} \frac{Ph}{V}. \tag{27}$$

Again, on putting $x = 1$ in (24),

$$V = \frac{2q}{3} \frac{h}{l}, \tag{28}$$

therefore

$$P_0 = \frac{\sqrt{(e)} \, F}{3l}. \tag{29}$$

Equations (24), (25), (28) and (29) completely describe conditions in a layer of given length and thickness. The subsequent behaviour of the layer is considered by remembering that $dh/dt = -V$, hence from (28),

$$dh/dt = -2qh/3l. \tag{30}$$

Further progress cannot be made until the behaviour of the length of the layer is known. This depends on the rate of dissipation of magnetic energy by Joule heating in the layer.

5. THE ELECTROMAGNETICS OF THE COLLISION LAYER

The problem will again be simplified by adopting Ohm's law in a medium moving with velocity $(u, v, 0)$:

$$j/\sigma = E + (vH_x - uH_y)/c, \tag{31}$$

where σ e.s.u. is the conductivity and E is the electric intensity in the z-direction. This neglects the effect of Hall current, but the reduction in conductivity due to collisions with neutral atoms as derived by Cowling[5] and by Piddington[6] could be allowed for by adopting the appropriate value of σ once more was known about conditions in the layer.

Maxwell's equation curl $\mathbf{E} = -\dfrac{1}{c}\dfrac{\partial \mathbf{H}}{\partial t}$ can be integrated and written

$$E = -\frac{1}{c}\frac{\partial A}{\partial t} + f(t), \tag{32}$$

where $f(t)$ is a function to be determined. By eliminating E between (31) and (32) and using the relation $j = -c\Delta^2 A/4\pi$,

$$\frac{\partial A}{\partial t} = -\frac{c^2}{4\pi\sigma}\Delta^2 A + cf(t) + (vH_x - uH_y). \tag{33}$$

In the photosphere A is independent of time by definition, there is no fluid velocity, and $\Delta^2 A/\sigma$ is supposed negligible. (33) therefore shows that $f(t) = 0$. At the neutral point N $H = 0$ and (33) shows that

$$dA_N/dt = -c^2[\Delta^2 A]_N/4\pi\sigma = -\lambda_6 c^2 a_N/4\pi\sigma hl, \tag{34}$$

where λ_6 is a dimensionless quantity of order unity.

The value of A_N effectively determines the length of the layer.

It is not possible to derive the precise relation on the simple analogy of the previous section. An empirical relation is therefore introduced in the form

$$A_N = A_N(0)\ (l/l_0)^n, \tag{35}$$

where n is a positive constant and $A_N(0)$ and l_0 are values at a given epoch $t = 0$ soon after the quasi-steady state is attained. (34) then shows that

$$\frac{dl}{dt} = -\frac{c^2}{4\pi\sigma nh} \tag{36}$$

taking $\lambda_6 = 1$ for definiteness. On dividing (36) and (30) and integrating,

$$l = l_0 \exp\{h_1(1/h_0 - 1/h)\}, \tag{37}$$

where h_0 is the initial value of h and

$$h_1 = \frac{3c^2}{8\pi\sigma nq}. \tag{38}$$

On substituting (37) into (30),

$$\frac{dh}{dt} = -\frac{2qh}{3l_0} \exp\{h_1(1/h - 1/h_0)\}. \tag{39}$$

Thus h may be derived implicitly at any time from

$$t = \frac{3l_0}{2q} e^{h_1/h_0} \int_{h_1/h_0}^{h_1/h} \frac{e^{-x}dx}{x}. \tag{40}$$

132

The layer vanishes after a time

$$\frac{3l_0}{2q} e^{h_1/h_0} \int_{h_1/h_0}^{\infty} \frac{e^{-x} dx}{x}.$$

The electric intensity at the neutral point is given by

$$E_N = -\frac{1}{c}\frac{dA_N}{dt} \sim \frac{cA_N(0)\, l^{n-1}}{4\pi\sigma h l_0^n} \sim \frac{cH_0}{4\pi\sigma h} \exp\{(n-1)\, h_1(1/h_0 - 1/h)\} \quad (41)$$

taking $A_N(0) = 0(H_0 l_0)$.

The rate of Joule heating is σE_N^2. If this is taken as typical of the whole layer, then the total rate of heating/cm of the layer in the z-direction is given by

$$J \sim \sigma E_N^2\, hl = \frac{c^2 H_0^2 l_0}{16\pi^2\sigma h} \exp\{(2n-1)\, h_1(1/h_0 - 1/h)\}. \quad (42)$$

If $2n \leqslant 1$ then $J \to \infty$ as $h \to 0$. We therefore take $2n > 1$, in which case J has a maximum value given by

$$J_{max} \sim \frac{H_0^2 l_0 nq}{6\pi(2n-1)} \exp\{(2n-1)\, h_1/h_0 - 1\}, \quad (43)$$

where
$$h = (2n-1)\, h_1. \quad (44)$$

This maximum occurs at time

$$t_1 = \frac{3l_0}{2q} e^{h_1/h_0} \int_{h_1/h_0}^{1/(2n-1)} \frac{e^{-x}\, dx}{x}, \quad (45)$$

and the time from maximum activity until the layer vanishes is given by

$$t_2 = \frac{3l_0}{2q} e^{h_1/h_0} \int_{1/(2n-1)}^{\infty} \frac{e^{-x}\, dx}{x}. \quad (46)$$

The electric intensity, in the z-direction, at maximum activity, is given by

$$E_{max} \sim \frac{2nqH_0}{3(2n-1)\, c} \exp\{(n-1)\, h_1/h_0 - (n-1)/(2n-1)\}. \quad (47)$$

6. APPLICATION TO A SOLAR FLARE

Consider a complex sunspot group with a neutral point near the base of the chromosphere, and suppose that the components of the group suffer relative displacements that break down equilibrium and produce a collision layer of length $l_0 = 10^9$ cm. Take $T = 10^4\,^{\circ}$K and $H_0 = 10^3$ gauss as further typical values. The effective value of σ, if collisions with neutral atoms are taken into account as in the work of Cowling [5] and Piddington [6], would be of the order of magnitude 10^8 e.s.u. It will be seen that the precise value has a negligible effect on both the rate of heating and the electric field. The results are not sensitive, in order of magnitude, to n provided this is not too near $1/2$. Take $n = 1$ for definiteness, and suppose that the flare operates along a line of length 10^{10} cm in the z-direction.

With the above values, $h_1 = 1\cdot 2 \times 10^6$ cm. The quasi-steady state sets in as soon as the layer becomes thin compared with its length. Hence, taking $h_0 = 10^8$ cm, $h_1/h_0 \ll 1$. The maximum rate of Joule heating in the whole flare is then practically independent of σ, and is given by

$$J_{\max} = 1\cdot 5 \times 10^{29} \text{ ergs/sec.}$$

The time of duration of the flare at maximum activity is given by $t \sim 10^4$ seconds, and the total energy output from the flare is therefore of order 10^{33} ergs. The electric intensity at maximum activity is given by

$$E_{\max} = 6 \text{ volts/cm;}$$

this value, also, does not depend on σ. The corresponding difference in potential between the ends of the flare is 6×10^{10} volts.

The above figures agree with the values associated with a large flare, to within the uncertainties of the theory.

7. CONCLUSIONS

The theory just developed reproduces some important features of solar flares. It gives a total radiation of energy and a duration of the right order of magnitude, and can account for the production of high-energy particles. It has been demonstrated conclusively that hydrostatic equilibrium can break down along a neutral line, and a narrow colliding layer set up in the ensuing motion. The theory cannot be accepted as definitive, however, until the hydrodynamic analogy introduced is replaced by a proper analysis of the dynamics of the colliding layer.

REFERENCES

[1] Cowling, T. G. *The Sun*, ed. G. P. Kuiper (Chicago, 1953), ch. 8.
[2] Dungey, J. W. *Phil. Mag.* **44**, 725, 1953.
[3] Picard, E. *Cahiers Scientifiques*, **5**, 1930, ch. 8.
[4] Prandtl, L. *Essentials of Fluid Dynamics* (Blackie, London, 1952), p. 266.
[5] Cowling, T. G. *Mon. Not. R. Astr. Soc.* **116**, 114, 1956
[6] Piddington, J. H. *Mon. Not. R. Astr. Soc.* **114**, 638 and 651, 1955.

THE NEUTRAL POINT DISCHARGE THEORY OF SOLAR FLARES.

A REPLY TO COWLING'S CRITICISM

J. W. DUNGEY*

Cavendish Laboratory, Cambridge, England

ABSTRACT

The discharge theory has been criticized by Professor Cowling in §5·3 of *The Sun*, ed. Kuiper (Chicago[1]). A discharge theory requires the thickness of the accelerating layer to be only 5 m, which he finds unacceptable. This criticism is not soundly based, because no lower limit for the layer width is obtained from observation.

Secondly, Cowling invokes Lenz's law: 'the effect of induced currents is always to oppose the change to which they are due.' Lenz's law needs verifying for conducting fluids, however, and though it is usually true, it is shown to be false when there is a neutral point. Since Lenz's law refers to electromagnetic induction, other effects such as the pressure gradient may be omitted in this test, and then a completely rigorous proof is possible: it is shown that in an ideal fluid which is perfectly conducting, perfectly compressible and inviscid, the current density at a neutral point must become infinite.

The pressure gradient is irrelevant to Lenz's law, but it usually opposes any motion and should also be discussed. A physical picture suggests that the pressure gradient will not stop a vortical motion, but merely retard it. The related problem of equilibrium between the pressure gradient and the electromagnetic force may be discussed mathematically for the special case with two-dimensional symmetry. It is found that equilibrium requires an infinite current density at the neutral point. This is a subsidiary argument in favour of discharges at neutral points.

I. INTRODUCTION

The suggestion that a solar flare results from an electrical discharge situated in the neighbourhood of a neutral point of the magnetic field was made by Giovanelli[2]. He had observed a large number of flares, and his suggestion was evoked by their position in the spot groups in which they occurred; he also supported his proposal by the observation that flares are more common in complex groups. Most observers agree that the flare

* Now at Department of Mathematics, King's College, Newcastle.

phenomenon seems to involve an electric discharge, but Cowling[1] has criticized any such possibility on theoretical grounds and his criticism has been widely accepted. The main purpose of this paper is to answer that criticism.

The defining feature of a discharge in this context is the existence of a large current density. The electrons at least must reach relativisitic energies and the order of magnitude is given by $j \sim nec$; furthermore, constriction may increase n to a value substantially larger than that in the surrounding gas. Then the relation c curl $\mathbf{H} \approx 4\pi j$ provides an estimate of the width b of the discharge, since $\frac{1}{2}b$ |curl $\mathbf{H}$| cannot exceed the value of H outside the discharge. Then $b \sim H/2\pi ne$ and with $H = 300$ gauss and $n = 10^9$ cm^{-3} this is 1 m. The width of a discharge must therefore be minute on the solar scale; at first sight this seems to be fatal to the the discharge theory, and Cowling finds it unacceptable. In section 4, however, it will be shown that this value of the width is not in conflict with any observation. Before discussing the observations, Cowling's theoretical criticism will be answered.

2. LENZ'S LAW

The fundamental difficulty raised by Cowling is Lenz's law: 'the effect of induced currents is always to oppose the changes to which they are due.' If this were true, the necessary large current density could never be built up under solar conditions. This law, however, does not have the same fundamental status as Maxwell's equations and has previously been applied only to electrical machines involving approximately uniform magnetic fields. It therefore needs verification before it can be generalized to hydrodynamics. Since the effects involved are just the induced electric field and the magnetic force density $j \times \mathbf{H}/c$ it is sufficient to consider a perfectly conducting fluid; in such a fluid the field moves with the material. Lenz's law can then be verified for currents which do not flow near any neutral point; for instance, the magnetic force of a twisted field tends to untwist it. It is found, however, that Lenz's law is reversed at a neutral point and this will now be explained.

Consider a neutral point N, where the lines of force in one plane have the form shown in Fig. 1. The limiting lines of force through N form an X and would be perpendicular, if there were no current flowing in the z-direction (normal to the paper). For the field of Fig. 1 there is a roughly uniform current in the z-direction, which contributes a field directed clockwise. The magnetic force therefore has the direction of the short arrows and tends to compress the material and field in the x-direction and stretch them in the y-direction. Since this motion reduces the acute angle between the limiting lines of force at N, it seems probable that it increases the current density. This has been verified mathematically (Dungey [3]). The pressure gradient can be omitted, since it is not involved in Lenz's law. N is chosen to be at rest initially and then does not move. The variables are the spatial gradients at N of $\mathbf{H}$ and of the velocity $\mathbf{u}$ and the mass density μ at N. The time derivatives of these variables are found to depend only on themselves, a very unusual situation in fluid dynamics. Then it is possible to work out what happens when a small perturbation is applied to a static state with $\mathbf{j} = 0$. It is only necessary to consider the signs of the variables to show that they all increase in magnitude indefinitely, and hence Lenz's law is reversed. The rate of growth of the discharge is proportional to the initial spatial gradient of $\mathbf{H}$.

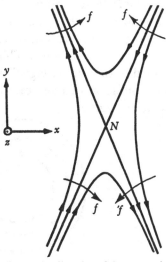

Fig. 1. The direction of the magnetic force, $\mathbf{f} = \dfrac{1}{c}\mathbf{j} \times \mathbf{H}$.

3. THE PRESSURE GRADIENT

Though the pressure gradient is not involved in Lenz's law, it needs to be investigated. No rigorous conclusion has been obtained, and Dr Sweet is at present studying the problem. The pressure gradient has a tendency to oppose the motion which produced it, but it does not by any means follow that it will stop the motion in this case. It would be more likely to stop the motion, if Lenz's law were true, so that the magnetic force were decreased by motion. With Lenz's law reversed there is a race between the two forces to build up fastest.

There is a more important reason why the pressure gradient should be unable to prevent a discharge. Since an increase of pressure must result from compression, a solenoidal motion does not build up a pressure gradient. Now it appears from Fig. 1 that a discharge can result from a solenoidal motion, and hence that it could occur even in an incompressible fluid.

Finally we may consider the possibility of equilibrium between the pressure gradient and the magnetic force. If the pressure gradient could prevent a discharge from occurring, a configuration of stable equilibrium should exist. The equation of equilibrium is $\mathbf{j} \times \mathbf{H} = c\nabla p$, which requires that $\mathbf{j} \times \mathbf{H}$ be irrotational. The general problem is complicated, but, when two-dimensional symmetry is imposed, the condition reduces to $(\mathbf{H}.\nabla)\mathbf{j} = 0$, and it is possible for real fields to approximate closely to two-dimensional symmetry. This problem has been studied (Dungey[31]) for a field of the 'figure eight' type shown in Fig. 2. The conclusion was that the condition of equilibrium requires an infinite current density at the neutral point N.

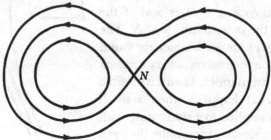

Fig. 2. A 'figure eight' field.

While a rigorous investigation is lacking, then, the indications are that the pressure gradient cannot prevent the discharge.

4. COMPARISON WITH OBSERVATIONS

It is well known that a large voltage can occur in a sunspot group; with a velocity of hundreds of km/sec, the voltage may be estimated as $\sim 10^{10}$ eV, though this could be out by an order of magnitude. It is also known that the energy of a spot field is sufficient to account for all the emissions of a flare (Kiepenheuer[4]). Since the discharge is very thin, it is desirable to check the number of particles accelerated. The discharge is thin in only one dimension and in both other directions extends over distances $\sim a$, determined by the scale of the spot field. The number of particles accelerated per second N is $n \cdot a \cdot b \cdot c$. The previous method of estimating b gives $nb \sim H/4\pi e$, so that N is independent of n. With $Ha \sim 10^{10}$ gauss cm, $N \sim 5 \cdot 10^{28}$ sec^{-1}, and if these were spread over a hemisphere at the earth's distance, they would be some twenty times as numerous as the normal cosmic rays. A thin discharge is therefore capable of accelerating sufficient particles.

The sudden onset of flares agrees with the discharge theory, because the time of growth for the discharge is not many seconds.

The discharge may be situated above the chromosphere. Some accelerated particles will be shot from the discharge towards the sun, and can produce the visual flare in a manner similar to the production of aurorae in the earth's atmosphere by incoming protons. The beam of accelerated particles is probably narrow in the same direction as the discharge, so that the luminous region should also be narrow; since each particle must have many collisions in the luminous region, however, it cannot be as narrow as the discharge. The true thickness of the luminous region could only be observed, if it were seen end on, and the tendency of flares to a somewhat filamentary form seems to agree with such a model. The brightening of parts of nearby prominences, which has occasionally been observed, could be explained by the impingement of the beam on the prominence, which has a large density; this suggests that the discharge occurs above the chromosphere.

The observed increases of cosmic ray intensity at the earth are compatible with a spectrum that is cut off sharply above a quite low energy. The increase occurs about half an hour after the flare, and increases are observed after most flares of magnitude 2 or more. This behaviour might be explained by a narrow beam of cosmic rays sweeping over a large range of directions, but the correct explanation may be more complicated.

Certain other features of flares may be accounted for by the bulk motion resulting from a discharge at a neutral point. The effect of the discharge is to 'reconnect' the lines of force at the neutral point, and this happens quickly. The 'reconnection' upsets the mechanical equilibrium in the neighbourhood in a way that can be visualized, if the lines of force are seen as strings. Then the mechanical disturbance will spread from the neutral point and may have energy comparable to the energy of the spot field in the solar atmosphere. This disturbance, characterized by a sudden onset, may account for several features: surge prominences and Doppler shifts, which are probably different aspects of the same phenomenon; the emission of a mechanical disturbance responsible for magnetic storms; the activation of prominences; the triggering of other neutral points, in whose neighbourhood the field is weak, resulting in multiple flares.

The emission of radio noise could result from either the high-energy particles or the disturbance in the plasma.

REFERENCES

[1] Cowling, T. G. *The Sun*, ed. Kuiper, (Chicago, 1953), p. 587.
[2] Giovanelli, R. G. *Mon. Not. R. Astr. Soc.* **107**, 338, 1947.
[3] Dungey, J. W. *Phil. Mag.* **44**, 725, 1953.
[4] Kiepenheuer, K. A. *The Sun*, ed. Kuiper, (Chicago, 1953), p. 393.

Cowling: My comments on self-induction were limited to increases in current due to changes in conductivity. Dungey's increased currents were not due to such changes.

May I make three tentative comments on Sweet's paper? First, it seems to provide a neutral line theory rather than a neutral point theory; the field considered as a two-dimensional one. Secondly, a disturbance at photospheric level can effect the field at a higher neutral point only by the transmission of magneto-hydrodynamic waves; the events are not sudden. Would differences in travel of such waves along neighbouring lines of force upset the mechanism proposed by the lack of synchronism resulting? Lastly, in a flare we do get an abnormal temperature which can be seen to rise. A flare would not be visible if it heated the gas up to the point where recombination of hydrogen did not occur; is there not some danger of a flare being invisible if its energy is produced in too narrow a layer?

Sweet: In reply to Professor Cowling's point concerning hydromagnetic waves, the spot pair need only approach each other slowly, and a theory of the forces which are transmitted upwards can be established without the introduction of waves. The limiting pressure to sustain such a quasi-hydrostatic equilibrium is first exceeded at the neutral point. Regarding the temperature in the layer, the density may be much higher, due to the compression at the neutral point, than in the surrounding chromosphere. The cooling by radiation in such a dense gas keeps the temperature low.

Dungey: The acceleration occurs only in a thin region. Giovanelli thinks that the accelerating region is well above the chromosphere and that the visible light is secondary, like that from an aurora.

Ferraro: Does the process not depend upon the rate of approach of the pair of binary sunspots? This rate does not occur in your formulae.

Sweet: The effect does not depend on the velocity of approach of the spots, provided this velocity is small compared to that of hydromagnetic waves. As the spots reach a certain separation the theorem shows that the gas becomes unstable at the neutral point. Equilibrium breaks down and the quasi-steady layer is set up. The spots need not move further.

SOME EFFECTS OF HYDROMAGNETIC
WAVES IN THE SOLAR ATMOSPHERE

J. H. PIDDINGTON

Division of Radiophysics, C.S.I.R.O., Sydney, Australia

ABSTRACT

A hitherto undisclosed source of absorption of hydromagnetic waves, due to the presence of neutral atoms, is investigated quantitatively. In the chromosphere the rate of absorption due to this effect may be 10^{10} times greater than that due to ordinary viscosity.

General solar heating, due to hydromagnetic waves caused by granules, is investigated and order-of-magnitude agreement with observational data is found.

The new effect is discussed in connexion with heating near sunspots and with flares.

I. THE RATE OF ABSORPTION OF
HYDROMAGNETIC WAVES

It has been shown [1, 2] that heating of gas by crossed electric and magnetic fields depends on the composite conductivity $\sigma_3 = \sigma_1 + \sigma_2^2/\sigma_1$, where σ_1 and σ_2 are the direct and Hall conductivities respectively. In a fully ionized gas like the solar corona σ_3 is very large and the rate of dissipation of electromagnetic energy very low.*

However, when neutral atoms are present in comparable numbers, even though the electron collision frequency is hardly affected (because of the small collision section of the atoms compared with ions), the value of σ_3 decreases by a large factor and the absorption increases correspondingly. The effect has been discussed in connexion with non-spot solar heating and with galactic hydromagnetic waves [3, 4].

When electrons collide much more frequently with ions than with atoms a partially ionized gas may be treated as a fully ionized gas co-existing with a neutral atom gas. The latter is only effective in exerting a viscous drag on the former which moves under the influence of the electro-magnetic field. For the shear waves with which we are concerned gas

---

* When the hydromagnetic waves are strong and irregular and some of the ions are moving much faster than the average, substantial dissipation of energy may occur by another process discussed below.

pressure gradients may be neglected and the momentum equations of the gases are

$$\frac{\partial \mathbf{v}}{\partial t} + (\mathbf{v} - \mathbf{v}')\,\frac{\eta}{\tau} + \frac{V^2}{H_0^2}\,(\mathbf{H}_0 \times \operatorname{curl} \mathbf{H}) = 0$$

and

$$\frac{\partial \mathbf{v}}{\partial t} = (\mathbf{v} - \mathbf{v}')\frac{1}{\tau},$$

where $\mathbf{v}$ and $\mathbf{v}'$ are the velocities of the plasma and neutral atom gases, η is the ratio of the mass densities of neutral atoms to plasma, τ is the collision period of a neutral atom with a heavy ion and $\mathbf{H}$ the magnetic field with a steady component $\mathbf{H}_0$; V is the hydromagnetic velocity $H_0(4\pi\rho)^{-\frac{1}{2}}$ where ρ is the density of the ion plasma alone. These equations are combined with the field equations for a moving fully-ionized gas (reference [1], Eq. (17)):

$$\nabla^2 \mathbf{H} - \sigma_2/\sigma_1 \frac{\partial}{\partial z} \operatorname{curl} \mathbf{H} = 4\pi\sigma_3 \left[\frac{\partial \mathbf{H}}{\partial t} - \operatorname{curl} (\mathbf{v} \times \mathbf{H}) \right],$$

where the z-axis coincides with $\mathbf{H}_0$.

The absorption coefficient of hydromagnetic waves described by these three equations takes a simple form when the wave angular frequency $\omega \ll \tau^{-1}$. We have

$$\kappa = -\frac{\omega^2 \tau \eta}{2SV(1+\eta)^{\frac{1}{2}}}, \tag{1}$$

where SV is the wave velocity, S having a value of $\cos \psi$ or unity for the 'ordinary' (O) and 'extraordinary' (E) waves respectively [5], ψ being the angle between $\mathbf{H}_0$ and the direction of wave propagation. In the middle chromosphere in a magnetic field of 100 gauss and with equal numbers of atoms and heavy ions this absorption rate is about 10^{10} times that due to ordinary Joule heating (neglecting neutral atoms) and much greater than that due to ordinary viscous effects.

Earlier theories of coronal heating, flares, etc., invoked large induction electric fields to provide acceleration. These theories have been shown untenable because the magnetic field is frozen into the plasma. It is not frozen into the neutral atoms, however, and its movement relative to this gas may result in a large induced electric field; hundreds of volts/cm is possible. It is this field which, according to the present theory, is responsible for the various phenomena. The movement of the ion plasma may be regarded as a Hall drift under the influence of the electric field. Rapid relative motion of the two gases results in collisions which cause heating, excitation and ionization with subsequent emission of quanta up to X-ray energies.

142

2. HEATING AWAY FROM SUNSPOTS

Alfvén[6] has suggested that some of the energy of the photospheric solar granules may be transferred upwards by shear waves and contribute to coronal heating. The available supply of kinetic energy is about 10^7 erg cm^{-2} sec^{-1} which may be 100 times or more that needed to maintain the corona in non-spot regions. The previous difficulty met by this theory was that the rate of absorption of wave energy appeared to be too low to cause much heating. It is now found that in a limited region where $\eta \sim 1$ this difficulty is overcome.

Upward transfer of energy must be by wave motion, the three possible waves being the O, E and S hydromagnetic waves (see reference[7]), the fourth wave, the space-charge electric wave, is not likely to be important at these frequencies. The O and E waves are the shear waves discussed above; the S wave is mainly longitudinal and approximates a sound wave in very weak fields. In a weak magnetic field, say 5 gauss, the velocities of these waves at the photosphere are about 0·2, 0·2, 10 km sec^{-1} respectively, which gives a measure of the relative effectiveness in energy transfer. Thus most of the energy is transferred initially by a sound wave as in the theory of Biermann[8] and others.

As the wave approaches a level of about 1500 km it tends to be strongly absorbed in the absence of a magnetic field. However, the field increases the coherence of the wave and also causes the weak shear waves to attain the same velocity as the strong S wave. There will then be a tendency to share the available energy equally between the waves as they have some electromagnetic vectors in common. At higher levels the O and E waves are slowly absorbed and also partially reflected due to a decreasing refractive index of the medium. According to the theory it is the energy provided by the absorption of these waves which heats part of the solar atmosphere.

A quantitative examination[3] suggests that a significant proportion of the original energy, in the frequency band $0\cdot1 < \omega < 0\cdot3$, may reach a level of 6000 km where $\eta \sim 1$. Waves of much lower frequency tend to be mainly reflected, those of much higher frequency to be absorbed at lower levels. At the level where $\eta \sim 1$ the ion and atom densities are about 3×10^{-9} cm^{-3} and $\tau \sim 1$ sec. Waves of frequency 0·2 rad sec^{-1} in a field of 5 gauss have an absorption coefficient of about 10^{-9} cm^{-1}. In a layer a few thousand km thick the energy released might be about 10^5 erg cm^{-2} sec^{-1}, perhaps sufficient to replenish the quiet or non-spot corona. At lower levels the heating per gas particle falls rapidly and in the corona itself the total heating is very small because of the dearth of neutral atoms.

A prediction of the neutral atom theory is that heating occurs, not in the body of the hot corona itself, but in the surface layers of the chromosphere and of prominence material. There is some evidence that this is the case.

3. HEATING NEAR SUNSPOTS AND FLARES

Current ideas of the origin, maintenance and decay of a sunspot magnetic field [9] suggest that there is irregular motion below the photosphere. The rate of development of a spot group indicates an upward motion of at least 0·03 km sec^{-1} from a depth of 20,000 km and more localized eddies may move much faster. These movements are communicated to the magnetic field and result in the formation of transverse and longitudinal hydromagnetic waves. The waves will tend to travel up the magnetic field and emerge either through the spot itself or through a non-spot region penetrated by a spot magnetic field.

Several varieties of hydromagnetic disturbance may travel up a cylindrical bundle of sunspot lines of force. Perhaps the most important form of shear wave is torsional, since this causes less disturbance (and so is less absorbed) in the surrounding medium than do waves involving lateral movement or expansion and contraction of the bundle. Torsional oscillations may be initiated by contact of the field with a vortex element in the gas at low levels. Longitudinal (S) waves may also be present and rise to the surface.

The Poynting flux of energy is given by

$$P = (H_o H_p v)/8\pi, \tag{2}$$

where H_p is the perturbation magnetic field and v the gas perturbation velocity given by

$$v = \frac{H_p}{H_o} \cdot V = \frac{H_p}{(4\pi\rho)^{\frac{1}{2}}}. \tag{3}$$

The total flux over a bundle of area A is $\int H_o H_p v \, dA/8\pi$ and, since $\int H_o \, dA$ is constant we have, for a uniformly expanding bundle, $H_p v = \text{const.}$ Hence $v \propto \rho^{-\frac{1}{4}}$ so that the disturbance becomes more violent as it rises.

A disturbance rising from a depth of 20,000 km to the photosphere experiences a density drop of $\sim 10^3$ so that an initial disturbance of say 0·1 km sec^{-1} becomes one of 0·6 km sec^{-1}. If the steady magnetic field is 2000 gauss, then $H_p \sim 70$ gauss and $P \sim 3 \times 10^8$ erg cm^{-2} sec^{-1}. This flow of energy, when absorbed by the process described above, may account at least in part for the very hot ($> 10^7$ °K), high-pressure regions whose existence above sunspots is inferred from both optical [10] and radio [11] observations.

As the hydromagnetic waves travel from the photosphere to the mid-chromosphere where $\eta \sim 1$ there is a further large increase in gas velocity to about 200 km sec^{-1}. The energy abstracted from the wave due to a collision between a heavy ion and an atom may be as high as 200 eV which is of the right order to account for the very high gas temperatures, emission of X-rays, violent movement of material and other observed and inferred effects. It may be noted that the perturbation magnetic field, H_p, associated with this very substantial gas velocity is small, only about 2 gauss in a total field of the order 1000 gauss.

Hydromagnetic waves with perturbation velocities of 10 km sec^{-1} at the photosphere will have perturbation magnetic fields of about 1000 gauss. These waves might be observable but should cause no very obvious effects at this level. However, on rising to a level where $\eta \sim 1$, the density has dropped to about 10^{-15} and the perturbation velocity risen to about 1000 km sec^{-1}. The results are likely to be catastrophic: there will in particular be violent heating of the surface layer of the chromospheric neutral atom gas. The heating is due to the violent beating of this gas by magnetic lines of force to which the ions are attached. If the waves are of angular frequency comparable with the neutral atom collision frequency then collision energies of thousands of electron volts are developed.

The mechanism is proposed as an explanation of the origin of flares.

A comparison of the observed properties of flares[12] with those anticipated from the mechanism shows fair agreement. In particular, flares grow near sunspots from pre-existent 'normal' bright hydrogen flocculi. Such flocculi constitute the surface layer of the neutral atom chromosphere and are bright because they are heated by the recurrent weak hydromagnetic disturbances. On occasions the heated surface layer may be on prominence material or on the underside of chromospheric clouds ('Hydrogen bombs'). The structure of flares is also related to the magnetic field distribution; the relationship is complex but sometimes the flares seem to extend along lines of force. Such effects are consistent with an origin in hydromagnetic waves, which are controlled by the field.

4. HYDROMAGNETIC WAVES IN THE CORONA

While the neutral atom mechanism may be adequate near the surface of the chromosphere (visible in line emission) it hardly seems capable of maintaining the corona, since the kinetic energy of the particles of hot gas would be dissipated in raising the gas to high coronal levels against the gravitational field. However, there are several factors which, with the aid of neutral atom heating, are likely to provide the observed hot gas at high levels.

There is considerable turbulence in the corona, particularly in the neighbourhood of sunspots. This would cause some mixing and raising of the gases. Strong hydromagnetic waves, and in particular shock hydromagnetic waves, instead of causing the gas only to oscillate back and forth, make it slide along the magnetic field and so provide electromagnetic energy to raise the gas to high levels.

There is another effect or perhaps series of effects which depends on the presence of a proportion of ions with velocities much greater than the average. The neutral atom mechanism should release such particles which will rise to 10^5 km or so without loosing a great proportion of their initial kinetic energy. When those particles have energies corresponding to a gas 'temperature' of less than 10^8 °K their collision periods are minutes or less and comparable with the wave periods. Under those circumstances the heating or accelerating mechanism may be described as a viscosity effect, the ordinary viscosity being enhanced by the presence of the fast particles.

We may use an expression for viscous absorption derived by van de Hulst[13] replacing the kinematic viscosity by the expression $\eta_F v_F^2 \tau_F$ where η_F is the proportion of fast particles in the total mass of gas, v_F is their velocity and τ_F their collision period. The absorption coefficient κ_F due to the viscous effects of those particles may then be compared with that due to neutral atoms (Eq. (1)) and we find

$$\frac{\kappa_F}{\kappa} = K \left(\frac{v_F}{v}\right)^2 \cdot \frac{\eta_F}{\eta} \cdot \frac{\tau_F}{\tau}, \tag{4}$$

where K is a constant of order unity and it is assumed $\eta \leqslant 1$. The expressions for κ and κ_F are similar as, in a way, are the mechanisms. In one case the particles causing absorption (neutral atoms) remain at rest while the ions are accelerated by the wave, in the other case the absorbing particles (fast ions) move rapidly from one part of the wave to another where the relative velocities are high. In either case a subsequent collision causes substantial loss of wave energy. In the corona in a magnetic field of 10–100 gauss and with say 10 % of ions at a 'temperature' of 10^7 °K the value of κ_F/κ does not fall far short of unity and allowing for the decreased density the heating effect may be important.

The viscous heating effect tends to increase with temperature until limited by the fact that the fast ions do not collide often enough to transfer their energy, which oscillates between the wave and the particles, this occurs when $\tau_F \gtrsim \omega^{-1}$. Meanwhile the third effect becomes operative, the fast particles collide not only with other particles but with strong hydro-magnetic waves; with the magnetic field itself. A single collision may increase the kinetic energy of the particle by many times its original value. Subsequently some particles will suffer collisions and degrade their kinetic energy to heat while others will become 'runaway' ions, suffering no collisions but, under suitable circumstances, gaining energy continously from the waves until some eventually become cosmic rays.

5. COSMIC RAYS AND RADIO BURSTS

The statistical gain of energy of ions by collisions with hydromagnetic wave packets was suggested by Fermi and reformulated by Thompson[14]. The results agree quantitatively for 'strong' waves, that is when $H_p \to H_0$ and are used to estimate the time taken to create cosmic rays in the corona.

Assume $H_p/H_0 = 0.2$ which is an order of magnitude greater than the flare value given above and may be expected for the exceptionally powerful disturbances which release cosmic rays. Hydromagnetic waves of length say 10^5 km in the corona then cause a gain of energy by a factor of exp. 1 in about 200 secs or a factor of 10^8 in 1 hr. The necessary injection energy is about 10^4 eV.

Strong flares are accompanied and followed by an outflow of gas moving at about 10^3 km sec^{-1}. This may be the remains of the longitudinal or S-type hydromagnetic wave which would be expected to accompany, at least in some degree, the transverse O- and E-type waves which are invoked to explain the flare. This flow of gas may be necessary to release the cosmic rays from the confines of spot magnetic field. On reaching a level of about 10^6 km or so in about $\frac{1}{2}$ hr it will so have stretched and attenuated the magnetic field that the cosmic rays may escape. The gas has in effect, changed the magnetic field from a typical dipolar form to a more-or-less radial form.

Same radio bursts are thought to originate in large-scale disturbances travelling out from the sun. Some such disturbances seem to travel out at about 10^3 km sec^{-1} and others at 10^5 km sec^{-1}, starting simultaneously from the vicinity of a flare[15]. The travelling sources of bursts need not be ion clouds, they could equally well be wave-packets and on this basis they have a simple explanation.

The slow travelling disturbance is the S-wave or the remains of the S-wave, probably degenerated into a shock wave directed largely along the magnetic field until it reaches a level where the field is too weak to control it. The fast travelling disturbance is perhaps the O or E hydromagnetic wave responsible for the flare. In the low corona in a magnetic field of 1000 gauss the velocity of these waves is about 10^5 km sec^{-1}. There will be a tendency for the velocity to remain steady as the field-strength and gas density both decrease outwards.

It seems that on the basis of a mild hydromagnetic disturbance below a sunspot, a number of observed effects may be explained.

REFERENCES

[1] Piddington, J. H. *Mon. Not. R. Astr. Soc.* **114**, 638, 1954.
[2] Cowling, T. G. 'The dissipation of magnetic energy in an ionized gas', *Mon. Not. R. Astr. Soc.* **116**, 114, 1956.
[3] Piddington, J. H. 'Solar atmospheric heating by hydromagnetic waves', *Mon. Not. R. Astr. Soc.* **116**, 314, 1956.
[4] Piddington, J. H. 'Galactic turbulence and the origins of cosmic rays and the galactic magnetic field', *Aust. J. Phys.* (in the press).
[5] Piddington, J. H. 'Hydromagnetic waves in ionized gas', *Mon. Not. R. Astr. Soc.* **115**, 671, 1955.
[6] Alfvén, H. *Cosmical Electrodynamics* (Oxford, Clarendon Press, 1950).
[7] Piddington, J. H. *Phil. Mag.* **46**, 1037, 1955.
[8] Biermann, L. *Z. Astrophys.* **25**, 161, 1948.
[9] Bullard, E. C. *Vistas in Astronomy*, vol. 1, ed. A. Beer, (London, Pergamon Press, 1955).
[10] Waldmeier, M. and Müller, H. *Z. Astrophys.* **27**, 58, 1950.
[11] Piddington, J. H. and Minnett, H. C. *Aust. J. Sci. Res.* A**4**, 131, 1951.
[12] Ellison, M. A. *Mon. Not. R. Astr. Soc.* **109**, 3, 1949.
[13] van de Hulst, H. C. *Problems of cosmical aerodynamics*, I.U.T.A.M. and I.A.V., Paris 1951; Central Air. Documents Office, Dayton, Ohio, 1951.
[14] Thompson, W. B. *Proc. Roy. Soc.* A, **233**, 402, 1955.
[15] Wild, J. P., Roberts, J. A. and Murray, J. D. *Nature, Lond.* **173**, 532, 1934.

Discussion

Schatzman: What is the direction of propagation of the flux of mechanical energy transported by the hydromagnetic waves?

Piddington: In a more or less uniform medium hydromagnetic waves may propagate freely in any direction and so transport energy freely in any direction. From a single small-scale disturbance a spherical wave would spread; from numerous granules an interference pattern would be formed.

Schatzman: What is the energy transmitted upwards in the chromosphere as a result of hydromagnetic waves compared to that transmitted by pure compression waves? As far as I know, the compression waves transport an energy sufficient for heating the chromosphere and the corona.

Piddington: Near the photosphere the speed of the shear waves is about 0·2 km/sec for a magnetic field of a few gauss. The speed of the compression waves is about 10 km/sec so that unless a particular disturbance greatly favoured the formation of shear waves these would initially play an unimportant part in energy transport. At higher levels, however, the situation is reversed.

Biermann: As was just pointed out, the velocity of hydromagnetic waves is, outside a spot region, quite small compared with the velocity of compression waves. Would that not indicate that outside spots the old picture of pressure waves is up to the higher levels of the chromosphere, as good an approximation as one might wish to have? Up to the limit just indicated only slight modifications should be expected from the interplay of the magnetic and shearing modes.

Piddington: Pressure waves would seem to play the main part up to about 1500 km. Above that level the shear waves are more efficient at transporting energy. There is evidence that the pressure waves would disintegrate, in which case the shear waves would be all-important as a medium of energy transport.

Biermann: This depends, of course, on the local value of the magnetic field and also on the hydrogen density.

Alfvén: Would not small charged grains, instead of neutral atoms, also be able to cause a heating of the corona?

Piddington: Yes, provided the charge and density of the grains is appropriate.

Sarabhai: One of the remarkable features of cosmic ray increases associated with flares is that the flare is often quite far removed from the central meridian. On 23 February 1956, the flare was about 70° from the central meridian and generated cosmic rays up to 30–40 BeV in energy. Is the angular distribution of ejection of energetic particles of this theory compatible with the observations?

Piddington: No statement can be made on the angular distribution.

SOME CONSIDERATIONS ON THERMAL CONDUCTION AND MAGNETIC FIELDS IN PROMINENCES

S. ROSSELAND, E. JENSEN AND E. TANDBERG-HANSSEN

Institute of Theoretical Astrophysics, Blindern, Oslo, Norway

ABSTRACT

Prominences which extend into the million degree temperature region of the corona will, in the absence of magnetic fields, be heated up to temperatures of the same order of magnitude in the course of at most a few hours. A magnetic field of reasonable magnitude inside the prominence, will, however, be sufficient to cut down thermal conduction and turbulence to such an extent that the long life of some prominences seems understandable.

I. INTRODUCTION

The temperature in prominences has been estimated by various authors[1] to lie around 4000° K. This low value indicates that the prominence in question is, practically speaking, in radiative equilibrium with the bulk of solar surface radiation. The majority of prominences extends well into the region of the corona which is credited with a temperature in the million degree range. We are hence confronted with a close juxtaposition of the matter in the temperature range a million degrees on one hand, and a few thousand degrees on the other. The drop in temperature of may be a million degrees takes place over a distance which may be less than a thousand kilometres. Let us further recall that the life period of prominences of the arched bridge type discussed at length by M. and Mme D'Azambuja[2] may run up into many months or even a year, without suffering serious deterioration until near the end of the period. The form of these filaments is that of a thin nearly vertical sheet, the thickness of which has been estimated by various investigators to range from 5000 to 10,000 km.

These facts suggest various ideas and hypotheses concerning the physical conditions maintaining long-lived prominences. Considering the admitted fact of the high thermal conductivity of the corona[3] it is puzzling to find lumps of cold matter like the prominences to exist there for weeks and months, as one would expect them to be broken up by coronal heating in the matter of a few minutes or even seconds.

We suggest that the solution of the puzzle lies in the proper recognition of the part played by magnetic fields in the maintenance of prominences. The magnetic field is long recognized as necessary for the understanding of *the form* of prominences, so that matter is restricted to flow along the magnetic-field lines. Such a field also suppresses turbulent convection of heat, which restricts the interchange of heat between a prominence and the corona to consist of ordinary molecular conduction. It will be shown in the following that the conductivity in the corona is so high—in the absence of magnetic fields—that prominences would not be expected to survive for more than a few minutes. For the understanding of the persistence of long-lived prominences their magnetic field is thus essential.

2. THERMAL CONDUCTIVITY IN THE CORONA

Let us first recall some elementary notions of the gas kinetic theory of conduction. The coefficient of the thermal conduction K is defined by the expression

$$\mathbf{F} = -K\nabla T \tag{1}$$

for the flux $\mathbf{F}$ of thermal energy, T being the absolute temperature. In the following we shall be interested in the case when the conduction of heat is mainly carried on by free electrons. This may be a poor approximation for the central part of a prominence, and will have to be amended in more refined calculations. For our present purpose it is, moreover, essential to keep the picture of the physical processes at work clearly in mind, and for this reason we base the considerations on the free path picture of elementary kinetic gas theory. The coefficient K is then defined by the expression

$$K = \tfrac{1}{2}kCNL, \tag{2}$$

where k is Boltzmann's constant while C, N and L represent the mean thermal velocity, mean number in unit volume, and mean free path of free electrons in the gas. For C the expression $(3kT/m)^{\frac{1}{2}}$ gives a sufficient approximation, m being the electronic mass. Assuming the gas to consist essentially of ionized hydrogen, the motion of the free electrons is mainly interfered with by the free protons, the number of which is approximately equal to N. The mean free path L of an electron is then defined by $1/N\pi a^2$ where a is the effective 'radius of collision' of a proton-electron encounter. Simple considerations, based on the Rutherford scattering formula leads to the expression $2e^2/3kT$ for a, $-e$ being the electronic charge. Combining these various terms we find for K the following expression

$$K = \frac{k(3kT)^{5/2}}{8\pi m^{1/2} e^4}. \tag{3}$$

The general statistical theory of conduction given by Chapman and Cowling[4] leads to nearly the same dependence of K on T as above, but to a slightly different numerical factor. For the following applications this difference is not essential.

To clear the picture as far as the corona is concerned we note that for the temperature $T = 10^6$ °K and a coronal electronic density $N = 5 \times 10^8$ cm^{-3} we find $C = 6800$ km/sec, and $L = 5200$ km. The average time spent on a free path in the corona becomes $L/C \approx 1$ sec.

In the corona proper the temperature is so high that the above expression for the conductivity should be a fair approximation. As the temperature goes down and approaches that of a prominence, the approximation becomes less good, partly because conduction by heavy particles will gradually have some effect, and partly because the free path will be cut down by collisions with neutral particles in addition to the influence of protons. We do not think it necessary, however, to amplify the theory in this direction for the purpose we have in mind. It seems probable that the speed of the heat wave is likely to be determined by values of the conductivity in the high temperature region rather than in the region of low temperatures, so that the exact expression of K for low values of T matters little.

When a magnetic field of strength H is present[5] the transverse thermal conductivity of completely ionized hydrogen will be cut down by a factor $(1 + \omega^2 \tau^2)^{-1}$, where $\omega = eH/mc$ is the magnetic gyro-frequency and τ the time spent by an electron on a free path $\tau = L/C$.

The realm of validity of the magnetic factor $(1 + \omega^2 \tau^2)^{-1}$ appears to be obscure. But it seems to be the best guide we have at the present time.

In many conduction problems it is possible to regard K either as a constant, or as a slowly varying function of space and time. In the present case conditions are different, in that the value of K as given by (3) varies by a factor a million when T drops from the coronal temperature 10^6 °K to a prominence temperature of say 4×10^3 °K. Such a variation must produce a tendency to build up steep temperature gradients in the surface region of the prominence. The consequent expansion of the heated region will, however, tend to reduce the rate of advance of the heat wave. Taken in full generality the problem is thus a very complicated one. But the fact that long-lived prominences seem to persist for a long time with only small changes of form makes it reasonable to use a static working model for our theoretical considerations.

The equation of heat transfer is then given by the simple equation

$$C_v \frac{\partial T}{\partial t} = \text{div} \ (K \nabla T), \tag{4}$$

where C_v is the specific heat per unit volume.

3. SOLUTION OF (4) WHEN K/C_v = CONSTANT

Eq. (4) has a simple solution for the case when K/C_v is constant. The fact that K varies by a factor about a million from the centre of the prominence to the corona then demands that C_v also varies by the same large factor. We imagine this to mean that the density increases outward with this factor, which of course is very far from the truth. However, the influence of this rapid increase of C_v with increasing temperature means a considerable slowing down of the progress of the heat wave into the prominence. As we are mainly interested in deriving upper limits for the life time of a prominence, the solution of this somewhat artificial problem will serve our end.

First of all we change the dependent variable from T to

$$U = \int K \, dT, \tag{5}$$

which, when introduced into (4) gives

$$\frac{\partial U}{\partial t} = \kappa \nabla^2 U; \quad \kappa = K/C_v; \quad C_v = 3Nk. \tag{6}$$

By our assumption κ is constant, Eq. (6) becomes a simple linear equation in U which may be solved in the conventional way by writing

$$U = \sum_s U_s (x, y, z) \, e^{-\lambda_s t}, \tag{7}$$

the functions U_s being functions of the space co-ordinates x, y, z only, and the λ_s being an infinite set of constants, determining the time scale of the problem. When (7) is introduced into (6) it follows that each U_s must satisfy the equation

$$\nabla^2 U_s + (\lambda_s/\kappa) \ U_s = 0. \tag{8}$$

It matters very little what kind of a model is adopted for our prominence, as it will be easily realized that models as different as an infinite plane sheet of thickness D or an infinite cylinder of diameter D will give comparable results for the time scale. In the first case the solution of (8) is

$$U_s = A_s \cos (\lambda_s/\kappa)^{1/2} x, \tag{9}$$

where A_s is a constant and x is the linear coordinate normal to the sheet. In the second case

$$U_s = B_s J_0 [(\lambda_s/\kappa)^{1/2} r], \tag{10}$$

where B_s is a constant and r the distance from the axis of the cylinder, while J_0 is a Bessel function of zero order. These functions are closely related to the circular ones, and their zeros tend asymptotically to those of a cosine for increasing values of the argument.

As suitable boundary values we may demand that at the time $t=0$ the temperature, and hence also U, shall be constant through the prominence, and rise abruptly to the coronal value at the boundary $(r=x=D/2)$. Adhering strictly to these conditions would be inconvenient, as it would make it necessary to handle an infinite and slowly converging series. It is better to date ones time from an instant after the coronal heating has had time to penetrate the skin of the prominence, so that the initial temperature distribution may be represented by a few terms only of the series (7).

Suppose that we consider the simplest case when the series (7) consists of one constant and one variable term, so that the solution is (for the case of the cylinder):

$$U = U(T_c) - [U(T_c) - U(T_0)] J_0((\lambda_1/\kappa)^{1/2} r) e^{-\lambda_1 t}, \tag{11}$$

where now

$$\lambda_1 = \kappa . \xi^2 (2/D)^2, \quad \xi = 2 \cdot 40.$$

Here ξ is the first root of the Bessel function J_0. Further are T_c and T_0 the temperatures in the corona and at the centre of the prominence at time $t=0$ respectively.

4. TIME SCALE OF THE PROMINENCE

The structures recognized as prominences have temperatures low enough for the material to show spectra of hydrogen and various metals in non-ionized states. Also, when seen projected on the disk, pictures in $H\alpha$-light show the filaments as dark, though with bright borders as a normal feature. This indicates that the bulk of the prominence is at a temperature comparable to or lower than that of the solar surface. The bright borders are naturally interpreted as the result of coronal heating. At a temperature much larger than that of the solar surface, say at 10,000 °K, the emissitivity of the prominence material would be expected to exceed that of the solar surface considerably, and make the prominence bright all over. To be on the safe side we may increase the limiting value to 20,000 °K, and state that a prominence heated at the centre to such a temperature is to be considered as invisible in the light of Hα. When the corresponding value of U is introduced on the left-hand side (11) and r is put equal to zero, this expression becomes an equation for the time $t = \theta$ during which a non-magnetic prominence embedded in the corona is likely to be visible.

For the case in hand, when K is proportional to $T^{5/2}$ and U, consequently, proportional to $T^{7/2}$, (11) assumes the form

$$T^{7/2} = T_c^{7/2} - (T_c^{7/2} - T_0^{7/2})\, e^{-\lambda_1\theta},$$

or solved with respect to $e^{-\lambda_1\theta}$:

$$e^{-\lambda_1\theta} = \frac{T_c^{7/2} - T^{7/2}}{T_c^{7/2} - T_0^{7/2}} \approx 1 - (T/T_c)^{7/2}.$$

This means that to a sufficient approximation

$$\lambda_1\theta = (T/T_c)^{7/2} = \left(\frac{20{,}000}{1{,}000{,}000}\right)^{7/2} \approx 10^{-6}. \tag{12}$$

The above method of deriving an expression for θ is easily generalized to the case when the series for U contains any number of terms, provided the exponential factors may legitimately be linearized. Using the data suggested by Öhman[6] $T_0 = 4000$ °K and $N = 3.2 \times 10^{10}$ cm^{-3}, and assuming the diameter of the cylinder to be $D = 10{,}000$ km, we find λ_1^{-1} equal to 4.6×10^7 sec, and by (12): $\theta = 46$ sec.

We do not mean to stress the meaning of this figure beyond the inference that by its method of derivation it is in the nature of an upper limit, and that it would have to be extended by a factor of at least 10,000 to provide a semblance of an explanation for the persistence of prominences.

We prefer to think that this extension is provided by the effect of the magnetic field of the prominence. To increase the time scale by a factor 10,000 it is sufficient to have a magnetic field present that makes $\omega^2\tau^2 \approx 10^4$, or $\omega\tau \approx 100$. In the corona we found τ to be of the order of 1 sec. In the surface region of the prominence where the density may be a hundred times higher, τ may be correspondingly less. But in any case only a gyro-frequency of the order $10^4 - 10^5$ has to be considered and this makes the corresponding magnetic field a small fraction of a gauss.

Suppose thus that the magnetic field prevents heat from leaking into the prominence from the corona. The temperature of the former is then determined on one hand by the balance of the energy absorbed from the light radiation of the sun, and the loss by light emission on the other. That prominences depend on ordinary solar light radiation for their thermal state is at least also suggested by the fact that they seem to be easily disrupted by adjacent flares.

One more remark: We have assumed that the unknown heating mechanism which is responsible for the high coronal temperature does not operate inside a prominence of the long-lived type we are considering here.

The research reported in this paper has been sponsored in part by The Geophysical Research Directorate of the Air Force Cambridge Research Center, Air Research and Development Command, U.S. Air Force, under Contract no. AF 61 (514)-651 C through its European Office at Brussels.

REFERENCES

[1] Unsöld, A. *Physik der Sternatmosphären* (Berlin-Göttingen-Heidelberg, 2nd ed., 1955), p. 684.
[2] d'Azambuja, L. and M. *Ann. Paris Obs.* (Meudon), **6**, no. 7, 1948.
[3] Woolley, R. v. d. R. and Allen, C. W. *Mon. Not. R. Astr. Soc.* **110**, 358, 1950.
[4] Chapman, S. and Cowling, T. G. *The Mathematical Theory of Non-Uniform Gases* (Cambridge University Press, 1939), p. 179.
[5] Spitzer, L. jr. *Physics of Fully Ionized Gases* (New York, 1956), p. 88.
[6] Öhman, Y. private communication in 1955, printed in *Ark. Astr.* **2**, 1, 1957.

Discussion

Cowling: One might consider the reverse problem. A prominence means condensing of coronal material. If a magnetic field is so successful in keeping material in, may it not be difficult to secure the condensation of coronal material into a prominence?

Jensen: The effect of motions has not been taken into account in this investigation, so I do not think that I can answer your question.

Parker: I have seen calculations which seemed to indicate that the enhanced radiation from a prominence due to its low temperature and high density, was sufficient to maintain the low temperature of the prominence immersed in the hot coronal gas without requiring any inhibiting magnetic fields.

Öhman: In my opinion a condensation of prominences from the hot corona gas is perhaps not so typical as we may expect, because coronal prominences do not show characteristics of high temperature when we first see them. They may start from low temperature objects instead. This would help us in overcoming this difficulty.

Gold: The condensation of material into the prominence requires one to suppose that the associated heat transport is radiated away. The luminosity which is observed is perfectly adequate to account for such a cooling.

Piddington: Dr Jensen has shown that the inhibition of thermal conduction across magnetic lines of force may be an important factor in the maintenance of solar prominences. In the corresponding case of electric fields and currents the Hall current plays a major part in causing current to flow across the electric field. This current may, in certain circumstances cause space-charge built up with a resultant potential electric field at right angles to the original field. Finally, this field causes Hall current in the direction of the original field and so apparently causes a large increase in the 'direct' conductivity (this is the σ_3 discussed earlier).

A similar effect should occur in the case of thermal conduction and even if it is on a much smaller scale may have to be considered in connexion with the energy balance in prominences of certain shapes.

Alfvén: This will mean that the thermal cross conductivity has the same position as the electric cross conductivity; and if one goes a little deeper into it, one should perhaps not speak of it at all!

Jensen: In the presented paper it is assumed that thermal conductivity is cut down by the magnetic field by the same factor as the electric 'cross conductivity'. I think this assumption is reasonable, even if no rigorous theory on thermal conduction in a magnetic field has as yet been established.

PART III

STELLAR MAGNETISM

STELLAR MAGNETIC FIELDS

HORACE W. BABCOCK

Mount Wilson and Palomar Observatories, Pasadena, California, U.S.A.

ABSTRACT

A report is given of a ten-year observational program directed toward the discovery and investigation of the magnetic fields of stars through the Zeeman effect in their spectra. The emphasis has been on the sharp-line stars of type A, of which ten have been found to show irregular magnetic fluctuations without reversal of polarity, ten others to fluctuate irregularly with occasional reversals of polarity, and six to show essentially periodic variations; of the latter group, four are large-amplitude reversers with periods near one week.

The search for and investigation of stellar magnetic fields, which was begun ten years ago, was based on the idea that strong coherent fields might result from rapid axial rotation. Some sort of dynamo process was envisaged; to bring this up to date, it would depend upon rotation in combination with an outer convective layer in which Coriolis forces play a part (Elsasser[1]). Rapid rotation is characteristic of the hotter stars (types B, A and early F) while convective zones prevail in the cooler stars, becoming thin and finally absent as we proceed from the F's to the hotter B-types. It is for the stars of types A and early F, then, that rapid rotation and convection both prevail, and it happens that among stars of just this group the most prominent magnetic effects have been observed. An alternative is that stellar fields are primeval, having been 'frozen in' when the stars were formed, and that they exist independently of any dynamo process.

The Zeeman effect is rather small, spectroscopically, and conditions must be specialized if it is to be observed at all in stellar spectra. The field must be strong, mainly longitudinal, and one polarity must predominate; further, line broadening from other sources, such as the Doppler effect, must be quite limited, else the lines will be too broad to permit detection of the small Zeeman displacements. Fig. 1 shows how the Doppler effect, for an average A-type star with an equatorial velocity of rotation, v_e of 115 km/sec, and a large axial inclination, results in spectrum line-widths of about 3 Å. For magnetic observations, we must select stars having lines little wider than one-tenth of this. Only a small proportion of all the hundreds of A-type stars brighter than the eighth magnitude have lines so sharp; for these, either the rotation is inherently slow or the inclination, i, to the line of sight is small. Fig. 2 is a plot of line-width, as ordinate, against spectral type, where each dot represents an observed star. The

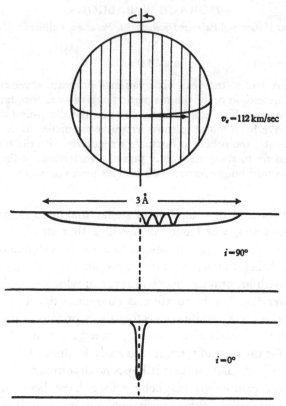

Fig. 1. Diagram illustrating the well-known Doppler broadening of spectrum lines due to axial rotation.

region of ultra-sharp lines in the lower part of the diagram has been rather completely observed, but for increasingly wider lines, additional stars could be found to populate the diagram. It is significant that of the twenty-one sharpest-line stars investigated, nineteen have been found to show a measurable magnetic field, and to display associated spectroscopic features such as abnormal line intensities. Only two of the twenty-one (γ Gem and 95 Leo) have 'normal' spectra without evidence of a coherent magnetic field.

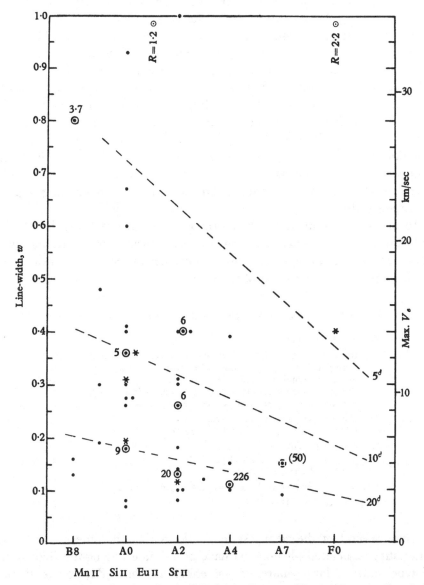

Fig. 2. The sharpest-line A-type stars plotted on a diagram of line-width, w, v. spectral type. Dots represent peculiar stars with magnetic fields; those encircled are periodic variables. Asterisks represent 'ordinary' stars. Of twenty-one stars having lines less than 0·2 Å wide, all but two have magnetic fields. The dashed lines indicate the minimum Doppler broadening corresponding to the rotational period in days.

If we regard these ultra-sharp-line stars as a statistical sample of normal, rapidly rotating A-type stars, selected for small axial inclination, then it would be a fair conclusion that a majority of the rapid rotators of whatever inclination also possess strong magnetic fields. The alternative is that the nineteen ultra-sharp-line magnetic stars are inherently slow rotators. This suggests the theory that they have lost angular momentum because of their strong magnetic field; this loss might have occurred through magnetic coupling between the star and a circumstellar gas cloud. But if this second alternative is correct, we have difficulty in answering this question: Statistically speaking, a small percentage of the hundreds of observable rapidly rotating A-type stars must have a small inclination and, therefore, sharp lines; if they are not included in our hard-found group of twenty-one ultra-sharp-line stars, then where are they?

In order to obtain as much information as possible from the observational work, a balance has been sought between the search for new magnetic stars and repeated observations for the study of variations. Table 1 gives a condensation of the results.

Table 1. *Number of magnetic stars with various characteristics*

Peculiar A-type stars (B 8, A 0, A 2, A 4, A 7, F 0)

Irregular magnetic fluctuations without reversal of polarity	10
Irregular magnetic fluctuations with reversal of polarity	10
Probably irregular, insufficiently observed	21
Periodic, large amplitude 'one week' reversers (early A-type)	4
Periodic, non-reversing, slower reversers (late A-type)	2

Metallic-line F-type stars. Fields relatively weak

HD 3883, 51 Sag, 63 Tau, 68 Tau, 16 Ori	5

Red Giants (Irregular fluctuations)

Type Mp: HD 4174, VV Cep, WY Gem	2
Type S: R Gem, HR 1105	2

RR Lyr (Cluster type variable, pulsating; rapid variations)	1
Others: AG Peg, HD 98088	2

Since several of the strongest magnetic fields and some of the most interesting associated phenomena have been found among the sharp-line A-type stars, a large share of the effort has been devoted to them; however, several other types, such as RR Lyrae, are receiving attention. Tests of classical cepheids have revealed no strong fields, although FF Aquilae deserves further investigation. There is no evidence that any star shows a constant magnetic field; for all that have been adequately observed, the fields appear to be variable. The peculiar stars of type A have been discussed by W. W. Morgan[2], the spectrum variables by A. J.

Deutsch [3, 4], and stellar magnetic fields by H. W. Babcock and T. G. Cowling [5]. It is hoped that a more complete report and discussion of the observations of stellar magnetic fields can be published shortly in the *Astrophysical Journal* [6].

REFERENCES

[1] Elsasser, W. M. *Rev. Mod. Phys.* **28**, 135, 1956.
[2] Morgan, W. W. *Publ. Yerkes Obs.* **7**, pt. 3, 1935.
[3] Deutsch, A. J. *Astrophys. J.* **105**, 283, 1947.
[4] Deutsch, A. J. *Publ. Astr. Soc. Pacif.* **68**, 92, 1956.
[5] Babcock, H. W. and Cowling, T. G. *Mon. Not. R. Astr. Soc.* **113**, 357, 1953.
[6] Babcock, H. W. 'A Catalog of Magnetic Stars', *Astrophys. J.*, Suppl. no. 30, 1958.

Discussion

Biermann: Do we have to understand that the contribution, say, of the relatively large following parts of a spot to the magnetic splitting of the lines, in the diagram we saw last, will effectively be dominant?

Babcock: The spots will not be dominant because they are dark and the field is concentrated in a very small area. The flux comes out from the spot and goes back to the photosphere over a very large area which is optically effective.

Schatzman: What is the limiting amplitude you can reach?

Babcock: Well, we can reach about seventh–eighth magnitude and perhaps the ninth if the dispersion is reduced.

Alfvén: In your picture showing the surface of a magnetic variable star you have some regions where the magnetic flux is directed outwards, and some regions where the flux goes inwards through the surface. The true average flux, integrated over the whole surface is, of course, zero. What comes out of your measurements? Do you measure the true average field (which should be zero) or do you measure a mean which differs from zero?

Babcock: We measure the mean longitudinal component of the magnetic field in the line of sight, weighted according to the surface brightness, and integrated over the visible hemisphere.

ASTROPHYSICAL APPLICATIONS OF SELECTIVE MAGNETIC ROTATION

Y. ÖHMAN

Stockholm Observatory, Saltsjöbaden, Sweden

When measuring the magnetic fields of sunspots the astronomer assumes that the magnetic field revealed by the inverse Zeeman effect is the same as if the splitting were produced by emission lines instead of absorption lines. No doubt this is in general a very fair approximation, but we have reason to remember sometimes that line absorption in the presence of magnetic fields is a very complicated process. In the immediate neighbourhood of absorption lines effects of magnetic rotation of the plane of polarization and magnetic double refraction may appear in the spectrum.

It is interesting to investigate whether such effects as magnetic rotation and magnetic double refraction may influence the measurements of solar fields. It is evident that no complication will appear if the magnetic field does not change its direction when we follow the line of sight through the gases contributing to the absorption line. But if considerable deviations of the field appear we may expect rather complicated effects which may influence our measurements.

One such effect which may be expected to appear sometimes is the possible disappearance of the central component, when we observe the Zeeman splitting of sunspots very near the solar limb and use for this purpose the conventional method of alternating strips. It is very well possible that longitudinal faint fields may in this case appear in the line of sight though the field of the spot itself has a direction perpendicular to the line of sight. In fact this would be expected on the general model of the magnetic field of sunspots [1]. When observing spots near the limb (Fig. 1) a device is generally attached to the spectrograph giving a number of strips where, say, the odd numbers are polarized parallel to the limb, and the even numbers perpendicular to this direction. In this way the simple Zeeman triplet shows the central component in one strip, the outer components next one, and so on.

Suppose now that a longitudinal field in the outer layers (Fig. 1), corresponding to a smaller optical depth, gives a certain rotation of the plane of polarization. The central component may then be rotated 90° by this longitudinal field, whereas the outer components in the Zeeman pattern will be more or less uninfluenced. When studying therefore the Zeeman splitting with the method of alternating strips we will see all three components in one strip and only a very weak central component or none in the perpendicular vibration.

The writer wants to draw attention to this possible effect. If the effect is found, it might give interesting possibilities in the study of external

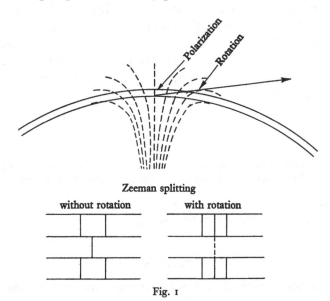

Fig. 1

fields of sunspots and the magnetic rotation of different lines, some of which cannot easily be excited in the laboratory.

When we observe instead of longitudinal Zeeman splitting, when sunspots are situated near the centre of the sun, it is evident that effects of magnetic rotation will have no influence on the measurements. If on the other hand magnetic double refraction is present it might influence the measurements in such cases when the two perpendicular fields in the line of sight have comparable strength. The primary components are otherwise too much displaced to be affected by this double refraction. Whether this effect may be of interest for the measurements of faint solar fields or not, is difficult to say without a thorough investigation.

It has been pointed out by me recently [2] that magnetic rotation may be found useful when studying the chromospheric structure of the sun in selected wave-lengths. In fact a solar monochromator based on the principle of magnetic rotation is very near at hand when considering the original observations of magnetic rotation made by Macaluso and Corbino.

When using a tube of length l between crossed polarizers in a magnetic field of strength $\mathbf{H}$ the intensity of the transmitted light of frequency ν is given by the following formula:

$$I_\nu = I_0 \sin^2 \chi_\nu \exp(-\kappa_\nu l), \qquad (1)$$

where I_0 is the intensity of frequency ν transmitted by the first polarizer, χ_ν the angle of rotation and κ_ν the absorption coefficient.

For the calculation of χ_ν and κ_ν reference is made to expressions given by W. Kuhn [3] and G. Stephenson [4]. As is well known χ_ν is proportional to H whereas both χ_ν and $\kappa_\nu l$ are proportional to the number of atoms of the absorbing vapour in the tube.

REFERENCES

[1] Alfvén, H. *Cosmical Electrodynamics* (Oxford University Press, 1950), p. 144.
[2] Öhman, Y. *The Observatory*, **79**, 89, 1956, and *Stockholms Observatoriums Annaler*, **19**, no. 4, 1956.
[3] Kuhn, W. *Math.-fys. Medd.* **7**, 12, 1926.
[4] Stephenson, G. *Proc. Phys. Soc.* A. **64**, 458, 1951.

Discussion

Dungey: Is there a method for detecting the component of the magnetic field at right-angle to the plane of your diagram?

Öhman: As I indicated, I do not think that the chances are too good to fit spots situated near the centre of the solar disk because this effect assumes that the light is plane polarized before it enters the gas where the rotation takes place. But there is an analogous effect which implies double refraction. The double refraction can affect the circular polarization. On the other hand in laboratory experiments the magnetic double refraction is not so clearly seen as the magnetic rotation. So I do not think this is a sensitive method of detecting fields.

THEORETICAL PROBLEMS OF
STELLAR MAGNETISM

LYMAN SPITZER, JR.

Princeton University Observatory, Princeton, N.J., U.S.A.

ABSTRACT

Relatively complete summaries of recent theoretical work in the field of stellar magnetism have been given by Cowling[1] and Elsasser[2, 3, 4]. For this reason the present survey does not aim at completeness, but is devoted instead to the following three outstanding problems in the field: (1) the formation of stars in the presence of an interstellar magnetic field; (2) the influence of convection currents on a stellar magnetic field; (3) the nature of magnetic variable stars. Study of these problems illustrates the type of difficulties encountered in theories of stellar magnetism.

I. STAR FORMATION IN A MAGNETIZED GAS

The observed polarization of starlight is probably caused by an interstellar magnetic field. The field strength, B, required to explain the observed polarization is about 10^{-5} gauss, if the Davis–Greenstein[5] theory of paramagnetic relaxation is accepted. However, a substantially lower field may be sufficient. Gorter suggested some time ago that ferromagnetic relaxation would be a much more dissipative influence than paramagnetic relaxation, provided we can assume, following Spitzer and Tukey[6], that an appreciable fraction of the grains are ferromagnetic. This suggestion has been analysed by J. Henry, who finds that a field of 10^{-7} gauss should suffice. However, B cannot be much less than 2×10^{-6} gauss if cosmic rays are assumed to be confined within the Galaxy by magnetic fields.

Another hypothesis which is now widely accepted is that stars are forming from the interstellar gas. Accretion, as compared' to direct condensation, may also be important, but in either case we may assume that the interstellar material is being brought together to form the young O, B and A stars of population type I.

It is therefore important to investigate the effects of a magnetic field on the contraction of an interstellar cloud. If the flux through the star is assumed constant, it is readily shown that the ratio of gravitational to magnetic forces remains constant during the contraction. If the condensation is large enough and dense enough to permit contraction to begin, magnetic forces will not prevent further contraction.

However, while condensations can develop, the minimum mass that can condense is large, greater than 10^3 suns for a field of 2×10^{-6} gauss or more. If the flux through the material remains constant there seems no way in which smaller stars can be formed from a big condensation, since the ratio of gravitational to magnetic pressure decreases as the mass decreases. In the case of the sun, for example, if the magnetic field were 2×10^{-6} gauss when the density of presolar matter was 10 atoms/cm^3, corresponding to a typical H I cloud, the present magnetic field at the center of the sun would be 10^{11} gauss, corresponding to a magnetic pressure about 10^3 times as great as the material pressure there. For the young stars of types O and B the discrepancy is almost as serious. If the stars are to form from interstellar clouds, either the initial field must be much less than 2×10^{-6} gauss, or there must be some way in which the magnetic flux through the material can be decreased.

Fortunately a powerful mechanism is available for reducing the flux through a contracting protostar. In a partially ionized gas the lines of force are attached to the electrons and positive ions, not to the neutral particles. The outwards force resulting from magnetic pressure is applied to the charged particles, and from these is transmitted to the neutral particles by frictional forces. These frictional forces can arise only from an outward drift of the electrons and ions relative to the neutral particles. The fewer electrons and ions present, the more rapid this outwards drift must be to transmit the same force. Since the relative ionization probably falls to a very low value in a protostar, owing to the extinction of ultraviolet radiation by the grains, this relative drift can be rapid. As a result, the lines of force, which have been stretched by the initial condensation of the cloud, pull their way out of the cloud, bringing the charged particles with them, and leaving a relatively low flux through the contracting cloud. Each line of force retains its identity in this process, but the magnetic energy decreases as the lines of force straighten, the energy being dissipated by the frictional force between the charged particles and the neutral atoms. The dissipation of energy occurs by the same process which Piddington[7] and Cowling (1955, informal communication) have proposed for the heating of solar flares.

A detailed analysis of this process, given by Mestel and Spitzer[8], will be summarized here. Let F be the outwards force on the electrons and positive ions in a cubic centimeter. Since F is produced by the gradient of the magnetic pressure, we have approximately, in c.g.s. units,

$$F = \frac{B^2}{8\pi R}, \tag{1}$$

where R is the linear dimension of the cloud. In a quasi-steady state, the force F will be balanced by the frictional force between positive ions and neutral atoms. If n_i and n_H are the particle densities of positive ions and neutral hydrogen atoms, respectively, and if v_D is the relative drift between them, the number of collisions/cm³ per sec will be $n_i n_H \overline{\sigma v}$, where v is the random relative velocity and σ is the collision cross-section. The momentum transferred at each collision will be approximately $m_H v_D$. Hence

$$F = n_i n_H \overline{\sigma v} \, m_H v_D. \tag{2}$$

Equating (1) and (2), we find that

$$v_D = \frac{B^2}{8\pi R n_i \, n_H \, \overline{\sigma v} \, m_H}. \tag{3}$$

Let us apply this result to a large cloud complex, some 30 pc in diameter, containing about 5×10^3 solar masses, that has contracted to a tenth of its original size, increasing n_H to 10^4/cm³, and B to 2×10^{-4} gauss. For the random mean velocity, v, we may take 10^5 cm/sec, about the value obtained from the virial theorem, and let σ equal 10^{-16} cm². We find

$$v_D = \frac{2 \cdot 2 \times 10^3}{n_i} \frac{\text{cm}}{\text{sec}}. \tag{4}$$

If the relative ionization were the same as in a normal H I cloud, with n_i/n_H equal to 5×10^{-4}, v_D would be about 4×10^2 cm/sec, much too slow to achieve any separation.

However, n_i/n_H must certainly be much reduced below its value in a normal interstellar cloud. The extinction through the cloud is increased by a factor of 100 in the contraction, and even in visible light will amount to some 20 magnitudes. Evidently n_i will be determined primarily by the rate of recombination. Radiative recombination is not very effective, reducing n_i by a factor of only 10^2 in 10^5 years. Recombination generally takes place by means of other, more rapid, processes, and in the present instance dissociative recombination and recombination by collision with grains should be more rapid than radiative recombination by a factor of at least 10^3. Thus n_i should be well below 10^{-4}/cm³, at the particular stage of contraction discussed above, and v_D should be at least 1 km/sec, even if B is reduced to 10^{-5} gauss.

Evidently the computed value of v_D is so large that the assumption of a quasi-steady state, made in deriving Eq. (3), becomes incorrect. For very small values of n_i the magnetic field nearly straightens itself out (or, if initial currents along the lines of force are present, at least nearly adjusts itself to a force-free condition) and the lines of force are then stationary as the gas contracts, the neutral atoms streaming by the electrons and positive ions; the small frictional force is offset by a slight, but roughly constant stretching of the lines of force.

We conclude that a magnetic field in interstellar space should not seriously interfere with the condensation of material to form new stars. One might expect some residual magnetic field to remain in a newly-born star, but it is difficult to predict the strength of such a field.

2. INFLUENCE OF FLUID MOTIONS ON MAGNETIC FIELDS

Internal motions are believed to be a characteristic of most stars—see the summary by Schwarzschild[9]. For stars later than type A a hydrogen convective zone is present at or below the surface, extending down in some stars by an appreciable fraction of the radius. For stars earlier than type F the central regions are assumed to be in convective equilibrium. There are probably very few stars which are in radiative equilibrium virtually throughout. In view of the tendency of lines of magnetic force to follow the material, fluid motions must have an intimate and powerful connexion with any magnetic properties of a star.

Much of the recent work on magnetic fields in moving fluids has been concerned with the dynamo theory of terrestrial magnetism. The objective of the research has been to show that fluid motions in a sphere can amplify an initial magnetic field and result in a finite external dipole field. The theoretical work, which has been summarized by Elsasser[3, 4] requires the assumption that the motions are not axially symmetric, since Cowling[10] has shown that a self-sustaining dynamo is impossible if the motions are axially symmetric. For reasons of mathematical simplicity these theories discuss the distortion of rather simple initial fields, and do not consider in detail the complicated magnetic effects produced by turbulence of all scales.

Actually, the interaction between turbulent motions and magnetic fields would seem to be a rather central problem of stellar magnetism. It is sometimes assumed that turbulence increases the effective resistivity of stellar material and accelerates the rate of decay of the external field.

More precise information on this point would be very helpful. Certain general theorems on this subject will be presented below. As we shall see, one may show that in an idealized case, with axial symmetry, an arbitrary field of fluid motions has only a relatively minor effect on the decay time.

We shall treat first the case of an infinite fluid cylinder, with the restriction that all quantities are independent of the axial co-ordinate, z, except for an imposed electrostatic potential which equals $-E_0 z$. This situation, which is mathematically simpler than the sphere, has been considered already by Sweet[11]. The fluid will be assumed incompressible and the resistivity, η, constant. The basic equation is Ohm's Law, which becomes

$$E_z + v_r\, B_\theta - v_\theta\, B_r = \eta j_z. \tag{5}$$

The r and θ components of Ohm's Law need not be considered. If we introduce the vector potential $\mathbf{A}$, only the component A_z appears in Eq. (5), and we have

$$\frac{\partial A_z}{\partial t} + v_r \frac{\partial A_z}{\partial r} + v_\theta \frac{\partial A_z}{r\partial \theta} = E_0 - \eta j_z. \tag{6}$$

We may express j_z in terms of A_z by the equation

$$j_z = -\frac{1}{4\pi}\, \nabla^2 A_z. \tag{7}$$

The quantities v_z and B_z need not be zero, but do not enter into the determination of A_z, B_r and B_θ. A line of force in the r, θ-plane is a line of constant A_z, and Eq. (6) describes the situation where arbitrary two-dimensional motions, in the r, θ-plane, bend, twist and distort a magnetic field which is also in the r, θ-plane.

Sweet [11] has demonstrated the important fact that when Eq. (6) is integrated over an area bounded by a line of force, the velocity terms cancel out. In a steady state $\partial A_z/\partial t$ vanishes, and we obtain

$$\overline{\eta j_z} = E_0, \tag{8}$$

where the average is taken over the surface enclosed by a line of force. On the other hand, by Stokes' theorem,

$$\frac{\eta}{4\pi} \oint \mathbf{B}.\,d\boldsymbol{\lambda} = \eta \int j_z\, dS = E_0 \int dS. \tag{9}$$

Since the lines of force are made very crinkly by small-scale motions, the ratio of the length of the line of force to the area enclosed will be much increased. Hence the mean value of B, without regard to direction, will be decreased by the motions. The mean value of B_θ will be decreased even further. From this result Sweet concludes that the effective resistivity of the fluid is much increased by the presence of fluid velocities.

While it is true that the internal field is decreased by the fluid motions, we shall show that the external field is not affected. The velocity terms in Eq. (6) also cancel out when this equation is integrated over a surface area bounded by a stream line of the flow; this is evident on integrating by parts, and utilizing the equation of continuity. In particular these velocity terms cancel out if Eq. (6) is integrated over the entire cross-sectional area of the cylinder, since the normal velocity v_r must vanish at the surface of the cylinder. Hence in the steady state, Eq. (8) is valid for the mean axial current in the cylinder, and if the total current is denoted by I, we have

$$I = \frac{\pi R^2 E_0}{\eta},$$

(10)

where R is the radius of the cylinder. Evidently the total current and the resultant external magnetic field are exactly the same as in the absence of fluid flow.

At first sight these two results, a weakened magnetic field inside and an unchanged field outside, seem contradictory. What actually happens is that in the bulk of the cylinder the magnetic field strength will be very low, and, by Eq. (7), j_z will also be very low. The total current I is concentrated in a thin layer close to the surface, where the effect of convection is much reduced. The ohmic dissipation ηj^2 is increased, but this additional dissipation is offset by the work done on the magnetic field by the fluid velocities. It is readily shown that the decay time of a cylindrical field of this sort is almost entirely unaffected by the presence of fluid motions of the type assumed here.

This same analysis may be applied to a sphere, provided we assume that all quantities are symmetric about the axis. If we introduce co-ordinates r, θ, and ϕ, the equation for the component A_ϕ of the magnetic vector potential is again found to be independent of v_ϕ, A_r and A_θ. Since there can obviously be no applied electrostatic potential in the ϕ direction, there is now no counterpart of the E_0 term in Eq. (6). The detailed analysis will be given elsewhere. Here we shall give simply the general result

$$\frac{\partial}{\partial t}\left\{\int A_\phi\, \rho r \sin\theta\, dV\right\} = -\int \eta\rho j_\phi\, r \sin\theta\, dV,$$

(11)

integrated over the entire volume, V, of the star; compressible flow has been assumed, and both the density, ρ, and the resistivity, η, must be assumed variable However, in a convective layer with γ equal to $5/3$, ρ varies as $T^{3/2}$; since η varies as $T^{-3/2}$ in an ionized gas, the product $\eta\rho$ is constant, and may be taken out of the integrand. We shall here consider the extreme case in which the star is in convective equilibrium throughout. The product $\eta\rho$ is now constant throughout the star, and the integral on the right-hand side of Eq. (11) may be expressed in terms of the external stellar dipole moment, $\mathcal{M}$. The dipole moment due to a current j_ϕ flowing across an area dS around a circle of radius a is $j_\phi \pi a^2 dS$. In the present instance, a is $r \sin\theta$, dS is $r\, dr\, d\theta$, and we obtain

$$\int j_\phi\, r \sin\theta\, dV = 2\mathcal{M}. \qquad (12)$$

The integral on the left-hand side of Eq. (11) may be expressed as a product of the dipole moment and a dimensionless parameter, ξ, which depends on the variation of A_ϕ throughout the star. If ρ and A_ϕ are both constant throughout, ξ is 3/4, while if A_ϕ is proportional to $1/r$, which it tends to be in the presence of strong axially symmetric convection, ξ equals unity. Evidently ξ does not vary enormously even for large changes in the spatial variation of A_ϕ. Hence it follows that $d\mathcal{M}/\mathcal{M}dt$, the relative rate of decrease of the dipole moment $\mathcal{M}$ is about the same for a completely convective star as for a star in which no fluid motion occurs. A convective layer should have an even smaller effect on the magnetic decay time.

It must be emphasized that this result is based on an idealized model, and does not necessarily apply to actual stars. The analysis is restricted to two-dimensional motions, which are not, of course, to be anticipated. A spherical bounding surface is assumed with $\eta\rho$ constant up to the boundary, and a vacuum outside; quite apart from irregularities at the surface (a point emphasized informally by Sweet), an actual star is surrounded by ionized gas, and the concept of an 'external' field must be used with caution.

However, it is not physically obvious why more general motions should lead to a faster decay rate, since they would not appear to bend the lines of force more drastically than the restricted motions assumed here. In the lack of a more general approach, there seems to be no very strong reason at the moment to doubt that the dipole fields of the sun and the stars could well be fossil, i.e. residual magnetic fields present originally in the material from which the star condensed.

3. THE NATURE OF MAGNETIC VARIABLE STARS

The theoretical explanation of magnetic variables is certainly one of the most exciting and challenging problems in astrophysics. In addition to the large and unexpected variation of magnetic field, the great apparent variation of chemical composition with phase in the outstanding spectrum variables is not readily explained.

The first task of theory is, of course, to explain the observed variation of the magnetic field; and, in particular, the conspicuous reversal of this field in a number of spectrum variables. We believe that during the short time of one period—only a few days—the lines of force are frozen in to the fluid. Hence, if the same material is responsible for the absorption of the spectrum lines throughout the oscillation, this material must change its orientation by a large angle to account for the reversal of the magnetic field. No one has yet succeeded in thinking up any specific mechanism which would produce such a large change in orientation. A non-rotating magnetically pulsating star, of the type first analyzed by Schwarzschild [12], might produce small fluctuations in a large magnetic field. Such a configuration can account, perhaps, for some of the irregular magnetic variables—Babcock's Group I [13]—but can apparently not produce the field reversals which are observed in about half of the magnetic variable stars.

We are thus forced to assume that at different phases of the outstanding magnetic variable stars we are looking at different regions, an assumption to which we would probably be forced in any case by the corresponding changes of chemical composition with phase. The simplest way in which a star can present different regions to the earth at different times is rotation, a mechanism which has been proposed for spectrum variables by A. Deutsch [14]. This is not the only way, however. One can imagine a non-rotating star whose surface is chiefly covered by two regions which have opposite magnetic polarity, and which oscillate in surface brightness. If the brightness oscillations are 180° out of phase in the two regions, effects somewhat similar to those observed might be produced. Non-radial oscillations with large amplitudes might conceivably produce such effects; however, such a model apparently gives the wrong sign for the 'cross-over effect' observed by Babcock [15] in the line profiles. In the absence of supporting data this model will not be considered further.

Rotation of a star with a magnetic axis at an angle to the rotational axis is certainly the simplest explanation of magnetic variables, and should be explored first before more complicated models are involved. According to Deutsch [14] such a theory provides a quantitative explanation of the line profiles in terms of stellar rotation. In addition, rotation of an oblique rotator provides a very natural explanation of the 'cross-over effect'.

On the other hand, uniform rotation of some constant configuration is definitely not consistent with the observational data. In the magnetic stars which reverse polarity irregularly—Babcock's Group 2[13]—no trace of a regular period has yet been found. Even in the spectrum variables, whose variations are, on the whole, periodic, the magnetic field changes are not strictly periodic. In α^2CVn, for example, the changes of magnetic field strength observed by Babcock and Burd[16] are not sufficiently regular to be accounted for by solid body rotation. Also, the variations in radial velocity seem to be non-harmonic.

The peculiar abundances observed in all magnetic variables, and the particular changes observed in spectrum variables, pose serious problems. It is tempting to assume that such abundance anomalies are produced by nuclear reactions at the surface. Intense hydromagnetic disturbances might be expected to accelerate ions to very high energies, and the magnetic field may well inhibit convection, thus preventing the mixing of the peculiar surface material with the normal stellar matter deeper in the star. As pointed out by Babcock[13] this hypothesis provides another argument against uniform rotation of an unchanging configuration, since solid-body rotation of a magnetized fluid would not be expected to accelerate many charged particles.

It has not been pointed out that intense hydromagnetic oscillations may well be inevitable in a so-called oblique rotator. Let us consider a star in which the axis of the dipole field makes an angle with the angular momentum of the star. The magnetic forces will have an effect on the density distribution of the star. In the absence of a magnetic field the surfaces of equal density are oblate spheroids, symmetrical about the axis of rotation. A magnetic force will tend to distort these spheroids, and the star will no longer be symmetrical about its axis of rotation. As a result, the star will not be rotating about a principal axis of inertia.

A solid body, rotating about an axis that is not one of the principal axes of inertia, undergoes a wobbling motion, in which the instantaneous axis of rotation moves about, while the total angular momentum remains constant both in direction and magnitude. This type of motion is accompanied by stresses in the solid. In a magnetized fluid these stresses will give rise to distortions which vary in time, resulting, presumably, in hydromagnetic waves and magnetic oscillations. In the body of the star, where the magnetic forces are relatively small, such oscillations should be almost inappreciable, but near the surface, where the magnetic energy becomes comparable to the material energy density, large hydromagnetic oscillations may be possible. Such oscillations will dissipate energy, of course, and may be expected in time to bring the magnetic axis into coincidence with the axis of rotation.

We are thus led to the following tentative picture of an idealized magnetic variable star. In a young star, newly formed from interstellar clouds, a fossil magnetic field will, in general, be present. The angle between the mean dipole field and the angular momentum will be arbitrary. In objects where the angle is appreciable, and where the rate of rotation is moderate, periodic changes of the magnetic field will be observed. Since the axis of rotation usually will not coincide with a principal axis of inertia, hydromagnetic oscillations will be superposed on the variations due to rotation. These oscillations may accelerate charged particles in the surface layers, and the transmutations produced by these particles may, perhaps, produce the differences in composition observed at the two magnetic poles. As the star grows older, the magnetic axis is brought into coincidence with the angular momentum vector, the magnetic variability and the hydromagnetic pulsations die away, fluid motions mix up the stellar material, eliminating the composition differences, and the star becomes more normal.

This working hypothesis can certainly not be regarded as established. It is presented to help stimulate further observations and additional theoretical work. Among the many questions raised by this hypothesis are the following. How can one account for Babcock's variable of Group 2, in which the polarity reverses irregularly? How can nuclear reactions produce different effects at the two magnetic poles? Why do differences in chemical composition occur in some magnetic variables, but are lacking in others, despite apparent similarities both in effective temperature and type of magnetic variation?

Certainly the answers to these questions will require modifications and changes in the simple hypothesis described above. The example of the solar atmosphere indicates that magnetic variations, both slow and rapid, may arise even when the axis of rotation agrees with the apparent dipole axis. It is evident that magnetic variable stars provide an exciting new field of astrophysics, whose detailed exploration and study has only just begun.

REFERENCES

[1] Cowling, T. G. Mon. Not. R. Astr. Soc. **113**, 371, 1953.
[2] Elsasser, W. M. Amer. J. Phys. **23**, 590, 1955.
[3] Elsasser, W. M. Amer. J. Phys. **24**, 85, 1956.
[4] Elsasser, W. M. Rev. Mod. Phys. **28**, 135, 1956.
[5] Davis, L. and Greenstein, J. Astrophys. J. **114**, 206, 1951.
[6] Spitzer, L. and Tukey, J. Astophys. J. **114**, 187, 1951.
[7] Piddington, J. H. p. 141 of this volume.

[8] Mestel, L. and Spitzer, L. *Mon. Not. R. Astr. Soc.* **116**, 503, 1956.
[9] Schwarzschild, M. *Structure and Evolution of the Stars* (Princeton University Press, 1958).
[10] Cowling, T. G. *Mon. Not. R. Astr. Soc.* **94**, 39, 1933.
[11] Sweet, P. A. *Mon. Not. R. Astr. Soc.* **110**, 69, 1950.
[12] Schwarzschild, M. *Ann. Astrophys.* **12**, 148, 1949.
[13] Babcock, H. W. p, 161 of this volume.
[14] Deutsch, A. J. *I.A.U. Transactions*, vol. 8 (Cambridge University Press, 1954), p. 801.
[15] Babcock, H. W. *Astrophys. J.* **114**, 1, 1951.
[16] Babcock, H. W. and Burd, S. *Astrophys. J.* **116**, 8, 1952.

Discussion

Burbidge: Do you believe, in your picture of the formation of a protostar in the gas containing a magnetic field, that the magnetic field can help you to get rid of the excess angular momentum?

Spitzer: Yes, I believe that the field may remove angular momentum.

Burbidge: I have no idea how nuclear reactions could produce different abundance anomalies on different parts of the stellar surface. On the other hand, do you know of any mechanism which could separate elements on the surface after they are produced?

Spitzer: No.

Gold: The picture of the contracting protostar may be severely complicated by rotation. If there is a relative rotation within the mass of the star, perhaps enhanced by the initial contraction, then a twisting of the magnetic field may happen. The result of this is that the ionized gas on the lines of force may not be subjected to an outward force, as was supposed, but even to an inward one due to the spiralling of the lines which are wound around the region considered.

Spitzer: The angular velocity of rotation of the galaxy corresponds to a period of about 10^8 years. The galaxy has to be contracted a lot more before it begins to twist up the field lines; I should think the contraction would take another 10^5 years.

Gold: But in galaxies you have differential velocities of the order of 8 km/sec and there will still be a possibility of twisting the field lines.

Spitzer: Did you state that the pressure will decrease?

Gold: Yes, if there is a twisting. I certainly agree that the main picture which you give has validity, but it may not be simple to deduce what the pressure is in case of a twisted magnetic field.

Spitzer: If the pressure actually decreases I think that the velocity would increase, but this cannot be true.

Gold: Well, in your case the force is directed outwards, but I do not think that your argument applies to the situation I am referring to.

Finally, I should like to point out that the problem of magnetic variable stars has some connexion with the investigation presented by Dr Davis the other day, i.e. the problem of the external flux from a simply connected body.

Schatzman: Is it possible to have orientation of interstellar particles by diamagnetic relaxation in the special case of graphite flakes? Cayrel and myself have shown that, due to their large optical and physical anisotropy, graphite flakes could well be the polarizing agent of interstellar space.

Spitzer: I do not think I can answer your question.

Schatzman: A difficulty for the formation of stars from protostars is the larger mass of the protostar if the protostar is to condense under its own weight. There is a way out of this difficulty. If the protostar is embedded in an H II region, the protostar can start contracting with a much smaller mass. In this connexion I would like to refer to the work of Ebert (Liège meeting, 1954) and to the beautiful photographs taken by Minkowsky which were shown at Cambridge in 1953. On these photographs, you can see several globules of small size, surrounded by the H II region. If the H II region pushes on the protostar, it can help it to contract considerably.

As you have brought here the explanation of many facts, would you not have also the hydromagnetic explanation of flare stars?

Spitzer: No, I do not think so.

Bostick: It is to be expected from the theoretical considerations by Kruskal and Schwarzschild that Taylor instability will set in when an ionized gas is being supported against a gravitational field by a magnetic field. This process is seen clearly in our experiments where plasmoids (essentially large-amplitude jets) projected almost radially inward across a magnetic field, spiral together to form a ring in the center. This process is obviously one where ionized matter can get from the outside to the inside across the magnetic field without carrying all the field lines along, and without having to wait for the plasma to diffuse across the lines. It should be pointed out that although in the laboratory experiments the central gravitational field is supplanted by inertial fields produced by the deceleration of the plasmoids, the resultant plasma ring has a certain amount of stability and furthermore has angular momentum and very probably a magnetic moment. Although as yet, no scaling factors from laboratory conditions to cosmical conditions have been given thorough consideration, it is nevertheless worth while to consider the effect of this Taylor instability process in star-production in regions where H II represents any appreciable fraction of the H I present.

Cowling: May I ask if the discussion of the second problem depended essentially on the assumption of incompressibility? I ask this in view of Walén's suggestion that as a consequence of compressibility, convection in the sun's surface layers might actually carry lines of force down below the solar surface.

Spitzer: The results for the sphere were derived for the compressible case.

Cowling: Is your second problem essentially a two-dimensional one?

Spitzer: Yes.

Mestel: The process discussed by Dr Spitzer and myself, by which the bulk of the contracting protostar can slip across the lines of force, is closely related to the 'ambipolar diffusion' of Dr Schlüter. The importance of this has recently been emphasized by Dr Piddington and Professor Cowling. We differ in that the drift of plasma through the neutral gas is not restricted to be small compared with the motion of the gas as a whole.

There is one aspect of the process which should be underlined. What is being destroyed by collisions between plasma and neutral gas is magnetic *energy*, not magnetic *flux*. If the ohmic field and the partial pressures are ignored, the magnetic field is frozen into the *plasma*; the motion of the system 'field plus plasma' is determined by the balance between the magnetic force and the friction between neutral gas and plasma. All the theorems of ordinary hydro-magnetics which follow from the identical motion of matter and magnetic field may be taken over by substituting the velocity of the plasma for the velocity of the gas as a whole. Thus, if we wish to destroy magnetic flux, we must employ ohmic dissipation to bring about relative motion of field and plasma.

PAPER 21

ON THE MAGNETIC FIELDS IN INTERSTELLAR SPACE AND IN NEBULAE*

G. A. SHAJN

Crimean Astrophysical Observatory, Crimea, U.S.S.R.

ABSTRACT

Some aspects of the problem of magnetic fields in the interstellar space and in the nebulae are discussed in this paper. Our observational basis are the numerous photographs of the nebulae in Hα and other rays, taken with high-speed 450 and 640 mm cameras.

The greatly elongated shape of many emission nebulae is interpreted as a result of three factors, the effect of the magnetic field, the macroscopic motions in these nebulae (including the tendency to expand) and the high electrical conductivity of matter. The expansion of nebulae is to be generally accompanied by the considerable decrease in brightness (roughly about d^{-5}), but if the expansion is only in one direction, owing to the presence of a regular magnetic field, the brightness decreases much more slowly (roughly about d^{-2}) and the nebulae remain visible for a longer time. The filamentary structure, which is very often inherent to the elongated nebulae, is, probably, an additional factor of a longer visibility of the nebulae under consideration. The great lengthening of nebulae may be reached in the period of the order of 10^6 years.

In the dark nebulae the matter seems to be electrically conductive and the macroscopic motions of the order of 1–2 km/sec are present there, so that there are reasons to suggest that the magnetic field is also responsible for the elongated shape of many dark nebulae. One may prove that the observed elongated shape of dark and emission nebulae could not be caused by differential galactic rotation.

Some correspondence between the apparent orientation of many elongated dark nebulae in Persei-Tauri, Cygni, Sagittar, Oph.a.o. on the one hand, and that of the local magnetic fields in the same regions on the other hand, as derived from the polarization observations, may be considered as a serious argument in favour of that the interstellar magnetic field is an important factor responsible for the elongated shape of nebulae.

An attempt to interpret the elongated emission nebulae as single arcs, or bars—the remains of gigantic peripheral nebulae—leads to the conclusion that this hardly holds true, at least with respect to several most elongated ones. However, when accounting both the frequent occurrence of the peripheral nebulae and the effect of magnetic fields, we get more opportunities to interpret the observed diversity of configurations of the nebulae, particularly the elongated nebulae.

* Presented by Dr A. B. Severny.

The structure of very elongated dark nebulae in Persei-Tauri, consisting of a number of mutually connected slightly elongated globules, seems to support the hypothesis of the magnetic field.

The high occurrence of the elongated dark nebulae oriented in addition predominantly at small and moderate angles with respect to the galactic equator, is interpreted as a result of a high efficiency of the controlling action of the interstellar magnetic field on the motion of matter in nebulae. We have to see in this very high occurrence of such nebulae also an argument in favour of the presence there of sensible macroscopic motions. The action of the magnetic field seems to be very efficient also with respect of emission nebulae, notwithstanding that the number of the observed elongated nebulae is not great.

It seems to be probable that in the emission nebulae there is an inner magnetic field of the order of 10^{-5} gauss, the variations being large within the observed nebulae, as well as within a single nebulae. Some correspondence between the structural features of the nebulae IC 1396 and Coal Sack and the orientation of polarization seems to be another fact in favour of the presence of an inner magnetic field. Practically, all bright emission nebulae, larger than $5'$ have filamentary characteristics and it is suggested in connexion with this that some nearly always operating factor, probably an inner magnetic field, is responsible for this widely spread phenomenon. The inner magnetic field does not seem generally to be a very important factor in the problem of stability of nebulae, although in some cases it may be very efficient in this relation in the more dense central and other regions.

The well-known strange peculiarity of planetary nebulae—the weakening of the brightness along the major axis, or at its ends, may be interpreted in the terms of the hypothesis of a magnetic field. There is some analogy in this with the uneven and systematic distribution of matter in some large peripheral nebulae, particularly in some shells, probably the remains of the very old supernovae or novae. One may suggest that in both cases the magnetic field of the central star, which had some influence on the primary conditions of the motion of the shells caused to some extent the observed peculiar distribution of matter and probably the magnetic field in the observed shells of the super-novae and novae in question, as well as in the planetary nebulae. The filamentary structure as revealed in the shells of former and sometimes in planetary nebulae is probably connected with the magnetic field.

The filamentary structure seems to be inherent not only to the majority of emission nebulae but also to a number of dust nebulae, both the reflexion and dark ones. The occurrence of the filamentary structure in the emission and dust nebulae consisting both of matter of high electrical conductivity and having at the same time such different temperature, $\sim 10,000°$ and ~ 100 °K, respectively, suggests that the structure in question is conditioned by the very nature of magneto-hydrodynamics.

The high occurrence of the elongated nebulae in the Galaxy and in the galaxies in general, oriented predominantly probably along the spiral arms, is to be considered as a fundamental characteristic of the spiral galaxies. One may suggest that the magnetic field is efficient enough to control to some extent the motion and distribution of diffuse matter in the Galaxy, hindering the dissipation of matter and contributing to the maintenance of the spiral structure of galaxies with respect to the gas-dust population as well as probably to some classes of stars. The diffuse matter dissipated in such a manner along the spiral arms may be responsible, at least partly, for the almost continuous emission background as revealed in the low latitudes in the last years.

While below 10° (by latitude) the elongated dark nebulae have predominantly small and moderate inclination to the galactic equator, there is no such tendency for small angles in the elongated nebulae within latitudes of 10–21°. When confronting this with small and very great dispersion in the angles θ' for the plane of vibration within latitudes from 0 to 10° and from 10 to 25°, respectively (as revealed from the polarization observations), a conclusion is made that the elongated shape of dark nebulae below 10° is conditioned by the known regular magnetic field, while within the latitudes 10–21° by the local fields.

I

Some points of the problem of magnetic fields in the interstellar space and in the bright and dark nebulae are discussed in this paper. Our observational basis are the numerous photographs of the nebulae in Hα and in other rays taken with high-speed 640 and 450 cameras of the Simeis observatory, in Crimea. Remarkable formations, systems of filaments, striae and strips were revealed on our photographs in 1950 and later. A striking example of such a system, as long as 11°, is met in Cygni. There are also other very elongated emission nebulae near ξ Persei, 10 Lacertae, η Canis Maj, α Orion and others. In 1952 we suggested that this phenomenon has a relation to the magnetic field and later on we obtained a number of independent arguments in favour of this supposition [1-7]. The strongly elongated shape of many emission on nebulae is interpreted as a result of three factors, the effect of the magnetic field, the macroscopic motions in these nebulae and the high electrical conductivity of matter. In fact, the study of the larger and brighter emission nebulae leads us to the conclusion that the preferential outward motion of matter seems to be an outstanding feature of some nebulae [8]. There are also very serious reasons to adopt from the theoretical point of view that the emission nebulae are expanding with sound velocity of 10–15 km/sec (L. Spitzer [9] was the first to suggest it). The phenomenon in question is very complicated, but there must be some tendency for expansion and dissipation in emission nebulae [10, 11]. Roughly speaking, the effect of the interstellar magnetic field is to restrain the motion of the nebular matter in the direction perpendicular to the magnetic lines of force and to let the matter move freely along these lines. The general expansion of nebulae must be accompanied by a decrease in brightness, approximately as D^{-5}, but if the expansion goes only in one direction (owing to the presence of a regular magnetic field), the brightness decreases much more slowly, roughly as D^{-2}, and the nebula must remain visible for a longer time. The filamentary structure, which is very often inherent to the elongated nebulae, is probably an additional factor of the longer visibility of the considered nebulae. Very elongated nebulae consist generally of a number of moderately elongated nebulae and filaments, giving an impression of a single very elongated nebula. The great length of a number of emission nebulae as revealed by observations may be reached probably in the period of the order of 10^6 years.

The matter of the dark nebulae seems also to be electrically conductive. Besides, the macroscopic motions of the order of 1–2 km/sec are present there, so that there are reasons to suggest that the magnetic field is also responsible for the shape of the numerous elongated nebulae.

One may prove that the observed elongated shape of dark and emission nebulae, at least in the majority of cases, could not be caused by the differential galactic rotation.

An attempt to interpret the elongated emission nebulae as single arcs, or bars—remnants of gigantic peripherical nebulae, leads to the conclusion that this hardly holds true, at least with respect to several most elongated ones. However, when taking into account both the frequent occurrence of the peripherical nebulae and the effect of magnetic fields, we get more opportunities to interpret the observed variety in the configurations of nebulae, particularly of the elongated ones [5].

2

There is a more direct argument in favour of the hypothesis of magnetic field as an important factor controlling, at least partly, the distribution and the motion of the diffuse matter [2, 3]. In fact, the hypothesis under consideration may be checked in some cases by comparing the orientation of the elongated nebulae with the results of observations of the polarization of the light of stars in the same region. Especially suitable for this purpose is the large region of ξ and ζ Persei: high latitude, polarization caused by absorption (probably in one cloud), the presence there of a number of remarkable elongated dark nebulae and of the emission nebula NGC 1499. In the region under consideration of nearly $30 \times 30°$ (Fig. 1) the contours of large dark nebulae $> 2°$ are drawn according to Becvar's Atlas, as well as all the known polarization data were taken according to Hiltner [12] and Hall and Mikesell [13]. While in low latitudes the plane of vibration nearly coincides with the direction of the galactic equator, in the higher latitudes ($b > 12°$) is brought out a local magnetic field, diverging largely from the magnetic field along the galactic equator. It seems to be remarkable that the angle under consideration $\theta' - 90° = -41 \pm 3°$ is nearly the same, as the mean angle of inclination to the galactic equator ($-35 \pm 5°$) for the six very elongated dark nebulae in this region. By the way, some of these elongated dark nebulae consist of a number of mutually connected slightly elongated globules, and it seems that such a very strange structure is easier to understand in the light of the magnetic field hypothesis.

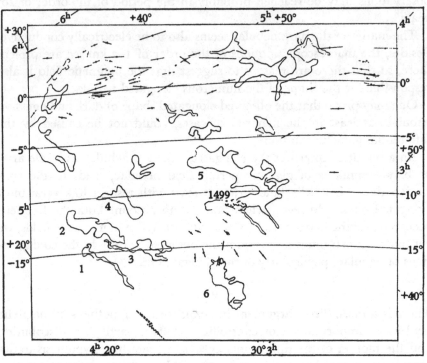

Fig. 1

One may add that the filaments of the emission nebula NGC 1499 in this region are slightly diverging from the galactic parallel also in the same direction. At last, attention has to be attracted to the fact that the very thin and long single bright filaments in the Pleiades, located on the SO border of Fig. 1, also oriented nearly in the same direction as the local magnetic field mentioned above. In connexion with this one may remember that Hall found recently that the final direction of short curved filaments (as though emanating from Merope) nearly coincide with the direction of magnetic field in this region [13].

186

Not entering here into details one may say that in higher latitudes ($b > 12°$) of this large region we have a complicated local magnetic field, which is responsible, on one hand, for the peculiar direction of the plane of vibration and, on the other hand, for the elongated shape of the nebulae.

The region in Sagittarius with its centre about 18^h 30^m–$13°$ is another illustration of the above idea (Plate I). In connexion with that the line of sight is oriented in Sagittarius approximately along the inner spiral arm, we have here as it is to be expected, a nearly occasional distribution of the plane of vibration. However, besides the enormous mass of dark matter along the galactic equator, there are outstanding dark nebulae, stretching from 18^h 30^m–$10°$ towards SO and O at an angle of inclination $\geqslant 60°$ to the galactic equator. The general picture of good agreement between the great inclination of these elongated nebulae and the direction of the plane of vibration here (vectors P, θ' being parallel and the polarization P very considerable) shows that in the region under consideration to the east from the galactic equator there is a local magnetic field oriented at an angle $\geqslant 60°$ to the regular galactic magnetic field, in harmony with the outstanding feature of the structure of the Galaxy in the region of Sagittarius [5].

The correspondence between the apparent orientations of the local magnetic fields, as derived from polarization observations, is also found in Oph, Mon, in two large regions of Cyg and others. For instance, we have a satisfactory agreement between the orientation of the well-known very elongated dark nebulae to the east from ρ Oph and the plane of vibration of the measured stars in this region. At last, we shall mention here one of the large regions in Cygni. As it is known there is in Cygni a lot of elongated filamentary emission nebulae with moderate and small inclination to the galactic equator. But further to the south-west, in the region around 20^h $5^m + 35°$ a considerable group of filamentary emission nebulae of similar structure, oriented nearly perpendicular to the galactic equator, were brought out. One may therefore suggest here a local field oriented in the same peculiar direction, which is observed in reality [3].

Additional examples of the correspondence under consideration might be given. When in many cases the direction of magnetic fields (as derived from the polarization observations) and the direction of the elongated nebulae differ but little from one another, and, on the other hand, both show a large inclination to the galactic equator, the conclusion seems to be almost inevitable that, namely, the interstellar magnetic field is responsible for the elongated shape of these nebulae. At the same time it is easy to understand that the correspondence under consideration may hold only under certain conditions.

3

In the low latitudes the general trend of the plane of vibration to be parallel to the galactic plane is expressed well enough (excluding the directions where the line of sight is oriented approximately along the spiral arms as Cygnus and Sagittarius, where naturally nearly occasional distribution of the position angles θ' is observed). However, this seems to be but the first approximation. In fact, appreciable systematic deviations from the trend in question are revealed for a number of regions in low latitudes. The observed data by Hiltner[12], Hall and Mikesell[13] and Smith[14] for all measured stars with $P > 0.3$ within $\pm 8°$ in latitude are plotted in Figs. 2 and 3. The deviations seem to be mostly real, since some of these regions are large and the number of measured stars in them is considerable. Such regions we have, for instance, in Monocer and Cygnus. There are also smaller areas, for which the plane of vibration differs appreciably from the direction of the galactic parallel. It is remarkable that, at least in a number of cases, such as in Cyg, Mon, Sgr and others, a correspondence between this peculiar direction of the plane of vibration and of the elongated nebulae in these regions was noticed. Therefore, there are very serious reasons to adopt that in the low latitudes, besides the general regular galactic magnetic field, there exist also local magnetic fields. This conclusion seems to be an essential feature of the interstellar magnetic field. One may suggest that a magnetic field of such character is conditioned by the complicated spiral structure of the Galaxy[6].

As it is known, the position angles θ' for the light of stars in latitudes $> 8°$ are distributed nearly at random. This may be interpreted as a serious indication of the local character of the interstellar magnetic field in the higher latitudes. Such fields, embracing sometimes tens or hundreds of square degrees, seem to be more or less uniform in direction as well as in strength. For instance, in the region ξ Persei, the plane of vibration makes an angle about $41°$ with the galactic parallel. If our representation of the cause of the elongated shape of nebulae is correct, one must expect the absence of the tendency of these nebulae to have small or moderate inclinations to the galactic equator, the tendency which does hold in the low galactic latitudes. In fact, the discussion of all available data shows that while below $10°$ (by latitude) the elongated dark nebulae have predominantly small and moderate inclinations to the galactic equator, no such tendency is observed for small angles in the elongated nebulae within latitudes from $10°$ to $21°$. When comparing this with small and very great dispersion of angles θ' for the plane of vibration below and above $8°$ of

Plate I

(*facing p.* 188)

galactic latitude respectively (as revealed from polarization observations) a conclusion is drawn that the elongated shape of dark nebulae below 10° is conditioned by the known regular galactic magnetic field and also partly by the local fields mentioned above, while in the higher latitudes ($b > 8°$) almost exclusively by the local fields[6].

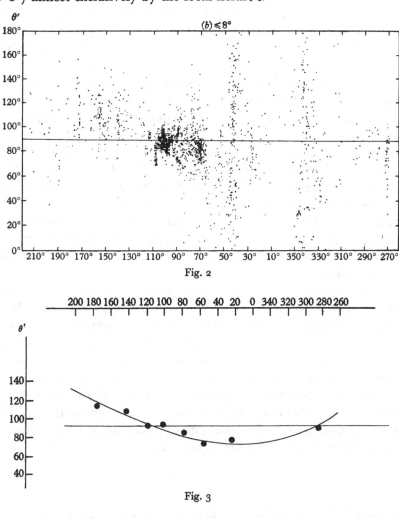

Fig. 2

Fig. 3

4

There is no doubt that there must exist in the nebulae an inner magnetic field. Particularly, this follows from the fact that the turbulence is the observed phenomenon in some nebulae, for instance, in the Orion nebula. When applying to the Orion nebula the relation $\frac{1}{2}\rho v^2 = H^2/8\pi$ one derives for the densest central part the strength H of the order of 10^{-4} gauss. It is probably smaller or much smaller in other nebulae. Generally it seems that the variations are large within the observed nebulae, as well as within a single nebula. Some correspondence between the structural feature of the nebulae IC 1396 and the Coal Sack and the orientation of the plane of vibration of the light of stars in these regions seems to be another observed fact in favour of the presence of an inner magnetic field [3, 14].

Attention must be paid to the fact that practically all brighter nebulae, larger than 5′, are of more or less filamentary character. It seems probable that some widespread and nearly always operating factor must be responsible for this phenomenon. It is suggested that the main cause of universal occurrence of filaments in nebulae is the inner magnetic field. Filamentary structure seems to be inherent not only to the majority of emission nebulae, but also to a number of dust nebulae, both the reflexion and the dark ones (where the filamentary structure may be discerned evidently only in rare cases). Occurrence of the filamentary structure in the emission and dust nebulae, consisting both of matter of high electrical conductivity and having at the same time such different temperatures, $\sim 10{,}000°$ and $\sim 100°$ K respectively, leads to the suggestion that the structure in question is questioned by the very nature of magneto-hydrodynamics.

A combination of such factors as the inner and outer magnetic fields, preferential outward motion, the turbulent motion and some hydrodynamical phenomena, connected with the encounters of single masses of gas and dust, may lead to a formation of such details, as the filaments, striae, rims, arcs and loops (especially in the outer parts of nebulae). An intimate connexion between the direction of the magnetic field and the direction of the filaments is suggested in the Crab nebulae [15].

Generally, the inner magnetic field does not ensure probably the stability of nebulae, although in some cases it may be very efficient (for instance, in the central dense part of the Orion nebula).

5

One may suggest apparent manifestation of the magnetic field in the emission nebulae, which is probably connected with the ancient novae or super-novae. Though there are large differences in the structure of these nebulae, it is quite evident that in addition to the ejection of matter and the turbulence, some other factor played there an important role. Such characteristics, as the systematic distribution and orientation of many details, sometimes the great curvature of the filaments and even the presence of closed loops, lead to the suggestion that this factor is the magnetic field. In the light of this idea we have discussed the filamentary nebula IC 443 (a radio source) [16], as well as two other filamentary nebulae S 147 and NGC 6960—NGC 6992 probably connected with ancient super-novae, or novae [3]. It seems to be probable that the magnetic field in question was carried out together with the matter ejected from supernovae and novae. One may suggest that the magnetic field of novae and super-novae influenced the primary conditions of the ejection of matter, and, on the other hand, caused to some extent the observed distribution of matter and of the magnetic line of force in the far environs of the stars (see also the paper by E. Mustel [17]). Though we deal here mainly with the inner magnetic field, it seems at the same time that the distribution of matter, as observed now in these nebulae, was influenced also to some degree by the controlling action of the interstellar magnetic field.

6

The well-known strange peculiarity of many planetary nebulae, namely, the weakening of the brightness along the major axis or at its ends, was not interpreted reasonably. A good example of a nebula belonging to such class is NGC 6818 (according to Curtis). There is some analogy in such structure with the uneven and systematic distribution of matter in some large peripherical emission nebulae, namely of some shells particularly, probably the mentioned above remnants of the ancient super-novae and novae. One may suggest that the peculiarity under consideration in the planetary nebulae may be explained in terms of the magnetic field hypothesis: in both cases the magnetic field of the central star influencing somewhat the primary conditions of the shell motions, caused to some extent the observed peculiar distribution of matter and probably the magnetic field in the observed shells of nebulae, connected with super-novae and novae in question, as well as in planetary nebulae [5]. The filamentary structure as revealed in the shells, of the former and sometimes in planetary nebulae, is probably connected with the magnetic field (section 4 above).

7

The high occurrence of elongated dark nebulae (oriented predominantly at small and moderate angles with respect to the galactic equator) is interpreted as a result of high efficiency of the controlling action of the interstellar magnetic field on the motion of matter in nebulae. At the same time we have to see in it also an argument in favour of the presence in the dark nebulae of appreciable macroscopic motions. The action of the magnetic field seems to be very efficient also in respect to emission nebulae, notwithstanding that the number of the observed elongated nebulae is, relatively, not great. The conclusion arrived at above must relate also to the extra-galactic spiral nebulae, where the greater and smaller bulks of dark matter often look, indeed, like elongated nebulae, predominantly oriented along the spiral arms. An efficient control of the motion of diffuse matter by the magnetic field means that the latter hinders to some extent the dissipation of matter and contributes in maintaining the spiral structures of galaxies with respect to the gas-dust population, as well as probably to some classes of stars.

REFERENCES

[1] Shajn, G. A. *A.J. U.S.S.R.* **32**, no. 2. 110, 1955.
[2] Shajn, G. A. *A.J. U.S.S.R.* **32**, no. 5, 838, 1955.
[3] Shajn, G. A. *A.J. U.S.S.R.* **32**, no. 6, 489, 1955.
[4] Shajn, G. A. *A.J. U.S.S.R.* **33**, no. 2, 210, 1956.
[5] Shajn, G. A. *A.J. U.S.S.R.* **33**, no. 3, 1956.
[6] Shajn, G. A. *A.J. U.S.S.R.* **33**, no. 4, 1956.
[7] Shajn, G. A. and Hase, V. *C.R. Acad. Sci. U.R.S.S.* **82**, 857, 1952.
[8] Shajn, G. A. and Hase, V. *Publ. Crim. Astroph. Obs.* **8**, 80, 1952.
[9] Spitzer, L. *Harv. Obs. Monogr.* **7**, 87, 1948.
[10] Kahn, F. *B.A.N.* **12**, no. 456, 1954.
[11] Oort, J. H. *B.A.N.* **12**, no. 455, 1954.
[12] Hiltner, W. A. *Astrophys. J.* **109**, 478, 1949; **114**, 241, 1951; **120**, 454, 1954.
[13] Hall, J. S. and Mikesell, A. H. *Publ. U.S. Nav. Obs.* **17**, pt. 1, 1950; *Les particules solides dans les astres* (1955), p. 543.
[14] Smith, E. V. P. *Sky and Telescope*, **14**, 8, 324, 1955.
[15] Shajn, G. A., Pikelner, S. and Ikhsanov, R. *A.J. U.S.S.R.* **32**, no. 5, 395, 1955.
[16] Shajn, G. A. and Hase, V. *C.R. Acad. Sci. U.R.S.S.*, **96**, 713, 1954.
[17] Mustel, E. *A.J. U.S.S.R.* **33**, no. 2, 182, 1956.

ON THE MAGNETIC FIELDS OF NOVAE AND SUPER-NOVAE

E. R. MUSTEL

Crimean Astrophysical Observatory, Crimea, U.S.S.R.

ABSTRACT

This paper is a development of the hypothesis, suggested earlier by the author, namely that novae possess a large general magnetic field. This hypothesis explains the following facts: (a) effects of the retardation of matter in novae before light maximum, (b) a preferential ejection of matter from novae in two diametrically opposite directions during an outburst, and (c) formation of rings and equatorial belts after light maximum, observed in N Aql 1918 and N Her 1934. These rings and belts are formed due to the fact that the general magnetic field of new stars deflects condensations of continuous ejection (the diffuse-enhanced and Orion spectra) towards the equator.

Magnetic fields inside novae must be 'tangled'. This explains (a), as well as the difference between cases (b) and (c).

The structure of envelopes, ejected by novae, must reflect the presence of tangled magnetic fields. According to G. A. Shajn this is confirmed for super-novae, the envelopes of which may be studied in detail.

Proceeding from the existing numerous spectroscopic data the author in a series of his papers [1, 2, 3, 4] came to the conclusion that in each nova, during its expansion before light maximum, considerable forces are present, which are directed to its centre and exceed gravity considerably in the case, when the mass of the nova is equal to the solar mass. In order to explain the origin of these forces a hypothesis was proposed by the author (in his early papers [1, 2]), in which large masses were postulated for novae. Later on, however, several facts showed that this hypothesis must be rejected, because of a number of difficulties. In connexion with this the author suggested [3] the hypothesis that magnetic fields are the principal source of retardation phenomena in the expanding nova before light maximum. As it is known (see, e.g. [5]) the conducting medium (ionized gas) becomes 'viscous', when magnetic fields are introduced. Moreover, the 'intertwining' (entanglement) of the magnetic lines of force, which is connected with the turbulent motions of ionized gases, increases this effect.

As a result of this the star, when being in a strongly turbulent state and possessing a magnetic field may be compared to a 'ball' of elastic threads.

There exist two possibilities in the problem of magnetic fields in novae: (a) the magnetic field of a nova is constant; it exists in the star before the outburst and after the latter, (b) the magnetic field arises during the outburst.

As we shall see further the first possibility is realized in novae, though the process of the outburst itself must lead to a strengthening of the magnetic field in the expanding star.

Let us now consider several arguments speaking in favour of the hypothesis about large magnetic fields in novae. It is natural to assume that the magnetic field of each nova is similar to the field of a dipole. This means that the influence of such a field on the motion of gases ejected from a nova must be non-isotropic.

First we shall consider the primary phase of the outbursts of the nova—the expansion of the latter before light maximum. If the magnetic field of the nova is of a dipolar character we must expect that the retardation of the moving (ionized) gases, ejected during the outburst, must be smallest in the polar directions, where the velocity vectors of gases and the magnetic lines of forces approximately coincide (Fig. 1). In other words, we must observe a preferential ejection of matter in two diametrically opposite directions.

This is fully confirmed by observations [6]. We have in mind:

(a) The well-known fact of the doubling of emission bands in the spectra of the majority of novae (see [6], Table 1; N Sgr 1954, which is not included in this table, also displayed double emission lines). In the cases in which there is no doubling of emission bands we may expect that the polar axis of the star is nearly perpendicular to the line of sight. Many observational facts convince us that the doubling of emission lines is not due to the phenomena of self-reversal and that we have two polar condensations (polar caps) in the expanding principal envelope, which are moving in two diametrically opposite directions;

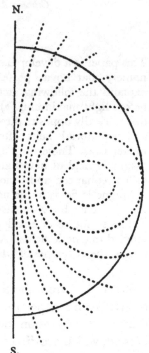

N.

S.

Fig. 1. A possible scheme of the distribution of the magnetic lines of force inside the nova.

(*b*) The so-called 'doubling' of novae, which was observed visually for N Pic 1925 and N Her 1934. In these cases two condensations in the expanding principal envelope, together with the nova itself, were located approximately on one common axis. In the spectra of these stars a doubling of emission lines was observed too, which confirms directly the conclusion that the doubling of emission lines in the spectra of other novae reflects also the presence of two condensations in the principal envelope of these stars (but not the presence of rings—toroids, which also give double emission lines).

It is of interest to note that the nebulosities which surrounded these two stars had an oval form; the just mentioned condensations were located on the large axis of the oval. Therefore the velocities of ejected matter were the greatest along the polar directions. This confirms also the statement that the retardation of ejected matter is the least in these directions.

Photographically the phenomenon of the 'doubling' was observed in N Aql 1918[7], where only the one condensation and a very dense one was visible; this corresponded to spectroscopic observations. The latter indicated that the 'violet' maximum in the emission bands of this star was noticeably stronger, than the 'red' one.

We have pointed out that in the polar directions the retardation of the moving ionized gases must be smallest. However, even in these directions it cannot be zero. As it has been shown by A. Kipper[8] magnetic lines of force in any star must be 'intertwined', if the magnetic field strength inside the star is greater than one gauss. This 'intertwinement' must play a great part in the phenomena of retardation. The significance of this must be greatly increased during the expansion of the star. The motion of the ejected ionized gases in the non-homogeneous magnetic dipolar field* of the star (this field decreases rapidly in strength from the centre of the star) must be accompanied by induced electric currents (due to the law of induction) and therefore by corresponding magnetic fields. And since the process of the expansion of the star is of a strongly turbulent character these magnetic fields must be partly 'intertwined'. It seems that these fields play the principal part in the general retardation phenomena in novae. The source of energy of this induced magnetic field is the energy of the outburst;† therefore the strength of this field may exceed very greatly the strength of the original dipolar field.

The presence of an 'intertwined' magnetic field in the expanding nova

---

* It seems that the source of this dipolar field is in the innermost parts of the star.

† The retardation phenomena in expanding novae show that the kinetic energy is being transformed here into the magnetic energy.

means that the retardation of matter must take place in all the radial directions, including the polar ones, though in the polar directions it must be the least. Moreover, the presence of such a field means that 'electromagnetic viscosity' exists not only in the radial directions, but in all other directions too. This explains why in spite of the divergence of magnetic lines of force we observe relatively compact, dense, polar condensations, 'polar caps', which produce double maxima in emission bands. Such a compactness was, e.g. strongly displayed in the case of N Aql 1918, see [7].

The viscosity in the process of expansion of the nova before light maximum is very important, because the expansion itself is a continuous

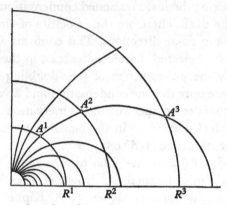

Fig. 2. The figure shows how the condensations, ejected after light maximum for the nova, are deflected to the equatorial plane of the star.

process. However, after light maximum this situation is radically changed. It is known [9, 10] that after light maximum a new ejection of matter begins from novae, but now in the form of separate gaseous condensations. These condensations, different in size, absorb the light from the star and produce the diffuse-enhanced and the Orion spectra of novae. The mean velocity of condensations which produce the diffuse-enhanced spectrum is approximately twice as great as the velocity of the principal envelope. The velocity of condensations, which produce the Orion spectrum is greater than the velocity corresponding to the diffuse-enhanced spectrum or is equal to it.

Since in this case an ejection of isolated condensations takes place, the role of electromagnetic viscosity must be negligible here, being essential only inside the condensations themselves. Therefore, these condensations must move independently one from another along the magnetic lines of force, see Fig. 2. This figure shows that the condensations ejected from the star after light maximum must be deflected to its equatorial plane. Therefore an equatorial belt must be formed in the principal envelope because the condensations have velocities greater than the velocity of the principal envelope and must overtake the latter after a certain lapse of time. An equatorial belt will be formed also in the case (N Aql 1918), when practically all the matter constituting the principal envelope is concentrated in its 'polar caps', while in the equatorial parts of this envelope there is practically no matter.

In the process of formation of the equatorial belt the following circumstances must be of great importance: (a) the remoteness of the principal envelope from the star; it changes with time, see Fig. 2; (b) the fact that at great distances from the star the influence of the dipolar magnetic field on the motion of condensations is very weak (magnetic field strength rapidly diminishes with distance).

Fig. 2 shows that the motion of condensations ejected from the equatorial parts of the star, is hindered due to a specific distribution of magnetic lines of force in these parts. When this factor is important, the amount of matter ejected in the equatorial directions, may be lowered. In these cases the equatorial belt will be divided into two parts, the 'southern' and the 'northern' ones. These parts may be called 'rings'. If (as it happens frequently), the velocities corresponding to the diffuse-enhanced and Orion spectra are different* condensations corresponding to both spectra will be deflected to the equatorial plane at a different rate (as in the case of a mass-spectrograph). This must lead to a rise of two pairs of rings, displaced symmetrically towards the equatorial plane.

These general considerations are confirmed by existing observational data on envelopes around N Aql 1918 and N Her 1934, see [6]. According to W. Baade [7, 11] the envelope of N Aql 1918 had a symmetrical system of rings and two polar caps. H. Weaver [12] has constructed a more exact model of the envelope, which surrounded N Aql 1918 and has found that actually both pairs of rings had velocities, which corresponded to the velocities obtained from the displacement of the absorption lines of the diffuse-enhanced and the Orion spectra. This shows that practically all the matter of the principal envelope is concentrated here in the polar caps and therefore condensations ejected after light maximum and deflected to the equatorial plane, were not retarded by the principal envelope.

It is of great importance that the equatorial rings around N Aql 1918 were characterized by strengthened emission in lines 6548 and 6584 (N II), while on the contrary the hydrogen emission was mainly concentrated in the polar caps [13]. This is in accordance with the fact that the

* The state of ionization and magnetization must also be different in both types of condensations.

Orion absorption spectrum of novae is characterized by strong absorption lines of nitrogen [9, 14].

In the same paper [6] a preliminary model (Fig. 3) of the envelope around N Her 1934 is constructed, which confirms also the above considerations. Existing spectroscopic observations of N Her 1934 and photographs of the nebula of this star in different wavelengths are used. An equatorial belt was also observed here in the same emission lines 6548 and 6584 (N II). The absence of separate rings in this case is easily explained by the fact that in the spectrum of N Her 1934 (in contradistinction of N Aql 1918) several systems of the diffuse-enhanced and the Orion spectra were observed. In connexion with this an approximately continuous belt of noticeable thickness was formed.

Additional facts on other novae confirm the above considerations [6].

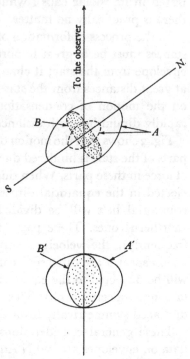

The problem of estimating the strength of magnetic fields which cause the retardation of gases in the expanding star before light maximum is very difficult. Three possibilities should be considered: (1) A magnetic field of a dipolar character. This field is produced in the innermost parts of the nova. (2) An 'intertwined' magnetic field existing in all the layers of the nova and even in its normal state [8], before the outbursts. Due to the expansion of the star the density of this field continuously decreases towards the light maximum. (3) A magnetic field which arises in the expanding new star before light maximum, due to the fact that the ionized masses of gas are moving in a non-homogeneous magnetic field of the dipole (see p. 196).

Fig. 3. The upper part of this drawing represents the structure of the envelope around N Her 1934, observed in the light of lines 6548 and 6584 (N II). A—is the principal envelope; B—is the equatorial belt. The lower part gives the projection of the envelope and of the equatorial belt on the plane perpendicular to the line of sight. B'—is the 'nitrogen belt' observed by W. Baade.

Probably this field (which is partly 'intertwined') plays the principal part in the process of retardation of the expanding new star. The strength of this field estimated according to the formula

$$\tfrac{1}{2}\rho v^2 \approx H^2/(8\pi) \tag{1}$$

amounted for N Aql 1918 to 2000 gauss near light maximum[6]. The strength of this field must be different for different novae. This may be concluded from the fact that 'masses', which were estimated by the author from the retardation effects, depend on the luminosity of the nova in its 'normal' state before the outburst[14]. The greater the luminosity of the nova is, the greater is its 'mass'. Of all the obtained masses the largest is the 'mass' of N Aql 1918, the smallest 'mass' is the one of N Her 1934. One may think that the same correlation must be true for the strength of magnetic fields, because these fields 'replace' now the 'large masses' of novae.

The question which field plays the principal part in deflecting the condensations towards the equator is very difficult to decide. It may be the original dipolar field of the nova. On the other hand it is also possible that the part played by the induced magnetic field is more important, of course if the latter is sufficiently strong. It is known that at large distances any magnetic field is similar to the dipolar field.

From the above considerations it follows that as a result of the outburst, magnetic fields which are present in the outer layers of the expanding nova are carried out from the nova together with the principal envelope. These fields must be partly 'intertwined'. If super-novae also possess large magnetic fields, the structure of their envelopes must also reflect the presence of magnetic fields, which are carried out from the star. This is fully confirmed by the recent results of G. A. Shajn[15]. He arrived at the conclusion that the presence of a systematic distribution of matter in the nebulosities IC 443, S 151 and in the system of fibrous nebulosities near the regions NGC 6960–NGC 6992, all these facts speak in favour of the hypothesis that the magnetic fields governing this distribution are carried out into the outer space together with the matter during the outbursts of super-novae. The same conclusion may be drawn from the study of the polarization of light from the Crab nebulae[16, 17, 18, 19].

Some authors do not agree with the idea that the origin of the magnetic field in the Crab nebula is connected with the magnetic field of the super-nova. For example, J. Oort and Th. Walraven[17] find that the total magnetic flux through the Crab nebula is extremely large. However, it is necessary to take into account that this magnetic flux must correspond not to the initial dipolar field of the star but to the field which arises during the outburst and which is of induction nature.* As I have already indicated,

* This field may also be connected with shock waves in the star during its outburst. The source of energy of these shock waves is the energy of the outburst. An initial magnetic field is needed in this case too.

the source of energy of this induced magnetic field is the energy of the outburst and therefore the strength of this field may exceed very greatly (by many orders) the strength of the original dipolar field.

The second problem is the problem of the diminishing of the field inside the nebula during its expansion. There are two possibilities here: (a) The ejected envelope (the nebula) is continuous in all directions. In this case the magnetic field strength decreases as r^{-2}, see Fig. 4 (a), (b) The envelope consists of separate condensations, and filaments, see Fig. 4 (b). In this case the magnetic field strength of these condensations will decrease

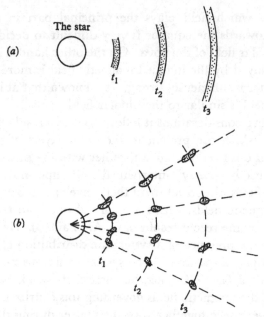

Fig. 4. Two possible types of expansion of the ejected envelope.

(during the expansion of the nebula) very slowly, especially if one takes into account the magnetic viscosity inside these condensations. One may think that in the Crab nebula the second case is realized because this nebula consists of separate filaments.

If novae and especially super-novae actually possess large magnetic fields, the above noted 'induction' mechanism may create in some way or other* particles of very high energies (cosmic rays), which may be released from the superficial layers of the expanding nova or super-nova. Owing to the pressure of magnetic fields in these layers the mechanism of

* For example, in the same way as in L. A. Artsimovich's experiments[20].

Fermi may become very effective. It seems that all these processes must be localized in separate small volumes and have the character of fluctuations. It is also necessary to point out that this 'induction mechanism' can act very long even after the outburst of nova or super-nova. It is well known that the ejection of matter from novae is observed during a period of many years [21], in several cases twenty, thirty years and even longer. One may think that super-novae, which are more unstable, eject matter during whole centuries. At the same time in all these cases we deal with the motion of ejected gaseous ionized condensations across the magnetic lines of force in a strong non-homogeneous field of the star (the same is true for a more short-lived process—the process of ejection of condensations, producing the diffuse-enhanced and the Orion spectra). If this 'induction mechanism', which may be accompanied by the acceleration mechanism of Fermi, is valid, then novae and super-novae may be the sources of high-energy particles lasting for decades and centuries.

The results of this paper lead to the conclusion that the magnetic fields, which secure the observed radio emission from the envelopes, ejected by super-novae [22] and the relativisitc particles (which are also needed for this radio emission) have their origin in the general magnetic fields of super-novae themselves.

From the above it follows that further investigation of novae and super-novae is needed. A more thorough study of envelopes ejected by novae and super-novae is especially important. This study must include spectroscopic material, as well as direct photographs in different wave-lengths. Another problem is also important—the study of light polarization in emission lines of H, He, N, C, which are produced by the novae themselves (not by their envelopes). These emission lines, which appear in the spectra of novae several months after light maximum, are produced close to the surface of the nova [21] and it is possible that the influence of the magnetic field on the emitting atoms H, He, N, C may involve some polarization due to Zeeman effect. It is also of great interest to investigate the polarization of light from 'old' novae and 'old' super-novae (excluding their nebulosities).

Finally we are going to discuss briefly the problem of the origin of magnetic fields of novae. Of course, this problem is a very complex one. However, it is necessary to point out that according to the recent article by M. Walker [23] N Her 1934 is an eclipsing variable with the shortest known period. Further, this is a reason to expect that in this binary the period of rotation of components in the system and the period of revolution coincide (as in the case of close binaries). Therefore N Her 1934 rotates very rapidly; in other words we have a very rapidly rotating dwarf. Besides it is necessary to note that the inclination of the orbit of N Her 1934 determined by M. Walker[23] and the inclination of the polar axis of N Her 1934 (obtained by the author, see Fig. 3) roughly coincides. At the same time the presence of magnetic fields is usually connected with rotation.

REFERENCES

[1] Mustel, E. R. *A.J. U.S.S.R.* **24**, 280, 1947.
[2] Mustel, E. R. *Publ. Crim. Astroph. Obs.* **4**, 152, 1949.
[3] Mustel, E. R. *Communications for the Symposium on Non-stable Stars* (Moscow, 1955), p. 130.
[4] Mustel, E. R. *Vistas in Astronomy*, vol. 2, ed. by A. Beer, Pergamon Press, London, 1956.
[5] Alfvén, H. *Cosmical Electrodynamics* (Oxford University Press, 1950).
[6] Mustel, E. R. *A.J. U.S.S.R.* **33**, 182, 1956.
[7] A communication in *Sky and Telescope*, **12**, 12, 1952.
[8] Kipper, A. J. *Trans of the 4th Conference on Problems of Cosmogony* (Moscow, 1955), p. 425.
[9] McLaughlin, D. B. *Publ. Obs. Michigan*, **8**, 149, 1943.
[10] Mustel, E. R. *A.J. U.S.S.R.* **24**, 97 and 155, 1947.
[11] *Problems of Cosmical Aerodynamics*, Proceedings of the Symposium held in Paris, Dayton, Ohio, 1951.
[12] Weaver, H. *I.A.U. Transactions*, vol. 9 (Dublin, 1955), Cambridge University Press, 1957, Report at the meeting of Commission 36a.
[13] Wyse, A. B. *Lick Obs. Bull.* **14**, pt. 3, 1939.
[14] Mustel, E. R. *Publ. Crim. Astroph. Obs.* **6**, 144, 1951.
[15] Shajn, G. A. *A.J. U.S.S.R.* **32**, 381, 1955.
[16] Dombrovsky, V. A. *C.R. Acad. Sci. U.S.S.R.* **94**, 1021, 1954.
[17] Oort, J. and Walraven, Th. *B.A.N.* **12**, 285, 1956.
[18] Shajn, G. A., Pikelner, S. B. and Ikhsanov, R. N. *A.J. U.S.S.R.* **32**, 395, 1955.
[19] Vashakidze, M. A. *A.C. U.S.S.R.* no. 147, 11, 1954.
[20] Artsimovich, L. A. *Nuclear Energy, U.S.S.R.* no. 3, 1956.
[21] McLaughlin, D. D. *Astrophys. J.* **117**, 279, 1953.
[22] Shklovsky, I. S. *A.J. U.S.S.R.* **30**, 15, 1953.
[23] Walker, M. *Astrophys. J.* **123**, 68, 1956.

Discussion

Spitzer: Would you explain again how the magnetic field of a super-nova is amplified by an explosion?

Mustel: The problem of strengthening of the field in hydromagnetics is a very complex one. In my communication I am considering the mechanism in which the strengthening of the field takes place, because of the fact that the ejected ionized gases are moving in the non-homogeneous dipolar field of the star. The source of this field is supposed to be located in the innermost parts of the nova. Of course this mechanism is connected with certain difficulties. It is possible that the non-homogeneity of the dipole field does not play any significant role here and that the strengthening of the field in the expanding star is connected only with the motion of ionized gases *across* the magnetic lines of force of the dipole (we do not speak here about the polar directions). In this case the strengthening of the field might follow from the fact that the gases crossing the magnetic lines of force are retarded, therefore the kinetic energy E of these gases is diminished. One may think that the lost energy ΔE must be mainly transformed into magnetic energy. A transformation of this energy ΔE into other kinds of energy is doubtful.

The decrease of velocity of the expanding gases in novae before light maximum follows from many spectroscopic facts (E. R. Mustel, *Symposium on Non-stable Stars*, 1957). This decrease is not accompanied by the heating of gases. The velocity of the reversing layer of N Her 1934 was diminishing all the time from the moment the star was discovered up to the moment of its light maximum (from 1300 km/sec to 300 km/sec) and the spectrum of the star changed from class B to class F.

In certain cases, however, it is possible (for example, in the case of the Crab nebula) that the magnetic field of the ejected envelope is actually the original magnetic field of the inner parts of the super-nova itself, this field being weakened by the nebula expansion. The magnetic flux in the superficial layers of the super-nova before the outburst cannot be very large, because this would violate the equilibrium of these layers.

In connexion with this the hypothesis (J. H. Oort and Th. Walraven, *Bull. Astr. Netherl.* **12**, 304, 1956), in which the phenomenon of the outburst of the super-nova arises from the equality between the magnetic energy of the internal parts of the star and the gravitational energy, is of great interest.

Schatzman: Have you to suppose that the ejection of matter at the beginning is spherically symmetric and that, later on, the magnetic field produces the rings and the polar caps, or is it necessary to suppose that the ejection of matter at the beginning was made of polar caps and rings?

Mustel: I suppose that the ejection of matter at the beginning and after light maximum is roughly spherically symmetric and that the magnetic field produces the rings and the polar caps. This supposition is quite justified. For example, the outburst of the nova is a sort of explosion and therefore the gases must be ejected practically in all directions. However, in reality we observe two polar caps. Further, the diffuse-enhanced and the Orion spectra were observed for each nova irrespective of the direction of its polar axis. This means that the ejection of corresponding condensations proceeds also in all directions. Nevertheless, after a lapse of time, we can see this matter only in the equatorial parts of the principal envelope and therefore the condensations are deflected to the equatorial plane of the envelope by magnetic fields. The most clear example is N Aql 1918. Its polar axis is near to the line of sight and in the spectrum of the star the strong diffuse-enhanced and the Orion absorption spectra were observed. However, after a certain lapse of time the corresponding condensations were concentrated in two pairs of equatoral rings. The identity of ejected matter was confirmed here by the equality of velocities inferred from displacement of both these spectra with the velocities inferred from the study of the corresponding equatorial rings.

INTERPRETATION OF THE POLARIZATION MEASUREMENTS OF THE DOUBLE CLUSTER IN PERSEUS

K. SERKOWSKI

Astronomical Observatory of the University of Warsaw, Poland

The stars in the Double Cluster in Perseus whose angular distance is small have similar polarization and similar colour excesses. Quantitative description of this effect is the purpose of this paper.

The observational data are taken from polarization measurements performed by Hiltner[1] and blue-yellow colour excesses obtained from data by Johnson and Morgan[2] and by Johnson and Hiltner[3]. The above investigations seem to be the most extensive ones; the measurements by other authors were not included in order not to destroy the uniformity of data.

The polarization is most conveniently described by the parameters Q, U, proportional to the Stokes parameters and defined by

$$Q = p \cos 2(\theta - \bar{\theta}),$$

$$U = p \sin 2(\theta - \bar{\theta}),$$

where p is the amount of polarization (in magnitudes) and θ is the position angle of the electric vector. For the cluster stars the mean square deviation of the position angle θ from the mean value $\bar{\theta}$ is only $\pm 4°$ and therefore $Q \simeq p$.

Within a radius of $2°$ from the centre of the Double Cluster the polarization of ninety-two stars was measured by Hiltner. These stars were connected in pairs in all possible combinations. The pairs were distributed in groups according to the angular distance ϕ of the stars in pairs. For every group the mean square of the difference in the parameters Q and U was computed for the stars in pairs. The values obtained in this way (given in Fig. 1 and Fig. 2) show that the mean square of the difference in every Stokes parameter for the close pairs is half that for the distant ones.

The mean squares of the differences in the colour excess E show still more

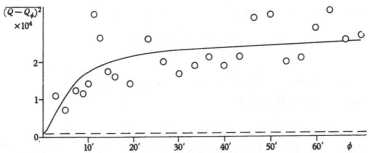

Fig. 1. The mean squares of differences in the Stokes parameter Q (amount of polarization) versus angular distance of the stars. The dashed line represents the errors of measurements, the solid line was computed from formula (8) for the micro-scale $s_{oq} = 1\cdot5$ pc.

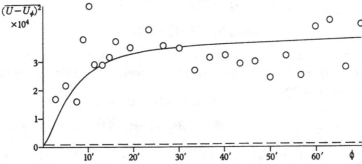

Fig. 2. The mean squares of differences in the Stokes parameter U.

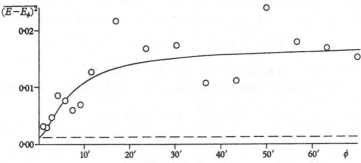

Fig. 3. The mean squares of differences in the colour excess. The solid line was computed for the micro-scale $s_{o\delta} = 1\cdot8$ pc.

distinctly the dependence on the angular distance of the stars forming the pair (Fig. 3).

Let H_x, H_y and H_z be the components of the galactic magnetic field in the co-ordinate system in which OY is in the direction of the light ray and OZ is perpendicular to the galactic plane. It is assumed after Davis [4] that the increase of the parameters Q, U, over the path ds is

$$dQ = c\rho(H_x^2 - H_z^2)\ ds,$$

$$dU = 2c\rho H_x H_z ds,$$

where ρ is the density of absorbing matter and $c = $ const. The similar increase of the colour excess E can be given by

$$dE = c'\rho ds \qquad c' = \text{const.}$$

Applying the method proposed by Chandrasekhar and Münch[5] the random functions $\delta(s)$, $q(s)$, $u(s)$, are defined by

$$\rho = \bar{\rho}[\mathrm{1} + \delta(s)], \qquad (\mathrm{1})$$

$$\rho(H_x^2 - H_z^2) = \bar{\rho}(\overline{H_x^2} - \overline{H_z^2})\ [\mathrm{1} + q(s)], \qquad (2)$$

$$2\rho H_x H_z = \bar{\rho}(\overline{H_x^2} - \overline{H_z^2})\ u(s), \qquad (3)$$

where s is a radius vector of given point in space. The assumption is made that $\delta(s)$, $q(s)$, $u(s)$, can be treated as homogeneous stochastical processes fulfilling the relations

$$\overline{\delta(s)} = \overline{q(s)} = \overline{u(s)} = \mathrm{o},$$

$$\overline{\delta^2(s)} = \alpha^2 = \text{const}, \quad \overline{q^2(s)} = \beta^2 = \text{const}, \quad \overline{u^2(s)} = \gamma^2 = \text{const},$$

$$\overline{\delta_1\delta_2} = \alpha^2 R(|\ s_1 - s_2\ |), \quad \overline{q_1 q_2} = \beta^2 R(|\ s_1 - s_2\ |), \quad \overline{u_1 u_2} = \gamma^2 R(|\ s_1 - s_2\ |),$$

where $|\ s_1 - s_2\ |$ is distance between the points in which the values of the functions are respectively δ_1 and δ_2, etc., and $R(|\ s_1 - s_2\ |)$ are the correlation functions which following Chandrasekhar and Münch are specified as

$$R(|\ s_1 - s_2\ |) = \exp\ (-|\ s_1 - s_2\ |/s_{ok}),$$

where instead of k should be put respectively indices δ, q, u.

From the preceding formulae by numerical integration was obtained

$$\overline{(E - \bar{E})^2} = 2\bar{E}^2\alpha^2 s_{o\delta}/S, \qquad (4)$$

$$\overline{(Q - \bar{Q})^2} = 2\bar{Q}^2\beta^2 s_{oq}/S, \qquad (5)$$

$$\overline{U^2} = 2\bar{Q}^2\gamma^2 s_{ou}/S, \qquad (6)$$

$$\overline{(E - E_\phi)^2} \simeq 2\overline{(E - \bar{E})^2}\ [\mathrm{1} - \exp\ (-a\phi S/s_{o\delta})], \qquad (7)$$

$$\overline{(Q - Q_\phi)^2} \simeq 2\overline{(Q - \bar{Q})^2}\ [\mathrm{1} - \exp\ (-a\phi S/s_{oq})], \qquad (8)$$

$$\overline{(U - U_\phi)^2} \simeq 2\overline{U^2}[\mathrm{1} - \exp\ (-a\phi S/s_{ou})], \qquad (9)$$

where $a = 0\cdot 23$ is constant resulting from the integration and $S = 2300$ pc is the distance of Double Cluster.

After eliminating the dependence of E and p on the galactic latitude (assumed to be linear) the observational data used in the present paper give

$$\bar{E} = 0^m 58, \quad \bar{Q} = 0^m 079,$$

$$\overline{(E-\bar{E})^2} = 9.10^{-3}, \quad \overline{(Q-\bar{Q})^2} = 1 \cdot 3.10^{-4}, \quad \overline{U^2} = 2 \cdot 0.10^{-4}.$$

All the three parameters E, p, U, are uncorrelated. Obviously, E and p are uncorrelated only when we are examining the stars in the narrow interval of the galactic latitude (e.g. $1°$) since both E and p are dependent on the galactic latitude. Further, the correlation between p and the absolute magnitude of the stars, suggested by Dombrowski[6] was not confirmed.

Applying the formula (7) to data plotted on Fig. 3 the micro-scale of the fluctuations in the density of absorbing matter $s_{o\delta} = 1 \cdot 8$ pc is obtained. The data given in Fig. 1 and Fig. 2 can be represented by the formulae (8), (9), with only one value of the micro-scale $s_{oq} = s_{ou} = 1 \cdot 5$ pc.

Formulae (4)–(6) when combined with the foregoing values give

$$\alpha^2 = \overline{(\rho - \bar{\rho})^2}/\bar{\rho}^2 = 15, \tag{10}$$

$$\beta^2 = 16, \tag{11}$$

$$\gamma^2 = 25. \tag{12}$$

Let us consider the large-scale homogeneous magnetic field on which are imposed fluctuations perpendicular to the mean direction of the field. In the region of the Double Cluster the mean direction of the magnetic field coincides approximately with the X-axis of the co-ordinate system formerly introduced. So it can be assumed that the fluctuations are isotropic in the YZ-plane and in the direction of the X-axis they disappear. In this case from the equations (1)–(3) we obtain

$$\beta^2 = \alpha^2 + 2(1 + \alpha^2) \, \overline{H_z^2}^2 / (\overline{H_x^2} - \overline{H_z^2})^2,$$

$$\gamma^2 = 4(1 + \alpha^2) \, \overline{H_x^2} \, \overline{H_z^2} / (\overline{H_x^2} - \overline{H_z^2})^2.$$

Substituting here the values (11), (12) found from the polarization data we obtain $\alpha^2 = 13$ which agrees sufficiently well with the value (10) resulting from the analysis of the colour excesses. In this way we obtain also the root-mean-square angular deviation of the lines of force from a uniform direction

$$\alpha_1 = (2\overline{H_z^2}/\overline{H_x^2})^{1/2} = 0 \cdot 71$$

about three times greater than the result obtained by Davis[4].

REFERENCES

[1] Hiltner, W. A. *Astrophys. J.* **114**, 241, 1951 and *Astrophys. J.*, **120**, 454, 1954.
[2] Johnson, H. L. and Morgan, W. W. *Astrophys. J.* **122**, 429, 1955.
[3] Johnson, H. L. and Hiltner, W. A. *Astrophys. J.* **123**, 267, 1956.
[4] Davis, L. *Vistas in Astronomy*, vol. 1, ed. by A. Beer, p. 336, (Pergamon Press, London, 1955).
[5] Chandrasekhar, S. and Münch, G. *Astrophys. J.* **115**, 103, 1952.
[6] Dombrowski, W. A. *Russ. Astr. J.* **30**, 603, 1953.

Discussion

Spitzer: How great are the probable errors of the obtained values of the micro-scale?

Serkowski: I think that the probable errors of $s_{o\delta}$ and s_{oq} are 30 % or somewhat more. The value of s_{ou} is much more inaccurate.

208

HARMONIC ANALYSIS OF THE PERIODIC SPECTRUM VARIABLES

ARMIN J. DEUTSCH

Mount Wilson and Palomar Observatories, Pasadena, California, U.S.A.

ABSTRACT

Certain stars are known to be periodic magnetic variables, and to show synchronous changes in line strength and radial velocity. The hypothesis has been made that the atmosphere of such a star is in rigid rotation, and that it is characterized by a permanent magnetic field and associated abundance irregularities. The magnetic potential and the local equivalent widths have been developed in spherical harmonics, and the Laplace coefficients of these expansions have been related to the Fourier coefficients of the observed curves. The theory has been applied to the star HD 125248 in an attempt to verify the original hypothesis and to map the magnetic fields and abundance anomalies over the stellar atmospheres.

Among the stars in which H. W. Babcock has found spectroscopic evidence of a general magnetic field [1], there are several that show periodic reversals of the observed field. In addition, some of these stars exhibit synchronous variations in the equivalent widths of certain absorption lines, and in the radial velocities indicated by these lines. These phenomena are illustrated for the star HD 125248 ($P = 9 \cdot 30$ days) in Figs. 1, 2 and 3.

Fig. 1 is taken from the work of Babcock [2]. He has converted the effective field H_e, as found from the observed Zeeman effect, into the polar field H_p, on the assumption that the field is that of a dipole viewed along the axis. The figure shows that different elements sometimes indicate significantly different values for H_p. Fig. 2 is based on 13 coudé spectrograms at $4 \cdot 5$ Å/mm, obtained by Babcock and Deutsch during one cycle in 1951 and one cycle in 1952. The figure rests on spectrophotometric measures of 127 different absorption lines chosen to be relatively free from blends. When the measured equivalent widths W of each line are expressed in units of its average equivalent width $\overline{W}$ at all observed phases, it is found that, within the errors of observation, all lines of a given element show nearly the same variation with phase. Moreover, among the nine elements discussed, each can be assigned to one of three groups, within each of which the intensity variation is the same. Fig. 2 shows how this assignment has been made. The mean points have been obtained by weighting each element in the group with the number of its lines that were measured.

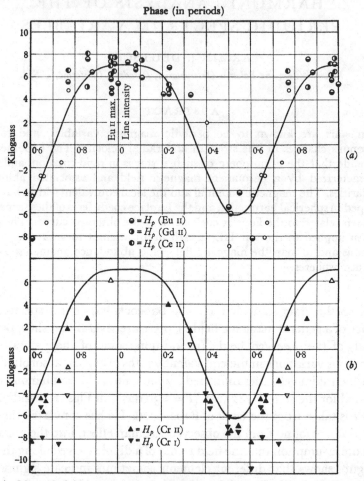

Fig. 1 (a). Magnetic field intensity, H_p, in kilogauss, for the rare earths, plotted against phase. Small open circles are of low weight. For comparison, the smooth curve for the Fe–Ti lines is also shown. Fig. 1 (b). Similar results for Cr I and Cr II. (Reproduced from the *Astrophysical Journal*.)

Fig. 3 is based on radial-velocity measures of the eight 1951 coudé spectrograms. As far as possible, all lines measured for intensity were also measured for velocity, and no others. With the possible exception of Sr II, for which only two lines could be measured, the same group behavior prevails in the radial velocities as in the equivalent widths. The details of the observations summarized in these figures will be discussed later in another publication. This is also true of the mathematical developments to be outlined below.

It has been proposed that the atmosphere of such a star as HD 125248 is

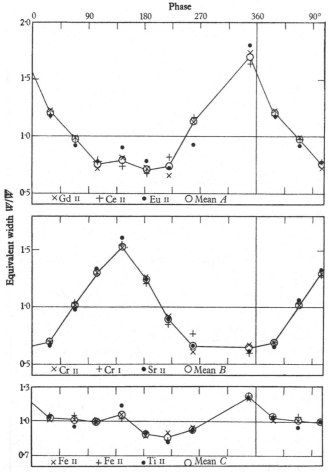

Fig. 2. Equivalent widths of 127 absorption lines. For each line, the measured equivalent width W has been divided by the average measured equivalent width $\overline{W}$ at all observed phases. For a given element at a given phase, the plotted point is the average value of $W/\overline{W}$ for all lines of that element.

spectroscopically non-homogeneous, and that it is in rigid rotation around an axis that is not a symmetry axis of the abundance irregularities or of the associated general magnetic field. The observed variations would then be attributed to the changing aspect of the star as it rotates; the observed period would be simply the period of rotation. The observations cannot all be satisfied with such a model if the abundance irregularities and/or magnetic field possess symmetry axes. It appears, however, that a satisfactory representation may be possible by a more general kind of rigid rotator. The evidence for this type of model has recently been summarized elsewhere [3].

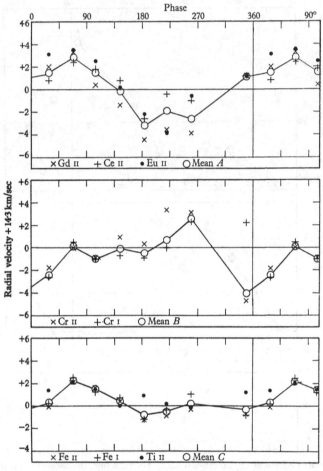

Fig. 3. Radial velocities from 127 absorption lines.
The systemic velocity is $V_0 = -14.3$ km/sec.

Fig. 4 represents a spherical star with the pole of rotation at P_0 and the subpolar point at S, the inclination of the rotational axis to the line of sight being χ. At an arbitrary point P on the surface, the polar distance is ψ and the azimuth v. The latter angle is measured from a meridian that rotates with the star, with $v = 0$ along the meridian that passes through S at the phase $\Phi = 0$. The angle between S and P is θ, and the azimuthal angle around S is ϕ, with $\phi = \pi$ at P_0.

We shall call $\xi(\psi, v)$ the distribution function for a given spectrum line. Its value is the local equivalent width of the line at P in light emergent normally, and its variation reflects the abundance irregularities over the surface of the star. In general, we may represent $\xi(\psi, v)$ as the real part of a Laplace series,

$$\Xi(\psi, v) = \langle W \rangle \sum_{n=0}^{\infty} \sum_{m=-n}^{n} A_n^m \exp(imv) P_n^{|m|}(\cos \psi), \qquad (1)$$

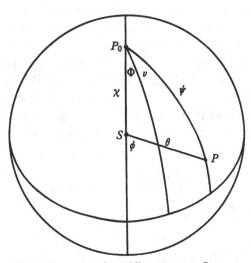

Fig. 4. Geometry of an oblique rotator. See text.

where $\langle W \rangle$ is the observed equivalent width averaged over the cycle of variation. To the extent that we can approximate the curve of growth by a straight line, it can be shown that all lines originating in a given group of elements will be characterized by the same set of Laplace coefficients A_n^m. Our first object will now be to compute the observed equivalent width W, in the integrated light from the visible hemisphere, as a function of the coefficients A_n^m and the phase Φ. We shall then compute the integrated radial velocity and Zeeman effect. Finally, we shall invert the argument and find the A_n^m from the observations of HD 125248.

We shall suppose that in the continuum the intensity is distributed over the stellar disk in the usual way, with the limb-darkening law

$$\Lambda = 1 - \mu + \mu \cos \theta. \qquad (2)$$

We shall also take account, in an approximate way, of the variation of equivalent width with angle of emergence at a given point. We suppose

213

that the local equivalent width in the direction θ can be written in the form $\xi(\psi, v) J(\theta)$, where

$$J = 1 - \kappa + \kappa \cos \theta. \tag{3}$$

The quantity κ is then a coefficient of line-weakening toward the limb, analogous to the coefficient μ of limb-darkening in the continuum.

In order to find the observed equivalent width W in the integrated light from the visible hemisphere, we must weight the local equivalent width ξJ with the local surface brightness Λ and the projected element of area at P, and then integrate over the visible hemisphere. Similarly, the observed radial velocity V (relative to the systemic radial velocity V_0) will be obtained by inserting the local radial velocity of rotation as an additional weighting factor, and integrating over the visible hemisphere. If V_e is the equatorial speed of rotation, the line-of-sight component of the local rotational velocity is $V_e \sin x \sin \theta \sin \phi$ at P. Again, to find the effective magnetic field, we must insert the line-of-sight component H_z of the local magnetic field $\mathbf{H}$ as the additional weighting factor. We then find that the observed quantities W, V and H_e can be taken as the real parts of the following expressions:

$$\mathscr{W} = \frac{\displaystyle\int_0^{\pi/2} \int_0^{2\pi} (\Xi J) \Lambda \sin \theta \cos \theta \, d\phi \, d\theta}{\displaystyle\int_0^{\pi/2} \int_0^{2\pi} \Lambda \sin \theta \cos \theta \, d\phi \, d\theta}, \tag{4}$$

$$\mathscr{V} = \frac{\displaystyle\int_0^{\pi/2} \int_0^{2\pi} (V_e \sin \chi \sin \theta \sin \phi)\,(\Xi J)\,\Lambda \sin \theta \cos \theta \, d\phi \, d\theta}{\displaystyle\int_0^{\pi/2} \int_0^{2\pi} (\xi J)\,\Lambda \sin \theta \cos \theta \, d\phi \, d\theta}, \tag{5}$$

$$\mathscr{H}_e = \frac{\dfrac{1}{2}\displaystyle\int_0^{\pi/2} \int_0^{2\pi} H_z[(\Xi + \tilde{\Xi})\,J]\,\Lambda \sin \theta \cos \theta \, d\phi \, d\theta}{\displaystyle\int_0^{\pi/2} \int_0^{2\pi} (\xi J)\,\Lambda \sin \theta \cos \theta \, d\phi \, d\theta}. \tag{6}$$

To evaluate these integrals, it is necessary to express the distribution function Ξ in the alternative Laplace series

$$\Xi(\psi, v) = \langle W \rangle \sum_{n=0}^{\infty} \sum_{m=-n}^{n} B_n^m \exp(im\phi)\, P_n^{|m|}(\cos \theta), \tag{7}$$

in which the coefficients B_n^m are functions of χ and Φ, and of the coefficients A_n^m. This transformation has recently been fully discussed by Satô[4].

With his results, it is possible to show that the integrals of interest can be put in the form of Fourier series in the phase, as follows:

$$\frac{\mathscr{W}}{\langle W\rangle}=\sum_{m=0}^{\infty} D_{-m} \exp\,(-im\Phi), \tag{8}$$

$$\left(\frac{W}{\langle W\rangle}\right)\mathscr{V}=(V_e \sin \chi)\sum_{m=0}^{\infty} E_{-m} \exp\,(-im\Phi), \tag{9}$$

$$\left(\frac{W}{\langle W\rangle}\right)\mathscr{H}_e=\sum_{m=-\infty}^{\infty} G_m \exp\,(im\Phi). \tag{10}$$

The Fourier coefficients in these expressions are related to the Laplace coefficients by the equations

$$D_{-m}=\sum_{n=0}^{\infty} \mathscr{P}_n^m A_n^m, \tag{11}$$

$$E_{-m}=\sum_{n=0}^{\infty} \mathscr{Q}_n^m A_n^m, \tag{12}$$

$$G_m=\sum_{n=1}^{\infty}\sum_{\alpha=0}^{\infty}\sum_{\beta=0}^{\infty} M_n^\beta (F_1 A_\alpha^{-m-\beta}+F_2 \tilde{A}_\alpha^{m+\beta}). \tag{13}$$

In these equations, the quantities $\mathscr{P}_n^m$, $\mathscr{Q}_n^m$, $F_1(n, \alpha, m, \beta)$, and $F_2(n, \alpha, m, \beta)$ depend only on the indicated indices, and upon the limb-coefficients μ and κ. They have been obtained explicitly by a term-by-term integration of Eqs. (4), (5) and (6), after substitution of Eq. (7) into these integrals. The coefficients M_n^β characterize the complex magnetic field $\mathscr{H}$, which we suppose can be obtained from a scalar potential,

$$\mathscr{d}=R\sum_{n=1}^{\infty}\sum_{\beta=0}^{n} (R/r)^{n+1} M_n^\beta \exp\,(i\beta v)\, P_n^\beta(\cos \psi), \tag{14}$$

where R is the stellar radius. In obtaining Eqs. (11), (12) and (13), it has been assumed that $A_n^m \equiv 0$ for $\dot{m}<0$. This assumption entails no loss of generality for the real distribution function $\xi(\psi, v)$.

The observed curves corresponding to Eqs. (8), (9) and (10) can be reasonably well represented by Fourier series of the second degree. Accordingly, we shall suppose that we can obtain an adequate representation of the large-scale structure of the field and the abundance irregularities by considering only terms to the second degree in the Laplace expansions for Ξ and $\mathscr{d}$. When this is done, we find that the total number of real Laplace coefficients required to specify the magnetic potential, and the distribution functions for elements of groups A, B and C, is 35. Additional free parameters that must be specified to determine the observed curves are the

215

inclination χ of the rotational axis to the line of sight; the equatorial velocity of rotation V_e; and the systemic velocity V_0. On the other hand, the second-degree Fourier representations of the observations summarized in Figs. (1), (2) and (3) give us directly a total of 42 Fourier coefficients. Through Eqs. (11), (12) and (13), it therefore becomes possible, at least in principle, to solve uniquely for the 35 Laplace coefficients and the additional unknown parameters χ, V_e, and V_0.

A first approximation to this solution has been carried out for HD 125248, with the results that are given in Table 1. As a check upon the solution, we must require that the radius of the star, as computed in the relation $R = PV_e/2\pi$, be appropriate for a main-sequence star near spectral type A 0. In addition, if we suppose that the variations in local equivalent width are due primarily to abundance irregularities and not to transfer differences over the surface of the star, we must obtain distribution functions that are non-negative over the whole surface. The first of these conditions is well-satisfied by the solution in Table 1. The computed radius is 2·5 ⊙, which is normal for a B 9 dwarf. Moreover, the solution yields $V_e \sin \chi = 6\cdot9$ km/sec,

Table 1. *Adopted constants for* HD 125248

Epoch of zero phase, $\Phi = 0$. JD 2430143·07. Period, $P = 9\cdot2983$ days.
Systemic velocity, $V_0 = -14\cdot3$ km/sec. Stellar radius, $R = 2\cdot5 R_\odot$.
Rotational velocity, $V_e = 13\cdot7$ km/sec. Inclination, $\chi = 30°$.
Limb coefficients: $\mu = 0\cdot62$ in the continuum; $\kappa = 0\cdot18$ in the lines.

Laplace coefficients

	a_0^0	a_1^0	a_1^1	α_1^1	a_2^0	a_2^1	α_2^1	a_2^2	α_2^2
A	2·50	−2·80	1·50	−0·90	1·36	−0·12	0·06	0·15	−0·24
B	2·75	−2·75	−0·30	0·28	0·06	−0·43	0·36	0·12	0·24
C	2·50	−2·50	0·60	−0·16	0·50	−0·05	0·09	0·12	−0·15

	m_0^0	m_1^0	m_1^1	μ_1^1	m_2^0	m_2^1	μ_2^1	m_2^2	μ_2^2
H (kilogauss)	0	5·20	2·80	−1·00	−0·50	0·40	0·00	−1·50	0·00

Fourier coefficients

	d_0	d_{-1}	δ_{-1}	d_{-2}	δ_{-2}	$\overline{W}/\langle W\rangle$	Elements
A	1	0·40	−0·25	0·04	−0·06	0·918	Eu II, Gd II, Ce II
B	1	−0·48	0·41	0·03	0·06	1·117	Cr I, Cr II, Sr II
C	1	0·16	0·03	0·03	−0·04	0·983	Fe I, Fe II, Ti II

	e_0	e_{-1}	ε_{-1}	e_{-2}	ε_2
A	0	0·176	0·289	0·087	0·054
B	0	−0·122	−0·139	−0·087	0·044
C	0	0·017	0·115	0·054	0·044

	g_0	$g_{-1}+g_1$	$\eta_{-1}-\eta_1$	$g_{-2}+g_2$	$\eta_{-2}-\eta_2$	
A	1·00	2·43	−1·05	−0·20	−0·14	
B	−0·61	1·57	−0·40	−0·38	0·04	(kilogauss)
C	0·59	1·97	−0·62	−0·27	0·03	

216

which is consistent with the widths of the relatively non-variable lines in the spectrum of this object[5]. The condition that ξ must be non-negative is not satisfied; for groups A and B, the distribution functions do go slightly negative over certain small areas. It would seem, however, that relatively small changes in some of the Laplace coefficients could remedy these defects in the solution, without appreciably changing its character.

In Figs. 5, 6 and 7, the observations are compared with the solution of Table 1. Only the mean points for each group of elements have been plotted on these figures. In the case of Fig. 6, the observed quantities differ slightly from those in Fig. 2, because the average equivalent widths $\overline{W}$ at the *observed* phases have been replaced by the averages $\langle W \rangle$ over the whole cycle.

The geometry of the resulting configuration is illustrated by the maps in Figs. 8, 9 and 10. The first two figures give the contours of the distribution functions, on an Aitoff equal-area projection. Fig. 9 also gives the contours of $|\mathbf{H}|$, the magnitude of the local magnetic field. The solution has the property that it will satisfy the observations equally well if the contours are reflected in the equator, together with the path of the subsolar point. Fig. 10 gives an orthographic projection in the plane of the sky. It illustrates the aspect changes with phase.

The probable errors of the observations in Figs. 5, 6 and 7 are such that many of the residuals from the computed curves are undoubtedly significant. These residuals could presumably be somewhat improved by a least squares adjustment of the Laplace coefficients to the observations. The method of solution that has been actually used is a crude one, and is not entirely systematic. On the other hand, some of the discrepancies between the computed curves and the observations may be attributable to the effects of the higher-order terms that have been neglected. In particular, the second-order Laplace coefficients are sufficiently large in some cases as to indicate that, in a systematic scheme of approximation, it would be necessary to consider certain higher-order terms that have been neglected in this treatment.

The solution that has been obtained here is further impaired by the neglect of the brightness variation that is known to occur over the surface of the star[6]. The amplitude of the observed light curve is only 0·05 magnitude, but this could imply relatively large local variations of brightness. If these are accompanied by temperature changes in the reversing layer, the assumption that all lines can be represented by a single linear curve of growth might also require modification.

Even without these complications, the solution must still be investigated with respect to its stability. In these circumstances, it seems reasonable to

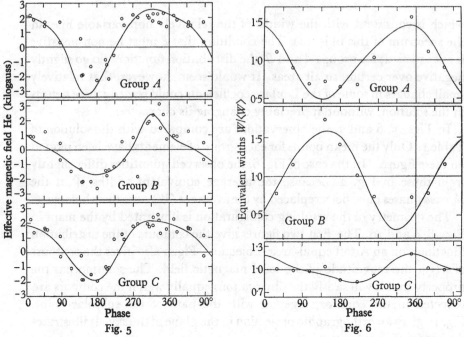

Fig. 5. Effective magnetic field, H_s, in kilogauss. From the measures of H. W. Babcock[2]. The smooth curves have been computed from the constants of Table 1.

Fig. 6. Equivalent widths of 127 absorption lines. The mean points are the same as in Fig. 2, except that a small correction has been applied to change $\overline{W}$ to $\langle W \rangle$, the equivalent width averaged over the whole cycle. The smooth curves have been computed from the constants of Table 1.

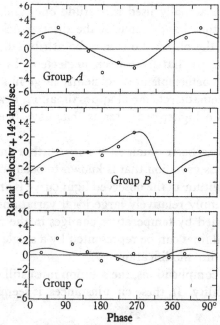

Fig. 7. Radial velocities from 127 absorption lines. The smooth curves have been computed from the constants of Table 1.

suppose that only the principal large-scale features of the solution can be considered as established. More detailed discussions, based on more extensive observations, may be indicated for the future.

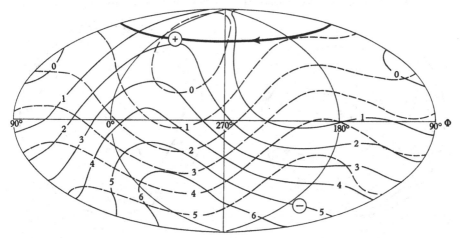

Fig. 8. Curves of constant local equivalent width for the lines of Group *A* (Eu II, Gd II, and Ce II; full curves) and Group *B* (Cr I, Cr II, and Sr II; broken curves). The subsolar point describes the heavy curve in the direction indicated; it lies at longitude $v = 0$ at phase $\Phi = 0$. The plus and minus signs mark the axis of the magnetic dipole. Aitoff equal-area projection.

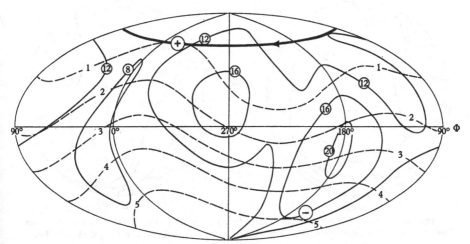

Fig. 9. The dashed curves are curves of constant local equivalent width for the lines of Group *C* (Fe I, Fe II, Ti II). The full curves are the contours of $|\,\mathbf{H}\,|$. The heavy curve and the plus and minus signs have the same significance as in Fig. 8.

An interesting feature of the solution is the situation of all three abundance maxima in the unobserved zone of the star. A calculation has shown that if this configuration were observed under an inclination of $\chi = 150°$, instead of 30°, all three groups of lines would be about four times stronger when averaged over the cycle. The amplitudes of the curves giving H_e and V as a function of phase would be comparable with the values actually observed. But the amplitudes of the curves for $W/\langle W \rangle$ would be less than

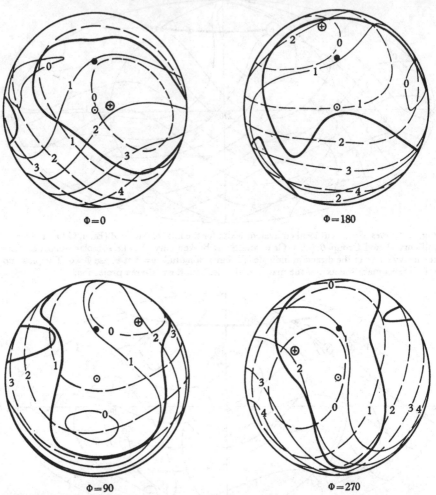

Fig. 10. Orthographic projections of the visible hemisphere on the plane of the sky at four phases. The light curves are the contours of Fig. 8. The poles of the dipole are indicated. The heavy curve is the locus where $H_s = 0$.

one-third as great, because of the increase in $\langle W \rangle$. It is possible that aspect effects of this kind could account for some of the magnetic variable stars that do not show conspicuous spectrum variation.

There is some evidence for secular changes in this star, in addition to the short-period effects that have been attributed here to mere aspect changes associated with rigid rotation. In all probability, these additional changes represent actual distortions of the configuration derived above. If so, they will have to be discussed within the theoretical framework of hydromagnetics. Such changes may also give rise to the acceleration phenomena that are required if nuclear processes occur in the atmosphere, as envisaged by Fowler and the Burbidges[7]. Meanwhile, we require a physical interpretation of the large-scale semi-rigid configuration derived above.

REFERENCES

[1] Babcock, H. W. This symposium, Paper 19.
[2] Babcock, H. W. *Astrophys. J.* **114**, 1, 1951.
[3] Deutsch, A. J. *Publ. A.S.P.* **68**, 92, 1956.
[4] Satô, Y. *Ball. Earthquake Research Inst., Tokyo*, **28**, 175, 1950.
[5] Deutsch, A. J. *I.A.U. Transactions*, vol. 8 (Rome, 1952), Cambridge University Press, 1955, p. 801.
[6] Stibbs, D. W. N. *Mon. Not. R. Astr. Soc.* **110**, 395, 1950.
[7] Burbidge, G. R. This volume, p. 222.

Discussion

Schatzman: I apologize for suggesting a slight increase in complications to the scheme of Dr Deutsch. It seems necessary to take into account the process of line formation in the expression of the weight functions used in numerator and denominator. In the more simple case of pulsating variable stars, Mlle Duquesne and myself have found that the ratio of the observed radial velocity to the material radial velocity is not 17/24 as usually assumed but is smaller by about 30 %.

Deutsch: I was aware that I have used a very approximate relation for the limb-weakening of lines, but in this context it seems adequate. It is surprising to find out how large the effect is for a scattering atmosphere on a Milne–Eddington model. It comes out that for an A-type star you can expect moderately strong lines to be something like 20 % weaker at the limb than they are at the centre of the disk.

NUCLEAR REACTIONS AND ELEMENT SYNTHESIS IN STELLAR ATMOSPHERES*

E. M. BURBIDGE, G. R. BURBIDGE and WILLIAM A. FOWLER

Mount Wilson and Palomar Observatories, and Kellogg Radiation Laboratory, California Institute of Technology, Pasadena, California, U.S.A.

ABSTRACT

A modified discussion of surface nuclear reactions in magnetic stars is given. The anomalous abundance effects found in magnetic stars are briefly described. It is suggested that the processes of particle acceleration are similar to those taking place in the solar atmosphere which give rise to the cosmic ray bursts observed by Wild, Roberts, and Murray, and to the solar component of cosmic radiation. Calculations of the rate of loss of energy following particle acceleration suggests that the duration of the hot spot is $1 \lesssim$ sec. It is estimated that in the region of acceleration (p, n) reactions will enable a ratio $n_n/n_p \simeq 10^{-2}$–10^{-3} to be built up. The majority of these neutrons will diffuse from the excited regions and form deuterium in the quiescent atmosphere. This deuterium will be continuously built up and re-acceleration will lead to the release of neutrons, some of which will be captured by the Fe group, eventually giving rise to the observed anomalous abundances of the heavy elements. Also the reaction $H(d, \gamma)$ He^3 may give rise to the formation of some He^3.

I. OBSERVATIONAL REASONS FOR CONSIDERING SURFACE NUCLEAR REACTIONS

As a class, the 'peculiar' A and F stars have long been known to show apparent anomalies in the spectrum lines of certain elements. It is also certain from the work of H. W. Babcock that there is a very good correlation between spectral peculiarities and the existence of a strong magnetic field. In fact, as far as known there is a one-to-one correlation between spectral peculiarities of this sort and the existence of a magnetic field. For a long time it was not known whether or not the spectral peculiarities were indications of real abundance anomalies in the stellar atmospheres, as they might have been the result of unusual ionization and excitation conditions. However, recent curve-of-growth analyses have shown that the

* Supported in part by the joint program of the Office of Naval Research and the U.S. Atomic Energy Commission.

abundance anomalies are real. The results for three 'peculiar A' stars, α^2 Canum Venaticorum (Burbidge and Burbidge[1]), HD 133029 (Burbidge and Burbidge[2]), and HD 151199 (Burbidge and Burbidge[3]) are shown in Table 1. The abundances were determined relative either to a standard star of about the same spectral type, or to the sun. Of these stars, the first is a well-known spectrum and magnetic variable, and the abundances which are given in the table are mean values, obtained through the period of variation. In some cases the variations in equivalent width of spectrum lines through the period are considerable, and lead to variations in the derived abundances by factors of 5 or, in the case of europium, even more. These represent changes which may be due to separation of elements in localized areas on the stellar surface (Deutsch[4]), perhaps connected with hydromagnetic phenomena through the magnetic variation, and we shall be concerned only with the enhancement of certain element abundances in the stellar atmosphere as a whole, so that these variations are outside the scope of our work. The second star in Table 1 is a magnetic variable, but apparently not a spectrum variable. The third star probably has a magnetic field, according to recent work by Babcock.

Table 1. *Abundances of the elements in three 'peculiar A' stars, relative to those in a normal star*

Element	α^2 CVn	HD 133029	HD 151199
Mg	0·4	1·4	1·2
Al	1·1	2·2	—
Si	10	25	1·3
Ca	0·02	0·05	2·6
Sc	0·7	—	—
Ti	2·6	2·3	—
V	1·3	3·2	—
Cr	5·2	10·3	1·8
Mn	16	15:	9
Fe	2·9	4·1	1·1
Ni	3·0	2	—
Sr	14	11·5	65
Y	20:	—	—
Zr	30	40	—
Ba	0·9	—	0·6
La	1020:	200:	—
Ce	400	190	—
Pr	1070	630	—
Nd	250	200	—
Sm	410	260	—
Eu	1910	640	130
Gd	810	340	—
Dy	760	460	—
Yb	—	—	—
Pb	1500	1500:	—

223

There are three possible explanations of the abundance anomalies:

(a) that the material in question has been accreted from the interstellar gas;

(b) that the material is a typical sample of material in the stellar core, where nuclear reactions have been taking place, and has been mixed to the surface;

(c) that the material has undergone nuclear reactions in the stellar atmosphere.

The first possibility can be dismissed immediately, since it would be applicable only if certain elements were accreted preferentially, or if the interstellar gas had these same abundance anomalies, both of which suggestions are highly improbable.

The second possibility might be the case if a source of neutrons were available in the stellar interior, since, as will be discussed later, the anomalies are mainly produced by neutron capture processes. Recent theoretical work on element synthesis has shown that to build heavy elements in this way the star must have reached an evolutionary stage at which it has a helium core and a hydrogen shell energy source. Stars at this stage must have reached the region of the red giants in the Hertzsprung–Russell diagram, or even a later evolutionary stage. On the other hand, there are a number of arguments which strongly suggest that the 'peculiar A' stars are relatively young, on or near the main sequence, and hence without large helium cores. For example, they occur in galactic clusters in which the main sequences extend up to the region in which they lie (Fowler, Burbidge and Burbidge[5]).

Thus we are left with possibility (c). If the process is a surface effect then the amount of nuclear activity which is demanded is directly related to the amount of mixing of the material in the stellar atmosphere with the material in the star's interior. A very interesting point is now brought to light. It seems probable that the normal main sequence A-type stars have very thin outer convective zones. The work of Rudkjöbing[6] and Vitense[7] shows that convection sets in up in the photosphere, but the convective zone will terminate where the temperature is under 20,000°, and the total thickness of the whole zone is then only $\sim 10^3$ km. Only in regions of great electromagnetic disturbance in the magnetic stars can we expect the mixing to greater depths to be important. The actual amount of the mixing will depend on the type of magnetic model that is assumed. It is of some interest to note that the models currently considered plausible for the sun by Bullard[8], by Parker[9], and by others, in which a toroidal field is convected to the surface to form sunspot pairs, is already in difficulties if it is applied to the magnetic stars, since they apparently have no deep convective envelopes. The total amount of material—the outer skin of the star—which must be processed by nuclear activity is somewhat uncertain. However, our original estimate that about $2 \cdot 5 \times 10^{28}$ g (corresponding to a total depth of 10^4 km with a mean density of 10^{-4} g/cm^3) has been contaminated by these processes still appears to be a reasonable one.

In previous work we supposed that the acceleration of ions by electromagnetic processes in stellar atmospheres gave rise to conditions which were suitable for element synthesis to take place. Two models were suggested, and it is a modification of these that we shall discuss here. In all cases, in order that synthesis of the observed heavy-element abundances can take place, a source of free neutrons must be available. Free neutrons can be produced only by (p, n), (α, n) and similar reactions. These involve threshold energies ~ 5 MeV for (α, n) reactions and proton energies in the range 3 MeV for C^{13} (p, n) to $18 \cdot 5$ MeV for C^{12} (p, n). Thus the problem of nuclear synthesis is that of accelerating a fairly large flux of protons and α particles to relatively low energies, and allowing them to interact locally. On the other hand, for the cosmic ray problem it is necessary to show that a fairly small flux of ions with energies $\geqslant 100$ MeV/nucleon is produced and ejected from a stellar atmosphere. The two processes are to some extent mutually exclusive at a particular level in the stellar atmosphere, since if a sufficiently large flux of high-energy protons and α particles is produced together with a low-energy component, spallation reactions of the high-energy particles will break up the nuclei which may have been built up in the neutron capture processes.

2. EVIDENCE CONCERNING PARTICLES ACCELERATED IN THE SOLAR ATMOSPHERE

In previous work we have suggested that the generation of the 'hot spots' in which neutrons are produced is due to a mechanism of the Swann type in which ions are accelerated in changing magnetic fields. Since our understanding of this mechanism is dependent on an adequate theory of magnetic stars, it is extremely difficult at present to estimate exactly how much magnetic activity is demanded to provide the energy for nuclear reactions. However, a number of observations of particles associated with solar activity suggest that an extrapolation from the sun to the 'peculiar A' magnetic stars is reasonable. The types of observation connected with the sun are then very significant.

Wild, Roberts, and Murray[10] have observed radio bursts which are sometimes associated with solar flares. They divide the bursts into three different types, each of which appears to be associated with a discrete range of velocities. Those associated with the highest velocities (type III bursts) are thought, from measurements of the frequency drift, to be caused by the coherent motion of fast particles upward in the solar atmosphere. These particles have velocities lying in the range $0 \cdot 1c$—$0 \cdot 3c$, which corresponds to proton energies in the range 5–50 MeV. These bursts have durations of 5–10 sec, and sometimes a cluster of bursts with a total duration of 1–2 min is observed.

Firor, Simpson, and Treiman[11] have made an attempt to estimate the total flux of particles emitted by the sun in the BeV range, using their observations of the low-energy solar cosmic-ray component. They find that the total flux is 2×10^{23} particles/sec/BeV at 4 BeV. Although there is some evidence for particles of much lower energies arising in the sun, both from auroral observations (Meinel and Fan[12]; Meinel[13]) and from balloon flights at the top of the earth's atmosphere (Meredith, van Allen, and Gottlieb[14]), no quantitative estimates of the average number of particles emitted has been made. However, if we assume that the particles accelerated follow a $1/E^3$ law (as is the case for an induction-type mechanism), then the results of Firor, Simpson, and Treiman suggest that about 10^{29} particles/sec are produced with energies in the range 4–14 MeV in the sun. Since only the particles which escape are observed by the Chicago group, this may be an underestimate.

Note added in proof

Recent results of Goldberg, Mohler and Müller (*Ap. J.* **127**, 302, 1958) show that deuterium probably builds up to approximately 10 % of the abundance of hydrogen in the material surrounding a solar flare such as that of 23 February 1956. This prompts us to abandon the conservative position taken in §5 below and to accept the possibility that some active regions do reach a temperature near 10^{10} degrees ($kT \sim 1$ MeV) with a neutron-to-proton ratio near 10 %. These neutrons diffuse into the surrounding material and can build up a 10 % concentration of deuterium in an amount of material equal to that in the active region. At the same time, each nucleus in the metal group will capture ~ 1 to 10 neutrons, if we assume that the surrounding material is raised in temperature to $\sim 10^5$ degrees where the metal group capture cross-sections become ~ 10 to 100 times that of hydrogen. In several processings of the surface material the observed over-abundances of the heavy elements will thus be synthesized. The deuterium will build up in the surface material to a figure at most equal to 10%, because it is destroyed in re-cycling even in low-temperature spots (10^7 to 10^9 degrees) by the $D(p, \gamma)$ He³, $D(d, n)$ He³, and $D(d, p)$ T^3 ($\beta^-\nu^-$) He³ reactions. An eventual 10% ratio for He³ to H in the surface material is not at all out of the question; this was the value which was obtained from the analysis by Burbidge and Burbidge (*Ap. J.* **124**, 655, 1956) of the magnetic star 21 Aquilae.

We have been encouraged in the above point of view by the detailed analysis of Parker (*Phy. Rev.* **107**, 830, 1957) of the acceleration mechansim of cosmic rays in solar flares. The E^{-5} power law for the cosmic-ray energy distribution which he finds theoretically agrees approximately with that found experimentally for the 23 February 1956 flare (Meyer, Parker and Simpson, *Phys. Rev.* **104**, 768, 1956). It confirms our belief that spallation processes at high energy will not counteract heavy element build-up, especially since the high-energy acceleration takes place only in regions of relatively low density ($\rho \lesssim 10^{-12}$). The hot spots which we have discussed must occur in the regions below the flares and may reach the high temperatures (10^{10} degrees) required for copious neutron production ($n/p \sim 10\%$) by some kind of an extensive magnetic 'pinch' effect.

3. DURATION OF THE 'HOT SPOTS'

We suppose that the generation of 'hot spots' begins with the acceleration of an appreciable fraction of protons and other less abundant ions. The protons, unless they interact very rapidly with other nuclei, will lose their energy primarily to the electrons in the gas, which will radiate both by 'Bremsstrahlung' processes and through acceleration in the magnetic field. At a density of about 10^{-7} g/cm³, the electrons will come into thermal equilibrium with the protons in times $\sim 10^{-5}$ seconds. The 'Bremsstrahlung' cross-section for the electrons, in the usual nomenclature (Heitler[15], ch. III), is given by

$$\sigma_{rad} = \frac{dE_{rad}}{dx} \frac{1}{nE} = \frac{5 \cdot 5 r_0^2}{137} \text{ for } E = 1 \text{ MeV}$$

$$= 3 \times 10^{-27} \text{cm}^2 = 0 \cdot 003 \text{ barn}. \tag{1}$$

The radiation lifetime for a 1 MeV electron under these conditions is given for σ_{rad} in barns by

$$\tau_{rad} = \frac{2 \cdot 2}{a\rho\sigma_{rad}v}, \tag{2}$$

where a = relative abundance of protons = 1, the density $\rho = 3 \times 10^{-8}$ to 3×10^{-7} g/cm³, and $v = 3 \times 10^{10}$ cm/sec. Thus $\tau_{rad} = 0 \cdot 1$ to 1 sec.

For the radiation in a magnetic field of strength H, if R is the radius of gyration of the electron, we have

$$\left(\frac{dE}{dt}\right)_{rad} = 3 \cdot 40 \times 10^{-16} H^4 R^2 \; \text{MeV/sec}$$
$$\simeq 10^{-8} H^2 \; \text{MeV/sec.} \qquad (3)$$

Thus for a 1-MeV electron, if $dH/dt \gg 10^4$ gauss/sec, then radiation in the magnetic field will be the predominant mode of energy loss, while for fields such that $dH/dt \ll 10^4$ gauss/sec, 'Bremsstrahlung' will predominate.

In general it seems, therefore, that the time scale for the 'hot spots' must be $\lesssim 1$ sec.

4. SIZES OF 'HOT SPOTS'

The dimensions of the regions in which particles are accelerated can be estimated theoretically only if the mechanism is fully understood. Failing this, we can appeal to the observational data on magnetic stars to place upper limits on the 'hot spot' dimension, as follows. We shall suppose that the sum of these areas present at any one time on the visible hemisphere of the star comprises a fraction of the disc characterized by the dimension a, and that these areas have a mean radiation temperature T_a. In following sections we shall consider accelerated proton fluxes which have some of the characteristics of a Maxwell–Boltzmann distribution of energies with $kT \sim 1$ MeV. The radiation losses from these particles are such that the radiation temperature $T_a \ll T$, but also it is probable that $T_a \gg T_{eff}$, where T_{eff}, the normal effective temperature of an Ao-type star, is 11,000°.

The Rayleigh–Jeans approximation to Planck's law can be used to derive the radiative flux from the hot areas, i.e.

$$I_\lambda = 2kcT\lambda^{-4}. \qquad (4)$$

This can be compared, at any wave-length λ, with the flux emitted by the undisturbed photosphere, calculated from Planck's law for a black body at 11,000°. Now we have the color measurements U-B and B-V by Provin[16] for a number of magnetic stars, where U, B, V are the stellar magnitudes on the photometric system of Johnson and Morgan, and represent filter-plate combinations with transmission maxima near the wave-lengths 3600, 4250, and 5300 Å respectively (Johnson[17]). Provin's colors for the 'peculiar A' stars in general indicate a slightly higher apparent temperature than that corresponding to the spectral type. If we postulate that this may be due, in part at least, to the presence of small 'hot spots' on the stellar surface, we may use Provin's color measurements, relative to the colors of a normal main sequence A o star, to derive an order of magnitude for the upper limit to the characteristic dimension of the 'hot spots' present at any one time. Using the measures for α^2 CVn, as a typical magnetic 'peculiar A' star, we find that for $kT_a = 0.1$, 1, and 10 KeV, $a = 3.8 \times 10^9$, 1.2×10^9, and 3.8×10^8 cm, respectively (we have taken the stellar radius to be $2R_\odot$, and have neglected limb-darkening and fore-shortening of spot areas).

Provin also gives measures of the colour and luminosity variation through the period of magnetic variation of α^2 CVn. From these we may estimate the total allowable *change* in 'hot spot' dimension a. For $kT_a = 0.1$, 1, and 10 KeV, we find that $\Delta a = 1.3 \times 10^9$, 4.0×10^8, and 1.3×10^8 cm, respectively.

5. THE PRODUCTION OF NEUTRONS

At sufficiently high particle temperatures, neutrons can be produced by a multitude of (p, n) and (α, n) reactions, while if very high radiation temperatures could be developed, large-scale neutron production by inverse beta-decay processes would be possible. The usual equations of statistical equilibrium (R. H. Fowler[18], Chandrasekhar and Henrich[19]) suggest that if both the particle and radiation temperatures were about 10^{10} degrees, then a ratio of protons to neutrons of about 6 would be reached at equilibrium. However, this would demand that the mass density of radiation $\simeq 10^3$ g/cm^3. If such an energy could be supplied by the magnetic field, the local magnetic intensity would have to reach a value of $\sim 5 \times 10^{12}$ gauss, which appears to be very unlikely. Under the reasonable conditions suggested above, in which only a particle flux of high intensity is present, the main contribution to neutron production will be (p, n) reactions on the light nuclei. It is reasonable to assume that a neutron-proton ratio of the order of 10^{-3} to 10^{-2} will be built up, on the basis that each light nucleus will interact with protons or α particles to produce several neutrons.

6. THE DIFFUSION OF NEUTRONS AND THE FORMATION OF DEUTERIUM

After the free neutrons are produced, they will interact with hydrogen, and eventually they will either form deuterium in the spots while a small proportion will be captured by the elements in the Fe abundance peak, synthesizing heavier elements directly, or they will escape from the spots to be moderated and form deuterium in the quiescent regions. The diffusion length L_D for neutrons can be calculated as follows. We have (Feld [20])

$$L_D = \left(\frac{l_{\text{trans}} \times l_{\text{cap}}}{3}\right)^{1/2},$$

where l_{trans} and l_{cap} are the transport length and the capture length, respectively. In hydrogen $\sigma_{\text{trans}} = \sigma_{\text{np}}/3$, where σ_{np} is the neutron-proton scattering cross-section. Thus

$$L_D = (l_{\text{np}} l_{\text{cap}})^{1/2}.$$

At energies of 1–2 MeV, $\sigma_{\text{np}} \simeq 3$ barns (Blatt and Weisskopf [21]), and σ_{cap} on the proton-rich isotopes of C, N, O, and Ne is $\sigma_{\text{geom}} \simeq 0.33$ barn.

Now

$$l = \frac{2.2}{a\rho\sigma},$$

where σ is in barns, and a is the abundance relative to hydrogen of the interacting nucleus. Since $a(\text{C, N, O, Ne}) \simeq 2 \times 10^{-3}$,

$$L_D\rho = 48.$$

If $\rho = 10^{-7}$ g/cm^3, then $L_D = 4.8 \times 10^8$ cm. Thus for spot radii $< 5 \times 10^8$ cm the majority of the neutrons moving in levels parallel to the star's surface will diffuse out of the hot areas before being captured, and will then be thermalized and form deuterium. If the total 'hot spot' area considered in §4 is an average area made up of a large number of smaller areas, then the neutrons will inevitably diffuse out before being captured.

The vertical diffusion problem is more complicated, but we have supposed that the depth of the 'hot spots' $\sim 10^8$ cm, and calculation of the scale height for an A-type star suggests that vertical diffusion of the neutrons both inward and outward away from the 'hot spot' will be possible. There is uncertainty, however, in the effective depth of the 'hot spot'.

It thus appears that following their production, the neutrons will escape from the accelerated regions into the quiescent atmosphere, where they will be thermalized and rapidly captured by protons to form deuterium which will be continuously built up on the surface until it in turn is depleted in the 'hot spots'.

7. THE SYNTHESIS OF THE HEAVY ELEMENTS AND He^3 BY REACTIONS BETWEEN PROTONS, NEUTRONS, DEUTERONS, AND THE METALS

In order to build the heavy elements to the over-abundances shown in Table 1, about 20 neutrons per Fe nucleus are demanded (Fowler *et al.* [5]). It is first necessary to decide whether the spots which are already required to provide the neutrons will be capable of synthesizing elements by means of the deuterium already present in material which has been produced in previous spot activity. If $kT \simeq 1$ MeV, we showed (Fowler et al. [5]) that if sufficient time were available the neutrons would eventually be captured by the metallic elements; the mean number N of neutrons captured per Fe nucleus was given by

$$N = 1 \cdot 5 \rho t \times 10^4.$$

At $\rho = 10^{-7}$ g/cm^3, if $t = 1$ sec, $N = 1 \cdot 5 \times 10^{-3}$, and in order that 20 neutrons should be captured per Fe nucleus, this would mean that each gram of material would have to be processed $\sim 10^4$ times. However, under these conditions further neutrons would be produced in each spot at the same time, and the final deuterium abundance in the surface would be about thirteen times the abundance of hydrogen. This result is completely in contradiction with observation, and it is clear, therefore, that a second type of 'hot spot' must be responsible for the synthesis. The characteristic of this second type of acceleration is that it must produce a flux of protons which contains practically no particles of high enough energy for neutron production to take place by (p, n) or (α, n) reactions. The neutrons are then produced only by the $H(\alpha, n)$ $2H$ reaction from the deutrons which have been produced in other regions. A Maxwell–Boltzmann distribution with $kT \sim 0 \cdot 2$ MeV or a flux having $N(E) \propto E^{-n}$ with a sharp cut-off near $E = 6$ MeV are possible cases to be considered. Thus in spots in which synthesis takes place, the reactions which we have to consider are those involving protons, neutrons, and deutrons: $H(d, \gamma)$ He^3, $H(d, n)$ $2H$, $H(n, \gamma)$ D, and the synthesizing reactions on the heavy elements which are either (n, γ) capture reactions, or (d, p) stripping reactions at low energy (Oppenheimer–Philipps processes).

The protons with very low energies will only transform the deuterium into He^3 by the $H(d, \gamma)$ He^3 reaction. The cross-section for this has been measured by Fowler, Lauritsen, and Tollestrup [22] to be given by

$$\sigma = 0 \cdot 74 E_p^{0 \cdot 72} \times 10^{-5} \text{ barn.}$$

Above the threshold for the $H(d, n)$ $2H$ reaction the cross-section is about $0 \cdot 1$ to 1 barn, so that the mean lifetime for the break-up of the deuteron $\tau_d = 0 \cdot 002$ to $0 \cdot 02$ sec, where we have put $\rho = 10^{-7}$ g/cm^3, $v = 3 \times 10^9$ cm/sec. On the other hand the cross-section for the $H(n, \gamma)$ D reaction is only about 10^{-4}–10^{-5} barn at these energies, so that the mean lifetime for the formation of deuterium is about 100–1000 sec. Thus in the duration of the spot, a large proportion of the deuterium will be disintegrated. Thus the majority of the heavy-element synthesis will take place through the capture of free neutrons, though there will be a very small contribution from Oppenheimer-

Philipps reactions, which have a cross-section ~ 0.1 barn with an energy dependence roughly proportional to E^2 in the energy range $2\text{ MeV} \leqslant E \leqslant 8$ MeV for the metallic elements (Peaslee [23]).

The cross-section for the (n, γ) reactions on heavy nuclei have been given in our previous paper, and we have shown earlier in this section that in order that about 20 neutrons per Fe nucleus should be captured on the average, each gram of material would have to be processed 10^4 times. The ratio r of the rates of the reactions which synthesize He^3 from deuterium and synthesize heavy elements by free neutron capture is given by

$$r = \frac{a_p a_D \displaystyle\int_{E_1}^{E_2} \sigma(d, \gamma)\, N(E_p)\, dE}{a_n a_{Fe} \displaystyle\int_{E_4}^{E_3} \sigma(n, \gamma)\, N(E_n)\, dE}. \tag{9}$$

Here we have supposed that $N(E)$ is a function representing the primary energy spectrum of accelerated particles. We shall suppose that the neutrons released by the $H(d, n)\, 2H$ reaction attain a similar energy spectrum. Also, since the neutrons are freed from the deuterium in a time which is short compared with the life-time of the spot, we can suppose that the number of neutrons present in the spectrum of accelerated particles is equivalent to the number of deuterons which would be present in the energy range above the neutron production threshold, $E_3 - E_4$. Thus we may put $a_n = a_D$ in Eq. (9).

An estimate of the numerical value of this ratio can be made by using the formula given by Fowler et al. [22] for $\sigma(d, \gamma)$ and by writing

$$\sigma(n, \gamma) = 4.8 \times 10^{-3} E^{-0.5}$$

(cf. Fowler et al. [5]). If we put $N(E) \sim E^{-1}$, E^{-2}, E^{-3}, and set $E_1 = 0.5$ MeV, $E_2 = E_3 = 2.5$ MeV, $E_4 = 6$ MeV, we find that $r \simeq 270$, 800, and 3000, respectively. Thus, in order that on the average each Fe nucleus shall capture about 20 neutrons, this mechanism demands that $20r$ neutrons in the form of deuterium should be available, and the majority of this will form He^3. The total abundance of He^3 which is in equilibrium with the over-abundant heavy elements will be given by

$$\frac{a He^3}{a_p} = 20r \times 2.2 \times 10^{-5},$$

and if $r \simeq 200$

$$\frac{a He^3}{a_p} \simeq 10^{-1}.$$

He^3 will be destroyed mainly by the He^3 (He^3, $2p$) He^4 reaction, but the effect of this will be small as compared with the $H(d, \gamma)\, He^3$ reaction. Thus it appears that the He^3 abundance may easily build up to a value very close to that of He^4 (the aHe^4/a_p ratio in normal stars $\sim 10^{-1}$). The a_D/a_p ratio (see note on p. 226) will be of the same order as the aHe^3/a_p ratio. The production of He^3 depends very sensitively on the form of the spectrum of the accelerated particles. If, for example, the spectrum has the Maxwell–Boltmann form, so that $N(E) \propto E^{1/2} \exp(-E/kT)$, a value of $kT \sim 0.2$ MeV will lead to practically all of the deuterium being converted into He^3, with very little synthesis of the heavy elements. In fact, unless the energy spectrum is fairly flat over the energy range which we have considered, the ratio of the amount of He^3 produced to that of the heavy elements will be very large and, to give the required production of the heavy elements, will lead to an amount of He^3 incompatible with the observed total abundance of He ($He^3 + He^4$). If synthesis took place following a burst of mono-energetic particles, then either only production of He^3, or only production of the heavy elements, would occur, depending on whether the particle energy was below or above the threshold for the $H(d, n)\, 2H$ reaction.

Thus a situation in which the ratio aHe^3/normal aHe^4 becomes greater than unity, and in which little synthesis of the heavy elements has occurred, may easily arise. In this case a magnetic star might be expected to show an apparent over-abundance of He, which would be detectable as He^3 if the effect of the isotopic shift on the spectrum lines could be measured. The magnetic star 21 Aquilae has moderately strong lines of He, and lines of Si II which are too strong to be compatible with them, although an abundance determination has not yet been carried out. Lines of Eu II and the other rare earths are not visible. The isotope shift between He^3 and He^4 should, if He^3 is present, be detectable by a wave-length shift of the spectrum lines $\lambda\lambda\,4388,\,4144$, and 4009, for which the difference between the two isotopes is $\Delta\lambda \simeq 0.3$ Å, relative to the lines at $\lambda\lambda\,4471,\,4121$, and 4026, for which $\Delta\lambda \simeq 0.1$ Å (Greenstein [24]; Fred et al. [25]). Preliminary results for 21 Aql, using plates obtained by H. W. Babcock and one obtained by Burbidge and Burbidge [26] indicate that such a shift may actually be present and may indicate an abundance of He^3 comparable to that of He^4.

8. THE AMOUNT OF ACTIVITY DEMANDED FOR SYNTHESIS

The amount of activity demanded on the stellar surface to produce the observed heavy-element anomalies can be estimated as follows. Since only a fraction of the neutrons may be available for heavy-element synthesis, the primary consideration is that enough neutrons should be produced to build He^3 through the sequence of events outlined in § 7. We have earlier estimated that the total amount of material which has been contaminated has a mass of $\sim 2 \cdot 5 \times 10^{28}$ g. Thus, if the He^3/H ratio is built up to about 10^{-1}, and if a similar ratio of D/H to that for He^3/H is assumed, the total number of neutrons which must be produced by (p, n) reactions is about 5×10^{51}. If the neutrons are produced mainly through the (p, n) reactions on light nuclei in spots in which the particle temperature is given by $kT = 1$ MeV, the sum of whose areas is such that the total effective radius is 10^9 cm, with a depth of 10^8 cm, and a mean density of 10^{-7} g/cc, then the total mass involved in each sum of spot areas present at any one time is about 3×10^{19} g, and the number of neutrons produced is $\sim 2 \times 10^{40}$. Thus, in order to produce the total neutrons required, about $2 \cdot 5 \times 10^{11}$ such spots or sets of spots are required. If the total time-scale available is $\sim 10^{16}$ sec (cf. Fowler et al. [5]), then the frequency of neutron-producing spots is about one (or one set) every 11 hr.

On the other hand, if the spots in which the synthesis takes place involve the same mass as those in which the neutrons are produced, then the number of spots demanded to process the surface material once is $\sim 10^9$. However, we have shown earlier that each gram of material must be processed about 10^4 times. Thus the total number of synthesizing spots must be about 10^{13}. Hence the frequency of these spots is about one every 10^3 sec.

Some comparison can be made between this activity and that on the sun. In § 2 we estimated, from the work of Firor et al. [11], that a representative number of particles in the MeV range was $\sim 10^{29}$/sec. Thus in 10^{16} sec about 10^{45} particles will have been accelerated. If the efficiency of production of neutrons is only about $\sim 1 \%$, this means that a magnetic star must have accelerated $\sim 5 \times 10^8$ times as many particles as the sun. It is difficult to devise theoretical methods of making comparisons, but if the activity is in any way related to the relative amounts of magnetic energy available, it is interesting to note that, since the mean solar field is ~ 1 gauss and that in magnetic stars is $\sim 10^4$ gauss, the ratio of surface magnetic energies in the two stars is $\sim 10^8$.

The estimates which have been made here are necessarily very uncertain. In particular, our concept of 'hot spots' may be superseded by models based on the particle fluxes responsible for the solar bursts observed by Wild et al. Thus it may be that the atmosphere of a magnetic star is continuously in eruption with large numbers of isolated small-scale flares taking place continuously over the surface.

9. CONCLUSION

The conditions for surface nuclear reactions which have been described differ from our previous ideas in a number of ways. In particular, by taking into account the radiation losses following particle acceleration, which we previously neglected, we have found that the time scale involved in discrete events is only of the order of seconds. The mode of neutron production, principally through (p, n) reactions on light nuclei, is the same as that proposed previously in our model (b), and the synthesis is still believed to take place through free neutron capture processes, but these will occur after the neutrons have been freed from the deuterium in which they customarily reside. Necessarily associated with this synthesis will be the production of He³ by the $H(d, \gamma)$ He³ reaction. The ratio between the two rates is such that we may find, in stars with anomalous heavy-element abundances, an amount of He³ equal to the normal He⁴ abundance. In equilibrium a similar amount of deuterium will also be present; it is still not certain what is the amount of deuterium on the star's surface which could definitely be detected observationally.

The details of the over-abundance ratios, and in particular the anomalies in certain elements such as Ba, still remain largely unexplained.

The mechanism of acceleration of the particles is still not understood in any detail, and we have been forced to appeal to the solar observations. Accelerations of the Swann type, the neutral-point type, or plasma mechanisms, may be effective.

Finally we wish to acknowledge the many helpful discussions on all aspects of this problem which we have had with Professor R. F. Christy.

REFERENCES

[1] Burbidge, G. R. and Burbidge, E. M. *Astrophys. J.* Suppl. **1**, 431, 1955.
[2] Burbidge, E. M. and Burbidge, G. R. *Astrophys. J.* **122**, 396, 1955.
[3] Burbidge, G. R. and Burbidge, E. M. *Astrophys. J.* **124**, 130, 1956.
[4] Deutsch, A. J. This symposium, Paper 24.
[5] Fowler, W. A., Burbidge, G. R. and Burbidge, E. M. *Astrophys. J.* Suppl. **2**, 167, 1955.
[6] Rudkjöbing, M. *Z. Astrophys.* **21**, 254, 1942.
[7] Vitense, E. *Z. Astrophys.* **28**, 81, 1951.
[8] Bullard, E. C. *Vistas in Astronomy*, ed. A. Beer (Pergamon Press, London, 1955).
[9] Parker, E. N. *Astrophys. J.* **122**, 129, 1955.
[10] Wild, J. P., Roberts, J. A. and Murray, J. D. *Nature, Lond.* **173**, 532, 1954.
[11] Firor, J. W., Simpson, J. A. and Treiman, S. B. *Phys. Rev.* **95**, 1015, 1954.
[12] Meinel, A. B. and Fan, C. Y. *Astrophys. J.* **115**, 330, 1952.
[13] Meinel, A. B. *Astrophys. J.* **118**, 205, 1953.
[14] Meredith, L. H., van Allen, J. A. and Gottlieb, M. B. *Phys. Rev.* **99**, 198, 1955.

[15] Heitler, W. *Quantum Theory of Radiation* (Oxford University Press, 1935), ch. 4.
[16] Provin, S. S. *Astrophys. J.* **118**, 489, 1953.
[17] Johnson, H. L. *Ann. Astrophys.* **18**, 292, 1955.
[18] Fowler, R. H. *Statistical Mechanics*, 2nd ed. (Cambridge University Press, 1936), ch. 16.
[19] Chandrasekhar, S. and Henrich, L. *Astrophys. J.* **95**, 288, 1942.
[20] Feld, B. I. *Experimental Nuclear Physics*, vol. II, ed. E. Segre (1953), pt. 7, p. 460.
[21] Blatt, J. M. and Weisskopf, V. F. *Theoretical Nuclear Physics* (John Wiley and Sons, New York, 1952), p. 70.
[22] Fowler, W. A., Lauritsen, C. C. and Tollestrup, A. V. *Phys. Rev.* **76**, 1767, 1949.
[23] Peaslee, D. C. *Phys. Rev.* **74**, 1001, 1948.
[24] Greenstein, J. L. *Astrophys. J.* **113**, 531, 1951.
[25] Fred, M., Tomkins, F. S., Brody, J. K. and Hamermesh, M. *Phys. Rev.* **82**, 406, 1951.
[26] Burbidge, E. M. and Burbidge, G. R., *Astrophys. J.* **124**, 655, 1956.

Discussion

Biermann: There are fairly strong reasons which indicate that the acceleration processes should take place in high layers of the stellar atmosphere. In contrast, the nuclear reactions take place in any case in rather deep layers, as considered by Dr Burbidge. Thereby the efficiency of the whole process is limited very severely by energy losses due to thermal electrons. The average energy per unit mass may necessarily be confined to the range 10–20 or 25 MeV on account of the preponderance of the spectrum above this energy range. Carrying out more detailed estimates (which may be found in a paper in the *Z. Astrophys.* **41**, 46, 1956), it is found that only if everything combines in the most favorable way can the process be both operative and effective in the sense of Fowler, Burbidge and Burbidge.

Burbidge: I agree that the problem is very difficult but I think this must be the most plausible one among the three possibilities that were discussed.

Spitzer: Is it really so difficult to get rid of the deuterium reaction? Heating the material up to moderate temperature—some 10 keV or more—will dispose of most of the deuterium by the D–D reaction, producing additional neutrons in the process. The cross-section of this process is known to be large.

Burbidge: I do not know but I think that since we only have a small amount of deuterium the rate of destruction by D–D reactions might be rather small.

Severny: Did you actually find He^3 in stellar spectra?

Burbidge: We believe so, although the difference between He^3 and He^4 lines is very small, and a distinction is difficult to make.

PART IV

SOLAR AND INTERPLANETARY
MAGNETIC FIELDS

PHOTOSPHERIC MAGNETIC FIELDS

H. W. BABCOCK AND H. D. BABCOCK

Mount Wilson and Palomar Observatories, Pasadena, California, U.S.A.

ABSTRACT

Since 1952 a total of 635 magnetograms of the sun have been obtained in a systematic investigation of weak magnetic fields in the photosphere. The frequent records give the location, polarity, and intensity of weak fields down to a fraction of 1 gauss, although with resolution limited to about 0·04 of the solar diameter.

Confirmation of previously reported results in 1954 comes from continuation of the series and from observations with a second, improved magnetograph on Mount Wilson. Three types of field pattern are found: (1) the poloidal field in high heliographic latitudes, consistently positive in the north, negative in the south, with intensity of the order of 1 gauss; (2) BM (bipolar magnetic) regions, often weak and extended, but which when strong are associated with plages, spots, flares, coronal emission, chromospheric fine structure, and filaments; and (3) UM (unipolar magnetic) regions, rather extended and weak, occurring in low latitudes, and associated in time with 27-day recurrent geomagnetic storms and cosmic-ray fluctuations. Attention is directed to the probable disposition of the magnetic flux in the high atmosphere and in interplanetary space, consistent with the observed magnetic areas and with the restriction div H = 0.

Alfvén has argued that the interpretation of the small Zeeman displacements is meaningless and irrelevant because the rather strong turbulent fields presumed to prevail in granules might be coupled systematically, in respect to magnetic polarity, with the intensity of the absorption lines used for measurement. But this would produce a bias, with a shift of zero point of magnetic intensity, for all observed fields on the disk, and no such bias is observed. The measurements, while limited in resolution, are on an absolute scale, and show, for the 'quiet sun', vast areas with only small random fields no greater than a few tenths of 1 gauss.

The solar magnetograph is an instrument for the measurement and recording of weak magnetic fields in the photosphere of the sun (Babcock[1]). It combines a powerful grating spectrograph with an oscillating analyzer and a sensitive photo-electric detector for measurements of very small Zeeman effects in the spectrum. It responds to $H \cos \gamma$, the component of the magnetic field in the line of sight, and by means of a scanning system with conformal recording, it maps the location, intensity, and polarity of magnetic fields on the disk of the sun. Spurious responses have been eliminated, so that the instrument, limited only by statistical noise, records fields down to a fraction of 1 gauss. The heliometric resolution on the disk is about 1' along the slit, so that the measurement when scanning is an average over a considerable area of the photosphere.

An automatic scanning system shifts the sun's image in a series of parallel traces across the slit of the spectrograph. The dispersion is 1 Å per 11 mm and the resolving power is 600,000. At the focus a double slit is placed on the line Fe I 5250.216. A sensitive, 2-tube, photo-electric detector, balanced to reject common-mode fluctuations, responds only to oscillating

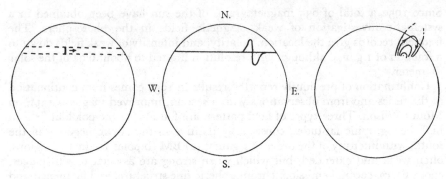

| Path of slit | Deflexion on a BM region | Inferred flux loops |

Fig. 1. Response trace of magnetograph as slit scans image.

Zeeman displacements of the line that result from the alternating sign of the electro-optic analyzer in front of the slit. A moving spot on the cathode-ray tube displays the magnetic effects conformally, and the record is photographed with a time exposure.

As we pointed out before (Babcock and Babcock[2]), an upper limit of the order of 2 gauss can be placed on the fields of individual granules by the observation that, with a slit fixed on a magnetically 'quiet' part of the image, admitting light from approximately 100 granules, the root-mean-square fluctuations are no greater than about 0·2 gauss. Hence it appears that equipartition in the sense expected by Alfvén[3, 4, 5] does not occur. For this reason, among others, we believe that the small Zeeman displacements that are measured are properly interpreted as representing the mean of coherent fields, averaged along the slit, apart from the effect of granules.

Fig. 1 shows schematically the type of response. The slit, having a length equivalent to 0·04 of the solar diameter, moves slowly across the image. The response on the cathode-ray tube is an upward deflexion if the field vector is toward the observer, and vice versa. Deflexions are proportional to the component of the field vector in the line of sight, up to several gauss. The trace shown is typical of a BM region. For such a region, we infer that the flux loops are somewhat as shown in the third part of the figure. BM regions commonly appear rather abruptly, and persist for several days, weeks, or months, with more or less associated activity on the disk and in the higher atmospheric layers.

Plate I shows a typical magnetogram with twenty-two parallel traces. The sensitivity here is such that a deflexion of one trace interval is equal to about 1 gauss. Other magnetograms show examples of BM regions and of a UM (unipolar magnetic) region, as well as deflexions characteristic of the poloidal or 'general' field at high heliometric latitudes.

When smaller surface elements are to be investigated, the desirable resolution in time is correspondingly increased, and records repeated at intervals of only a few minutes would be appropriate. But magnetograms taken at a rate of only one a day are adequate for a rough analysis of the magnetic regions of large and moderate size for which the daily changes are often rather minor. Altogether, more than 650 magnetograms have been obtained during the last four years. The data on the distribution, intensity, and polarity of the magnetic flux as it passes through the photosphere, together with the consequences of the fact that div $\mathbf{H} = 0$, enable us to infer a good deal about the disposition of the lines of force above (and perhaps even below) the photosphere. Thus, in examining the magnetograms, one is usually visualizing the magnetic lines of force and attempting to correlate them with material motions and with other phenomena observed either optically or in the radio spectrum.

We can usually regard the highly conducting material of the sun, within any given volume, as identified with the magnetic flux within that volume. In a sense, the magnetic flux serves as a 'tracer', or tag, for the material, enabling us to follow its movements. For example, the appearance of a bipolar magnetic field on the sun where there was none before indicates that material formerly submerged has been brought to the surface, carrying its magnetic flux with it. The lines of force at the surface are comparatively free to loop upward into the region of the corona. Spreading and weakening of the surface fields with conservation of total flux suggests that the associated material is extending itself over a larger portion of the sun's surface, and that the flux loops are rising higher. Magnetic flux and material motions on a large and moderate scale (from $0 \cdot 3 R$ down to $0 \cdot 03 R$ or less) may thus be traced. The extension of similar observations to yet smaller turbulent elements is a problem of technique that is capable of much further development.

The observational evidence of the magnetic records is now firm in its major respects, and must be taken into account in the development of theories of the sun's magnetic field, both internal and external.

The principal classes of magnetic patterns observed are those of the main poloidal or 'general' field, the BM (bipolar magnetic) regions, and the UM (unipolar magnetic) regions. Flux patterns of these three types are illustrated schematically in Fig. 2.

1. The general field. Evidence for this is usually found only in high heliographic latitudes, roughly above ± 60°, although the limits are rather variable and indefinite. Polarity has been consistently positive in the

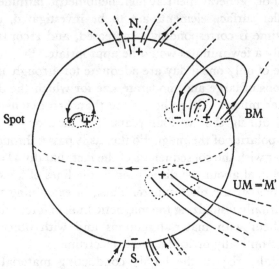

Fig. 2. Schematic representations of magnetic lines of force of (a) the general field, near the poles, (b) two BM regions, one with a spot, and (c) a UM region. Above UM regions, the lines of force are probably carried out to great distances by energetic corpuscular streams expelled from these regions; the lines of force presumably return through widespread areas in which the field is below the limit of detectability.

north, negative in the south. The fine structure is variable. Mean field intensity is of the order of 1 gauss. The coronal features known as polar tufts arise from the same regions, and presumably delineate the lines of force of the general field.

2. BM fields. These are the strongest features of the magnetic records. There are two contiguous areas of opposite polarity. A great diversity is apparent in field intensity, in total flux, and in area. The BM regions often extend towards the poles far beyond the zones of latitude in which sunspots are common. Spots are often found within BM regions, but many BM regions do not contain spots. The preceding and following

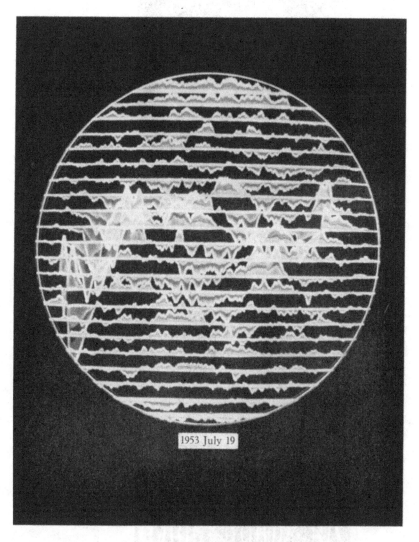

1953 July 19

Plate I. Typical magnetograms, showing extensive weak fields.

(*facing p.* 242)

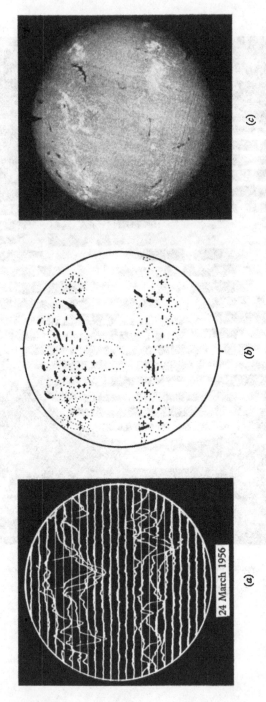

(c)

(b)

(a)

24 March 1956

Plate II. The magnetogram (a) has been converted to a diagram (b) showing magnetic areas of positive and of negative polarity. From the H spectroheliogram (c) the stable filaments have been transferred to diagram (b) to show their approximate position relative to the magnetic areas.

parts of BM regions obey the laws of magnetic polarity found by Hale to apply to spot groups. The regions originate as if loops of a submerged toroidal field were brought to the surface. The more prominent and concentrated regions, in their early stages, usually develop spots, but the expanding BM fields persist long after the spots have disappeared. Whereever the average field of a BM region is stronger than about 2 gauss, bright Ca II plages appear on the spectroheliograms. A less sensitive indicator of the field is bright Hα. However, on spectroheliograms of good quality, the small, fine, light and dark Hα flocculi that form the fine structure of the chromosphere, and which suggest a random pattern in regions free of any extended magnetic field, take on a coherence of alignment if the mean field is 1 gauss or more. With increasing age, the BM regions tend to expand gradually and sometimes asymmetrically as the field intensity diminishes. At this stage, stable filaments—seen in the light of Hα—may appear above the surface. These filaments or prominences are usually disposed either across a BM region so as to divide it into two parts of opposite polarity, or they tend to delineate the border of such a region, preferentially on the poleward side, toward which the region is slowly expanding (Plate II). Such bordering filaments are frequent at the present stage of the solar cycle, when the expanding BM regions seem to push the prominences before them into quite high latitudes. Several BM regions are often found to merge into a complicated pattern. Some of the 'softer' BM regions have weak fields and are very extensive, up to about $0.3R$; they probably do not develop spots or other marked characteristics of active regions.

There is a general correspondence between the arching lines of force, that, by inference, rise high into the solar atmosphere, joining the positive and negative parts of BM regions, and the coronal regions showing the bright green (λ 5303) and red (λ 6374) radiations reported by the coronagraph observers.

3. UM regions. These are rather extensive regions of only one polarity. The strongest mean fields have been about 2 gauss; areas are of the order of one-tenth of the disk. They appear unrelated to other surface features, and it is not apparent where the emergent magnetic flux returns to the sun. The UM regions are transitory, but the outstanding example could be identified on eight successive rotations of the sun. It occurred in 1953 as the last sunspot cycle reached its terminal stages. We have suggested that, as a class, the UM regions are to be identified with the heretofore hypothetical 'M' regions of Bartels. The best UM regions of 1953, on its central meridian passages, preceded by about 3 days the onset of a series

of 27-day recurrent geomagnetic storms (Babcock and Babcock[2]). The CM passages of the same region coincided with a series of 27-day recurrent fluctuations in cosmic ray intensity (Simpson, Babcock and Babcock[6]). We infer that the magnetic flux emerging from UM regions forms a coherent bundle (identifiable with a coronal streamer) extending outward into interplanetary space sufficiently far to intercept the earth on occasion. Presumably the magnetic lines of force identified with the streamers are carried out to a great distance by energetic particles expelled from the photosphere.

As we have remarked before, a number of solar phenomena can be qualitatively related to solar magnetic fields on the assumption that corpuscular emission from the photosphere occurs preferentially in regions where there is a coherent magnetic field, whether near the poles, or in the lower-latitude BM and UM regions. Above the photosphere the ionized corpuscular streams are guided by the lines of force or distort them, depending upon the relative magnetic and kinetic energies; they condense, if trapped in sufficient quantity above BM regions, to form the dark filaments. Above UM regions the corpuscular streams are presumed largely to overbalance in energy the associated fields, thus carrying the fields in extended bundles to great distances.

While most of the principal results described here have been reported earlier (Babcock and Babcock[2]), they have gained added weight through the continued accumulation of data since 1954. Furthermore, the existence, polarity, and order of magnitude of the weak general field have been confirmed by a second magnetograph incorporating certain technical improvements and operating with the advantage of the superior sky of Mount Wilson as compared to Pasadena.

REFERENCES

[1] Babcock, H. W. *Astrophys. J.* **118**, 387, 1953.
[2] Babcock, H. W. and Babcock, H. D. *Astrophys. J.* **121**, 349, 1955.
[3] Alfvén, H. *Nature, Lond.* **168**, 1036, 1951.
[4] Alfvén, H. *Ark. Fys.* **4**, no. 24, 1952.
[5] Alfvén, H. *Tellus*, **8**, 1, 1956.
[6] Simpson, J. A., Babcock, H. W. and Babcock, H. D. *Phys. Rev.* **98**, 1402, 1955.

Discussion

Parker: Do you think it might be possible to use your method also to obtain some idea of the magnetic configuration in the vicinity of a solar flare?

Babcock: I think this is much more difficult. As a rule we can get only one or two magnetograms a day. Flares have a short lifetime, and their adequate observation would require improved resolution.

Schlüter: The correlation between the magnetic field and filaments (quiescent prominences)—showing that filaments occur only where the vertical component of the magnetic field disappears—this correlation is exactly that which I had predicted on the basis of my theory of filaments which I have described in the paper by Kippenhahn and myself (the paper will soon be published). Already in 1954 I tried to verify these predictions using the records then available and kindly made accessible to me by Dr Babcock. At that time, however, the geometrical resolution of the magnetograph was not quite sufficient to give an unambiguous check.

Lehnert: Just let me make some comments on the coupling between a turbulent magnetic field and the density fluctuations. Alfvén has argued that, due to this coupling, the measurements may not give the proper value of the magnetic field at the solar surface. Have I grasped it correctly that the resolution of your instrument corresponds to a region containing about 100 granules?

Babcock: Yes, when the slit is held stationary on the image.

Lehnert: Then, let us suppose that the total magnetic field, $\mathbf{B} = \mathbf{B_0} + \mathbf{b}$, in such a region, D, consists of a homogeneous general field, $\mathbf{B_0}$, and a turbulent field, $\mathbf{b}$. The probability of picking out a sub-region inside D with the field $\mathbf{b}$ in the range $db_x\,db_y\,db_z$ is defined by $f\,db_x\,db_y\,db_z$, where f is a normalized distribution function. Further, for the sake of simplicity, we do not include temperature fluctuations and assume that only the density $\rho = \rho(B_x, B_y, B_z)$ is coupled with the magnetic field. The total contribution to the measured Zeeman effect gives a measured mean of the magnetic field

$$\overline{B_z} = \frac{1}{\rho_0} \iiint_{-\infty}^{+\infty} \rho B_z\, f\, db_x\, db_y\, db_z,$$

where the z-axis has been chosen along the observation line and ρ_0 is the density in absence of turbulent fluctuations. Introducing the excess density $\rho' = \rho - \rho_0$ we may also write

$$\overline{B_z} = B_{0z} + \frac{1}{\rho_0} \iiint_{-\infty}^{+\infty} (\rho' B_z + \rho_0 b_z)\, f\, db_x\, db_y\, db_z.$$

There is no doubt that the measurements give the proper value, $\overline{B_z} \approx B_{0z}$, when the field B_0 is strong (as in sunspots) and $|\rho'/\rho_0| \ll 1$ as well as $|b/B_0| \ll 1$. On the other hand, if the general field $\mathbf{B_0}$ disappears it is easily seen that also B_z disappears. Consequently, I should think that your measurements show that there *exists* a general magnetic field of the sun, regardless of Alfvén's mechanism being of importance or not. But, what I should like to point out is that it has so far not been proved that the effect of turbulent coupling can be neglected.

Babcock: If this systematic coupling is significant, would not a bias be observed when the magnetic field is measured over the whole solar surface?

Lehnert: Not necessarily. Suppose that the measurements are compared in two regions, D_I and D_{II}, on the solar surface, equally distant from the equatorial plane and situated on the same meridian circle as in Fig. 3. The general magnetic field is assumed to be symmetric around a vertical axis in the figure. When comparing the measurements in D_I and D_{II} the observer is imagined to rotate around the line of sight when observing D_{II}. The rotation is carried out such as to bring the general field in D_{II} in a direction anti-parallel to that in D_I (open arrow).

Now, if the current density is reversed at every point within the whole configuration the density distribution can be assumed to be unchanged. Thus, the density has the property

$$\rho(B_x, B_y, B_z) = \rho(-B_x, -B_y, -B_z).$$

Further, the turbulent field $\mathbf{b}$ is assumed to be axisymmetric with respect to the magnetic field direction, i.e.

$$f = f[\hat{\mathbf{B}}_0 \cdot \mathbf{b}, |\hat{\mathbf{B}}_0 \times (\mathbf{b} \times \hat{\mathbf{B}}_0)|]; \quad \hat{\mathbf{B}}_0 = \text{unit vector of } \mathbf{B}_0.$$

We have to distinguish between two cases, namely:

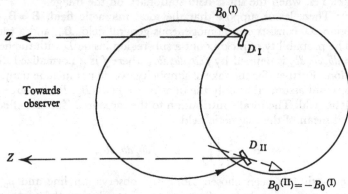

Fig. 3. Measurement of the Zeeman effect in two regions, D_I and D_{II}, of the sun equally distant from the equatorial plane and situated on the same meridian circle. The observer is situated approximately in the equatorial plane. The general magnetic field $\mathbf{B}_0$ is assumed to have a dipole-like shape.

1. The turbulent intensity is the same in D_I as in D_{II}. This applies, e.g. when the turbulence is *homogeneous* all over the solar surface. In such a case the spectral functions f_I and f_{II} in D_I and D_{II} are equal $(f_I = f_{II} = f)$. The proper fields are $\mathbf{B}_0^{(I)} = \mathbf{B}_0$ and $\mathbf{B}_0^{(II)} = -\mathbf{B}_0$, respectively, and the measured fields are $\overline{B_z^{(I)}}$ and

$$\overline{B_z^{(II)}} = -\frac{1}{\rho_0} \iiint_{+\infty}^{-\infty} \rho(-B_{0x} - b'_x, -B_{0y} - b'_y, -B_{0z} - b'_z)(-B_{0z} - b'_z)$$

$$\times f[-\hat{\mathbf{B}}_0 \cdot (-\mathbf{b}'), |-\hat{\mathbf{B}}_0 \times (\mathbf{b}' \times \hat{\mathbf{B}}_0)|] \, db'_x \, db'_y \, db'_z = -\overline{B_z^{(I)}},$$

where we have introduced $\mathbf{b}' = -\mathbf{b}$. Consequently, no bias exists in the homogeneous case.

2. However, if the distribution of the turbulent magnetic intensity over the solar surface is *inhomogeneous* in the sense that $f_I \neq f_{II}$ it also follows that $\overline{B_z^{(I)}} \neq -\overline{B_z^{(II)}}$, and a bias exists.

To sum up, the non-existence of a bias may be explained in two ways. Either does it indicate that Alfvén's mechanism is unimportant, or it implies that the intensity of the turbulent magnetic field is distributed homogeneously over the solar surface.

Alfvén: Professor Cowling has already touched the problem in his introduction yesterday when he said he was inclined to accept observational evidence and that I was questioning observational facts. The observational facts, which Babcock has obtained in an admirable way, consist of a line displacement—that is the observational fact—and I have not at all questioned that. But in order to have the right to put 'magnetic field' instead of 'line displacement' into the resulting diagram one has to use Zeeman effect theory, i.e. the behavior of a single atom, and one also has to use magneto-turbulent theory of the photosphere. Only by using these two theories is it possible to infer anything about magnetic fields. Because a magneto-turbulent theory does not exist at present it is impossible to infer anything about magnetic fields. What Lehnert has pointed out now shows that such a theory is very complicated and we are very far from being able to say anything definitely. I am not at all trying to sketch such a theory but just let me point out the following. Suppose that a general magnetic field of say 10 gauss or perhaps even less is superimposed by a turbulent field of 200 gauss. In some regions the turbulent field is, roughly speaking, in the direction of the external field, in other regions it opposes the external field. Even a very weak coupling with density or intensity of spectral lines could result in a pronounced systematic effect, provided that the turbulent magnetic field is strong enough.

Yesterday, in the discussion on stellar magnetic fields Dr Babcock pointed out that the measured average field could differ from the true average. A systematic coupling of this type could be effective.

Cowling: To explain the discrepancy between the fields estimated by Babcock and those required by Alfvén, it is not sufficient simply to show that turbulent fields may affect the field estimated from observation. In Alfvén's case it is necessary also to show that the effect is regularly to reduce the observed field and not to increase it, i.e. that the sign 'goes the right way'. Objections to the observed value based on the inadequacy of theories of magneto-turbulence cut two ways. I personally do not find the arguments in favor of a turbulent magnetic field of order 200 gauss in the photosphere altogether convincing.

Lehnert: I should like to emphasize that in my comments I have said nothing about the sign of the difference $\bar{B}_z - B_{0z}$. The only thing I want to point out is that this difference hardly vanishes identically and that a non-existence of a bias cannot be used as an argument against Alfvén's hypothesis.

ON MERIDIONAL CIRCULATIONS IN STELLAR CONVECTIVE ZONES

L. BIERMANN

Max Planck Institut für Physik, Göttingen, Germany

ABSTRACT

It is shown in outline by a discussion of the equation of motion in all its components and by taking regard of the geometrical properties of turbulent exchange, that in general the state of pure rotation (without meridional circulations) is not a possible stationary state of motion for convective zones of stars.

It would not be possible, in the time available to me to give a complete survey of existing sunspot theories, to which so much effort has been devoted in recent years by a number of competent workers. Instead I will rather aim at grouping and discussing these theories from a point of view, suggested by the theory of stellar structure which will be developed below. In doing so, I shall make use of certain results concerning the structure of thermal convection and on meridional circulations in convective zones in stars. Only part of these results have been published; more recent ones are outlined in a preliminary note, which has been distributed to this symposium.

I start then by indicating which conceptions regarding the internal structure of the sun will be used as a basis for our discussion.

1. The hydrogen convection zone is regarded as deep, that is to say, of a depth of the order of $0 \cdot 1$ or $0 \cdot 2 R_\odot$, and of nearly adiabatic structure, such that

$$\Delta \frac{d \ln T}{d \ln P} \equiv \frac{d \ln T}{d \ln P} - \left(\frac{d \ln T}{d \ln P} \right)_{ad} \ll 1 \qquad \left\{ \begin{array}{l} R_\odot = \text{solar radius} \\ T = \text{temperature} \\ P = \text{pressure,} \end{array} \right.$$

(but > 0) everywhere, except near the photosphere, by several powers of 10. The convective transport of energy is assumed to dominate; for the detailed picture reference is made to the work of Mrs Boehm-Vitense[1] and to my own work[2,3].

2. For the main part of the interior it is assumed that the radiative transport of energy leads to a sub-diabatic temperature gradient.

3. The existence of a central convective zone, due to the contribution of the carbon cycle to the energy production, has become doubtful; if it exists at all, its instability cannot be pronounced, most of the energy sources being outside.

A crucial point for the hydromagnetics of sunspots and of the solar cycle, in particular for the interpretation of the migration of the activity zones on the surface, is the state of rotation and of large-scale circulation. On the surface we observe pronounced differential rotation such that

$$\frac{\partial \omega}{\partial \phi} < 0 \quad \begin{cases} \omega = \text{angular velocity of rotation} \\ \phi = \text{latitude,} \end{cases}$$

and there are at least observational indications for $\partial \omega / \partial r$ being < 0, where r is the distance from the centre (U. Becker[4]).

From the theory of internal structure, one has, for the radiative deep interior, the theorem of von Zeipel–Vogt–Eddington, according to which in general the state of large-scale motion of a star with non-vanishing angular momentum of rotation, must be one of differential rotation with superposed meridional circulations (of velocity v_m), such that the isobaric and the isothermic surface will generally not coincide. Their angle as well as the relative temperature difference between axis and equator on isobaric surfaces, which I denote by

$$\Delta^* \ln T,$$

will in general be of the order of (g acceleration of gravity, ϵ energy production in erg. g^{-1} sec^{-1})

$$\frac{\omega^2 r}{g} \quad (\approx 2 \cdot 10^{-5} \text{ on the solar surface})$$

and v_m of the order of $(\omega^2 r/g)$ (ϵ/g), that is to say exceedingly slow. Also the initial conditions may still be reflected in the actual law of rotation of the sun. If magnetic fields are present, one has furthermore the familar law of isorotation, of Ferraro[5] and Alfvén[6].

In a convective zone one would, at first sight, expect $\Delta^* \ln T = 0$ owing to the efficiency of the turbulent exchange ('Austausch'), and likewise $\nabla \omega \approx 0$ for the same reason. On the other hand we observe at the surface $\nabla \omega \neq 0$, from which Bjerknes[7] and Randers[8] have concluded that a meridional circulation of the order of 1 m/sec must be present in order to keep stationary this state of rotation in spite of the observed turbulence of the surface layers. The meridional circulation, in the scheme of Bjerknes and Randers, had to be supposed to be maintained by a given $\Delta^* \ln T > 0$ (the sign being fixed by the direction of the required circulation, towards the equator near the surface).

I am now going to discuss an analogue to the von Zeipel–Vogt–Eddington theorem, which, in contrast, applies to convective zones in rotating stars. The starting point is a property of turbulent exchange due to thermal instability, which as such has been discussed already long ago, but has received little attention, that is to say its nonisotropy[9]. Since the primary acceleration in thermal convection is anti-parallel to the direction of gravity, whereas the motions in the plane perpendicular to that direction are only necessitated by the conservation of mass, it is obvious that the turbulence of largest scale under such conditions cannot be isotropic. This follows also from a consideration of the respective properties of large-scale motion and of turbulent exchange in the earth's atmosphere, where the conservation of angular momentum in large-scale motions is the basis for some well-known laws of the winds. It follows, that the turbulent exchange, which strictly speaking is a tensor, for the transport of angular momentum of rotation, cannot be reasonably approximated, by one scalar coefficient. The simplest approximation expressing the non-isotropic character of large-scale turbulence, is a combination of the two scalar coefficients, one of which (A_1) may be said to denote the isotropic part, whereas the other one (A_2) expresses a monotropic transport of angular momentum, in the direction of gravity. The flux of angular momentum of rotation, in this approximation, is thus thought of as being made up by two parts, one of which is analogous to that due to ordinary viscosity proportional to A_1 and to the gradient of the angular velocity of rotation, the other one being essentially the product of the monotropic exchange coefficient A_2 and of the radial derivative of the angular momentum of rotation per unit mass. In the absence of meridional circulations, it would follow, that ω can depend only on the distance r from the centre, its rate of change with r being given by

$$\frac{d \ln \omega}{d \ln r} \approx \frac{-2 A_2}{A_1 + A_2}.$$

For the discussion which follows we assume, as is true for the sun, $\omega^2 r/g \ll 1$ and disregard the contribution of the pressure of radiation and of turbulence to the total pressure gradient. We fix our attention to the momentum balance in a meridional plane. Writing the equations in question in such a form that all terms have the dimension of accelerations, we see by taking the curl that in a stationary state (cf. Randers[8])

$$\text{curl} \ [\boldsymbol{\omega}[\boldsymbol{\omega}r]] = \text{curl} \ \left(\frac{1}{\rho} \nabla P\right) + \text{terms only due to } v_m \ (\text{e.g. } A_1 \ \Delta^2 \mathbf{v}_m).$$

Using only the equation of state (by an expansion effectively in spherical harmonics) it may be shown that

$$\left|\text{curl} \ \frac{1}{\rho} \nabla P\right| \approx \left(\frac{g}{r} \ \Delta^* \ln T\right) \sin 2\phi$$

$$+ \text{terms of higher order} \equiv \omega^2 \left(\frac{g}{r\omega^2} \ \Delta^* \ln T\right) \sin 2\phi + \dots$$

The expression in brackets is either of order unity, in the general case of baroclinic rotation, or $\ll 1$, the special case of almost barotropic rotation ($\Delta^* \ln T \ll \omega^2 r/g$). The ratio g/r varies along r approximately as $1/r^3$.

Independent information regarding $\Delta^* \ln T$ may be gained from the equation of energy transfer by turbulence

$$H_k = \text{const.} \ Pv_t \ \Delta \frac{d \ln T}{d \ln P} \ \text{erg. cm}^{-2} \ \text{sec}^{-1} \ (v_t = \text{velocity of large-scale turbulence}).$$

The non-dimensional constant is of the order unity, if c_p is not increased by partial ionization, or if the effectivity of the transport is not diminished in some way (see below).

Apart from the alternative last mentioned, $\Delta^* \ln (H_k/Pv_t)$ would again be of order $\omega^2 r/g$, and hence $((d \ln T/d \ln P)_{ad}$ being the same on axis and equator)

$$\Delta^* \Delta \frac{d \ln T}{d \ln P} = \Delta^* \frac{d \ln T}{d \ln P} \ \text{of order} \ \frac{\omega^2 r}{g} \ \Delta \frac{d \ln T}{d \ln P},$$

that is to say, small compared with unity to a higher order, owing to the smallness of $\Delta \frac{d \ln T}{d \ln P}$ itself. Integrating over $d \ln P$, the change of $\Delta^* \ln T$ along r is obtained, which is still $\ll \omega^2 r/g$.

For the case of pure rotation ($v_m = 0$) the argument given above ($\omega \equiv \omega(r)$ only), leads to

$$|\text{curl} \ [\boldsymbol{\omega} \ [\boldsymbol{\omega}r]] | \approx -\omega^2 \frac{d \ln \omega}{d \ln r} \sin 2\phi + \text{terms of higher order with}$$

$d (\ln \omega)/d (\ln r)$ of order -1.

The boundary condition at the inner boundary of the convective zone, which is least favourable for meridional circulations in the zone, is obviously $\Delta^* \ln T = 0$. Since $\Delta \ln P$ along r in the convective zone is only of order of 10 it would follow that

$$\Delta^* \ln T \text{ of order } 10 \frac{\omega^2 r}{g} \Delta \frac{d \ln T}{d \ln P},$$

that is

$$\ll \omega^2 r/g$$

throughout the whole convective zone. On the other hand, assuming for the argument the absence of such circulations it would follow that the curl of $[\omega[\omega r]]$ is of the order ω^2 itself.

Hence the curl of $\frac{1}{\rho} \nabla P$ would be small compared with ω^2, but that of $[\omega[\omega r]]$ of order ω^2. It follows, from the contradiction thus reached, that $v_m = 0$ is no possible stationary solution of the momentum equations. It may be concluded, then, that the normal state of motion of the convective zone in a rotating star is such that meridional circulations regulate together with the turbulent exchange the flow of momentum in such a way that the law of rotation is stationary, which in this case must be expected to lead to a variation of ω, both with latitude and with depth. Since the depth is not very small compared with the solar radius, these relative variations would not be expected to be relatively small, as is actually observed to be the case for the variation of ω with latitude and is indicated by the afore-mentioned observations for the variation with depth.

Further discussion, which cannot be presented here, shows that the effect of partial ionization and/or the local diminishing of the effectivity of transport—by the stabilizing action of rotation (Cowling [10]; Biermann [11]) or by magnetic fields (Biermann [12]; Chandrasekhar [13, 14]) must not be disregarded for the actual law of rotation and for the geometry of the circulations, which therefore might be not quite simple. From the analogy of the earth's atmosphere one might expect in the equatorial regions considerable circulations with likewise considerable variations of ω, whereas in the polar regions almost solid body rotation without meridional circulations would be expected. As in the earlier reasoning of Bjerknes and Randers, the observed state of differential rotation together with the isotropic part of turbulent exchange would lead to a meridional velocity of the order of 1 m/sec.

The magnetic observations, as discussed by Babcock at this symposium—which I choose to take at their face value—indicate that in the polar regions, down to a distance of about $45°$, an apparently permanent poloidal magnetic field is present, with an intensity of the order of 1 gauss. This field may well be a relict of a primordial magnetic field of the sun. On the other hand, in the middle and low latitudes the observations strongly indicate the presence only of a local field of essentially toroidal shape as indicated by the BM regions and the sunspots. Both seem to be similar in their general properties and with respect to the total magnetic flux, the lifetime of the magnetic fields brought to the surface being apparently considerably longer than that of visible sunspots. The main difference between a BM region and a bipolar spot appears to be the concentration of the magnetic flux in the photosphere, as to which references is made to the paper of Schlüter and Temesváry to be discussed later. Owing to the long lifetime of magnetic fields even in the surface regions of the sun, the magnetic fields connected with BM regions and spots must be supposed to be brought to the surface by convection, as was pointed out long ago by Cowling[15]. Hence the theoretical picture of differential rotation and meridional circulations as discussed before, and the magnetic observations, seem to fit quite well, in the sense, that the near axis regions seem to be the seat of the permanent poloidal field, the law of rotation conforming to the law of Ferraro and Alfvén, whereas in the low latitudes the meridional circulations do not seem to permit any permanent poloidal component, but strongly favour magnetic fields of toroidal character as long-lived features.

Turning now to the several schemes proposed for understanding the cycle of solar activity and the properties of sunspots, I will start by indicating, as a central problem, the question of whether the change of location of the activity belts with the cycle is due to mass motion or due to a change of the physical state, that is to say to a wave-type phenomenon.

The picture proposed by Alfvén[16, 17, 18, 19] and Walén[20] about ten years ago, and similarly some other schemes proposed by more recent workers, agree in assuming a poloidal magnetic field also in low latitudes, along which disturbances may be propagated giving rise to travelling or to standing waves. Now I do not want to repeat the arguments advanced already earlier, e.g. by Cowling against these schemes. What I want to emphasize is, that, apart from the low value of the magnetic intensity now observed, the theoretical results concerning circulations presented here, again seem to disfavour strongly any possibility that stationary poloidal magnetic fields exist in the regions in question. One might argue, of course, that the influence of magnetic fields on the deceleration of meridional circulations has not been taken account of in the analysis above; but it is seen without much difficulty, that although magnetic fields being able to restore the balance of momentum can easily be constructed mathematically, their properties are such, that they should not be expected theoretically, and there is no observational indication whatsoever of their presence. We infer that on the sun in these regions the magnetic fields should adjust themselves to the circulations maintained by the conditions discussed above (and not vice versa).

We next turn to the scheme advanced already long ago from hydrodynamic considerations by Bjerknes [7, 21]. Replacing Bjerknes's vortex girdles around the axis by a toroidal magnetic field, it is seen at once that the general pattern is compatible with the state of rotation and of meridional circulations derived above. This does not mean, of course, that the theory has been proved to be correct. There is, e.g. lacking a proof that an arrangement of two girdles of opposite direction in each hemisphere is a stable one, and that it is continuously reinforced in some way. What I would like to indicate is only that the general background of this scheme is in better harmony with what one should expect from the theory of internal structure, than one probably used to think.

There are two points which I would like to note in passing. Magnetic girdles of purely toroidal structure would be dynamically unstable, since in a sense their specific weight is smaller than that of the layers in which they are embedded. This is probably one main cause for the appearance of BM regions and spots on the surface as suggested by Elsasser and Parker [22]; but to secure dynamical stability over long periods of time, the magnetic fields must be of such a structure that their specific weight is effectively equal to that of the regions to which they belong. Hence fields of the type studied by Lundquist [23], Lüst and Schlüter [24], called force-free-fields, are suggested.

The second remark is this. There has been a tendency to set aside the theoretical possibility that impressed electro-motive forces, play a role in creating and maintaining toroidal magnetic fields around the axis of rotation. Now it may easily be shown, from my work in 1949 [25] that the magnetic flux, created by such forces in the inner parts of the hydrogen convection zone, has roughly the order of magnitude, observed in big sunspot groups and BM regions. Furthermore this mechanism seems to be a very natural one to create complete magnetic girdles around the axis. But obviously more questions remain open here than have been answered.

Lastly I would like to discuss briefly the scheme recently proposed by Parker [26]. This scheme has several features in common with the general pictures of the solar magnetic fields we had discussed earlier. In particular, according to Parker, toroidal fields are the main feature of the low latitude regions, poloidal fields dominating in the polar regions. Parker assumes that the toroidal field is continuously reinforced from the poloidal field by differential rotation, and the poloidal field vice versa from the toroidal field by cyclonic motion. In this connexion Parker notices the theoretical possibility of migratory dynamo waves in the convective zone, which he identifies with the migration of the activity belts. Now from his expression for the velocity of this migration, it is found that the latter depends only on the velocity of differential rotation and on that of the cyclonic motion (not on the magnetic field intensity) and with the most probable choice of these velocities there is a discrepancy in the velocity of migration of roughly three powers of 10. On the other hand, no account has been taken in Parker's picture as published, of any meridional circulation; a possible line of further development of this work is to be desired. From what has been said earlier, it may be concluded, that the observed migration of the activity belts should be at least partially connected with the circulations postulated by the theoretical considerations presented above.

In discussing the theory of turbulence in stellar atmosphere last year in Dublin (cf. *I.A.U. Transactions*, vol. 9), I emphasized the conclusion, that the various aspects of solar activity are probably all based on the turbulence of the hydrogen convection zone. This is also in harmony with what has been discussed today; but the essential role of the law of rotation in determining, e.g., the pattern of the spot activity has perhaps become somewhat clearer by the present discussion.

In closing, I would like to emphasize that this discussion was mainly based on one theoretical argument which is believed to be of some bearing, but it is obvious that in a complex problem like this some other aspects, to which earlier writers have given closer attention, have to be reconsidered also (cf. Temesváry[27]) with further references).

REFERENCES

[1] Boehm-Vitense, E. Z. Astrophys. **32**, 135, 1953.
[2] Biermann, L. Astr. Nachr. **264**, 361 and 395, 1938.
[3] Biermann, L. Z. Astrophys. **21**, 320, 1942.
[4] Becker, U. Z. Astrophys. **37**, 47, 1955.
[5] Ferraro, V. C. A. Mon. Not. R. Astro. Soc. **97**, 458, 1937.
[6] Alfvén, H. Ark. Mat. Astr. Fys. **28**A, no. 6, 1942.
[7] Bjerknes, V. Astrophys. Norveg. **2**, 263, §11, 1937.
[8] Randers, G. Astrophys. J. **94**, 109, 1941.
[9] Biermann, L. Z. Astrophys. **28**, 304, 1951.
[10] Cowling, T. G. Mon. Not. R. Astr. Soc. **105**, 166, 1945.
[11] Biermann, L. Z. Astrophys. **25**, 135, 1948.
[12] Biermann, L. Vjschr. Astr. Ges., Lpz. **76**, 194, 1941.
[13] Chandrasekhar, S. Phil. Mag. **43**, 501, 1952.
[14] Chandrasekhar, S. Phil. Mag. **45**, 1177, 1954.

[15] Cowling, T. G. *Mon. Not. R. Astr. Soc.* **106**, 218, 1946.
[16] Alfvén, H. *Ark. Mat. Astr. Fys.* **29**A, no. 12, 1943.
[17] Alfvén, H. *Mon. Not. R. Astr. Soc.* **105**, 3 and 382, 1945.
[18] Alfvén, H. *Ark. Mat. Astr. Fys.* **34**A, no. 23, 1948.
[19] Alfvén, H. *Tellus*, **8**, 274, 1956.
[20] Walén, C. *Ark. Mat. Astr. Fys.* **30**A, no. 15, 1944; **31**B, no. 3, 1944; **33**A, no. 18, 1946.
[21] Bjerknes, V. *Astrophys. J.* **64**, 93, 1926.
[22] Elsasser, W. M. and Parker, E. *Astrophys. J.* **121**, 491, 1955.
[23] Lundquist, S. *Ark. Fys.* **2**, 361, 1950.
[24] Lüst, R. and Schlüter, A. *Z. Astrophys.* **34**, 263, 1954.
[25] Biermann, L. *Z. Naturf.* **5**A, 65, 1950.
[26] Parker, E. *Astrophys. J.* **122**, 293, 1955.
[27] Temesváry, St. *Z. Naturf.* **7**A, 103, 1952.

Discussion

Alfvén: The main difference between my opinion and Professor Biermann's is that the source of energy of the sunspots, according to my point of view, is the centre of the sun and the energy goes out as waves to the solar surface where they create solar activity, sunspots and perhaps also prominences at higher latitudes. There could, of course, be given many arguments against this view and at least the same number in favour of it; I think that in this, as in many other cases, the final decision could only come from observations. This theory results in a correlation between the sunspots of the northern hemisphere and the sunspots of the southern hemisphere. The correlation has been checked independently by Galvenius and Wold by statistical methods and by Whittle and they have found a correlation which supports the theory. I have myself made another correlation analysis which also seems to support it. It would be very interesting to discuss the objections against it because, again, the decision of problems like this cannot be given by theoretical arguments alone because we know too little about the interior of the sun.

Tuominen: The convective zone is thin as compared to the solar radius. Is it therefore possible that the velocity of the meridional circulation in it does not change its sign between the equator and the pole? If the circulation takes place in different directions at different latitudes, it will be difficult to see the connexion between the motion of the sunspot belts and the meridional circulation in the convective zone.

Biermann: I am not aware of any necessary contradiction in this case, since the sunspot belts are confined to latitudes $< 40°$; the direction of the meridional circulation might be different at higher latitudes, as suggested, e.g. by the displacement of the zones of prominences, and of the coronal isophotes, with phase in the cycle of solar activity. In the inner parts of the convective zone in the spot latitudes the velocity v_m will depend on the detailed structure of the velocity field, which we do not know. I am doubtful whether the simplified picture which I have discussed today contains as yet the whole truth, and I indicated earlier that certainly wave-type phenomena also have to be thought of. But the structure of the circulation itself might be rather complex, on account of the complex character of the physical effects which should be relevant.

Tuominen: In the terrestrial atmosphere the circulation is in different directions at different latitudes and it may be so also in the sun.

Biermann: Yes, that is what I would expect. But I think that the system of circulation might be more complex than the corresponding system in the earth's atmosphere.

Tuominen: But then it should have nothing to do with the motion of sunspots in a belt?

Biermann: O, yes, it should. The motion of the sunspot zone and that of an individual spot are not necessarily identical. I would like to emphasize that in my picture the meridional circulation is an essential element of the migration of the spot belts in the cycle of solar activity, but that contributary causes might be important for the motion of individual spots.

Schatzman: With reference to the existence of the inner convective zones, referred to by Professor Alfvén, the important factor is the effective power of temperature. This power is usually supposed to be 4, but if one takes into account the effect of electron screening in thermonuclear energy generation rate one finds that the effective power is slightly smaller, may be 3, which reduces considerably the likelihood of the existence of a central convective zone.

Tayler: The existence or non-existence of a small convective core has very small effect on the observed solar properties and does not seem to effect the mechanism proposed by Professor Biermann. It would, however, be critical to the Alfvén theory.

Biermann: I mentioned the central convective zone exclusively on account of its (postulated) relation to some sunspot theories; I perfectly agree that its existence or otherwise does not essentially affect the internal structure in its overall features.

THE MAGNETIC FIELDS OF SUNSPOTS AND THE EVERSHED EFFECT

E. JENSEN

Institute of Theoretical Astrophysics, Blindern, Oslo, Norway

ABSTRACT

The observational data on the magnetic fields and the physical parameters in sunspots indicate in a qualitative way how the magnetic lines of force run relative to the isobaric surfaces. If matter is confined to move only along the lines of force it is shown that for sufficient tilt between these lines and the isobars, matter will be accelerated outward along the lines of force. The flow corresponding to this forced convection works as a cooling cap for the core of the spot. It is indicated how a stationary state may be reached with the outward velocity adapted to the temperature difference between the spot and the photosphere.

It is the purpose of this note to suggest a self-maintaining mechanism, which at the same time provides the driving force of the Evershed-motion and maintains the low temperature in the upper layers of the spots. We shall deal mainly with the outer layers of a fully developed stationary sunspot, leaving questions as to the formation and development of the spots open.

As is generally accepted, the electrical conductivity, even in the relatively cool spot region, is sufficiently high for matter to be effectively 'glued' to the magnetic lines of force (for example [1]). With magnetic fields of the order observed in sunspots this has the consequence that in a stationary spot the motion takes place along the magnetic lines of force, as suggested by Hoyle [2]. We shall make this assumption in the following. It should be remarked, however, that this assumption does not fit the observations very well. According to Kinman [3] no vertical component of the Evershed-motion is observed at the boundary between the umbra and the penumbra. In one of the spots observed by him the mean error is given as 0·26 km/sec. and the measured horizontal velocity component at this point is 1 km/sec. Since no vertical velocity was observed we may conclude that the magnetic lines of force are here inclined at least 75° to the vertical. On the other hand, according to the Mount Wilson observations [4] this angle is only about 40°, if a linear penumbra-umbra ratio of 2·3 is assumed. This discrepancy may perhaps be explained as a level effect. Considering this possibility and the many sources of error involved both in the measurement of velocities and magnetic fields we shall stick to our assumption that the motion follows the magnetic lines of force.

In Michard's empirical model of a large sunspot[5], not only is the temperature of the spot found to be lower than in the surroundings, the density is also less at the same level. Both effects contribute towards a smaller gas pressure in the spot region. As has been emphasized by

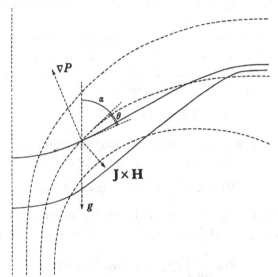

Fig. 1. Schematic drawing of the magnetic lines of force (dotted curves) and the isobars (solid curves) of a sunspot.

Sweet[6], this model shows departures from hydrostatic equilibrium. The gas pressure at a given level turns out to be too small to support the weight of the matter above. Near the centre of the spot the electromagnetic force has, however, no vertical component. Thus the discrepancy cannot be explained by taking this force into account. Further observational data are required to decide whether this effect is real.

Even if the physical conditions in sunspots are not known with great accuracy, we may consider sunspots as low-pressure regions. For such a region to be in equilibrium with a magnetic field, certain conditions concerning the orientation of the isobaric surfaces relative to the magnetic lines of force must be fulfilled. The situation is somewhat similar to that shown in Fig. 1, where the magnetic field is supposed to be axially symmetrical and with the axis vertical. In the central region the isobaric surfaces are perpendicular to the magnetic field and gravity. Here the electromagnetic force on matter is probably quite small and has no vertical component. Further out this force is larger, and here the isobaric surfaces are also tilted relative to the direction of gravity.

The three forces which determine the equilibrium are: the gradient of the gas pressure, the electromagnetic force, $\mathbf{J} \times \mathbf{H}$ (where $\mathbf{H}$ denotes the magnetic field strength, and $\mathbf{J} = (1/4\pi)$ curl $\mathbf{H}$ the current density) and gravity. Since matter may move freely only along the magnetic lines of force, the important quantity is the resulting force-component along the direction of the magnetic field.

Let α denote the angle between the magnetic lines of force and the vertical at a given point, and $\pi/2 - \Theta$ the angle between the gas-pressure gradient ∇P and the field vector $\mathbf{H}$. Then the force component F directed outward in the direction of the lines of force, is

$$F = |\nabla P| \sin \Theta - \rho |g| \cos \alpha. \tag{1}$$

In the direction perpendicular to the magnetic lines of force of a fully developed spot we assume the forces to be in equilibrium, thus,

$$|\nabla P| \cos \Theta - \rho |g| \sin \alpha - JH = 0, \tag{2}$$

where $\mathbf{J}$ is assumed to be at right angles to $\mathbf{H}$. When $\rho |g|$ is eliminated between the Eqs. (1) and (2) the following expression for F is obtained,

$$F = \frac{|\nabla P|}{\sin \alpha} \{\beta \cos \alpha - \cos (\alpha + \Theta)\}, \tag{3}$$

where we have introduced the parameter $\beta = JH/\nabla|P|$, the ratio between the magnetic force on matter perpendicular to the lines of force and the absolute value of the gas-pressure gradient. This parameter, β, will attain its maximum value somewhere outside the spot centre, probably near the border between the umbra and the penumbra, and approach zero in the outer region of the penumbra.

The condition for equilibrium is that F is everywhere equal to zero. If F is positive, we get an outward directed acceleration, if it is negative, matter will fall into the low-pressure region.

When motions are taking place, other additional forces will act on the moving material, and somewhat disturb this simple picture. First of all inertial effects will come into play. These will be most important where the curvature of the lines of force is largest. The dissipative effects of viscosity must also be considered.

Any specification of the quantitative dependence of β on α must necessarily be only rough guesswork. The most probable value of the limiting angle Θ_1 making F vanish, seems to be of the order of 15–30° in the penumbra region of the spots.

A static equilibrium with the tilt everywhere attaining the value Θ_1 is clearly not a stable one. As shown by Walén[7] at least the upper layers of the cool core of the spot when exposed to the radiation from the hotter surroundings, would be heated to photospheric temperature in a time very much shorter than the lifetime of a spot. Thus some effective cooling mechanism must be operating.

From the previous considerations, we may see what will happen to a spot with Θ equal to Θ_1 when its temperature is increased. First of all the gas pressure will increase in the spot region, thus the isobaric surfaces are partly levelled out, making the low-pressure region less marked. Provided the magnetic field is not altered significantly, this means that the angle Θ is increased above the equilibrium value, thus making F positive. This force will set the matter moving outwards and upwards, sucking material along the magnetic lines of force from the deeper layers. This forced convection will, since the upper layers of a spot are in stable radiative equilibrium, lead to a cooling of the moving matter relative to the surroundings. Another effect may be of importance here, that is the cooling which results when conductive matter is forced through a strongly divergent magnetic field. The material which is set into motion will thus act as a cooling cap for the umbra regions of the spot. After the motion has started protecting the spot from the incoming radiation, the temperature in the spot will decrease, leading to a corresponding decrease in the tilt. In a stable spot where conditions to a good approximation are stationary, we can imagine that the tilt (and thus the force F) is adjusted to the dissipative forces. The stationary velocity of the Evershed-motion will depend upon the temperature difference to be maintained.

The largest spots are also the coolest (cf. [5], and the Evershed-velocity is found to increase with spot size (cf. [8] and [9])). This would seem to confirm our hypothesis. However, the efficiency of the cooling depends upon many parameters besides the velocity. The thickness of the moving layer, the temperature gradient within this layer and the topography of the magnetic field all enter.

In the discussion above we have tacitly assumed that some mechanism is at work in deeper layers which cuts down the energy flux in the spot region. Inhibition of convection by the magnetic field, according to Biermann's idea[10] would seem the most plausible.

The research reported in this paper has been supported in part by The Geophysical Research Directorate of the Air Force Cambridge Research Center, Air Research and Development Command, U.S. Air Force, under Contract No. AF 61(514)–651 C through its European office at Brussels.

REFERENCES

[1] Unsöld, A. *Physik der Sternatmosphären* (Berlin-Göttingen-Heidelberg, 2nd ed. 1955), p. 594.

[2] Hoyle, F. *Some Recent Researches in Solar Physics* (Cambridge University Press, 1949), p. 4.

[3] Kinman, T. D. *Mon. Not. R. Astr. Soc.* **112**, 425, 1952.

[4] Hale, G. E. and Nicholson, S. B. *Magnetic Observations of Sunspots* (Washington, D.C., 1938).

[5] Michard, R. *Ann. Astrophys.* **16**, 217, 1953.

[6] Sweet, P. A. *Vistas in Astronomy*, vol. 1, ed. A. Beer (1955), p. 675.

[7] Walén, C. *Ark. Mat. Astr. Fys.* **30** A, no. 15, 1944.

[8] Michard, R. *Ann. Astrophys.* **14**, 101, 1951.

[9] Kinman, T. D. *Mon. Not. R. Astr. Soc.* **113**, 613, 1953.

[10] Biermann, L. *Vjschr. Astr. Ges., Lpz.* **76**, 194, 1941.

THE INTERNAL CONSTITUTION OF SUNSPOTS

A. SCHLÜTER AND S. TEMESVÁRY

Max-Planck-Institut für Physik, Göttingen, Germany

ABSTRACT

The constitution of stationary single sunspots of circular shape is considered. Account is taken of the mechanical effects of the magnetic field, including those which arise from the curvature of the lines of force. To make the system of magneto-hydrostatic equations manageable, it is assumed that the relative distribution of the vertical component of the magnetic field is the same across the flux-tube of the spot in all depths. Preliminary results indicate that suppression of convective energy transport by the magnetic field in those depths in which ionization of hydrogen takes place, will give the essential observable properties of sunspots, relatively independent on the asumptions about the physical processes in greater depths. There the physical properties of matter can deviate but very little from those of the indisturbed hydrogen convection zone.

I. ASSUMPTIONS

The assumptions made in this attempt to determine the relation between the constitution of sunspots and their magnetic field divide into two groups; those which seem obvious, if we deal with a simple picture of a sunspot (group A), and those which determine the particular model to be considered (group B). The latter are not necessarily true, but are suggested for a first attempt by their simplicity.

A 1. We consider a circularly shaped single spot (H 1 in the Zurich classification). Therefore cylindrical symmetry around the z-axis of a co-ordinate system r, ϕ, z is assumed, the z-axis being perpendicular to the sun's surface, which is taken to be plane, the positive direction of z pointing toward the centre of the sun. On the axis (i.e. for $r = 0$) the electric current and the r-component of the field H_r vanish, and the vertical component H_z has there a maximum.

A 2. The spot is assumed to be in a quasi-static equilibrium, that means, all derivatives with respect to time are put equal to zero. This seems a fair approximation for relatively long-living single spots.

A 3. The magnetic field strength and its derivatives vanish for $z = -\infty$ (infinite distance above the sun) and for $r = \infty$.

B 1. All material motions in the spot are neglected.

B 2. The magnetic field has no torsion, that means H_ϕ vanishes everywhere, each line of force lying in a plane perpendicular to the sun's surface.

B 3. We assume that the relative distributon of magnetic flux through a horizontal cross-section of the sunspot is everywhere geometrically similar.

2. FORMULATION OF THE SIMILARITY-ASSUMPTION

The ratio of the vertical component $H_z(z, r)$ to the central intensity $H(z)$ at the same depth z is assumed to depend on z only by a scale factor $\zeta(z)$, or

$$H_z(z, r) = H(z) \cdot D(\alpha)/D(o); \quad \alpha = \zeta(z) \cdot r. \tag{1}$$

Then $1/\zeta(z)$ describes the dependence on depth of the diameter of a flux-tube constituting the magnetic field of the sunspot, and the function $D(\alpha)$ determines the shape. Due to the continuity of the lines of force (i.e. div $\mathbf{H} = o$) H_z at homologous points (i.e. α fixed) and $H(z) = H_z(z, o)$ in particular, have to be larger where the tube is more constricted and vice versa, they vary in fact like $\zeta^2(z)$, so that with a suitable normalization of D we may write for the field on the axis

$$H(z) = \zeta^2(z) \cdot D(o) \tag{2}$$

and generally for the vertical component

$$H_z(z, r) = \zeta^2(z) \cdot D(\alpha); \quad \alpha = \zeta(z) \cdot r. \tag{3}$$

Using once more the equation div $\mathbf{H} = o$, we find the component of the field perpendicular to the axis of the field-tube to be given by:

$$H_r(z, r) = -(d\zeta/dz) \cdot \alpha D(\alpha). \tag{4}$$

The condition of similarity of flux-distribution therefore implies that the ratio of the field components increases linearly with the distance from the axis

$$\frac{H_r}{H_z} = -r \frac{d \ln \zeta}{dz}. \tag{5}$$

The quantity α introduced so far as the distance from the axis scaled down by ζ, has a simple meaning: The equation

$$\alpha(\zeta, r) = \text{const}$$

264

determines the lines of force, i.e. α is constant along the lines of force. For the magnetic flux through a horizontal circle of radius r we obtain

$$2\pi \int_0^r H_z r\, dr = 2\pi \int_0^\alpha D(\alpha)\, d\alpha. \tag{6}$$

The magnetic flux between a given line of force and the axis is therefore independent of depth, as it has to be.

3. THE DIFFERENTIAL EQUATION FOR THE MAGNETO-HYDROSTATIC EQUILIBRIUM

According to the assumptions (A 1) and (B 1) we have

$$\frac{1}{4\pi}\, \mathbf{H}\cdot\mathrm{curl}\, \mathbf{H} = \rho\mathbf{g} - \mathrm{grad}\, p, \tag{7}$$

ρ being the density, $\mathbf{g}$ the gravitational acceleration and p the gas pressure. The r-component of (7) is

$$H_z\left(\frac{\partial H_r}{\partial z} - \frac{\partial H_z}{\partial r}\right) = 4\pi\, \frac{\partial p}{\partial r},$$

or using a dash to denote the derivative with respect to the depth z

$$\alpha D^2 \zeta\zeta'' + \frac{1}{2}\frac{d}{d\alpha}\,(\alpha D)^2 . \zeta'^2 + \frac{1}{2}\frac{d}{d\alpha}\,(D^2) . \zeta^4 = -4\pi\, \frac{dp}{d\alpha}.$$

This gives, integrated over r (that is over α) from the axis to infinity,

$$\zeta\zeta'' \int_0^\infty \alpha D^2(\alpha)\, d\alpha - \tfrac{1}{2}\zeta^4 D^2(0) = -4\pi\Delta p,$$

because αD vanishes for $\alpha = 0$ as well as for $\alpha = \infty$, with

$$\Delta p = p(\infty, z) - p(0, z),$$

and the abbreviations

$$H = \zeta^2 D(0) = y^2 \quad \text{and} \quad f = \frac{2}{D(0)}\int_0^\infty \alpha D^2(\alpha)\, d\alpha \tag{8}$$

we obtain
$$fyy'' - y^4 + 8\pi\Delta p = 0; \tag{9}$$

the basic differential equation of the magnetic field.

The z-component of (7) gives for the pressure difference between the axis and the undisturbed layers of the same depth:

$$(\Delta p)' = g\Delta\rho, \tag{10}$$

(where $\Delta\rho = \rho(\infty, z) - \rho(0, z)$) because the magnetic force vanishes for $r = 0$ and $r = \infty$. Due to our assumption of constancy of shape we need only to know the physical state of the matter along the axis of the field-tube

265

and in great distances from this tube and our partial differential equation (7) reduces to an ordinary one with the depth as the only independent variable. The term y^4 in Eq. (9) represents the effect of the magnetic pressure across the lines of force, while the term fyy'' expresses the effect of the Maxwell–Faraday tension along the lines of force, which comes into play only where these lines are curved. The usual neglect of this term is inadmissible, particularly above the photosphere where Δp certainly becomes insignificantly small. Before entering into the more complete discussion of the constitution of the field-tube we consider some formal properties of Eq. (9).

$y'' = 0$ is a solution of this equation. if

$$y^4/8\pi = H^2/8\pi = \Delta p \qquad (11)$$

and at the same time $\qquad y' = \text{const} = a, \qquad (12)$

say, that gives
$$\left.\begin{array}{l} y = a(z - z_0), \\ 8\pi\Delta p = a^4(z - z_0)^4, \\ 2\pi g\Delta\rho = a^4(z - z_0)^3, \end{array}\right\} \quad a \neq 0. \qquad (13)$$

It is interesting to note that configurations can exist, where the tension along the lines of force do not contribute to the lateral forces even though the lines of force are not straight (i.e. $a \neq 0$); the practical importance of these configurations must however be quite limited since they possess an essential singularity at the point $z = z_0$, where the diameter of the flux-tube ($\alpha 1/y$) becomes infinite. Yet, the case $a = 0$ with $\Delta p = \text{const}$; $\Delta\rho = 0$ may be a good approximation wherever the flux-tube is sufficiently slender, this is, however, at least near the photosphere, certainly no good approximation.

If y', $y \neq 0$ we may multiply Eq. (9) by y'/y and integrate once to obtain

$$fy'^2 = \tfrac{1}{2}y^4 - 16\pi \int \Delta p\, \frac{y'}{y}\, dz + \text{const}$$

$$= \tfrac{1}{2}y^4 + 16\pi g \int \Delta\rho \ln y\, dz + \text{const}. \qquad (14)$$

This form shows that above the sun's surface ($z < 0$) where to a very good approximation $\Delta p = 0$, the condition of vanishing field strength for infinite distance leads to the first-order equation

$$fy'^2 = \tfrac{1}{2}y^4. \qquad (15)$$

The solutions of (9) for the interior ($z > 0$) have to be adapted to this

266

condition. The only solutions of (15) which are regular for all $z < 0$ are of the form

$$H = y^2 = \frac{2f}{(z-z_0)^2}, \tag{16}$$

where $z_0 > 0$ is the point in which the field strength would become infinite, if the solution were continued into the interior. The equation of the lines of force is

$$r = \text{const } (z - z_0). \tag{17}$$

This is the equation of straight lines which intersect the axis $r = 0$ at z_0. The field therefore falls off like that of a monopole, which is an obvious consequence of our assumptions.

4. PARTICULAR DISTRIBUTIONS OF MAGNETIC FLUX

By the Eqs. (6) and (8) the constant f is related to the total magnetic flux F by:

$$\pi f / F = \int_0^\infty \alpha D^2(\alpha) \, d\alpha \Big/ D(0) \int_0^\infty \alpha D(\alpha) \, d\alpha. \tag{18}$$

We study this relation by considering some examples for the shape-function $D(\alpha)$. We remark, that this quotient is invariant against the substitution $c_1 D(c_2 \alpha)$ for $D(\alpha)$, c_1 and c_2 being constants.

1. If we assume a gaussian shape for the horizontal distribution of magnetic flux, namely

$$D(\alpha) = D(0) \, e^{-\alpha^2}, \tag{19}$$

we obtain by Eq. (18)

$$F = 2\pi f. \tag{20}$$

2. The observations of flux-distribution at the surface of sunspots lead to the well-known Broxon formula, which gives

$$H_z = H_0 \left(1 - \frac{r^2}{r_0^2} \right) \cos \left(\frac{\pi}{2} \frac{r}{r_0} \right), \tag{21}$$

where H_0 is the central field strength and r_0 the radius of the penumbra, and further very closely

$$F = 2\pi \int_0^\infty H_z r \, dr \approx H_0 r_0^2. \tag{22}$$

Putting

$$\alpha = \frac{\pi}{2} \frac{r}{r_0},$$

we obtain

$$D(\alpha) = D(0) \left(1 - \frac{4}{\pi^2} \alpha^2 \right) \cos \alpha \quad \text{for} \quad \alpha < \frac{\pi}{2} \tag{23}$$

267

and
$$D(\alpha) = 0 \quad \text{for} \quad \alpha \geqslant \frac{\pi}{2}. \tag{24}$$

So we obtain from Eq. (18) $F = 1 \cdot 70 \pi f.$ (25)

Considering the uncertainty of the Broxon formula (cf. W. Mattig[1] and G. Thiessen[2]) the difference between Eqs. (20) and (25) seems to be unimportant.

3. We have assumed H_z to have everywhere the same sign and have neglected the returning lines of force, while it is more probable that in reality the total magnetic flux must equal zero. To estimate the influence of these returning lines of force we assume that we have two concentric tubes, both with constant H_z, the outer tube of k-fold radius representing the (unobserved) returning lines of force. That means:

$$D(\alpha) = \begin{cases} D(0) & \text{for} \quad 0 \leqslant \alpha < 1, \\ -D(0)/(k^2 - 1) & \text{for} \quad 1 \leqslant \alpha < k, \\ 0 & \text{for} \quad k \leqslant \alpha. \end{cases}$$

The observed flux has then to be taken as the flux $F(1)$ of the inner tube ($\alpha < 1$) for which we get

$$F(1) = \left(1 - \frac{1}{k^2}\right) \pi f.$$

The deviation due to the returning lines is expressed by the term with $1/k^2$ and will be of the order of a few per cent, if k is of the order 10 or more.

The examples given show the relation between f and the flux-distribution. If we want to apply this to the field of actual sunspots, we have first to test the applicability of the underlying assumption of similarity. We recall, that we have no *a priori* reason as to why this similarity should hold, other than this being the simplest possibility. We can test this hypothesis only at the surface, where both H_z and H_r are observable. According to Eq. (5) the ratio of H_r/H_z should theoretically increase linearly with r. The observations are usually quoted as showing this ratio to vary like $tg(\pi r/2r_0)$ where $r_0 =$ radius of the sunspot. Fig. 1 (adapted from S. Chapman[3]) shows, however, that the observations are as well compatible with a linear increase as with the quoted relation. Assuming now the linear law to hold sufficiently well, also the increase with depth of the field on the axis follows from Eq. (5). It results in

$$dH/dz \approx 4H/r_0$$

$$\approx 0 \cdot 8 \text{ gauss/km} \tag{26}$$

(with $H \approx 2,000$ gauss; $r_0 \approx 10,000$ km) in fair agreement with the value $0 \cdot 5$ gauss/km found previously by Houtgast and van Sluiters[4].

5. THE ENERGY TRANSPORT IN THE SPOT

The problem remains of determining the function $\Delta p(z)$. We obtain a necessary condition on Δp by integrating the basic differential Eq. (9) over z, between z_1 and z_2, say:

$$\int_{z_1}^{z_2} (fy'^2 + y^4)\, dz + \frac{f}{2}\{y^{2'}(z_1) - y^{2'}(z_2)\} = 8\pi \int_{z_1}^{z_2} \Delta p\, dz. \qquad (27)$$

If we choose for z_1 a sufficiently great height above the sun, then $y^{2'}(z_1)$ disappears. So does $y^{2'}(z_2)$ if we choose such a great depth z_2 that the field

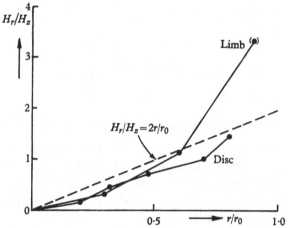

Fig. 1. The ratio between the radial and vertical components of the magnetic field in a sunspot plotted as a function of r/r_0. Full curves indicate observed values.

either is constant, or has disappeared there. It then follows that the mean value of Δp must be positive and must even be larger than would follow by just considering the 'magnetic pressure' $H^2/8\pi = y^4/8\pi$.

On the axis, hydrostatic equilibrium is not affected by the magnetic field, nor is, of course, the equation of state. Therefore the only possibility is that the temperature is different, and since the temperature is controlled by the transport of energy, this transport must be affected by the presence of the magnetic field of the spot. This can conceivably be the case where the energy is ordinarily transported not by radiation but by convection. So one is lead to the picture first proposed by L. Biermann[5]: The convective transport of energy in the hydrogen convection zone is hindered by the magnetic field of the sunspot. This causes the temperature to drop faster in the outward direction, thereby producing a positive Δp, which then is needed to keep the lines of force together and so to make the existence of the sunspot field possible. At the same time the reduction of the effective thermal conductivity deflects a part of the energy flow out of the sunspot-tube, thus causing the lowering of the effective temperature at the surface.

Since, as yet, a theory of the influence of a magnetic field on convection and energy-transport in a zone of great thermal instability does not exist, the effect of the sunspot-field in the hydrogen convection zone cannot be asserted with sufficient certainty. One may, however, surmise that a necessary condition for convection to be effectively suppressed is gained by using the scale height as the characteristic length in Walén's stability criterion [6], which then reads:

Stability occurs if

$$\left(\frac{d \ln T}{d \ln p}\right)_{\text{radiative}} - \frac{H^2}{4\pi p} < \left(\frac{d \ln T}{d \ln p}\right)_{\text{adiabatic}}. \tag{28}$$

6. NUMERICAL INTEGRATIONS

Lacking safe theoretical guidance, one has now to make an assumption about the energy transport below the sunspot, then one has to evaluate Δp by numerical integration and to solve the differential Eq. (9) for the magnetic field. Since this part of the work is still in progress, we shall give only a brief sketch of our attempts.

1. First we supposed that the suppression of convection inside the spot-tube is complete and that the energy is there transported by radiation and in vertical direction only. With these assumptions the temperature increases with depth much faster inside the tube than outside it, particularly so in the hydrogen-ionization zone. With the so determined Δp we tried to find solutions for the magnetic field. However, it turned out, that with no reasonable assumptions on y and y' at the surface a solution for $y(z)$ existed which did not become irregular within a depth of a few scale heights. From these numerical attempts it followed that at least in greater depths the pressure difference Δp must be very small compared to p for a regular solution of Eq. (9) to exist.

2. To gain an impression what Δp should look like, we then assumed a reasonable shape for the vertical dependence of the magnetic field and determined by Eq. (9) Δp and thereby $\Delta \rho$ and ΔT. We so circumvented the difficulty of determining the energy transport mechanism. Two models were considered in particular.

(i)
$$H = y^2 = 2fd^{-2}\left[\ln\left\{\exp\left(\frac{1}{d}\sqrt{(2f/H_c)}\right) + \exp\left(-\frac{z-z_0}{d}\right)\right\}\right]^{-2},$$

$$f, d, H_c, z_0 = \text{const.} \tag{29}$$

For $z \to +\infty$ the field becomes constant $(H \to H_c)$, and for $z \to -\infty$ the solution (16) is approached.

(ii)
$$H = y^2 = \frac{2f}{a^2 + (z - z_0)^2}, \tag{30}$$

$$a, z_0 = \text{const.}$$

Here the magnetic field drops off also as z goes to $+\infty$. The pressure difference is given by

$$4\pi\Delta p = \frac{3a^2 f^2}{[a^2 + (z - z_0)^2]^3} \tag{31}$$

and falls rapidly with increasing distance from the point of maximal field strength $z = z_0$, so that the solution (16) is approximated.

A number of plausible parameters were tried in these models. In every case, it turned out that in a depth well within the hydrogen convection zone the internal constitution of the flux-tube was almost indistinguishable from that of the undisturbed layers.

3. The first attempts have shown, that some modification has to be made in the assumption of pure radiative transport of the observed energy flux in the spot-tube. The easiest explanation would be that the suppression of convective energy transport is far from being complete. Since, however, every detailed assumption on the effectiveness of the convective transport mechanism under the conditions prevailing in the spot-tube would be completely arbitrary, we considered an alternative solution. In addition, the stability criterion (28) showed no reason for instability in the examples considered by us.

If the magnetic field is much stronger in great depths than it is near the surface, then the flux-tube is there correspondingly thinner and horizontal flow of energy may become appreciable. This might be a mechanism which causes the desired similarity in the thermodynamic state of the gas in the tube and of the gas outside it. To obtain a model containing the horizontal energy transport, we assume that the temperature across the flux-tube is given by a gaussian distribution the half-width of which corresponds to the diameter of the flux-tube

$$T(r, z) = T(\infty, z) - \Delta T(z)\exp\left(-\frac{y^2(z)\, r^2}{2f}\right). \tag{32}$$

271

Since we assume radiative equilibrium we have the equations:

$$\text{div } \mathbf{F} = 0, \quad \mathbf{F} = \text{radiative energy flux}, \tag{33}$$

$$\mathbf{F} = -\sigma \text{ grad } T, \quad \sigma = \text{thermal conductivity (due to radiative transport).} \tag{34}$$

Similar as before we fulfil these equations only on the axis ($r = 0$), taking σ and the molecular weight μ to depend only on z (not on r). So we arrive at the following system:

$$\left.\begin{array}{l} fyy'' = y^4 - 8\pi\Delta p, \\[4pt] (\Delta p)' = g\Delta\rho, \\[4pt] (\Delta T)' = \sigma^{-1}\Delta F, \\[4pt] (\Delta F)' = \dfrac{2}{f}\,\sigma y^2 \Delta T + F'(\infty, z), \end{array}\right\} \tag{35}$$

where all quantities (except the constants g and f) depend on depth only. $F(\infty, z)$ denotes the part of the energy transported vertically by radiation in the undisturbed layers.

These equations, together with the equation of state and suitable boundary conditions suffice to determine the magnetic field (i.e. $y(z)$) and the difference in constitution between the flux-tube and the normal layers of the sun. If one starts at the surface (at $z = 0$, say), then the values of y and y' are provided by the magnetic measurements (together with the value of f), while the values of ΔF and ΔT are determined by the observed defect in brightness of the sunspot. The value of Δp is not directly observable since it depends essentially on the geometric depression of the spot which is very hard to determine.

For our attempts to solve these equations numerically, we used for the constitution of the undisturbed layers a model, the dates of which were kindly placed at our disposal by Mrs Böhm-Vitense[7].

It turned out, that the solutions of Eq. (35) were unstable to a degree, which we had not anticipated, the range of initial conditions for which the solution could be continued to a depth of more than a few thousand kilometres, being extremely small. This then means, that the condition that the solution should be regular down to a depth where the ionization of hydrogen is complete, determines the constitution of the sunspot for all higher layers, if only f is given. (Only at this depth the horizontal flux of energy becomes important, if at all.) This means that we can not infer anything about the constitution at greater depths, except that we know the difference between the spot and the normal layers to be quite small, as we pointed out earlier. These results are to be considered as preliminary, since the work described here is still in progress. In particular, we are attempting to obtain a consistent model for the constitution of the decisive part of the sunspot-tube from the photosphere down to the hydrogen ionization zone.

REFERENCES

[1] Mattig, W. *Z. Astrophys.* **31**, 273, 1953.
[2] Thiessen, G. *Naturwissenschaften*, **40**, 218, 1953.
[3] Chapman, S. *Mon. Not. R. Astr. Soc.* **103**, 117, 1943.
[4] Houtgast, J. and van Sluiters, A. *B.A.N.* **10**, 325, 1948.
[5] Biermann, L. *Vjschr. Astr. Ges., Lpz.* **76**, 194, 1941.
[6] Walén, C. *On the Vibratory Rotation of the Sun* (Stockholm, 1949), cf. *The Sun*, ed. G. P. Kuiper (Chicago, 1953), ch. 8 (T. G. Cowling), pp. 561–5.
[7] Vitense, E. *Z. Astrophys.* **32**, 135, 1953.

Discussion on Papers 28 and 29

Bostick: One mechanism for the cooling of sunspots is that provided by the magnetic evaporation of hot ions and electrons by the presence at the sunspot of the divergent magnetic field which acts like a magnetic vacuum pump. It can be calculated that a hydrogen ion and electron pair with an energy $W_\perp$ greater than 3 eV transverse to the sunspot magnetic field will feel an upward force $(W_\perp/H)$ grad H which is greater than the gravitation force on the ion. Accordingly, the hot ion pairs will be evaporated from the sunspot, leaving the cooler ones behind. This mechanism is expected to be operating strongly in regions where the Larmor frequency of the electrons exceeds the collision frequency. This mechanism can also conceivably explain the penumbra of the sunspot, because we expect the thickness of the penumbra to be representative of the thickness of the current sheet producing the magnetic field of the spot. Within this current sheet grad H in the vertical direction is less and we therefore expect less evaporation cooling in the penumbra. The sharp boundary of the penumbra is presumably produced by the edge of the current sheet where the cooling mechanism disappears completely.

The fact that solar prominences are streamers, frequently from one sunspot to another, suggests that the same mechanism is operative here as in the laboratory (Bostick, W. H., *Phys. Rev.* **104**, 1191, 1956), where we have produced streamers in a geometrically similar magnetic field generated by a horseshoe magnet. The important fact brought forward by the laboratory experiments is that for streamer formation it is necessary to have not only the magnetic field from north to south but to have a current flowing between north and south as well. Hence, we infer that the probable distribution of magnetic field, currents, and velocity in a whirl ring producing the spot pair is helical, i.e. the field lines form helices around the central circle of the ring. This helical distribution of the magnetic field in the ring will cause a current to flow from north to south or vice versa when the ring intersects the surface of the sun (see also page 113 of this volume).

Alfvén: This is a very interesting idea you put forward but it is necessary to check in detail whether the similarity transformation from your plasmoid to the conditions in the sun could be made. I look forward with great interest to future results in this field.

Dungey: There is some circumstantial evidence that the sunspot fields are twisted. I wish someone would look for this in the Zeeman effect.

Biermann: I would like to add here one remark which should have been made in my report. A magnetic field of the kind I discussed can be stationary (against drifting towards the surface, owing to smaller density) only if it is approximately of the type called 'force-free' field by Lüst and Schlüter. Hence, it would seem to be worth while to look after screw type geometrical features also of the magnetic spot fields on the surface.

THE MAGNETIC FIELD IN THE CORONA

T. GOLD

Harvard College Observatory, Cambridge, Mass., U.S.A.

ABSTRACT

The shapes of coronal features are discussed, on the basis that they are indicative of the configurations of the magnetic fields. Some considerations about the strength of the fields are advanced, and the forces that must be responsible for streamers and for the quiet corona are discussed. The magnetic fields would seem to dominate over all inertial hydrodynamic forces, leaving only such magnetic configurations as can be set up by variations of the gas pressure.

Several reasons can be given why the shapes of the corona can be taken to be indicative of a magnetic field. Indeed the coronal streamers and the polar plumes gave the first suggestion of the existence of the general magnetic field of the sun. These reasons can be classified as follows:

1. Many features look as if they trace out lines of force. What is meant by that, of course, is that they represent a pattern of lines that could be lines of force produced largely, though not entirely, by currents in the sun. They look like the external field of a magnet, distorted, but not beyond recognition (Plates I and II).

2. The shapes are of a sort that could not be expected from hydro-dynamical motions on the rotating sun. There are no indications of eddies or of cyclons, but instead there are straight elongated features which one would not easily associate with pure hydrodynamic motions.

3. Coronal features appear to take part in the differential solar rotation of the surface underneath them. Petri has measured the rotation periods of photometrically recognizable coronal features and these measurements show the existence of approximately the correct differential rotation. Also the plumes and other features never show any loops which would result from the projected appearance of a spiral shape. Long-lived individual features must therefore be thought of as being in solid body rotation, and anchored in the photosphere below them.

4. There appear to exist substantial horizontal gradients of density and pressure (with the temperature positively correlated), and such differences

would disappear quickly without a magnetic field. The observed velocities that are normally present are too low to lead to such pressure differences.

5. There exists a strong connexion between coronal shapes and the magnetic activity in the photosphere.

Even if one did not accept that the coronal lines are indicators of the magnetic field, one can still use them to indicate the absence of very different shapes of magnetic fields of significant strength. A particular type of field evidently not possessed by the sun should be mentioned here. A star condensing out of a large mass of magnetized gas would cause a curious deformation of the field: the star will no doubt rotate after contracting from the gas mass to which some lines of force remain connected. Such lines must, therefore, get twisted around one another in the vicinity of the poles before reaching out to the large gas masses in which they remain anchored. There is not the slightest tendency for such twisted knots in the corona (Fig. 1). It appears therefore that the fields of the rotating sun are not anchored in distant gas masses, but belong entirely to the sun.

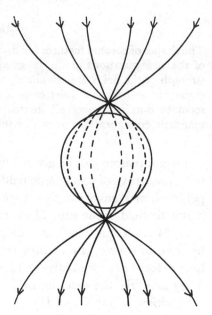

Fig. 1. Twisted knots.

Now let us deal with actual attempts to explain some of the features in magneto-hydrodynamical terms. For the large broad streamers there are two basic types of explanation possible. Either they really are streamers in which the outward flow of material is essential for maintaining the density distribution that is observed, or they may represent shapes which are in magneto-hydrostatic equilibrium and which, in the case of a perfect conductor, would require no flow at all. For the first explanation one has to meet van de Hulst's point that hydrodynamic continuity implies that a great acceleration of the material must occur as it rises from the solar surface.

Observations do not really seem to allow one to exclude such accelerations, and Schlüter has given a mechanism which would provide it. If any gas were made available low down in the corona with the magnetization much less than that of its surroundings, it would act like a diamagnet and

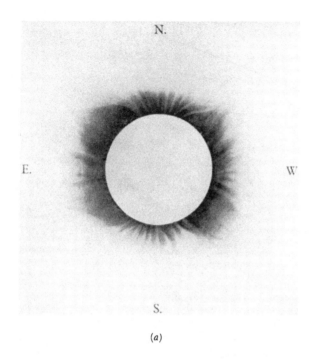

(a)

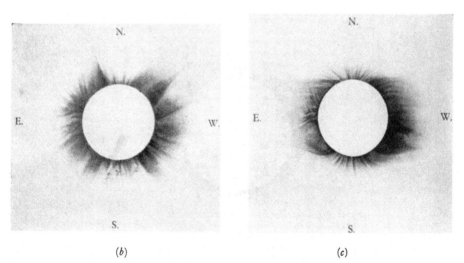

(b)   (c)

Plate I. Shapes of the corona. (a) 22 January 1898. (b) 3 January 1908.
(c) 18 May 1901.

(facing p. 276)

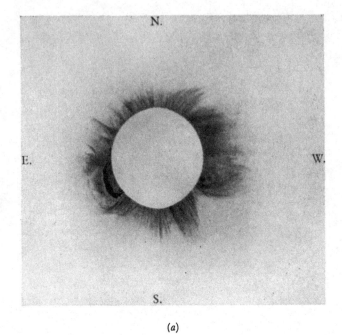

(a)

(b)

Plate II. Shapes of the corona. (a) 29 May 1919, (b) 30 August 1905.

would be expelled by pushing the existing field apart and finding its way out in between. In fact one can give a general reason why Schlüter's process should be expected on the surfaces of stars. A chaotic magnetic field would possess an average strength varying with the two-thirds power of the density in the case of expansion or contraction of the gas. If a star forming from tenuous interstellar gas possessed the appropriate field strength at each level, then the Schlüter instability on the surface would be avoided. Material brought from one level to another would bring with it the correct amount of magnetic field for the new surroundings. But over long periods of time ohmic dissipation of the fields must be considered. In places where the gradient of the density is high ohmic dissipation will in general mean that in the higher density region the fields will be weakened and in the adjoining lower density regions they will be strengthened, as compared with those appropriate to the density. The surface of the photosphere is a region where this effect would be most pronounced and where any material would become unstable against expulsion as soon as its field strength has been diminished by a certain margin as a result of ohmic decay. It is interesting to think that there is a mechanism here for the continuous emission of matter from stars.

The alternative possibility is to regard the coronal streamers as approximations to magneto-hydrostatically stable equilibrium shapes. Using the approximation of infinite conductivity and the usual notation, the equation

$$-\operatorname{grad} p + \rho \mathbf{g} + \frac{1}{4\pi} \operatorname{curl} \mathbf{H} \times \mathbf{H} = 0$$

would have to be satisfied. What configurations can satisfy this equation for the boundary conditions that the lines of force are anchored on the surface does not yet seem to have received any consideration. It is likely that stable shapes of overlying arches exist, and a stability condition will have to be satisfied according to which the field strength in the center has to diminish sufficiently rapidly with height (Fig. 2). It would be extremely valuable if such configurations with the appropriate boundary condition on the surface could be calculated by electronic computing machines, if only to see whether the observed shapes are at all of the right kind. In particular one would like to know whether as narrow a structure of arches can be expected as occurs commonly in coronal streamers. If such magneto-hydrostatic stable shapes exist, then they would really be expected to occur in the corona, provided the magnetic field is strong enough. This explanation would then take precedence over any other requiring similarly strong fields.

In the 1954 eclipse the field was sufficiently regular to make it worth while to discuss the departures it possessed from a dipole field. Fig. 3 demonstrates the type of departure and it will be seen that the angle of the lines at the photosphere, as well as their position far out in equatorial regions, corresponds simply to that deformation of a dipole field where the

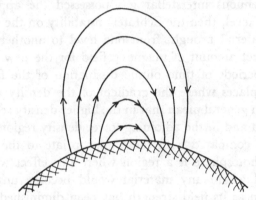

Fig. 2. A magneto-hydrostatic configuration anchored on the solar surface.

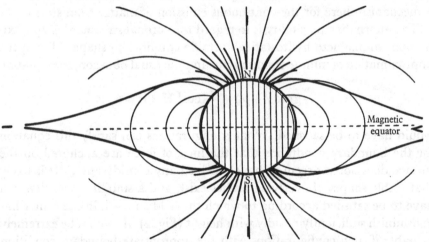

Fig. 3. Magnetic field in the corona during the 1954 eclipse (thick lines). The dipole field component is indicated for comparison (thin lines).

lines of force in the equatorial plane are stretched outwards. An extra source of heat or any other outward force in equatorial regions acting on the coronal gas would produce this type of deformation.

Now it is worth while noting the orders of magnitude of the quantities concerned. Although the magnetic and hydrodynamical forces would depend in detail upon the configurations and the flow, the comparison of the gas pressure variations, the stagnation pressure appropriate to the velocities that are reported (Waldmeier), and the magnetic pressure, may be of interest. The variations of the gas pressure at the same level appear to be of the same order as the gas pressure. At a distance of $1/10$ radius above the photosphere one has the gas pressure $p \approx 2 \times 10^{-2}$ dynes/cm², the stagnation pressure $\frac{1}{2}\rho v^2$ (at 10 km/sec) $\approx 10^{-4}$ dynes/cm² and for $H = 1$ oersted $H^2/8\pi = 4 \times 10^{-2}$ dynes/cm². Such values would allow of the magneto-hydrostatic interpretation. The velocities of flow would only introduce insignificant variations of pressure, while the magnetic field could exercise a decisive influence. If stable configurations exist then one must expect any variation in the heat supply to set them up. But with fields much weaker than 1 oersted one could not expect the large density variations at the same level.

Recent radio observations by Conway in Cambridge appear to place an upper limit on the mean field strength over a hemisphere of about 2 or 3 oersted. This conclusion depends upon the absence of circular polarization. If these considerations turn out to be correct, it will thus be possible to bracket the strength of the coronal field very closely, and to compare it with values inferred from other types of observation. At present photospheric fields are inferred by Babcock from his magnetograph tracings.

Discussion

Righini: I want to point out that we have observational evidence that the poles of the sun, as indicated by the polar plumes of the corona, are not antipodal but there is a difference of a few degrees.

Schatzman: I would like to mention that Trellis at the Pic du Midi had made an extensive study on the coronal streamers. He has studied the motions of the coronal streamers in longitude and latitude and found observations of the coronal lines, the distribution of density and temperature inside the coronal streamers.

Alfvén: I would like to ask Dr Gold about the argument for putting a maximum value of two gauss in the corona. Was that in the inner corona?

Gold: The field of one or two gauss is estimated as the limiting value of the surface field, assuming it to tail off into the corona more or less like a dipole field. A stronger field in the corona of opposite sense in the two hemispheres would have produced a measurable amount of the two senses of circular polarization.

Alfvén: Polarizarization of radio waves?

Gold: Yes.

Alfvén: I see, but does that take account of the fine structure of the corona?

Gold: No, that is on the basis of a distributed field.

Alfvén: I am rather surprised that it is possible to accept such a limit if you are not very sure about the fine structure of the corona.

Gold: It is quite true that the surface flux could be compressed into narrow bundles and then guided through the coronal material, leaving most of it unmagnetized. Then the flux could be anything you like.

Alfvén: Oh, yes.

Schlüter: I completely agree with Dr Gold as to the importance of getting as many solutions of the magneto-hydrodynamic and magneto-hydrostatic equations as possible. In most actual cases the problem is essentially three-dimensional and even a numerical solution on the largest electronic computers is very difficult. Finding a solution, however, is only part of the problem—one has also always to establish the stability behaviour of it. Obviously, this is even more difficult.

van de Hulst: I think what Dr Gold said in the beginning of his paper some-what defeats the argument at the end. He said it seemed as if most of the currents would be in the sun. If that is true, of course, the magnetic pressure vanishes outside. I see nothing wrong with your picture excluding all magneto-hydrostatic data.

Gold: But I do. The shapes which are commonly seen in the corona just cannot be made with currents entirely in the sun. It is not a curl-free field.

Cowling: The shape of the polar plumes of the corona appears to suggest that the magnetic field near the equator is small, since the plumes bend over towards the equator more than the lines of force of a dipole.

ON THE THEORY OF IONIZED
INTERSTELLAR GAS MOTION

J. P. TERLETZKY

Moscow University, Moscow, U.S.S.R.

ABSTRACT

It is pointed out that the model of mutually penetrating ideal charged and neutral gases can be used with particular success to describe the motion of an extremely rarefied plasma. As a simplest instance of application of such a model, the acceleration of totally ionized gases in crossed electric and magnetic fields is considered. It is shown that the general solution of the problem in two limiting cases is either magneto-hydrodynamic or the free charge solution.

Magneto-hydrodynamic equations afford a sufficiently exhaustive way for the description of interstellar gas motions, if the mean free path l of electrons and ions is small as compared with the radius of curvature, $R = PC/He$, of the electron trajectory in a given magnetic field moving with mean thermal velocity. Contrariwise, if $l > R$ new phenomena are originating, which may partly be accounted for by introducing an anisotropic electric conductivity and other additional terms. Much more complicated equations are obtained in this case. However, some phenomena remain unaccounted for in them, namely those connected with the probable temperature difference between electrons and ions, or with ionization and recombination.

The phenomena occurring in extremely rarefied gases can be accounted for as exhaustively as possible, if a system of kinetic equations for electrons, ions and neutral molecules is used for these gases. The equations must be solved together with equations of the electromagnetic field, caused by the electronic and ionic currents and charges. However, the solution of these equations is a rather complicated task.

A much more simple model of a rarefied plasma, in which the phenomena connected with the peculiarities of electronic and ionic motions in the magnetic field at $l \gtrsim R$ are taking place, is the system of mutually penetrating ideal gases of electrons, ions and neutral molecules, where hydrodynamical equations are used for each gas (taking into account the mutual friction of gases). This system of equations must be solved together with the electrodynamical equations, like in an ordinary magneto-hydrodynamical case. This model was used successfully for investigations of electro-acoustical waves in a plasma, originating in gaseous discharges [1]. It may also be used for calculating the ionic and electronic beams in a plasma. Such a model was also applied by Schlüter [2] in his deductions of magneto-hydro-dynamical equations for a plasma.

The main aim of the present communication is to direct attention to the fact that the model of mutually penetrating ideal gases can be used with particular success to describe the motion of an extremely rarefied plasma. An example of such a plasma is the interstellar ionized medium.

Let us consider the phenomenon of acceleration of totally ionized gases in crossed electric and magnetic fields as a simplest application of a mutually penetrating gas model.

This problem was considered earlier [3] from the point of view of magneto-hydrodynamical equations. If the electric field $\mathbf{E}$ and the magnetic field $\mathbf{H}$ are to be assumed as homogeneous and constant (i.e. if the field caused by the plasma is neglected) then for $\mathbf{E}$ directed along the x-axis and $\mathbf{H}$ along the z-axis the magneto-hydrodynamical equations may be given the following particular solution

$$U_x = U_x^0 e^{-t/\tau},$$

$$U_y = U_y^0 + \left(U_y^0 + c\frac{E}{H}\right)(e^{-t/\tau} - 1)$$

and
$$U_z = U_z^0, \quad \tau = \frac{\sigma c^2}{\lambda H^2}, \qquad (1)$$

where U_x, U_y, U_z—are the velocity components, U_x^0, U_y^0, U_z^0—their initial values, σ the medium density, λ the electric conductivity and c the velocity of light.

From the point of view of the model of mutually penetrating gases this problem may be solved for a totally ionized plasma by means of the following equations:

$$\sigma_+\left\{\frac{\partial \mathbf{U}^+}{\partial t} + (\mathbf{U}^+.\nabla)\,\mathbf{U}^+\right\} = -\nabla P_+ + \frac{e}{m_+}\sigma_+\left\{\mathbf{E} + \frac{1}{c}[\mathbf{U}^+\mathbf{H}]\right\} - \alpha\sigma_+\sigma_-(\mathbf{U}^+ - \mathbf{U}^-),$$

$$\sigma_-\left\{\frac{\partial \mathbf{U}^-}{\partial t} + (\mathbf{U}^-.\nabla)\,\mathbf{U}^-\right\} = -\nabla P_- - \frac{e}{m_-}\sigma_-\left\{\mathbf{E} + \frac{1}{c}[\mathbf{U}^-\mathbf{H}]\right\} - \alpha\sigma_+\sigma_-(\mathbf{U}^- - \mathbf{U}^+),$$

$$\operatorname{div}(\sigma_+\mathbf{U}^+) + \frac{\partial\sigma^+}{\partial t} = 0, \quad \operatorname{div}(\sigma_-\mathbf{U}^-) + \frac{\partial\sigma^-}{\partial t} = 0, \quad P_+ = \frac{kT_+}{m_+}\sigma_+, \qquad (2)$$

$$P_- = \frac{kT_-}{m_-}\sigma_-, \quad \operatorname{curl}\mathbf{H} = \frac{1}{c}\frac{\partial\mathbf{E}}{\partial t} + \frac{4\pi}{c}e\left(\frac{\sigma^+}{m_+}\mathbf{U}^+ - \frac{\sigma^-}{m_-}\mathbf{U}^-\right),$$

$$\operatorname{curl}\mathbf{E} = -\frac{1}{c}\frac{\partial\mathbf{H}}{\partial t}, \quad \operatorname{div}\mathbf{E} = 4\pi e\left(\frac{\sigma^+}{m_+} - \frac{\sigma^-}{m_-}\right), \quad \text{and} \quad \operatorname{div}\mathbf{H} = 0.$$

Taking **E** and **H** as constant and directed correspondingly along the x- and z-axis, as in the preceding problem, we obtain for a particular case, when $\mathbf{U}^+$ and $\mathbf{U}^-$ depend only on time, the following system of equations

$$
\begin{aligned}
\dot{U}_x^+ &= \frac{e}{m_+}\left(E + \frac{U_y^+}{c}H\right) - \alpha\sigma_-(U_x^+ - U_x^-), \\[4pt]
\dot{U}_y^+ &= -\frac{e}{m_+}\frac{U_x^+}{c}H - \alpha\sigma_-(U_y^+ - U_y^-), \\[4pt]
\dot{U}_x^- &= -\frac{e}{m_-}\left(E + \frac{U_y^-}{e}H\right) - \alpha\sigma_+(U_x^- - U_x^+)
\end{aligned}
\tag{3}
$$

and
$$
\dot{U}_y^- = \frac{e}{m_+}\frac{U_x^-}{c}H - \alpha\sigma_+(U_y^- - U_y^+).
$$

The solution of the latter system of linear equations is not difficult, but we shall not give it here. We shall only consider two limiting cases.

For a limiting case, when the density of the plasma has sufficiently high values, an expression for the mean velocity $(m_+\mathbf{U}^+ + m_-\mathbf{U}^-)/(m_+ + m_-)$ is obtained, which coincides with (1).

For another limiting case, when the density is infinitesimally small (the terms containing the coefficient α in Eq. (3) may be neglected) a solution of the following type is arrived at:

$$
\begin{aligned}
U_x^+ &= U_o^+ \sin\Omega_+ t, \quad U_y^+ = U_o^+ \cos\Omega_+ t - c\frac{E}{H}, \\[4pt]
U_x^- &= U_o^- \sin\Omega_- t, \quad U_y^- = U_o^- \cos\Omega_- t - c\frac{H}{E},
\end{aligned}
\tag{4}
$$

where
$$
\Omega_+ = \frac{e}{m_+ c}H, \quad \Omega_- = -\frac{e}{m_- c}H.
$$

This solution coincides with the one for free electrons in crossed fields.

Thus, the system (2) of equations describes the plasma fairly well, both in the case of high and low densities.

System (2) of equations may be easily generalized, either for the case of incompletely ionized gases, in which the process of ionization and recombination is taking place[1], or in the case of relativistic velocities. The system of relativistic hydrodynamical equations of mutually penetrating gases may be successfully applied not only to the interstellar medium, but may also be used for calculations of various accelerators.

REFERENCES

[1] Konjukov, M. V. and Terletzky, J. P. *J. Exp. Theor. Phys.* **27**, 542, 1954; **29**, 874, 1955.
[2] Schlüter, A. *Z. Naturf.* **5**A, 72, 1950.
[3] Kolpakov, P. E. and Terletzky, J. P. *C.R. Acad. Sci. U.R.S.S.* **76**, 185, 1951.

INTERPLANETARY MAGNETIC FIELD

H. ALFVÉN

Royal Institute of Technology, Stockholm, Sweden

ABSTRACT

The magnetic field outside the sun (interplanetary magnetic field) is certainly far from a dipole field. A model of it has recently been suggested (*Tellus*, **8**, 1, 1956). This model is in agreement with cosmic ray and magnetic storm data and reconcilable with the coronal ray structure. Some possibilities to check the model are discussed.

I. INTRODUCTION

The properties of interplanetary space has been only a minor chapter in astrophysics. This is due to the fact that except for the zodiacal light there has been no astronomical phenomenon which we have associated with interplanetary matter. With the rising interest for electromagnetic phenomena in astrophysics, the situation is now changing. As the degree of ionization of the interplanetary gas is high, the condition for magneto-hydrodynamic coupling is well satisfied in interplanetary space (see e.g. Lehnert, p. 54 of this volume). Hence the motion of the interplanetary gas will in general produce electromagnetic phenomena, which may affect the ionosphere (magnetic storms and aurora) and the cosmic radiation (intensity variations). As the motions induce electric fields the electromagnetic state of interplanetary space is a reflexion of the state of gas motion, so that we have a possible way to study the interplanetary 'winds' with electromagnetic methods. We may also say that the electromagnetic state of interplanetary space is regulated by the interplanetary 'meteorological' situation, i.e. the state of gas motion, pressure conditions, etc.

A considerable part of the subjects discussed at this symposium converges into the problems of interplanetary space. A general study of magneto-hydrodynamics is of course fundamental even to this field. Solar physics is important because the interplanetary space may be considered as the extreme outskirts of the solar atmosphere, and its state is certainly regulated to a large extent by the solar activity. Some ionospheric phenomena, especially magnetic storms and aurorae, will certainly be of great value for the understanding of the motions of interplanetary matter. Even if we still disagree fundamentally about the mechanism underlying these phenomena, we do agree that they are caused by the emission of ionized beams or clouds from the sun. Last but not least, the cosmic rays which reach the earth have travelled through interplanetary space and hence may act as probes of its electromagnetic state. If correctly interpreted the variations in primary cosmic radiation will no doubt be an extremely valuable tool in the exploration of interplanetary space.

Hence we find that several different fields of research coalesce and we can trace the contours of a new field of research which may be called *interplanetary meteorology*, concerned with the winds in interplanetary space and hence also with its electromagnetic state. This field is still in a very early stage of its development and quite different ways of approach have been tried—and should be tried.

I intend to discuss a model of the magnetic field in interplanetary space which at the same time is a model of the state of gas motion and hence also of the electric field. This model is essentially the same as has been used long ago in the electric field theory of aurorae and magnetic storms[1] and more recently also in a theory of cosmic ray variations and the local generation of cosmic radiation[2]. The model is only meant to be a first approximation to the real conditions and to represent, at most, an average state.

Its essential properties are shown in Figs. 1, 2 and 3. The model has been discussed in detail in *Tellus*[3].

The background of the model is the following. In the absence of electric currents in space the solar magnetic field should be a dipole field at least at a large distance from the sun. According to Ferraro's theorem it is likely that the magnetic axis coincides with the rotational axis of the sun. The solar activity introduces disturbances of the dipole field, and especially the emission of beams in the equatorial region will tend to 'blow up' the field, so that the lines of force are drawn out. Within a beam the field—if frozen in—should fall as r^{-1}. Outside the beam (or beams) the field recovers but rather slowly, the time constant being of the order of a month or a year[3]. The decrease of the field with increasing solar distance should be more rapid than inside the beam but less rapid than in a dipole field. In the model the field outside the beam is assumed to decrease in average as r^{-2}.

2. PROPERTIES OF THE MODEL

The most important properties of this model are:

1. *The field lines agree with the visual structure of the solar corona.* Independently, Gold[4] has concluded that the fine structure in the corona

should be interpreted as indicating the direction of the magnetic field even near the equator.

2. The field permits *cosmic rays produced by a solar flare* to go almost directly out to the earth. Their path follows a line of force which according to Figs. 1 and 2 may go almost radially. When they arrive to the earth part of them come from the direction of the sun, but there is also a considerable scattering.

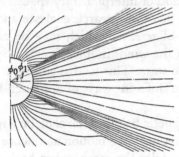

Fig. 1. The solar magnetic field inside a beam (*Tellus*, **8**, 8, 1956).

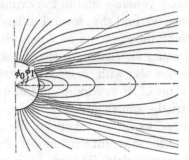

Fig. 2. The solar magnetic field outside a beam (*Tellus*, **8**, 8, 1956).

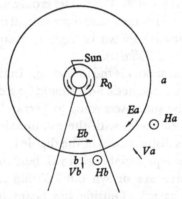

Fig. 3. Field model in the equatorial plane (*Tellus*, **6**, 232, 1954).

3. It gives an *electric field* inside the beam which according to the electric field theory and Malmfors–Block's experiment is essential in order to explain the *magnetic storms and aurorae*.

4. The model also gives a *cosmic ray storm effect and a 27-day variation* of cosmic radiation. These effects will be discussed later.

5. The model is in *apparent conflict* with the Zeeman effect measurements of the photospheric field if these measurements are interpreted in the

conventional way. However, as there is no theory of the Zeeman effect in a turbulent atmosphere, this conflict could not be regarded as serious [5, 6, 7]. On the other hand the model is in agreement with the value of the solar magnetic field derived from the magneto-hydrodynamic theory of sunspots.

3. INFLUENCE OF SOLAR ROTATION

The solar rotation introduces a modification of the model. When ionized matter is ejected in a beam, the time of travel from the sun to the earth is not long enough to make the matter rotate with the same angular velocity as the sun. This is in reality a consequence of the fact that the velocity of emission is higher than the magneto-hydrodynamic velocity. The result is that the angular momentum is conserved so that the beam gets the same geometrical shape as in Chapman–Ferraro's theory.

On the other hand, outside the beams the flow is directed inwards and much slower. There may be time enough to establish a state of rotation with the same angular velocity as the sun. Sandström [8] has found that after application of Brunberg–Dattner's correction the tangential component of the diurnal variation is remarkably constant. This very interesting result seems to indicate that isorotation is established in interplanetary space at the solar distance of the earth's orbit.

It is possible that the magnetic field transfers momentum from the sun to its environment. This will cause a drag so that a tangential component is introduced.

4. FORBUSH STORM EFFECT AND THE 27-DAY RECURRENCY

It is of special interest to see how the Forbush storm effect and the 27-day recurrency is accounted for by the model. The influence of a beam on cosmic rays moving in the equatorial plane has been treated by Brunberg and Dattner [9]. Recently Block has started detailed calculations of the orbits of cosmic rays in the model field. A detailed theory cannot be worked out before these calculations are ready. However, already now it is clear that the effect is rather complicated and at least three different phenomena are of importance. This could be seen from a discussion of the motion in the equatorial plane of particles with so low energy that the radius of curvature of their paths is small compared to the distance sun–earth.

Fig. 4 shows the orbit of such a particle. When it passes a beam, (which contrary to Swann's opinion it could do even if the radius of curvature is small compared to the thickness of the beam) it will be influenced in the following way:

1. *Electric field effect.* The particle is decelerated by the electric field. This will cause a continuous decrease dI/dt in the intensity of cosmic radiation as long as the earth is inside the beam.

2. *Magnetic field effect.* Due to the difference in magnetic field inside and outside the beam the particle will drift outwards (from the sun) or inwards when it enters the beam. If outside the beam the intensity of cosmic radiation is a function of the solar distance, the intensity inside the beam will be different from outside the beam. It may be either higher or lower.

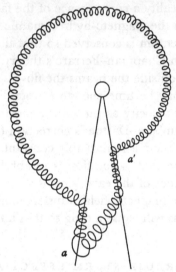

Fig. 4. Path of a high-energy particle in the accelerator model with crossed electric and magnetic fields (*Tellus*, 6, 232, 1954).

3. *Capture effect.* Cosmic ray particles may be captured inside the beam which also means that certain orbits cannot be entered by particles from outside. When the beam moves outwards the captured particles are decelerated in the decreasing magnetic field in the beam. The present model does not give capture in the equatorial plane but probably in three-dimensional orbits. Further, a small disturbance in the magnetic field of the beam is enough to produce a capture of orbits in the equatorial plane.

If these three effects are superimposed the result may be shown in Fig. 5. If the K_p index is considered to be a measure of the electric field of a beam, we get a correlation between cosmic ray intensity and K_p index, which agrees with Kane's results[10]. However, it is in conflict with an

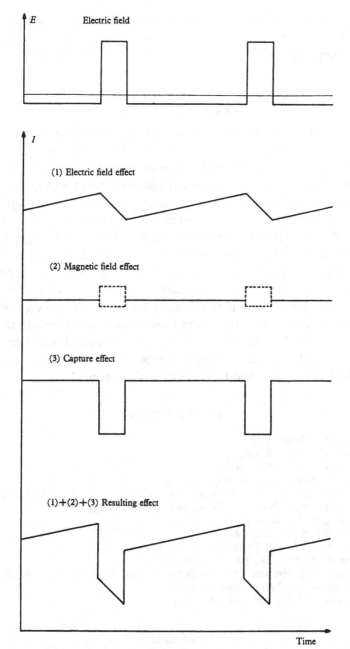

Fig. 5. When the sun rotates, the earth will be situated alternatively in a positive electric field (inside the beam) or in a small negative field (outside the beam). The cosmic rays will be affected in three different ways (1, 2 and 3 in the figure) which together give the resulting effect.

investigation by van Heerden and Thambyahpillai [11]. Venkatesan [12] will report about a similar investigation which supports Kane's results and gives an indication of the existence of at least effects (1) and (3).

5. POSSIBLE TURBULENCE IN THE MAGNETIC FIELD

Our model is designed in order to give a first approximation of the magnetic field in interplanetary space, but it is quite natural that the field should be superimposed by turbulent fields. It is an open question how important the turbulence is. I do not know any observed effect which calls for the assumption of turbulence in interplanetary space and in my opinion its importance has often been exaggerated. Interplanetary space may be considered as an extrapolation of the corona. Coronal photographs show a fine structure in the corona which may very well go far out in inter-planetary space, but there is no indication of a large-scale turbulence. Although a small-scale turbulence, associated with the granulation, may exist, it seems more difficult to assume a turbulence with large elements. Especially, I cannot see any reason to introduce an extremely turbulent beam or cloud as Morrison [13] and others do. It is much easier to capture cosmic rays and to produce forbidden orbits in a rather regular field like the field of our model than in violently turbulent fields.

REFERENCES

[1] Alfvén, H. *Tellus*, **7**, 50, 1955.
[2] Alfvén, H. *Tellus*, **6**, 232, 1954.
[3] Alfvén, H. *Tellus*, **8**, 1, 1956.
[4] Gold, T. This symposium, Paper 30.
[5] Alfvén, H. *Nature, Lond.* **168**, 1036, 1951; *Ark. Fys.* **4**, 407, 1952; *Tellus*, **8**, 1, 1956.
[6] Alfvén, H. and Lehnert, B. *Nature, Lond.* **178**, 1339, 1956.
[7] Lehnert, B. Comment given on p. 245.
[8] Sandström, A. E. *Tellus*, **8**, 18, 1956.
[9] Brunberg, E. Å. and Dattner, A. *Tellus*, **6**, 254, 1954.
[10] Kane, R. P. *Phys. Rev.* **98**, 130, 1955.
[11] Heerden, I. J. van and Thambyahpillai, T. *Phil. Mag.* [7], **46**, 1238, 1955.
[12] Venkatesan, D. This symposium, Paper 37.
[13] Morrison, P. *Phys. Rev.* **101**, 1397, 1956.

Discussion

Parker: I would like to know how the close approach to isotropy (within $\pm 0.5 \%$) following the onset of a Forbush decrease of 10% (presumably produced by a beam) can be reconciled with a large cosmic ray decrease of 10%. The 10% effect of a beam is essentially entirely anisotropic.

Alfvén: In absence of the electric field the cosmic radiation would become isotropic very quickly because of scattering, so it is only the electric field, or we can say the 'wind', which produces the anisotropy. The anisotropy during the storm becomes larger. In a paper presented at the Guanajuato conference Sandström has made a division of the cosmic ray variation into one tangential and one radial component. What comes out is that the latter is independent of the K_p but the former is a function of the K_p. The variation in the anisotropy is in good agreement with the model I have presented. What happens during the storm is, to the first approximation, that the voltage of the earth is changed when it passes across an electric field and that changes the whole cosmic ray intensity. When the voltage is changed the intensity is changed but not the anisotropy.

Singer: I want to make two remarks to Alfvén and to some extent also to van de Hulst (paper No. 1). One has to do with the winds in interplanetary space, I think this is a nice way to put it. I do not know any method of detecting these winds but I will report briefly that one can calculate the equilibrium charges of dust particles in interplanetary space and this charge comes out to be very highly positive. It is controlled merely by the solar photo-electric effect and also by the electron density and the electron temperature in the region of the dust particles. One of the interesting consequences of this is that the winds, such as are produced by corpuscular streams, will tend to carry these dust particles far up, away from the sun. There might be a way of observing the 'sweeping out', for example by studying the outer corona of the sun before and after solar activity.

Alfvén: The interplanetary dust is surely important here. It is of interest to see to what extent it moves with the beam and of course it will be best studied from a satellite.

Singer: The other remark has to do with the emission of the beam. The omitted beam will be a supersonic jet and the density is high, 200–800/cm³. Then, turbulence will be set up. This could perhaps explain the absence of low-energy cosmic ray particles and why they reappear during low solar activity. Lumps with large magnetic field and high density created in this way will act as very effective sinks for low-energy particles (see paper to appear in *Phys. Rev.*). This cosmic ray effect would, in addition, provide a method of observing the onset and decay of the turbulence.

Alfvén: A crucial thing here is the importance of the shock-waves. The model here is a stationary one. It should only be taken as a first approximation and there are some obvious difficulties with this model which you see immediately.

It is natural that there should be turbulence in interplanetary space. But the importance of this is difficult to estimate. From the structure of the solar corona and from the rather long regular pattern which Gold has shown us, you do not get the impression that the large-scale turbulence is very important. In my model of the magnetic field it is neglected, but it is quite reasonable that we should introduce some turbulence at a later point. On the other hand, it is quite likely that the field has a fine structure. The structure of the corona may very well penetrate outwards. However, at present I can see no meaning in introducing a turbulent field and, as far as I know, there is no phenomenon which calls for the introduction of turbulence, so I think we could forget it at present. The extremely turbulent clouds which Morrison has proposed do not produce any effect which could not be produced much easier with the present model. For example, in the beam you could get captured orbits which give the same decrease as is obtained with these turbulent clouds, but with the beam you get them with much lower magnetic fields and in a much easier way.

Gold: The cosmic ray flare effect does not allow us to suppose that a very particular position on the sun is required to let the particles come to the earth, as the very greatest flares are known to be responsible; if an improbable condition of position has to be satisfied then it would be unlikely that the few largest flares would have satisfied this.

Further, the very great extension, as far as the earth, of coronal streamers is also made a difficult concept because of the solar rotation. Some of these streamers are long lived, and it would be necessary to suppose the gas at the earth to go around once in 27 days. At some distance this situation must clearly break down.

Alfvén: I think that this model works out to some earth radii. It is likely that matter at the earth's orbit takes part in the rotation, but this requires a much lower density in the interplanetary space than what is usually assumed.

PART V

ELECTROMAGNETIC STATE IN INTERPLANETARY SPACE

A. THEORIES OF MAGNETIC STORMS

B. COSMIC RAY METHODS OF EXPLORING INTERPLANETARY SPACE

A. THEORIES OF MAGNETIC STORMS

PAPER 33

THE PRESENT STATE OF THE CORPUSCULAR THEORY OF MAGNETIC STORMS

V. C. A. FERRARO

Queen Mary College, University of London, England

ABSTRACT

The evidence in favour of a corpuscular theory of magnetic storms is briefly reviewed and reasons given for believing that the stream must be neutral but ionized and carry no appreciable current. It is shown that under suitable conditions the stream is able to pass freely through a solar magnetic field; the stream may also be able to carry away with it a part of this field. However, because of geometrical broadening of the stream during its passage from the sun to the earth, the magnetic field imprisoned in the gas may be wellnigh unobservable near the earth.

The nature, composition and dimensions of the stream near the earth are discussed and it is concluded that on arrival the stream will present very nearly a plane surface to the earth if undistorted by the magnetic field.

Because of its large dimensions, the stream will behave as if it were perfectly conducting. During its advance in the earth's magnetic field the currents induced in the stream will therefore be practically confined to the surface. The action of the magnetic field on this current is to retard the surface of the stream which being highly distortible will become hollowed out. Since the stream surface is impervious to the interpenetration of the magnetic tubes of force, these will be compressed in the hollow space. The intensity of the magnetic field is thereby increased and this increase is identified with the beginning of the first phase of a magnetic storm. This increase will be sudden, as observed, owing to the rapid approach of the stream to the earth.

The distortion of the stream surface is discussed and it is pointed out that two horns will develop on the surface, one north and the other south of the geomagnetic equator. Matter pouring through these two horns will find its way to the polar regions.

The main phase of a magnetic storm seems most simply explained as due to a westward ring-current flowing round the earth in its equatorial plane. Under suitable conditions such a ring-current would be stable if once set up. The mode of formation of the ring is, however, largely conjectural. The possibility that the main phase may be of atmospheric origin is also briefly considered. It is shown that matter passing through the two horns to the polar regions could supply the energy necessary for the setting up of the field during the main phase. The magnetic evidence in favour of such a hypothesis, however, seems wanting.

I. INTRODUCTION

The corpuscular theory of magnetic storms and aurorae was first proposed by Birkeland in 1896[1]. He suggested that both these phenomena are due to charged particles emitted from the sun and that these particles are guided towards the polar regions by the geomagnetic field. This hypothesis found support in his experiments in which cathode-rays projected towards a magnetized sphere were seen to impinge along two zones, one around each pole of the sphere.

These experiments led Störmer[2] to study mathematically the motion of a single charged particle in the magnetic field of a dipole—which is a good approximation to the geomagnetic field except near the surface. The trajectories derived by Störmer showed many analogies with the forms of the aurora; nevertheless the theory is unsatisfactory in so far as it takes no account of the electrostatic forces which arise when many particles are present. These forces far exceed the deflecting force on a single charge by the geomagnetic field. In a cloud consisting of charges of one sign only the mutual electrostatic repulsion of its parts would disperse the particles to a negligible density long before their arrival at the distance of the earth.

To overcome this destructive criticism, first directed by Schuster[3] against one-signed corpuscular theories Lindemann (Lord Cherwell as he became later) in 1919 put forward[4] the suggestion that the streams emitted by the sun are neutral though ionized. He discussed at some length the process of emission, the speed attained by the particles of the stream and showed that the number of recombinations of ions and electrons to be expected during the passage of the stream from the sun to the earth was negligible. But he did not consider the phenomena which would develop during the advance of the stream in the geomagnetic field. Chapman and Ferraro[5] made the first attempt to develop a theory of magnetic storms based on the neutral stream hypothesis, and this will be described here. An entirely different theory, also based on the neutral stream hypothesis, was later proposed by Alfvén[6,7]. But because of the neglect of the powerful electrostatic forces between the ions and electrons, the bearing of his theory on magnetic storms, like that of Birkeland–Störmer, is uncertain.

The evidence for a corpuscular theory of magnetic storms and aurorae had long been adduced by Maunder[8] and Chree[9] on the basis of the 27-day recurrence tendency shown by these phenomena. But this evidence was indirect and not indisputable until Meinel[10] showed from his observations of the auroral spectrum that high-speed protons were entering the upper atmosphere along the magnetic lines of force of the earth's field.

2. THE NEUTRAL IONIZED GAS

Observations of the solar atmosphere shows that from time to time gas is accelerated away from its surface. The long coronal streamers seen at the time of a solar eclipse is also taken as evidence of solar corpuscular emission. But as yet we do not know from what level in the solar atmosphere the neutral gas is emitted nor the mechanism of emission.

Great magnetic storms, however, are closely associated with intense solar flares. About a day after the occurence of such a flare the magnetic storm begins suddenly. This time lag is usually interpreted as indicating that the neutral gas travels from the sun to the earth with a speed of 1000–2000 km/sec. Since magnetic storms seldom occur if the flare is more than about 45° from the centre of the disc it is supposed that the angle of emission of the stream is large, ranging from about 40–50°.

Moderate and weak storms show a tendency—not shared by great storms —to recur after a period of a solar rotation of about 27 days. This recurrence tendency was interpreted by Maunder[8] as indicating a continued emission, lasting for a month or more, of gas from particular disturbed regions of the sun. Bartels has labelled these M regions. They are not always associated with sunspots.

Because these moderate storms do not begin abruptly it is difficult to infer the speed of M region streams. The evidence suggest that it is lower than for flare streams and lies probably in the range 500–1000 km/sec.

The gas is likely to be typical solar atmospheric gas, that is, mainly hydrogen atoms (mostly ionized) with a small admixture of other elements, notably Ca^+. The density and temperature are unknown: if the gas is emitted from the lower chromosphere the temperature is about 6000 °K, and the density of the order of 10^{10}/cc. If the emission take place from a higher level the temperature will be correspondingly higher and the densit lower.

During its passage from the sun to the earth the stream will expand because of geometrical broadening and because of thermal velocities of the particles in the gas. For a gas at 6000 °K the velocity of expansion of the surface on this account is 11 km/sec; for the temperature of the corona it is 130 km/sec. If unimpeded during its passage from the sun to the earth, the stream would expand to linear dimensions of the order of 10^6 km whatever the original size. The expansion due to geometrical broadening (produced by the divergent directions of emission at the sun's surface) will be still greater. For a conical angle of 20° the breadth of the stream at the distance of the earth is 5.4×10^7 km. For a flare-burst stream it is still greater. This expansion will reduce the density by several powers of ten in the neighbourhood of the earth, depending on the original size and mass of the stream.

As seen from the earth the undisturbed stream surface will appear to be nearly plane. For a flare-burst stream this surface may be normal to the direction of travel of the particles in the stream. For an M region stream, the longitudinal surface of the stream is inclined to this direction (which is nearly radial from the sun) so that the stream overtakes the earth in its orbit once every 27 days with a speed of about 4×10^7 cm/sec. If the streaming velocity of the particles is 10^8 cm/sec the stream surface will be inclined to the sun-earth line at an angle of about 20°.

3. INFLUENCE OF SOLAR AND INTERPLANETARY MAGNETIC FIELDS

Several suggestions have been put forward as regards the mode of emission of solar streams. Milne[11] supposed that it was due to the action of selective radiation pressure on certain chromospheric gases and showed that this process could accelerate the particles affected to speeds of about 1600 km/sec. This accords well with the estimate derived from solar terrestrial relationships. Kahn[12], however, questioned whether the available radiation would suffice to accelerate the particles away from the sun in sufficient numbers. Kiepenheuer[13] applied the Milne mechanism to solar flares and showed that it could account for the emission of Ca^+ ions in sufficient numbers if one makes reasonable assumptions about the energy output of the flare. Unfortunately, the mechanism is inapplicable to hydrogen atoms which are by far the most important constituent of the solar streams. Alfvén[6] has attributed the emission to the presence of a general solar magnetic field.

But, it is difficult to be certain of the influence of solar magnetic fields during the period of emission of the gas. Unless the gas is emitted along the direction of the magnetic lines of force of the sun's field, the latter will tend to hinder the passage of the gas outwards. If the gas be polarized by the magnetic field so that the resulting electric field exactly balances the electromagnetic deflecting force on the charges, then provided the energy density of the stream is sufficiently large the gas will be able to pass freely outwards through the magnetic field. Let ρ and $\mathbf{v}$ be the density and velocity of the stream at any point and $\mathbf{H}$ the intensity of the magnetic field. Then the electric field $\mathbf{E}$ required to balance the magnetic deflecting force on the charge is $-\mathbf{v} \times \mathbf{H}/c$, where c is the speed of light. The condition that the stream should be able to pass through the magnetic field unhindered is that the kinetic energy density $\frac{1}{2}\rho v^2$ should be large compared with the electrostatic energy $E^2/8\pi$ or $v^2H^2/8\pi c^2$, that is, ρ must be large compared with $H^2/4\pi c^2$. This is a lower limit. It seems more likely that the polarization electric field will be unable to entirely balance the magnetic deflecting force on the charges. In this case the magnetic lines of force of the solar field will be carried along with the gas and if this is able to escape from the sun it may well carry with it a part of the solar magnetic field—general or sunspot. This will be the case if the kinetic gas pressure ρv^2 much exceeds the magnetic pressure of the tubes of force, $H^2/8\pi$, that is, if ρ is large compared with $H^2/8\pi v^2$. This is a more stringent condition unless the velocity of the gas approaches the speed of light. Considering, for example, a neutral cloud of ionized gas moving outwards with a speed of 1000 km/sec the cloud will be able to escape through a solar magnetic field of the order of one gauss if ρ exceeds 4×10^6 protons/cc. Considerably higher densities or velocities would be needed to enable the gas to escape from the magnetic field of sunspots.

If the gas is able to take with it a part of the solar magnetic field and the gas and the magnetic field imprisoned within it move into a region where the magnetic field is weaker, the surface of the gas will expand because of the tendency of the tubes of force to swell out. The velocity of expansion is likely to be of the order of $(H^2/8\pi\rho)^{1/2}$ and so less than the streaming velocity of the gas and probably greater than the thermal velocities. Thus the gas will inevitably expand to the dimensions mentioned earlier. The intensity of the magnetic field imprisoned in the gas will be greatly reduced by the time the gas reaches the earth's orbit. We may estimate this reduction as follows: since the lines of force are frozen in the gas, the magnetic flux through an open surface consisting of fluid particles will be conserved. Using also the equation of continuity of mass we find that the magnetic field will be reduced in the ratio $(\rho/\rho_0).(l/l_0)$ approximately, where l denotes a typical length and the suffix o refers to the values of ρ and l at emission. Since $\rho/\rho_0 \sim (l_0/l)^3$, the reduction in the intensity of the magnetic field of the gas is likely to be of the order of $(l_0/l)^2$. As a numerical illustration, suppose that the linear dimensions of the gas at emission are of the order of 10^4 km. Near the earth the breadth of the stream is likely to exceed 10^7 km. Thus the magnetic field in the gas will be reduced by a factor of a million at least and may well be unobservable near the earth.

An interplanetary magnetic field may also affect the advance of a solar stream towards the earth. Supposing that the magnetic lines of force of the interplanetary field are nearly perpendicular to the direction of travel of the stream, electric currents will be induced in the surface of the stream which will prevent the magnetic tubes of force penetrating into the gas. These will be pushed forward by the stream surface and the magnetic pressure exerted by the tubes of force will tend to retard the surface of the stream and render it very sharp long before the stream comes under the influence of the earth's field. The reduction in speed of the stream surface will become appreciable only if the intensity of the interplanetary magnetic field much exceeds 10^{-4} gauss. Estimates based on cosmic ray considerations do not indicate a field greater than this.

Before leaving the subject of the constitution of corpuscular streams, reference must be made to a recent attempt by Bennett and Hulburt[14] to revive the Birkeland–Störmer auroral theory by proposing a new variant of the neutral stream hypothesis. In this it is supposed that a conical beam of approximately equal numbers of fast moving ions and electrons are emitted from the sun. As it moves through the corona, the electrons in the stream have the smaller momentum and thus undergo a greater longitudinal retardation than the positive ions during encounters with the coronal particles and so tend to lag behind. Hence the stream carries an electric current. But because electrons are lost from the beam in the process, it acquires a positive charge which is then supposed to be neutralized by slower interplanetary electrons. Because of the inhibiting effects of self-induction, however, the current appears to be limited[15] to the value mu/e, where m is the electronic mass and u the velocity of the ions. Unless there are several such streams formed, the maximum current is of the order of 100 amperes and this is too small to be of interest.

4. THE AVERAGE CHARACTERISTICS OF MAGNETIC STORMS

During a magnetic storm the magnetic effects in middle and low latitudes are characterized by a sudden, world-wide increase in the horizontal force within the period of a minute. This rise is maintained for a few hours afterwards and constitutes the *first phase* of the storm. Thereafter the horizontal force decreases to a minimum some 8–15 hr after the beginning of the storm, the minimum below the normal exceeding the initial rise above normal. This period is the *main phase* of the storm. The field then returns to its normal value before the onset of the storm at a rate which becomes progressively slower and may last for many days.

Over the polar regions the disturbance is far more intense than in the middle regions of the earth. The evidence points to the existence of a system of electric currents flowing in the atmosphere which includes narrow concentrated filaments called electrojets flowing along the auroral zones. The flow of current is nearly eastward over one half of each of the two auroral zones lying on the post-meridian hemisphere, and westward over the other half. The current is completed in the atmosphere mainly across the polar cap. The height of the auroral currents estimated from magnetic data seems to be of the order of 100–150 km; the height of the polar cap currents is unknown.

Whilst there can be little doubt that in the polar regions a part of the current-system flows in the atmosphere, the same seems unlikely to be true of the rest of the current-system. The magnetic data points rather to an extra-terrestrial system of currents as the cause of geomagnetic disturbance in middle regions.

5. THE THEORY OF THE FIRST PHASE OF A MAGNETIC STORM

We next consider what happens when a neutral ionized stream advances into the earth's magnetic field. The velocity of 1000 km/sec inferred from solar-terrestrial relationships is about one hundred times the sonic velocity. The motion is therefore hypersonic and can be deduced from simple Newtonian considerations. The density of the gas near the earth appears sufficiently low for the effects of collisions to be neglected.

The gas may be treated as one of infinite electrical conductivity provided that the linear dimensions of the stream are large compared with the skin-depth $d = (4\pi n e^2/m)^{-1/2}$, where n denotes the number density of the gas, e and m the electronic charge and mass. This condition is amply satisfied since for solar stream d is of the order of a kilometre at most. It follows therefore that during its advance in the earth's magnetic field the stream will be shielded from the geomagnetic field by surface electric currents induced by the field. The surface of the stream will be sharply defined since it offers resistance to the interpenetrations of the tubes of force of the earth's field.

The magnetic tubes of force exert a pressure of amount $H^2/8\pi$ over the surface of the gas; since this is compressible it will yield to this pressure and a hollow will be carved out by the magnetic field. In the equatorial plane the section of the hollow will be roughly parabolic in shape except far out where little distortion of the stream surface takes place. The apex of the hollow will eventually be brought to rest. The distortion of the stream surface elsewhere is more difficult to determine. In the early stages of the motion, when the stream is far away from the earth, the surface of the gas can be considered as plane and the form of the current lines are as shown in

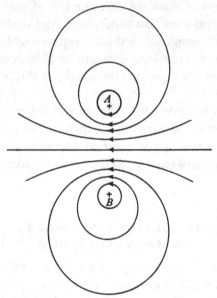

Fig. 1. Current-lines of the electric current-system induced in the front surface of the stream considered to be plane, parallel to the earth's dipole axis and normal to the direction of travel of the particles.

Fig. 1. The current intensity vanishes at the foci A and B and at these points the gas is not retarded; hence two horns will emerge and as the gas advances and its surface becomes distorted the position of the foci will change. It is difficult to infer the further development of these two horns, but, as Chapman and Ferraro pointed out in their original papers [5], matter passing through the horns seems likely to find its way to the polar regions. It seems unlikely that the gas emerging from the two 'horns' will gain speed and energy. The regions of the atmosphere most likely to be affected by this gas is the F-layer in the polar regions.

As the hollow in the stream deepens during the advance of the stream, more tubes of force are crowded together in it. The magnetic field in the hollow is thereby increased and would produce at the earth's surface a rise in the horizontal force. Chapman and Ferraro identified this increase as the beginning of the first phase of a magnetic storm. As long as the surface of the stream remains sensibly plane, the magnetic field of the induced currents adds to the earth's field a field equal to that of the image dipole of the earth. When the stream surface becomes highly distorted the rise in the magnetic field will be somewhat greater than the value calculated from the image dipole. It is possible to estimate the rate of diminution of velocity of the vertex of the hollow and hence the duration of the initial rise of the magnetic intensity during the first phase. The writer has shown [16] that it is of the order of a minute for an initial rise of about 20γ, as is, in fact, observed. The minimum distance Za (a being the earth's radius, Z a numerical factor) from the earth attained by the vertex can be found by equating the kinetic gas pressure ρv^2 to the magnetic pressure $H^2/8\pi$ on the stream surface or $H_0^2/8\pi Z$, where H_0 is the value of the horizontal field at the equator. Thus $Z = (H_0^2/8\pi\rho v^2)^{1/6}$ and so is very insensitive to changes in energy density. Taking the streaming velocity to be 1000 km/sec, the density of the stream at large distances from the earth necessary to produce an initial rise of 20γ is inferred to be from 1 to 100/cc; and the minimum distance of approach about $5a$.

6. THE RING-CURRENT AND THE THEORY OF THE MAIN PHASE

During and after the main phase of a magnetic storm the earth is surrounded by an external field which is nearly uniform and directed from north to south. The most direct explanation, and one which is supported by the magnetic records, is that this field is produced by a westward electric current flowing in a ring encircling the earth. This suggestion was first made by Störmer in a form which is untenable. It was revived by Schmidt [17] in 1924 purely on the evidence of the magnetic data. He thought that the ring was always present being re-enforced from time to time and that it decayed during periods of magnetic calm. He suggested that the ring is electrically neutral with the positive ions circulating westward and the electrons eastward. He gave no details as regards its size or the speed of the particles in it.

Chapman and Ferraro also attempted to explain the main phase of a magnetic storm on the hypothesis of a westward ring-current. They indicated how such a ring might be formed during a magnetic storm but their suggestion was little more than a qualitative sketch. They discussed the relative equilibrium, stability and decay of the ring and showed that there was no difficulty in accounting for the continued existence of the ring for many days once established. Chapman and Ferraro supposed that the ions and electrons would flow round the earth in the same sense and with very nearly the same speed. They pointed out that for any continuing ring the current must be westward because the radial acceleration towards the centre of the earth, necessary to maintain the motion, must be supplied by the ponderomotive force which the earth's magnetic field exerts on the current. Such a current would diminish the horizontal force at the surface of the earth as is observed during the main phase of a magnetic storm.

Alfvén[18] has questioned whether the ring would, in fact, be stable. He correctly states that the circular orbit of an isolated charge in the geomagnetic field is unstable since in the equatorial plane the field decreases more rapidly than the inverse square law, but concludes that the same is likely to be true for a larger assembly of charges. This conclusion is incorrect as is borne out by the discussion of an idealized problem devised by Chapman and Ferraro[19] to examine the radial stability of the ring-current. In this the ring-current is replaced by a cylindrical sheet of ionized gas and the earth's field is replaced by a unidirectional field parallel to the earth's axis and varying inversely as the cube of the distance from the axis. Let H be the intensity of this field, r the radial distance from the axis of the sheet, and write

$$H = H_0(a/r)^3, \tag{1}$$

where a is the radius of the earth and H_0 the value of the earth's field at the equator. Let $\pm Q$ be the charges carried by the superposed ionic and electronic sheets, per unit length, v_i and v_e the azimuthal velocities of the ions and electrons in the sheet, respectively, and m_i, m_e their respective masses. Define the azimuthal mass-velocity v and the differential velocity v' of the ions and electrons by the equations

$$mv = m_i v_i + m_e v_e, \quad v' = v_i - v_e, \tag{2}$$

where $m = m_i + m_e$. Then it can be shown[19] that the following equations hold

$$v = \frac{K}{r}, \quad v' = \frac{K'}{r} + \frac{eH_0 a^3}{m'' cr^2}, \quad m'' = m' + \frac{eQ}{c^2}, \tag{3}$$

where $m' = m_i m_e/m$ is the reduced mass of the charges and r the radius of the sheet. The first equation expresses the conservation of angular momentum of the sheet about its axis. The second determines the currents induced in the sheet by its motion across the magnetic field. In addition we have the equation of radial motion

$$m\ddot{r} = \frac{mK^2 + m''K'^2}{r^3} + \frac{3eK'H_0 a^3}{cr^4} + \frac{2e^2 H_0^2 a^6}{m'' c^2 r^5}. \tag{4}$$

The current per unit length of the sheet is Qv' and the magnetic field, H', which it produces within the sheet is equal to $2Qv'/cr$. Outside the sheet there is no additional field. It is convenient to write

$$mK^2 = \mu m''K'^2, \quad r = aZ, \quad p = -eH_0 a^2/(m''cK'), \tag{5}$$

where Z, μ and p are pure numbers; (4) may then be rewritten

$$\ddot{Z} = \frac{m''K'^2}{ma^4Z^5} \{(\mu+1)\, Z^2 - 3pZ + 2p^2\}. \tag{6}$$

We can interpret this as the equation of rectilinear motion of a particle in the field of force derived from the potential

$$U = \frac{m''K'^2}{2ma^4Z^4} \{(\mu+1)\, Z^2 - 2pZ + p^2\}. \tag{7}$$

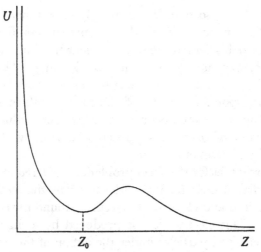

Fig. 2. Potential energy function for the radial motion of an infinite cylindrical sheet of ions and electrons in a unidirectional magnetic field whose intensity decreases inversely as the cube of the distance from the axis of the sheet. The position of stable relative equilibrium occurs at the minimum Z_0.

The positions of relative equilibrium of the sheet are given by the stationary values of U if they exist. These occur for values of Z given by

$$(\mu+1)\, Z^2 - 3pZ + 2p^2 = 0 \tag{8}$$

and the roots of this equation are real if $\mu < \frac{1}{8}$. This may be interpreted as an upper limit to the angular momentum of the sheet. When the positions of relative equilibrium exist the potential function has the form shown in Fig. 2. The stable position occurs for the smaller of the two values of Z for which U is stationary and a stable ring-current could be set up in this position. Its radius and the current within it would depend on the constants

K, K', and Q, always provided that $\mu < \frac{1}{8}$. The magnetic field H' ($= 2Qv'/cr$) produced by the current in the sheet can be expressed in terms of Z and p only (provided that $m'' \gg m'$ which is likely to be true in all cases of interest). Substituting for K' and v' from (3) and (5) in the expression for H' quoted we find

$$H' = (2H_0/p^3)\left[\left(\frac{p}{Z}\right)^3 - \left(\frac{p}{Z}\right)^2\right]$$ (9)

approximately, since $m'' \gg m'$ implies that $m''c^2 \sim eQ$, by (3). From (8) the positions of stable equilibrium, $Z = Z_0$ say, is given by

$$\frac{p}{Z_0} = \frac{3 + \sqrt{(1-8\mu)}}{4}.$$ (10)

Clearly $3/4 < p/Z_0 < 1$, so that $H' < 0$ by (9), as was mentioned earlier in this section. The magnetic field of this current would thus produce a decrease in the earth's field at the surface. Near the limit of radial stability the roots of (8) become equal so that $\mu = \frac{1}{8}$ and $p = \frac{3}{4}Z_0$. In this case $H' = -2H_0^3/(3Z_0 3)$. For a moderate disturbance of about 30γ at the earth's surface the corresponding radius of the sheet would have to be about 8 earth radii. Chapman and Ferraro showed that no undue demands were made on the value of Q or on the differential velocity of the charges of opposite sign, in all cases of interest.

In a ring-current, as for the sheet problem considered above, it is necessary that a radial electric field should act across the section of the ring and enable the ions and electrons to circulate round the earth together in the same sense. This electric field is produced by a slight separation of the oppositely charged particles under the action of the nearly equal and opposite forces acting on them. As was shown by Martyn [20] the maximum potential difference across the section may be of the order of four million volts. The polarization charges induced on the surface will thus be repelled from the surface along the magnetic lines of force and impinge on the polar regions with energies of the order of a million volts. This energy would suffice to account for the observed auroral penetration and occasion auroral luminescence.

7. THE ORIGIN OF THE WESTWARD CURRENT SYSTEM

The increase in the horizontal force produced by the currents induced in the surface of the stream (which is identified with the first phase of a magnetic storm) will be maintained so long as there is matter pouring into the current-layer from the shielded regions of the stream. If the depth of the stream is taken to be 5×10^{12} cm, as seems likely, this increase could persist for as long as half a day if the speed of the particles is taken to be 1000 km/sec. Thereafter the horizontal force would rapidly return to its normal value before the onset of the storm, without reversal. Chapman and Ferraro therefore supposed that the ring-current would be set up whilst the earth was still enveloped in the stream. They suggested that soon after the hollow space becomes stationary the ring would begin to be formed by positive charges spiralling away from the walls of the hollow facing the anti-meridian side of the earth and bridging the gap at the back of the hollow. They thought that the current in this secondary stream, though feeble at first, would grow steadily. In order to bridge the gap the spiral radius of the ions, if these be protons, would have to be at least comparable with the breadth of the hollow. Taking this as 10 earth radii and the speed of the ions as 1000 km/sec the ions would begin to flow across the gap at a distance of about 60 earth radii. Although the current in the secondary stream would be westward, an estimate of its intensity suggests that it is too small to be of interest in this connexion.

It is possible that the westward current may be produced by the outward motion of ionized gas across the earth's magnetic field, as in the case of the first phase. Such an expansion of gas, atmospheric or interplanetary, may, under suitable conditions, give rise to the observed diminution of the field over the earth. A theory of this type in which the ionized gas is atmospheric has been advocated by Oliver Wulf and Vestine. An atmospheric origin of geomagnetic disturbance was suggested by Schuster long ago, but it has always been maintained that the observations lent little support to this hypothesis since there appears to be no noticeable difference in the mean intensity of the storm variations over the dark and sunlit hemispheres. An exception appears to be the station at Huancayo where the amplitude of sudden commencements, and possibly the first phase, are enhanced during daylight. Another criticism which may be directed against a purely atmospheric theory is that it would be difficult to account for the world-wide character of storms and the fact that occasionally marked similarities appear in the magnetograms at widely separated stations.

Nevertheless, if a substantial part of the material passing through the two 'horns' mentioned in section 5 finds its way into the earth's atmosphere, mainly over the polar regions, the rate and amount of energy supplied by this material suffice to account for the observed rate of increase of the magnetic energy during the main phase. The energy is supplied at the rate of $\frac{1}{2}\rho V^3$, where ρ is the density and V the velocity of the gas. Taking

$V = 1000$ km/sec* this is equal to $0.8N$ ergs/cm^2/sec, where N is the number density of the gas. Supposing that this energy is absorbed along the two sunlit halves of the auroral zones, and taking their mean radius and breadth as 23° and 6° respectively, the energy supplied per second is of the order of $10^{17}N$ ergs/sec. Chapman and Bartels[21] give the rate of increase of magnetic energy during a storm as 2×10^{18} ergs/sec. If one-tenth, say, of the energy added to the atmosphere were converted into magnetic energy the required density of the incoming gas would have to be about 200/cc, which is not excessive. To account for the world-wide character of magnetic storms this energy absorbed would have to be quickly redistributed over the whole atmosphere. This seems doubtful but the possibility cannot be ruled out.

REFERENCES

[1] Birkeland, K. *Arch. Sci. Phys. Geneva*, **4**, 497, 1896.
[2] Störmer, C. *Arch. Sci. Phys.*, *Geneva*, **24**, 5, 113, 221 and 317, 1907; **32**, 33 and 163, 1911; **35**, 483, 1913.
[3] Schuster, A. *Proc. Roy. Soc.* A, **85**, 44, 1911.
[4] Lindemann (Lord Cherwell). *Phil. Mag.* **38**, 669, 1919.
[5] Chapman, S. and Ferraro, V. C. A. *Terr. Mag.* **36**, 77 and 171, 1931; **37**, 147 and 421, 1932; **38**, 79, 1933; **45**, 245, 1940.
[6] Alfvén, H. *K. Svenska Vet. Akad. Handl.* **18**, 1939.
[7] Alfvén, H. *Cosmical Electrodynamics* (Oxford University Press, 1950), ch. 6.
[8] Maunder, E. W. *Mon. Not. R. Astr. Soc.* **64**, 205, 1904; **65**, 2, 538 and 666, 1905; **76**, 63, 1916.
[9] Chree, C. *Proc. Roy. Soc.* A, **101**, 368, 1922.
[10] Meinel, A. B. *Astrophys. J.* **111**, 555, 1950; **113**, 50, 1951.
[11] Milne, E. A. *Mon. Not. R. Astr. Soc.* **86**, 459, 1926.
[12] Kahn, F. D. *Mon. Not. R. Astr. Soc.* **109**, 324, 1949; **110**, 477, 1950.
[13] Kiepenheuer, K. O. *Geophys. Res.* **57**, 113, 1952.
[14] Bennett, W. H. and Hulburt, E. O. *Phys. Rev.* **95**, 315, 1954; *J. Atmos. Terr. Phys.* **5**, 211, 1954.
[15] Ferraro, V. C. A. *Ind. J. Meteor. Geophys.* **5**, 157, 1954.
[16] Ferraro, V. C. A. *J. Geophys. Res.* **57**, 15, 1952.
[17] Schmidt, A. *Z. Geophys.* **1**, 3, 1924.
[18] Alfvén, H. *Tellus*, **7**, 50, 1955.
[19] Chapman, S. *Terr. Mag.* **46**, 1, 1941.
[20] Martyn, D. F. *Nature, Lond.*, **167**, 92, 1951.
[21] Chapman, S. and Bartels, J. *Geomagnetism*, vol. 2, p. 897 (Oxford Clarendon Press, 1940).

Discussion

Alfvén: There are a few questions I want to ask here. First, why do you assume that the magnetic field decreases like $1/r^2$ when you have a beam of this type and you have a magnetic field with a component perpendicular to the

---

* We have seen that there is little likelihood of the matter gaining speed or energy during its approach to the earth.

plane? The beam widens as $1/r$ and the magnetic field, assuming a constant velocity, should also decrease as $1/r$ which will give you a much higher magnetic field at the earth.

Ferraro: I rather think that there is a velocity spectrum along the stream as well and that the density decreases as $1/r^3$, not as $1/r^2$. I think that is much nearer the truth.

Alfvén: You assume a constant emission, and that the beam should be accelerated?

Ferraro: Yes.

Alfvén: But if it moves with constant velocity you have the magnetic field $H=H_0 r_0/r$. If it doubles its velocity when it moves outwards the field is $H=H_0 r_0/(2r)$, but not proportional to $1/r^2$.

Parker: I think Dr Ferraro is suggesting that the beam contains internal thermal motions which result in lateral accelerations in the beam, thereby decreasing the magnetic field more rapidly than $1/r$.

Alfvén: That cannot be correct in a model where a beam is assumed to go out radially, i.e. the way in which it is usually presented.

Parker: I do not think that this assumption is necessary.

Ferraro: My one feeling is that the inverse square law of the distribution of density is much too optimistic. I think the decrease should be more than this.

Alfvén: The second question is about the density. Let us assume, again, a radial emission such that the density n in the beam is proportional to v/r, where v is a constant. Then, you can calculate an upper limit for the density from the condition that the beam could not be denser and brighter than what can be observed in the corona. You could get an upper limit to the number of electrons in the beam, because you see the electrons in the corona by scattered light. If you consider the values of the density in the corona at different distances from the sun, you find that you have a high value near the solar surface but at a distance of about 4 or 5 solar radii you come to a minimum value of the constant v. If from the value you calculate the density at the earth's orbit you obtain about 20 particles/cm³ which is an upper limit under the assumption that all the light of the corona is due to the beam. Under more reasonable assumptions you may come to a value of 2 particles/cm³. This is a value which is much lower than the minimum value which is needed in order to explain the hollow and other effects in your theory. If you want to make the beam decrease with an expansion such as Parker mentioned this brings down the density value still more.

Ferraro: Oh yes, if you assume an expansion according to a $1/r^3$-law and assume 10^6 particles in the beam you would come to a density of $1/$cm³.

Alfvén: But do you not need $1000/$cm³?

Ferraro: Yes, but this is for a great storm. I think that in our papers we estimated a density of 100–$200/$cm³ to be necessary for a moderate storm.

Alfvén: But then you would see it, would you not?

Ferraro: Yes, but you really do not know from what level in the corona this emission takes place.

Alfvén: Perhaps not, but this argument will break down only in the case where you assume the beam to be emitted from more than 5–10 solar radii in the corona.

If it is emitted from that part of the corona which you could observe, the density could not be so high.

Ferraro: No, but why does it come from the corona?

Alfvén: Where does it come from otherwise?

Ferraro: We really do not know that.

Alfvén: Of course, you can let it be produced somewhere in the interplanetary space. That is all right. But then it has no connexion at all with the sun. You mentioned that the beam was supersonic but that collisions were negligible. Is that possible?

Singer: Would not the high density modify the present picture? If you take account of the fact that the cross-section is pretty large, would the picture really work at a density of $1000/cm^3$? (The mean free path is less than 0.01 a.u.)

Ferraro: I am sorry I cannot answer this offhand.

Singer: Could you explain from your theory why the initial phase should have a duration of 8 hr?

Ferraro: I think the argument has been turned around here. The duration of the initial phase is used to estimate the depth of the stream.

Singer: Can you explain why the decay time of the ring-current is about 1–2 days?

Ferraro: Yes, this is due to the very great electromagnetic inertia of the ring.

Singer: What determines the time constant of the discharge in the auroral zones?

Ferraro: The time of leakage of the charges.

Lehnert: Have you investigated the stability of the ring-current in the axial direction?

Ferraro: No. That case we have not considered; it is very difficult mathematically.

Alfvén: The ring-current is introduced in order to explain equatorial disturbances. Now, the disturbances which you observe are not equally large at the day and night sides of the earth. Consequently, if you introduce a ring-current to explain equatorial disturbances it must be eccentric. This gives you a new type of instability because the forces cannot be balanced as far as I can see.

Ferraro: I think that this is a special case of the radial displacements which we have treated in our stability considerations.

Spitzer: Has anybody investigated the instability in the hollow; you have two gases of different density with a heavy one at the top?

Ferraro: I am afraid not. The problem is difficult enough already. It might be fairly stable and not break up into tongues.

Gold: It is important to consider the explanation of individual storms, not the mean effect. In any individual case there may be a factor of 3 or 4 in the magnitude of the movements at different longitudes in low latitudes, although they may be simultaneous. A distant ring-current can therefore frequently only explain a quarter of the effect, and the rest has to be more local, though still synchronized for some other reason. The ring-current at something like 4 radii would thus explain only so little that it is hardly worth invoking.

Ferraro: I think this is true but on the other hand it would be very difficult to explain the world-wide character of these disturbances. One way in which one could account for the differences in intensity at different stations would be to take into account the magnetic effects of possible current systems which actually could be present in the atmosphere. I think, however, that it would be extremely difficult to explain the total magnetic storms by currents flowing in the earth's upper atmosphere solely. I cannot see why we should necessarily reject the hypothesis of a ring-current simply because the intensities at various stations may differ by a factor of 2 or 3.

Dungey: I have a suggestion that the Chapman–Ferraro surface is unstable with respect to the formation of surface waves; it is just like wind over water. I think that this happens all the time. Due to the orbital motion of the earth the waves will travel eastward by day and westward by night. I suggest that this can be used to get one quantity referring to the 'winds' which Professor Alfvén has mentioned (*Proc. Ionosphere Conference*; Phys. Soc. (London, 1955), p. 229).

Block: I want to ask Dr Ferraro about the stability of the ring-current. I think we will agree that one single particle is unstable in the geomagnetic field.

Ferraro: Yes.

Block: Therefore you must need a minimum number of particles in the ring in order to get it stable. Have you ever calculated this minimum density?

Ferraro: It is in our paper of 1941. This density agrees quite well with that needed for the explanation of the first phase of a magnetic storm.

Block: When the ring-current is formed from the beginning, the density must be lower than this minimum density and then it is very difficult to understand how it can ever be formed at all, when it is not stable in the beginning of the formation.

Ferraro: These problems are very difficult and we have never claimed to show or to give a mechanism of how the ring-current is set up. On the other hand Professor Chapman has suggested an alternative way in which the ring-current can be formed. It is in a paper he has published in the *Indian Journal of Meteorology and Geophysics*. The idea is that some of the debris of the stream combine to form a ring-current in a way somewhat like the formation of a plasmoid of four jets that Dr Bostick has mentioned.

PAPER 34

THE PRESENT STATE OF THE ELECTRIC FIELD THEORY OF MAGNETIC STORMS AND AURORAE

L. BLOCK

Royal Institute of Technology, Stockholm, Sweden

ABSTRACT

The main features of the electric field theory are outlined. The theory should be considered as a first approximation. The validity of the approximations and assumptions introduced is discussed.

Some model experiments on the theory are described. It is impossible to construct an entirely true model of nature in the laboratory. The similarities and differences between nature and model are discussed.

The mechanism of the model seems to be described very well by the theory. Some experimental results, which because of the complicated phenomena cannot be predicted by the theory, are compared with observations in nature. As far as we can see at present, the agreement between nature and model is astonishingly good.

I. INTRODUCTION

During the last few years the electric field theory of aurorae and magnetic storms has made some progress, both theoretically and experimentally by model experiments performed at this institute. In this paper I will briefly present the results of this work and discuss it in some detail, and also point out some problems, still doubtful or unsolved.

Aurorae and magnetic storms must be consequences of the electromagnetic state in space around the earth. If, therefore, we assume some special electromagnetic state as a probable cause of the phenomena, and then work out the consequences of this electromagnetic state, we may judge about the correctness of our assumption.

2. THEORY

The electric field theory is based on the assumption, that an electric field exists in the surroundings of the earth. How this field is produced is immaterial, but it may be an effect of a cosmical wind in interplanetary space, e.g. a beam from the sun. The normal direction of the field should be from the evening side towards the morning side of the earth in order to agree with the diurnal variation of aurorae and magnetic disturbances during a storm.

Starting from this basic assumption the main problem is to calculate what happens to charged particles acted on by this electric field in the presence also of the geomagnetic field. It is certainly impossible to do this exactly, and, therefore, it is necessary to simplify the problem by physically sound approximations. At the present state of our knowledge about electrical discharges in magnetic fields it is very difficult to judge the soundness of the approximations made in the theory. The following approximations and assumptions are made:

1. The electric field is as a first approximation taken to be homogeneous and perpendicular to the magnetic dipole axis of the earth. This means that all space charges are neglected.

2. The geomagnetic field is approximated by a dipole field.

3. An interplanetary magnetic field exists ($\approx 10^{-5}$ gauss), and it is homogeneous and parallel to the geomagnetic dipole axis. (This is only essential for the theory of the initial phase.)

Having made these approximations it is possible to calculate the electronic and ionic drift orbits in the magnetic equatorial plane. These orbits are seen in Fig. 1. The earth's centre is at the origin of the co-ordinate system. The real particle orbits are trochoids along the drift orbits. At the great circle, indicated in the figure, ($R \approx 30$ earth radii) the interplanetary and geomagnetic fields are approximately equal. Because of the gradient of the magnetic field in this region the particles from the sun are retarded, and the inertial forces deflect them in such a way that they produce an eastward ring-current, which causes the increase of the geomagnetic field at lower latitudes, observed during the initial phase of a storm. Nearer the earth the inhomogeneity of the magnetic field gives a drift motion of the electrons around the earth, producing a westward current responsible for the general decrease of the geomagnetic field during the main phase of a storm. There will also be a forbidden space (mean radius about 7 earth radii), which neither the electrons nor the ions can enter. All this is described in more detail by Alfvén [1, 2].

So far the calculations of the orbits in the equatorial plane are quite rigorous, provided the density of the beam is infinitely low, so that all space charges can be neglected. However, this is by no means the case, and therefore we must in some way account for the effects of the space charges, and also for the motion of the particles above and below the equatorial plane. The following assumptions are supposed to account for this.

4. Particles not moving entirely in the equatorial plane are oscillating through this plane along the magnetic-field lines from north to south and back again, at the same time as they drift perpendicular to the field in orbits similar to those calculated for particles in the equatorial plane. These orbits are, therefore, of fundamental importance and may be considered as a framework for the motion of all particles.

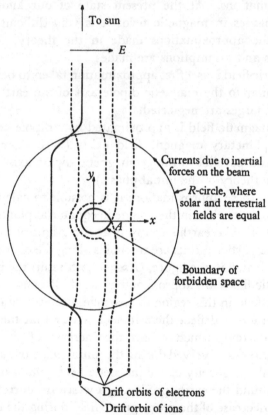

Fig. 1. Currents and particle orbits in the geomagnetic equatorial plane with an electric field perpendicular to the geomagnetic dipole axis.

5. Space charges will certainly be produced, but they will discharge along the magnetic-field lines towards the ionosphere. It is assumed, that the space charges will never be so large, that the motion shown in Fig. 1 is fundamentally changed.

Most space charges are produced at the boundary of the forbidden space and at the R-circle, where the interplanetary and geomagnetic fields are approximately equal (see Fig. 1). These space charges will discharge towards the auroral zones, so that the main auroral zones form the projection of the boundary of the forbidden space along the magnetic-field lines upon the earth's surface, and the inner auroral zones are the projections of the R-circle.

It is difficult to judge about the validity of the assumptions of the theory. Certainly there are sufficiently many charged particles to form considerable space charges, if sufficient charge separation occurs somewhere. This would make approximation 1 entirely invalid. Mathematically, it can be expressed as

$$e.n \gg \mathrm{div}\ \epsilon_0\ E; \tag{1}$$

$n =$ number density of charged particles,

$e =$ electronic charge,

$\epsilon_0 =$ dielectric constant of vacuum,

$E =$ electric field.

Eq. (1) is always fulfilled in cosmical physics. But we do not know anything about to what extent charges will be separated. If assumption 5 is correct, charge separation is largely compensated by the currents along the field lines.

Another phenomenon, which is likewise not at all understood at present, are the instabilities of a plasma in a magnetic field. It is well known from investigations by Åström[3]; Massey and collaborators[4] and Webster[5], that a plasma is unstable in a magnetic field, so that a bunching of the particles takes place, and this makes it easier for the charged particles to diffuse perpendicular to the magnetic-field lines.

In particular the experiments by Webster are interesting. He produced an electron beam, shaped like a hollow cylinder. Parallel to the axis of the cylinder he applied a magnetic field. The electrons moved along the field lines. Plate I (a) shows the beam near the cathode and Plate I (b) far from the cathode. It is seen that the beam is broken up very strongly.

Recently Bostick[6] has investigated the properties of such plasma bunches, called plasmoids. We do not know whether they will affect the properties of the auroral discharge as outlined in the electric field theory.

Thus, it may be fair to say, that at the present state of our knowledge it is in principle impossible to treat the problem by ordinary theoretical methods. The best we can do is to account for the phenomena, which we believe to be most important, and to check the results by comparison with the observational results in nature, and as far as possible, by model experiments in the laboratory.

3. MODEL EXPERIMENTS

The theory has been simulated by model experiments in the laboratory by Malmfors[7] and by Block[8].

In a vacuum chamber an electric field is applied between two condenser plates. In this electric field a terrella is placed. The terrella is magnetized by a coil inside it, with the magnetic dipole axis usually perpendicular to the electric field. By some ionizing device a gaseous discharge is started around the terrella. The discharge may be self-sustained (glow discharge, Plate II (a)) or non-self-sustained (dark discharge, Plate II (b)) where the ionizing agent must be in continuous operation. The surface of the terrella is covered by fluorescent material, so that one can see where the particles impinge.

There is no doubt, that the general character of the model discharge is in agreement with the theory. Luminous eccentric ring-shaped auroral zones appear (Plate II), and their latitudes vary with the magnetic and electric field strengths as predicted by the theory. The current system is in essential agreement with the theory. There exist, however, some differences between nature and model, which will be discussed now.

According to the similarity laws of gaseous discharges with different linear dimensions the magnetic and electric field strengths must be increased by the same factor as the linear dimensions are decreased. Since the earth is about 10^8 times greater than the terrella, the magnetic field of the terrella should be 6×10^7 gauss at the poles, which is impossible to obtain. However, it can be shown (Block[9]), that the drift orbits of the electrons and ions are properly scaled down in the model experiments, although the radius of curvature of the circular motion superimposed on the drift motion is comparatively much larger in the model than in nature (Fig. 2). It is probable, however, that this incorrect scaling down of the radius of curvature is of minor importance. The main reason for this is, that as soon as the magnetic field has reached the value where auroral rings appear, the general character of the discharge is unchanged even if the magnetic field is increased by a factor 10.

Eq. (1) is certainly fulfilled in the glow discharge but not in the dark discharge. In both cases, however, the general character of the discharge agrees with the theory, and this favours the opinion, that even if sufficiently many charged particles are available, the charge separation will not be too great.

The pressure in the dark discharge is a few times 10^{-4} mm, which means that the mean free path of the electrons is longer than the linear dimensions of the forbidden space by a factor 3 or so. In nature the corresponding

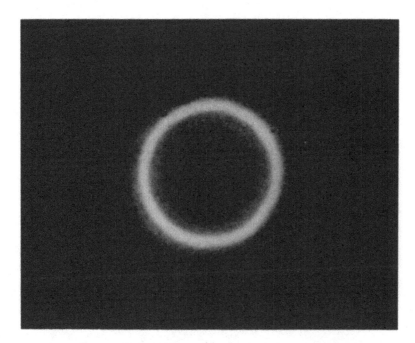

(a)

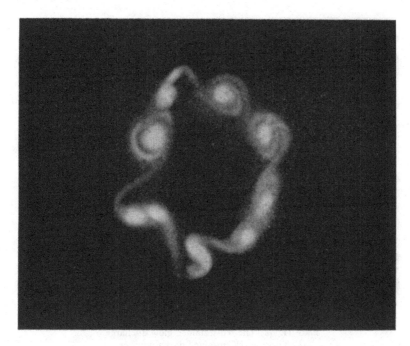

(b)

Plate I. Webster's hollow electron beam, (a) 1 cm from the cathode,
(b) 8·5 cm from the cathode (from H. F. Webster[5]).

(facing p. 316)

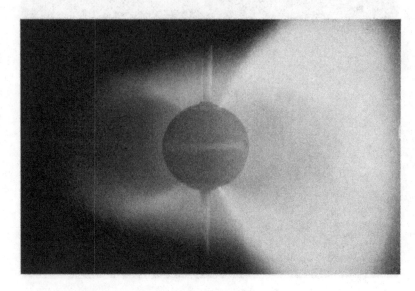

(a)

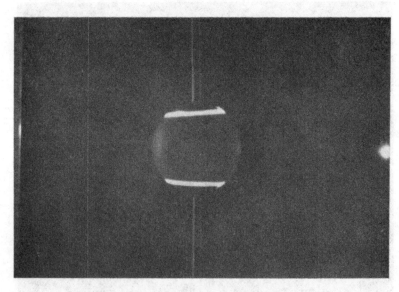

(b)

Plate II. (a) The self-sustained glow discharge.
(b) The non-self-sustained dark discharge.

factor is several powers of 10. In the glow discharge, however, the mean free path of the electrons is smaller than the forbidden space. Thus, the dark discharge is more like nature as far as the pressure is concerned, but considering the number of charged particles—Eq. (1)—the glow discharge is a far better model of nature.

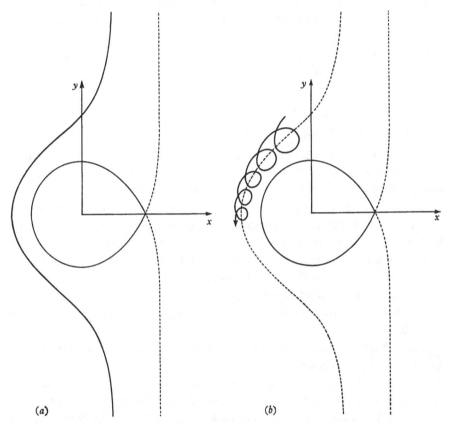

Fig. 2. Comparison of electron orbits in nature (a), and in the model (b).

The most serious disagreement between nature and model is probably that the ions can move rather freely in the model without being affected appreciably by the magnetic field of the terrella. This means, that the positive space charge accumulation predicted by the theory on the day side of the terrella will not take place, so the ion current in the auroral zone will be directed parallel to the electric field instead of from the day side towards the night side. This is also observed in the model.

It is observed in the experiment, that as soon as the magnetic field has become so strong that auroral rings are formed, a noise suddenly appears.

The spectrum of the noise extends at least from some 10 kc/s up to more than 5 Mc/s. This noise may be due to something like the previously mentioned plasmoids or bunches of charged particles.

The complicated pattern of aurorae with many different auroral forms may very well indicate the existence of some sort of particle bunching in space outside the earth. The particles in the beam from the sun may form bunches, expanding and stretching along the magnetic lines of force with one end appearing as aurora in the ionosphere. The motion of the bunches across the magnetic field may account for the rapidly moving auroral forms, so frequently observed in nature.

4. SOME SPECIAL PROBLEMS

(a) *The density of the beam*

We will now consider the particle density of the beam from the sun. There are several reasons to believe that the density of the beam could not exceed about 10 particles/cm³ at the earth's orbit (Alfvén[10]), and a lower limit of the density seems to be 10^{-1} cm^{-3} for an average magnetic storm. There are three independent ways of obtaining a lower limit. The highest of these three different limits thus obtained, must certainly be chosen as a real lower limit. We may consider

    (i)   the currents in the auroral zones,

    (ii) the luminosity of the aurorae,

    (iii) the time rate of producing the magnetic energy of a storm disturbance.

These considerations, in particular that of the luminosity of aurorae, are certainly very uncertain.

The currents in the auroral zones must be closed by currents along the magnetic field lines from the equatorial plane. The auroral particles are picked up from the beam over a cross-section which may be considered as a rectangle of area A with one side equal to the size of the forbidden zone, 10^{10} cm, and the other side may be 10^9 cm, or a little more than one earth radius. Then the current will be

    $nevA = I \approx 10^6$ amps,

    $n =$ lower limit of number density of charged particles in the beam,

    $e = 1 \cdot 6 \times 10^{-19}$ coulombs,

    $v = 2 \times 10^8$ cm/s,

    $A = 10^{19}$ cm².

This gives $n = 3 \times 10^{-3}$ cm^{-3}.

Thus the number of particles impinging in the auroral zones is

$$N = nvA = 3 \times 10^{-3} \times 2 \times 10^8 \times 10^{19} = 6 \times 10^{24} \text{ particles/sec.}$$

If the total area of the auroral zones is equal to two 1000 km wide, ring-shaped zones at a mean latitude of 23° from the poles or

$$\pi \times 6 \cdot 4 \times 10^8 \times \sin 23° \times 10^8 = 10^{17} \text{ cm}^2,$$

we get about 10^8 particles/cm² sec.

If 10^{-3} of the energy of all these particles is converted into visible light, it would be equivalent to 10^9–10^{10} photons of oxygen green lines ($\lambda = 5577$ Å) per cm² and sec, covering the whole area of the auroral zones at the same time. This corresponds to auroral international brightness I–II (Seaton[11]). Thus, this number of particles is sufficient to produce the visible light of aurorae.

The magnetic energy of the S_D-field must be taken from the kinetic energy of the beam. Across an area A, perpendicular to the motion of the beam, there is passing a kinetic energy/sec amounting to

$$nAv \cdot \frac{mv^2}{2} = P.$$

Assuming the magnetic energy to be due to a dipole M, parallel to the earth's dipole, at a distance r from the earth, so that the magnetic disturbance field on the earth is ΔB, we get the mutual energy of the two dipoles (the energy necessary for moving M from infinity)

$$W_m = \Delta B \cdot M_0,$$

where $M_0 =$ the earth's dipole moment.

The time T to produce this energy is obviously

$$T = \frac{W_m}{P} = \frac{\Delta B \cdot M_0}{nA \cdot \dfrac{m}{2} \cdot v^3},$$

or
$$nT = 5\Delta B \text{ sec/cm}^3$$

if the above-mentioned values of A and v are used, if $m =$ proton mass and if ΔB is measured in gammas.

Applying this formula to a sudden commencement with $\Delta B = 10\gamma$, we get, e.g. $n = 10^{-1}$ particles/cm³ and $T = 500$ sec, and for the main phase with $\Delta B = 100\gamma$ and $n = 10^{-1}$, T becomes 5000 sec.

It can thus be concluded, that a particle density in the beam of 0·1 per cc is sufficient for a moderate storm.

(b) The energy of the auroral particles

The energy of the particles impinging in the auroral zones may be expected to be of the order of eV, where V is the voltage difference across the forbidden space in the equatorial plane. This is probably about 100 kV in nature, and in the model it is a few kV. The energy of the auroral particles in the model has been measured by putting a spherical grid around the terrella and supplying a variable voltage between the grid and the terrella (Fig. 3). When the terrella is so negative with respect to

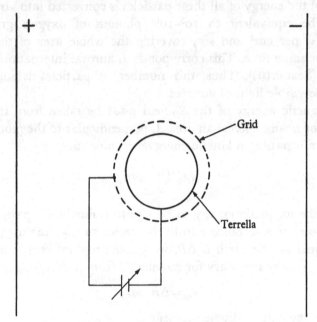

Fig. 3. Arrangement for measuring the energy of the particles impinging on the terrella.

the grid, that the voltage difference is about half of the voltage across the forbidden space, the ionization suddenly increases very much outside the auroral rings of the terrella. This must be due to Barkhausen oscillations of the electrons in and out through the grid. If the voltage between the two condenser plates is increased a little, the electrons gain more energy so that they can reach the terrella and be absorbed there. Thus, the intense ionization vanishes again.

(c) The eccentricity of the auroral zones

As seen from Plate II the experiments give eccentric auroral zones. The direction of the eccentricity indicates the direction of the electric field. By observing the eccentricity during a particular magnetic storm, it should

(a)

(b)

(c)

Plate III. The dipole axis of the terrella is tilted 23·5° at different directions, corresponding to different seasons of the year: (a) September, (b) December, (c) March.

(facing p. 320)

thus be possible to derive the direction of the electric field causing this particular storm.

We define the eccentricity as

$$\epsilon = 1 - \frac{\text{Minimum polar distance}}{\text{Maximum polar distance}}.$$

The experiment gives $\epsilon_{exp} = 0 \cdot 10 - 0 \cdot 20$.

The theory gives $\epsilon_{th} = 0 \cdot 29$.

The difference may be explained by space charges, which change the shape of the forbidden space in the equatorial plane.

The eccentricity of the auroral zones should cause a double periodicity in the diurnal variation of the vertical component, ΔZ, of the S_D-field at magnetic observatories situated at auroral latitudes (Alfvén[1], p. 197). Consulting Vestine et al.[12] one finds, that such a double periodicity is detected at stations between $70 \cdot 8°$ N (Julianehaab) and at least $67 \cdot 1°$ N (Tromsö) or possibly $64 \cdot 5°$ N (College, Fairbanks). This means that the eccentricity of the northern auroral zone may be between $0 \cdot 16$ and $0 \cdot 25$.

(d) *Diurnal and seasonal variation of auroral frequency*

It is seen from Plate II (b), that all points in the auroral zones are equally illuminated in the experiment. However, this is only true when the dipole axis of the terrella is exactly perpendicular to the electric field and the motion of the beam. If the dipole axis is tilted only a few degrees, the auroral zones are divided into strongly luminous spots, separated from each other by areas of weak luminosity. This is certainly in better agreement with nature.

By tilting the dipole axis $23 \cdot 5°$ at different directions, corresponding to different seasons, the photos in Plate III have been obtained. They are all taken from the 'night side' of the terrella, so the 'morning side' is to the right. It is seen that at the solstices, there are two bright spots, one before and one after midnight. Thus, we should have two maxima of auroral frequency during the winter nights. At the equinox in September, on the other hand, we have only one bright spot at about midnight in the southern auroral zone. In March there is only one bright spot in the northern auroral zone.

It is well known from observations in nature (see, e.g. Vegard[13]) that there are two maxima of auroral frequency, one before and one after midnight. But this is an average over the whole year, and it might very well be only one maximum in some seasons. In fact, the observations by Carlheim-Gyllenskiöld[14] during the first polar year 1882–3 indicate that there occurred only one maximum at midnight at the equinoxes. The diagrams

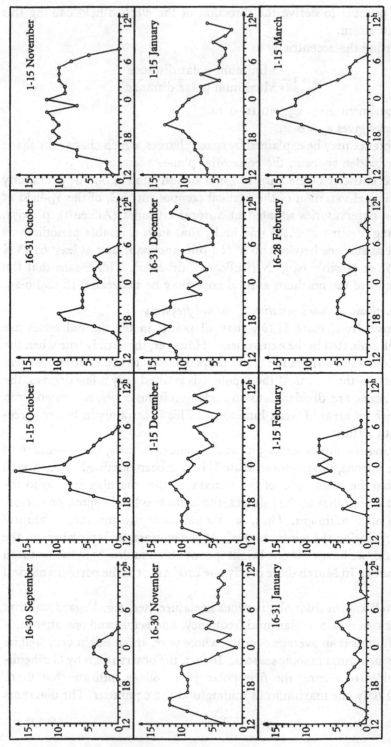

Fig. 4. Hourly total number of aurorae observed each half month during the winter 1882–3 by Carlheim-Gyllenskiöld[14] at Cap Thordsen, Spitzbergen.

in Fig. 4 show the hourly total number of aurorae observed by Carlheim-Gyllenskiöld during each half month of the winter 1882–3.

In the model experiments it is also observed, that the current of the discharge at the 'equinoxes' exceeds the current at the 'solstices' by about 25 %. This may have something to do with the well-known fact that magnetic disturbances are more frequent at the equinoxes than at the solstices.

It is astonishing, that the agreement between nature and the model is so good, although it is not yet proved, that some of the agreements are not more or less accidental.

In Table 1 a summary is given of some differences and similarities between nature and the model discharges.

Table 1. *Properties of discharges in nature and in model experiments*

	Nature	Glow discharge	Dark discharge
$en > \operatorname{div} \epsilon_0 E$	Yes	Yes	No
Mean free path > L = size of forbidden zone	Yes	No	Yes
Plasma instabilities	Many auroral forms indicate instabilities	Noise observed	Noise observed
Ion current in auroral zones	Perpendicular to el. field	Unknown	Parallel to el. field
ρ/L = trochoidal radius of curvature comp. with size of forb. zone	10^{-4}	10^{-2}	5×10^{-2}
Eccentricity	0·16–0·25	0·10–0·15	0·15–0·20
Seasonal var. of magn. activity	Magn. dist. more frequent at equinoxes	Discharge current at 'equinoxes' 25 % greater than at 'solstices'	
Seasonal and diurnal var. of aurorae	Two max. during winter night, possibly one in spring and autumn	Unknown	Two max. in winter, one in spring, no def. in autumn

REFERENCES

[1] Alfvén, H. *Cosmical Electrodynamics* (Oxford University Press, 1950), ch. VI.
[2] Alfvén, H. *Tellus*, 7, 50, 1955.
[3] Åström, E. *Experimental Investigation on an Electron Gas in a Magnetic Field*, Transactions of the Royal Institute of Technology, Stockholm, no. 22, 1948, p. 70.
[4] Massey, H. S. W., Bohm, D., Burhop, E. H. S. and Williams, R. M. *The Characteristics of Electrical Discharges in Magnetic Fields*, ed. A. Guthrie and R. K. Wakerling, (McGraw Hill, 1949), ch. 9.
[5] Webster, H. F. *J. Appl. Phys.* 26, 1386, 1955.
[6] Bostick, W. H. *Experimental study of ionized matter projected across a magnetic field*, Univ. of California Radiation Lab., UCRL-4695, 1956.

[7] Malmfors, K. G. *Ark. Mat. Astr. Fysik*, **34**B, no. 1, 1946.
[8] Block, L. *Tellus*, **7**, 65, 1955.
[9] Block, L. *Tellus*, **8**, 234, 1956.
[10] Alfvén, H. *Tellus*, **9**, 92, 1957.
[11] Seaton, M. J. *J. Atmos. Terr. Phys.* **4**, 285, 1953.
[12] Vestine, E. H., Lange, I., Laporte, L. and Scott, W. E. *The geomagnetic field, its description and analysis*, Carn. Inst. Publ., no. 580 (Washington D.C., 1947).
[13] Vegard, L. *Handbuch der Experimentalphysik*, xxv: 1, (Leipzig, 1928), p. 404.
[14] Carlheim-Gyllenskiöld, *Exploration Intern. Polaires 1882–3. Exp. Suédoise. Aurore boréale*, Kungl. Vet. Akad. (Stockholm, 1886).

Discussion

Singer: In our models, when we try to explain such complicated phenomena as magnetic storms and aurorae, we start with an idealized situation. That means that we take some observational facts as relevant and reject others. The problem is that we all differ on what is relevant or not relevant. As an example, the electric field theory considers the diurnal variations of importance. From my point of view I consider the fact that the sudden commencement currents flow mostly in the atmosphere as very relevant, but Ferraro does not. I also consider very significant the fact that the sudden commencement is preceded by a reverse commencement. This is particularly shown in one of the slides which Ferraro demonstrated.

Now, I want to ask Dr Block: for what gas densities around the earth does your model work?

Block: I think that it works at an extremely low density but there must be an upper limit, i.e. that the mean free path cannot be orders of magnitude smaller than the distance for the beam to proceed as far as to the earth. It would lose too much energy by collisions with the interplanetary gas. Very close to the earth the density may be higher.

Singer: I believe that the initial phase of a magnetic storm is explained by the ring-current at 30 earth-radii from the earth, according to the electric field theory.

Block: Yes, partly. But I think we must also take into account the jet streams in the ionosphere.

Singer: How does one explain the delay between the sudden commencement and the main phase in the electric field theory?

Block: This is not worked out very much in detail. There are certain differential equations which are very difficult to solve, but some theoretical calculations indicate that the delay should not be smaller than about 10 min; this estimate is very rough, of course.

Singer: It should be a few hours.

Block: Yes, but this is not very well explained.

Alfvén: I think that Block's experiments and the whole theory in its present development concentrates on the main phase; concerning the initial phase there are some recent attempts to study this phenomenon. In the electric field theory the main phase is caused by the formation of space charges at the boundary of a forbidden region. The energy of the incoming particles is taken from the electric field (Fig. 1). But then there is also, as shown in Fig. 1, a current produced at a very large distance of about 30 radii which is due to the braking of the beam when it comes into the earth's magnetic field. Here the earth's magnetic field equals the interplanetary field and the drift motion of the particles is affected by inertia forces. This would correspond to an inner auroral zone at a polar distance of about 5° whereas the outer auroral zone is at 23°. It would be very important to look for the existence of this inner auroral zone. I was very glad to hear at the arctic conference here in Stockholm in May that research in the Soviet Union by Dr Nikolski has given very good evidence for the existence of such a zone.

The effect of the application of an electric field in the region around the earth is not confined to the phenomena we have studied all of which derive from the properties of the beam. The electric field should also set into motion the ions which exist already in the neighbourhood of the earth. This is a point where Singer's question is very important. The motion of the already existing ions should produce currents in the upper atmosphere. We should also study the currents produced by the electric field directly in the upper atmosphere.

Ferraro: I should like to make a remark to Dr Block about the model experiments. Chapman has drawn the attention to what is the right scale when moving from nature to laboratory. Professor Cowling drew attention to this also yesterday.

Now, as regards the R-circle (see Fig. 1) which is supposed to produce the first phase, I am not quite sure I understand how this comes about. One thing that puzzled me was the effect that the displacement of the positive ions in the regions where the earth's magnetic field and the interplanetary field are equal was greater than that of the electrons. Is that a line current?

Alfvén: No. This is a space current and it comes out straightforward from the assumptions which Block stated very clearly. You just calculate the motions of electrons and ions in an electric field and a combined magnetic field of the earth and interplanetary space.

Ferraro: Could you tell me quite briefly how it comes about that the displacement of positive ions is greater than that of the electrons? I should have thought that in the regions where the interplanetary and terrestrial fields are equal, the gradient is very near to zero.

Block: I think Dr Ferraro has misunderstood us here. The gradient of the magnetic field is not zero because we assume that the interplanetary and terrestrial field have the same direction. The deflexion of the particles is caused by inertia forces.

Ferraro: I thought that the gradient in the earth's magnetic field was responsible for the spiralling and streaming motions of the electrons around the earth so as to produce a westward current. But does not this produce an eastward current?

Alfvén: This is an inertia effect of the ions. If you have an ion which drifts into an increasing magnetic field you brake the translational velocity and this produces a drift.

Ferraro: Then somewhere you get a reverse of that drift?

Alfvén: Yes.

Ferraro: Does not this cause a discontinuity?

Alfvén: This is due to the braking of the velocity. We can take it as a transformation of kinetic energy into field energy. It is necessary to have such a displacement; it comes out through straightforward calculations.

Lovell: In connexion with Dr Block's simulation of the diurnal effects a comment on the recent radio echo results may be of interest. In this work the radio echoes scattered from the ionized auroral regions are recorded, and it is

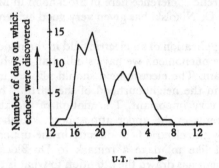

Fig. 5. Diurnal variation in the rate of occurrence of aurorae as determined by the radio echo technique.

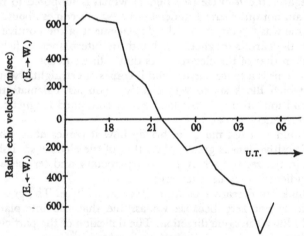

Fig. 6. Diurnal variation in drift speeds of aurorae as measured by the radio echo technique.

possible to determine the range, speed of drift movements and the nature of the reflecting agency independent of daylight or sky conditions. The diurnal variation is given in Fig. 5 which shows two main peaks at about 18^h and 02^h with a minimum at 22^h. The most significant features of this minimum may be listed as follows:

1. The minimum is associated with a change in drift motions of the reflecting regions as shown in Fig. 6. At the first maximum the drifts reach 600 m/sec

326

east to west; at the minimum the drift is zero and reaches over 600 m/sec west to east at the time of the second maximum.

2. The variation of the ΔV component of the earth's magnetic field follows this drift curve closely.

3. There is a marked change in the type of radio echoes observed, those in the early maximum being mainly diffuse, whereas after the minimum the echo structure is predominantly discrete. In one notable case observed on 25/26 September 1951 these changes in echo structure were closely correlated with a change in the appearance of a visual aurora from a stable arc to pulsating rays and diffuse patches.

4. The drift motions determined from radio star scintillation observations are normally associated with the F-region and show reversals in direction at midnight. When observed in the auroral zone these drifts show a partial reversal in direction at the time of the minimum in Fig. 5.

These data have been obtained at Jodrell Bank (geomagnetic latitude 56°) during the years of sunspot minimum (1949–54), the reflecting regions being about 50 km northwest of the station. Some of this information has been published by Bullough, K. and Kaiser, T. R. (*J. Atmos. Terr. Phys.* **5**, 189, 1954, and **6**, 198, 1955).

Block: Do the aurorae move across the sky in a certain direction?

Lovell: We do not know if the visible aurorae moves. What we detect is the ionization in the aurorae and this shows a drift and a reversal.

Lowes: Do you have to have the interplanetary field in the same direction as the geomagnetic field in order that the mechanism shall work? Is it possible for it to work with an interplanetary magnetic field in the opposite direction?

Block: The interplanetary magnetic field is only necessary for the theory of the initial phase. For the main phase the interplanetary field is not essential. What is essential is that an electric field exists, directed from the evening side towards the morning side of the earth.

Cowling: Does the main phase of a magnetic storm persist only while the ionized stream is flowing continuously past the earth?

Block: For the main phase it is only essential that an electric field exists. The direction of this field is determined by the diurnal variation of the aurorae and associated magnetic disturbances. This direction can be explained by a beam moving from the sun in an interplanetary magnetic field of the same direction as that of the geomagnetic field. If you can produce an electric field in some other way it is all right.

Cowling: In such a case the stream must persist for about 3 or 4 days?

Block: Yes.

Alfvén: If a discharge of this type stops at once there are after effects. If you switch off the electric field the particles will no longer move in eccentric orbits but in circular orbits around the earth. The eccentricity of the forbidden region is produced by the electric field. If you take away the electric field you have still ions and electrons present and they will drift around in circles until they are absorbed by some mechanism. But at the same moment as you switch off the electric field the eccentricity of these orbits, and also the eccentricity of the auroral zone, will become zero.

Cowling: Will you at the same time have essentially the same ring-current?

Alfvén: No, the ring-current in this case is due to electrons moving in tro-chodial orbits. It is well known from experiments that such a ring-current is stable and all the objections against the stability of the currents in Chapman–Ferraro's theory are inapplicable here.

Singer: How do you explain the acceleration of the protons impinging on the auroral zones?

Block: One possible explanation is as follows. The earth may be considered as a probe in a gaseous discharge. It will be charged by electrons to a negative potential approximately equal to that at the point A in Fig. 1 at the boundary of the forbidden space around the earth in the equatorial plane. Then, the resulting potential differences between the other points of the forbidden space boundary and the earth will accelerate the protons.

Other mechanisms are also conceivable, e.g. plasma instabilities and bunching of protons in the equatorial plane. The electric fields of these positively charged bunches may accelerate protons towards the auroral zones. This may perhaps explain the highly unstable and fluctuating auroral forms.

A NEW MODEL OF MAGNETIC
STORMS AND AURORAE

S. F. SINGER

Physics Department, University of Maryland, U.S.A.

ABSTRACT

Three topics are discussed dealing with interplanetary phenomena. They are: (i) sudden commencement of magnetic storms[1]; (ii) main phase of magnetic storms[2]; (iii) cosmic ray effects associated with solar corpuscular emission[3].

To explain the sudden commencement(SC) of magnetic storms, the reverse sudden commencement (SC*), and the pre-SC disturbances, we invoke the following model: The solar eruption produces a shock-wave which arrives at the earth 22–34 hr later. *High velocity particles* having a smaller interaction precede the shock-wave and cause the pre-SC bay-like disturbances at high latitudes. The shock-wave itself is retarded by the body forces produced by the geomagnetic field, but speeds up as it enters the auroral zones. In pushing out lines of force it creates the polar SC* events. Charge separation in the shock-wave produces the driving force for the SC currents which flow in the atmosphere (in accordance with Vestine's analysis).

The storm decrease is produced by the high velocity particles following the shock-wave (up to 9 hr later) which enter because of field perturbations into the normally inaccessible Störmer regions around the dipole. Here they are trapped and will drift producing the ring-current which gives rise to the storm decrease. Particles with small pitch angle, however, can reach the earth's atmosphere and contribute to aurora, the air-glow, and ionospheric ionization. These particles are replenished by perturbations produced by solar influences having a 27-day recurrence. Many other particles are absorbed or scattered out of the trapping regions so that their number diminishes rapidly in a day or so, as does the magnetic storm decrease.

The model thus attempts to explain for the first time the cause of the reverse sudden commencement events (SC*), the atmospheric nature of SC, the delay between SC and the main phase, the formation and decay of the ring-current. A by-product is auroral particle acceleration by a shock-wave[4].

New experimental tests are suggested by the model: (i) Acoustic observations with balloons to look for the shock-wave penetrating into the atmosphere in the auroral zones. (ii) Observations with rockets or satellites to establish the location of the SC and main phase currents. (iii) Measurements of the nature and energy of the auroral particles[5].

It is suggested that the cosmic ray decrease occurring with magnetic storms

(*Forbush events*), as well as the 27-*day decreases* of cosmic ray intensity, are modulation effects produced primarily by the deceleration of cosmic rays in interplanetary space due to the expansion of turbulent gas clouds from the sun. The detailed mechanism depends on a statistical decrease of the initially high turbulent fields and can therefore be called an 'inverse Swann effect' or 'inverse Fermi effect'. The cosmic ray intensity variation during the *solar cycle* is accounted for as the cumulative effect of this mechanism which operates in connexion with emission of solar gas. In this way it is possible also to account for the decrease lasting six months observed by Forbush starting in February 1946. Some experimental tests are suggested to discriminate between different theories for the origin of cosmic ray time variations [3].

REFERENCES

[1] Singer, S. F. *Trans. Amer. Geophys. Union*, **38**, no. 2, 175, 1957.
[2] Singer, S. F. *Nuovo Cimento*, Suppl. II, 1957.
[3] Singer, S. F. *Phys. Rev.* 1957 (in the press).
[4] Singer, S. F. *A new model of magnetic storms and aurorae*, Phys. Dept. Techn. Rep. no. 48 (University of Maryland, 1956).
[5] Singer, S. F. *Cosmic ray time variations produced by deceleration in interplanetary space*, Phys. Dept. Techn. Rep. no. 50 (University of Maryland, 1956).

Discussion

Ferraro: One general comment which I should like to make in connexion with both Alfvén's theory and Singer's is that they have only considered the motion of a single particle in the external electric and magnetic field. Because of the neglect of the interaction between the particles of the stream it is difficult to be sure of the bearing of their results on the actual phenomena.

As regards the anomalous SC and initial phase amplitudes at Huancayo, whilst this is undoubtedly an atmospheric effect, it is not necessarily adverse to the hypothesis that the sudden commencements are due to sudden increase in the earth's magnetic field outside the earth. There will be, undoubtedly, ionospheric currents induced by this impulse and these may produce local variations because of the non-uniform conductivity of the ionosphere.

Forbush: Vestine and Forbush found that the sudden commencements at Huancayo were larger on days with larger diurnal variation. How does this influence the induced effects if the larger diurnal variation arises from greater ionospheric conductivity?

Ferraro: Huancayo is a very abnormal station. I cannot see that there is necessarily anything extreme in the observed effect.

Singer: Concerning the situation of the main phase I think that the interaction between particles is taken into account when you consider the interaction between currents. There is a decrease in the magnetic field gradient which guarantees the stability.

Then we come to the second problem concerning the sudden commencement. The theory of the sudden commencement should also explain the reversed sudden commencement, i.e. the decrease in the magnetic field which immediately precedes the sudden commencement increase. This decrease should be very pronounced in the auroral regions. It may be a strong argument for the hypothesis that shock-waves penetrate to the auroral zones and not to the equator.

Now let me make some comments on the question put forward by Dr Forbush. The theory has to explain why the sudden commencement currents are enhanced on the day side of the earth and why this enhancement is large at Huancayo. There is nothing strange about this station, except that it is situated near the geomagnetic equator. The sudden commencement enhancement is increased by a factor of about 8 at Huancayo as compared to other stations which are not situated near the geomagnetic equator. Further, a comparison of the diurnal variations at Huancayo and at Cheltenham also shows that the sudden commencement currents are large at Huancayo. I think that this situation is explained by the fact that the electric conductivity at the geomagnetic equator is high in the direction parallel with the equator. This is a phenomenon well known from the investigations of the electrojet. But a high ionospheric conductivity would shield, i.e. cancel, an external current, not reinforce it.

Ferraro: What you say about the conductivity at Huancayo is quite true. However, one should not exclude other possibilities of explanation for such a complicated phenomenon as this. Chapman has, e.g. suggested that it might be due to return currents from the polar regions. This could reduce the effect of the currents which produce the 'normal' sudden commencement.

Parker: I remember having read somewhere that in auroral latitudes during years of solar activity there are auroral displays nearly every night. Does this imply that in years of solar activity nearly the entire orbit of the earth is flooded with beams and/or shock-waves, etc.?

Alfvén: If an electric field is the cause of magnetic storms and aurorae, then there is almost always an electric field in interstellar space giving a weak aurora or a strong aurora, according to the strength of the field.

B. COSMIC RAY METHODS OF EXPLORING INTERPLANETARY SPACE

PAPER 36

THE 27-DAY VARIATION IN COSMIC RAY INTENSITY AND IN GEOMAGNETIC ACTIVITY

SCOTT E. FORBUSH

Department of Terrestrial Magnetism, Carnegie Institute of Washington, Washington D.C., U.S.A.

ABSTRACT

The amplitude of the average 27-day wave in cosmic ray intensity, at Huancayo, Peru, and its phase relative to that for the 27-day wave in international magnetic character figure (ICF) is determined from results of harmonic analysis of data for each of 246 intervals (or solar rotations) of 27 days. From these data, the variability of which is essential for tests of statistical significance, the amplitude of the average 27-day wave in cosmic ray intensity and its phase relative to that in geomagnetic activity is determined for each of three groups of solar rotations selected according to the average of the amplitudes of the 27-day waves in magnetic activity. A fourth group contained only 27-day intervals in which large cosmic ray decreases occurred. Relative to that in magnetic activity, the phase of the 27-day wave in cosmic ray intensity is found for the averages, to be the same for the four groups.

The maxima of the average cosmic ray waves occur about 1·5 days after the minima of the corresponding waves in ICF. In general, the amplitude of the average 27-day wave in cosmic ray intensity, in the co-ordinate system in which its phase is relative to that of the 27-day wave in ICF tends to be greater for the selected groups of rotations with larger average ICF amplitudes. For most years near sunspot minimum the amplitude of the 27-day cosmic ray wave does not differ significantly from zero.

Bartels found for 27-day waves in ICF the effective number of statistically independent 27-day waves for N successive solar rotations to be $N/3$; the number found for cosmic ray intensity is $N/2$. Thus, on the average the 27-day recurrence tendency is less for cosmic ray intensity than for magnetic activity.

1. INTRODUCTION

Probably all of the established variations in cosmic ray intensity are in some way connected with solar activity. The large sudden decreases of cosmic ray intensity occur during magnetic storms although storms without decreases in cosmic ray intensity often occur[1]. The cosmic ray intensity averaged for the five magnetically disturbed days of each month is nearly always less than that for the five magnetically quiet days of the month. These facts imply a close connexion between the mechanism responsible for magnetic storms and geomagnetic activity and that for many of the variations in cosmic ray intensity. Morrison[2] has proposed one ingenious mechanism to explain many of the variations of cosmic ray intensity based on the hypothesis that the solar streams responsible for magnetic storms and magnetic activity generally contain a low density of cosmic rays, so that a decrease of cosmic ray intensity may be observed when the earth is inside the stream. The relation between the 27-day waves in magnetic activity and those in cosmic ray intensity should provide information of value to theories for the time variations.

Since our first[3] investigation of the relation between the 27-day variations in magnetic activity and cosmic ray intensity, the amount of data has greatly increased. As a measure of geomagnetic activity the international character figure[4] is used. This measure is chosen mainly because Bartels, in a famous paper[5] made a classic investigation of the waves in magnetic activity based on daily international magnetic character (ICF) figures for 378 solar rotations of 27 days each, beginning 11 January 1906.

2. DATA AND METHOD OF ANALYSIS

The basic ICF data were the daily mean ICF figures for each Greenwich day for the period 13 June 1936 through December 1955. The 27-day waves for the ICF data were determined by a six ordinate scheme for harmonic analysis using differences[6] to eliminate non-cyclic change. The six ordinates used were means for alternate sequences of 4 and 5 days for each 27-day interval.

The cosmic ray data from Compton–Bennett meters[1] at Huancayo (starting 13 June 1936) and at Cheltenham (starting 6 April 1937) were similarly analysed to provide, for each station, a set of harmonic coefficients: A and B for the 27-day waves; similarly, the analysis of the ICF data provided another set a and b, for corresponding 27-day intervals. In the 27-day harmonic dial for ICF each pair of coefficients a and b defines

333

the end-point of a vector **c**, the length, $c = \sqrt{(a^2 + b^2)}$, of which is the amplitude of the 27-day wave. The time of maximum of the wave is indicated by the number on the scale on the periphery of the harmonic dial [5] to which the vector points.

To determine the phase of 27-day waves in cosmic ray intensity relative to that for ICF each pair of coefficients A and B for cosmic ray intensity, CRI, is transformed (by rotation of the axes) to the pair A_R and B_R referred to a set of axes in which the vector **c** for ICF for the same 27-day interval has its time of maximum at zero. This co-ordinate system is designated the rotated co-ordinate system (RCS), or by a subscript R, to distinguish if from the original co-ordinate system (OCS).

3. STATISTICAL PARAMETERS FOR THE VARIABILITY OF THE 27-DAY WAVES

Fig. 1 shows the harmonic dials for the 27-day waves in cosmic ray intensity in the rotated co-ordinate system (RCS) for three different ranges in amplitude for the 27-day waves in ICF. The starred point on the vertical axis, in each of the three dials, indicates by its distance from the origin the average of the ICF amplitudes for that group. Table 1 indicates the number of rotations, n, for each of the three dials, together with the average of the ICF amplitudes, $\bar{A}_R$, $\bar{B}_R$, and the amplitude $\bar{C}_R$ of the average CRI waves in the RCS. The average $\bar{C}_R$ vector is shown for each dial.

The means, standard deviations, and other parameters, determined for the distributions, in the rotated co-ordinate system, for Fig. 1 (as well as for Figs. 2 and 3) are indicated in Table 1. In Table 1 the standard deviations, s_{A_R} and s_{B_R} of A_R and B_R are about their means (in Figs. 1, 2 and 3, A_R is plotted on the vertical axis and B_R on the horizontal axis). It will be noted from the first four rows of Table 1 that the standard deviation, s_{B_R}, of B_R is in every case somewhat greater than that of A_R which would suggest that the two-dimensional frequency distributions for the points in Figs. 1, 2 and 3 may be slightly elliptical. However, using the results from line four of Table 1 derived from 230 rotations, the ratio of the variance of B_R to that for A_R is 1·31. If samples of 230 statistically independent pairs A_R, B_R are drawn from populations with the same variance, then in about 5 % of such samples the ratio of the variance would equal or exceed 1·31. Thus, the difference in the variances is barely statistically significant at the 5 % level and the distributions may for practical purposes be regarded as circularly symmetric, with

$$M = \{S_{A_R}^2 + S_{B_R}^2\}^{1/2}$$

334

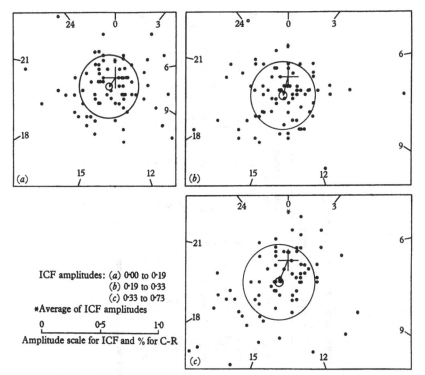

ICF amplitudes: (a) 0·00 to 0·19
(b) 0·19 to 0·33
(c) 0·33 to 0·73
*Average of ICF amplitudes

0   0·5   1·0

Amplitude scale for ICF and % for C-R

Fig. 1. Harmonic dials 27-day waves cosmic ray intensity (C–R), Huancayo (1936–54) phases relative to those for waves in international character figure (ICF) rotations with large C–R storm effects excluded.

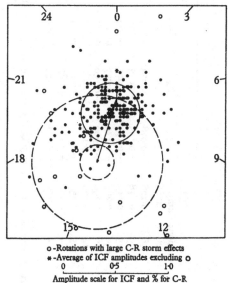

o -Rotations with large C-R storm effects
* -Average of ICF amplitudes excluding o

0   0·5   1·0

Amplitude scale for ICF and % for C-R

Fig. 2. 27-Day harmonic dial using data pooled from Fig. 1 (a), (b) and (c) and including 16 rotations with large C–R storm effects.

335

Table 1. *Data for 27-day harmonic dials in Figs. 1, 2 and 3*

For cosmic ray intensity

Fig. no.	No. of rotations n	ICF amplitude	A_R %	B_R %	C_R %	s_{A_R} %	s_{B_R} %	M %	ρ %	Points inside ρ	Points outside ρ	κ	$P = e^{-\kappa^2}$	T days
1 (a)*	76	0·126	−0·076	−0·053	0·093	0·195	0·233	0·304	0·253	41	35	2·64	1 × 10⁻³	2·6
1 (b)*	79	0·248	−0·155	−0·043	0·161	0·207	0·258	0·329	0·274	45	34	4·36	8 × 10⁻⁹	1·2
1 (c)*	75	0·419	−0·173	−0·077	0·189	0·248	0·264	0·362	0·301	43	32	4·54	1 × 10⁻⁹	1·8
2*	230	0·262	−0·135	−0·057	0·147	0·219	0·252	0·234	0·278	129	101	6·70	3 × 10⁻²⁰	1·7
2†	16	0·290	−0·600	−0·190	0·629	0·642	0·447	0·782	0·651	7	9	3·27	3 × 10⁻⁵	1·3
3 (a)‡	14	0·300	−0·154	−0·066	0·167	0·312	0·347	0·467	0·389	8	6	1·61	8 × 10⁻²	1·8
3 (a)§	75	0·254	−0·024	−0·010	0·026	0·127	0·155	0·200	0·167	40	35	0·12	9·8 × 10⁻¹	1·8
3 (b)*	82	0·258	−0·199	−0·080	0·214	0·250	0·304	0·394	0·328	46	36	4·93	4 × 10⁻¹¹	1·6
3 (b)†	12	0·270	−0·534	−0·154	0·556	0·693	0·449	0·827	0·689	5	7	2·33	5 × 10⁻³	1·2

* Excluding intervals with large C-R storm effects.　† Only intervals with large C-R storm effects.　‡ 1952 only.　§ All except 1952.

as the parameter governing the distribution. The radius of the so-called probable error (p.e.) circle is given [5] by 0·833 M. From Table 1 it will be seen that in most cases a few more points lie inside the circles than outside, but the difference in no case is statistically significant, based on the χ-square test.

A more important point is whether in the rotated dial the points for successive rotations are statistically independent. To test this point the

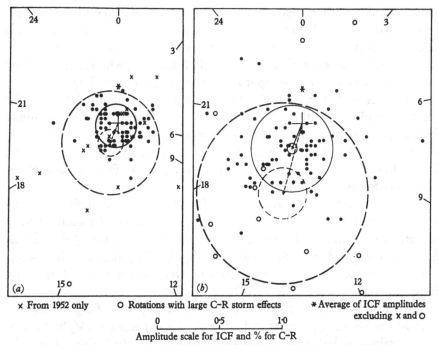

x From 1952 only   O Rotations with large C-R storm effects   *Average of ICF amplitudes excluding x and O

0   0·5   1·0

Amplitude scale for ICF and % for C-R

Fig. 3. 27-Day harmonic dials C–R intensity Huancayo, Peru, phases relative to those for ICF: (a) near SS minimum, (b) near SS maximum.

sample of 230 rotations (plotted in Fig. 2 omitting the open circles) was used. The standard deviation, $s(1)$, of individual values of B_R from the mean was computed as well as the standard deviation, $s(h)$, for means of h chronologically successive values of B_R with $h = 5$, 10, and 15. The characteristic [5] $c(h) = s(h)\sqrt{(h)}/s(1)$ was determined for the above values of h. The resulting values of $c(h)$ were essentially constant and independent of h. Thus [5] these successive values of B_R are statistically independent, and their standard deviation $s(n)$ for means of n is reliably given by $s(n) = s(1)/\sqrt{(n)}$.

A similar test for successive values of A_R (plotted as the vertical co-ordinate in Fig. 2) showed successive values of A_R were definitely not statistically independent, since for these, the characteristic $c(h)$ increased with $\sqrt{(h)}$ with no indication of reaching an asymptotic value (such as is indicated for example in Fig. 6). A plot of means of A_R for 5 successive, non-overlapping rotations, showed these means tended definitely to cluster in groups I and II, around two quite different average values which were: -0.193% and -0.025% respectively for 150 and 75 rotations. The smaller values of A tended to occur roughly near the two sunspot minima and the others near the maxima. This fact is also indicated by the data in rows two and three from the bottom of Table 1, from which it is evident that the average cosmic ray vector for the 75 rotations near sunspot minimum does not differ significantly from zero; whereas the average for the 82 rotations near sunspot maximum is eight times larger.

The value of the variance of the A_R in group I about the mean for group I, pooled with the variance of the A_R in group II about the mean for group II resulted in a standard deviation (s.d.) of 0.206% for A_R. It will be noted that this is only slightly less than the s.d. of 0.219% for A_R in line four of Table 1 for which the deviations were from a single mean (-0.135) for all 230 rotations.

Next, the standard deviation for deviations of A_R (for single rotations) was computed with the deviation of each A_R in group I measured from the mean (-0.193%) for group I, and with the deviation of each A_R in group II measured from the mean (-0.025%) for group II. The successive individuals in this set of deviations were found to be statistically independent.

Since successive values of B_R (from one mean) and of A_R (using two means) are statistically independent, then the expectancy, M, for successive rotations will also be statistically independent. In Table 1 it will be noted that s_{A_R} and s_{B_R} are smaller for years near sunspot minimum than for years near sunspot maximum (see second and third rows from bottom of Table 1). This indicates that M is also less near sunspot minimum. Thus the waves for the 230 rotations (for example) summarized in line four of Table 1 comprise a sample composed of two sub-samples, one from each of two populations having different means and variances. For this reason the distribution derived from the single parameter, M, provides only an approximation to the actual distribution. This accounts for the tendency for slightly more than half the points falling inside the p.e. circles. Nevertheless, the distribution of means from such composite samples, one part of which is from one population with a certain mean and variance and the remainder of which is from another population with different mean and

338

variance, will be governed by a single M, provided the individuals from the two populations are statistically independent. Consequently, the expectancy for means of n vectors in Table 1, is obtained from the expectancy M for single vectors on division by $\sqrt{(n)}$.

4. TESTS FOR STATISTICAL SIGNIFICANCE OF AVERAGE WAVES AND RESULTS

From the results of the preceding discussion the expectancy for the averages of n waves is obtained from M, the expectancy for single waves or vectors by dividing M by $\sqrt{(n)}$. Then the ratio, κ, of the amplitude of the average vector (for example $\bar{C}_R$ in Table 1) to the expectancy[5] for the average is $\kappa = \bar{C}_R \sqrt{n}/M$. The probability, P, of obtaining an average vector as large or larger than $\bar{C}_R$, from a population with $\bar{C}_R = 0$, is given[5] by $P = e^{-\kappa^2}$.

These values of P are given in Table 1 for different samples, the harmonic dials for which are plotted in the indicated figures. The tabulated values of P indicate only two average vectors $\bar{C}_R$, both in Fig. 3 (a), one from 14 rotations for 1952 only, and the other from 75 rotations for the years 1942, 1943, 1944, 1945, 1953 and 1954, which are too small to be regarded as differing significantly from zero. Incidentally, the sample for Fig. 3 (b) with 82 rotations is derived from the years 1937, 1938, 1939, 1946, 1947, 1948 and 1949. For the remaining samples the average vector $\bar{C}_R$ is large enough to be quite definitely statistically significant. The small circles centered on the end-points of the average vectors in Figs. 1, 2 and 3 are the probable error circles for these averages. Finally, the last column of Table 1 gives the time of maximum of the 27-day CRI wave in days after the minimum of the ICF wave. To these values of T, 0·2 day should be added since daily means of ICF are for G.M.T. days and the daily mean CRI values are for 75° W.M.T. days.

Thus, on the average, the maxima and minima of the 27-day waves in CRI occur respectively about 1·9 days after the minima and maxima of the ICF waves. Earlier results[4] from a smaller sample of only 34 rotations indicated the average CRI wave to be essentially opposite in phase to the ICF wave.

Using Chree's method Simpson[7] finds that the peaks or selected *maxima* from curves of daily mean neutron intensity (during 19 months, 1 May 1951 to 30 November 1952) tend to occur about 1 *day after* a minimum in magnetic activity. To this extent his results are in approximate agreement with those obtained herein. He also found, however, that the same neutron peaks were followed after about 1 day by a maximum in magnetic activity. This differs from the result for the 27-day waves.

During those magnetic storms with large decreases in cosmic ray intensity, the beginning of the cosmic ray decrease generally occurs[1,8] near the start of the main phase of the magnetic storm. The minimum cosmic ray intensity during the storm usually coincides with the minimum of H, the northward geomagnetic component at the equator. It is most likely that the minimum of H coincides closely with the maximum of geomagnetic activity as measured by K indices[8] or by ICF. Thus, for the

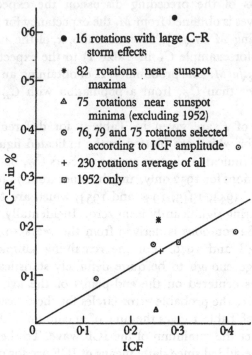

Fig. 4. Amplitude of 27-day waves in cosmic ray intensity (C–R) as function of average character-figure amplitudes (ICF)

large C–R decreases associated with some geomagnetic storms, the minimum CR intensity probably coincides with the maximum geomagnetic activity, whereas, on the average, the maximum of the 27-day waves in CRI occurs about 2 days after the minimum for ICF. This difference may possibly indicate that in geomagnetic storms the solar stream strikes the earth head on, giving rise to a maximum magnetic activity and a minimum of cosmic ray intensity within hours after the sudden commencement of the storm.

Fig. 4 indicates a plot of the amplitude, $\bar{C}_R$, of the average vector for CRI in the RCS, as a function of the ICF amplitudes (from data in Table 1).

In a rough way the former increases with the latter excepting for rotations with large C–R storm effects for which $\bar{C}_R$, relative to the ICF amplitude, is several times greater than for any other group. Also $\bar{C}_R$ for years near sunspot minimum is relatively much smaller than for any other group.

5. THE DEGREE OF QUASIPERSISTENCE OF 27-DAY WAVES IN COSMIC RAY INTENSITY AND IN INTERNATIONAL MAGNETIC CHARACTER FIGURE

Fig. 5 shows a summation harmonic dial for 27-day waves in CRI for Huancayo from 264 rotations (vectors in the original non-rotated co-

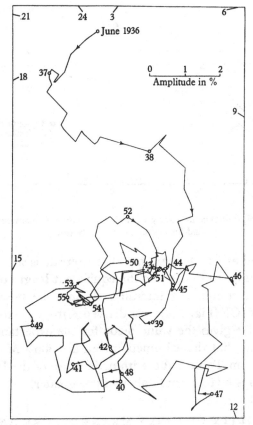

Fig. 5. Summation dial 27-day cosmic ray waves, Huancayo 1936–55.

ordinate system). Several stretches in the same direction indicating quasi-persistence are evident in the intervals: 1937–8, 1938–9, 1946–7 and 1949–50. Bartels[5] determined the degree of quasipersistence for 27-day

341

waves in ICF. His results are shown by the upper curve (a) of Fig. 6. With $m(h)$ equal to the expectancy (or two dimensional standard deviation) for means of h successive vectors $c(h) = m(h) \sqrt{(h)}$ and $c(1) = m(1)$. $m(1)$ is simply the r.m.s. amplitude for single vectors. For statistically independent vectors $c(h)/c(1) = 1$ for all h. With increasing values of $\sqrt{(h)}$ for ICF the characteristic $c(h)/c(1) = 1$ approaches the asymptotic value of 1·74 indicating that the effective number of statistically independent vectors [5] in a sample of N is $N/1·74^2 = 3·0$, or that the equivalent length of sequences is 3·0 rotations. For the 27-day waves in cosmic ray intensity

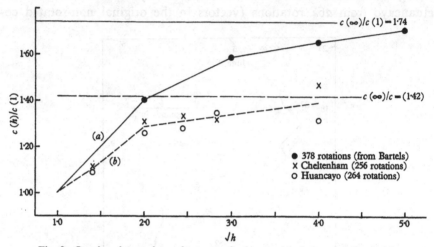

Fig. 6. Quasipersistence in 27-day waves for international character figure (a), and for cosmic ray intensity (b).

at Huancayo and at Cheltenham the characteristic is shown in Fig. 6 (b) with asymptote 1·41 so that 2·0 is the equivalent length of sequences for 27-day waves in cosmic ray intensity. It may be noted that Bartels characteristic for ICF (Fig. 6 (a)) was derived [6] from 378 rotations starting 11 June 1906. He gives the value 0·262 for the expectancy M of single vectors and shows that the arithmetic mean is 0·886 × M or 0·232 in the units of ICF. From Table 1 the arithmetic mean of the ICF amplitudes for 246 rotations is 0·264 or only about 14 % greater.

6. VARIABILITY OF DAILY MEANS BEFORE AND AFTER REMOVING 27-DAY WAVES IN COSMIC RAY INTENSITY

Table 2 summarizes for Huancayo the pooled values for each year of the standard deviation s_d of CRI daily means from monthly means. The third column of Table 2 gives the r.m.s. value, n, of the amplitudes of all the

27-day CRI waves in each year. Since the c.d. of daily means from the monthly means will differ little when pooled for the year from the yearly pooled value of c.d. of daily means from the means for 27-day intervals, then the standard deviation s_k for the residuals with the 27-day waves deducted [5] is closely approximated by: $s_k = (s_d^2 - n^2/2)^{1/2}$. A plot of s_d as function of s_k from the values in Table 2 shows that $s_k = 0.78\, s_d$.

Table 2. *Standard deviation of daily mean CRI at Huancayo before and after removing 27-day waves; harmonic coefficients for yearly mean 27-day wave*

Year	s.d. of daily means from monthly means pooled for each year (%)	r.m.s. C-R amplitude (%)	s.d. of residuals after removing C-R wave (%)	$\bar{A}_R$ (%)	$\bar{B}_R$ (%)	$\bar{C}_R$ (%)
1937	0·43	0·35	0·35	−0·09	−0·05	0·10
1938	0·67	0·61	0·51	−0·32	−0·07	0·33
1939	0·49	0·37	0·42	−0·20	−0·02	0·20
1940	0·38	0·39	0·26	−0·15	−0·06	0·16
1941	0·37	0·31	0·30	−0·21	−0·09	0·23
1942	0·39	0·41	0·26	−0·09	−0·06	0·11
1943	0·25	0·17	0·22	+0·03	−0·03	0·04
1944	0·24	0·16	0·21	−0·08	−0·01	0·08
1945	0·29	0·24	0·24	+0·04	−0·10	0·11
1946	0·83	0·69	0·68	−0·17	+0·03	0·17
1947	0·61	0·57	0·45	−0·30	−0·23	0·38
1948	0·52	0·50	0·38	−0·20	−0·15	0·25
1949	0·50	0·44	0·39	−0·10	−0·07	0·12
1950	0·38	0·36	0·28	−0·13	−0·08	0·15
1951	0·49	0·52	0·33	−0·27	−0·01	0·27
1952	0·52	0·48	0·39	−0·15	−0·07	0·17
1953	0·28	0·21	0·24	+0·01	+0·06	0·06
1954	0·21	0·15	0·18	−0·06	+0·08	0·10
1955	0·31	0·26	0·25	—	—	—

Finally, in Table 2 the harmonic coefficients A_R and B_R for CRI at Huancayo are tabulated together with the amplitude $\bar{C}_R$ of the average CRI wave.

7. ACKNOWLEDGMENT

It is a pleasure to acknowledge the invaluable contribution of Miss Isabelle Lange whose expert assistance made possible this investigation.

REFERENCES

[1] Forbush, Scott E. *J. Geophys. Res.* **59**, no. 4, 525–42, 1954.
[2] Morrison, P. *Phys. Rev.* **101**, no. 4, 1397–404, 1956.
[3] Forbush, Scott E. Transactions Washington Meeting, 1939; Internat. Union Geod. Geophys. Assoc. Terr. Mag. Electr., Bull. no. 11, 438–52, 1940.
[4] Chapman, S. and Bartels, J. *Geomagnetism* (Oxford, Clarendon Press, 1940).
[5] Bartels, J. *J. Geophys. Res.* **40**, no. 1, 1–60, 1935.
[6] Bartels, J. *Beitr. Geophys.* **28**, 1–10, 1930.
[7] Simpson, J. *Phys. Rev.* **94**, 426–40, 1954.
[8] Bartels, J. Presented at the Symposium on Geophysics, April 1956. (To be published in *Proceedings National Academy of Sciences*.)

CORRELATION BETWEEN COSMIC RAY INTENSITY AND GEOMAGNETIC ACTIVITY

D. VENKATESAN
Royal Institute of Technology, Stockholm, Sweden

ABSTRACT

The Chree method of analysis has been adopted for the analysis of the Ionization Chamber data for Huancayo, Cheltenham and Godhavn for 1946 and for the former two stations for 1945. The same procedure is adopted for the planetary index K_p also.

The cosmic ray minimum (or maximum) precedes the minimum (or maximum) of K_p by about 4–5 days. It is also observed that the relative decrease in cosmic ray intensity per day, $-\Delta I/(I.\Delta t)$, follows the changes in K_p in a general way, and hence the electric field as would be expected from the consideration of the theory of emission of beams of particles from the sun with the associated frozen magnetic field and the electric field arising due to polarization.

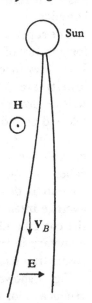

The concept of a storm-producing beam consisting of ionized rarified gas ejected from the sun and reaching the earth in about a day, was first introduced by Schuster (see Fig. 1). The beam originates from the sun where magnetic fields exist. If this is the case, as pointed out by Alfvén[1] the field would be frozen in the beam, since the conductivity of the beam is large. The beam carries the field as far out as the earth. Due to the motion with the velocity V_B ($= 2 \times 10^8$ cm/sec) the beam becomes electrically polarized, the electric field being given by the equation

$$\mathbf{E} = -\frac{1}{c}.\mathbf{V}_B \times \mathbf{H}.$$

Fig. 1. Storm producing beam, emitted from the sun (equatorial plane).

The voltage V across the beam is given by $V = E.B$, where B is the breadth of the beam. Cosmic ray particles with energy V_0 on passing the beam change their energy

345

by V. If ΔI is the change in intensity, the relative decrease in intensity per day is given by the equation:

$$-\frac{\Delta I}{I} = k\frac{V}{V_0} = kEB/V_0 = \frac{k.E}{V_0} \cdot \frac{2\pi R}{27} \cdot \Delta t,$$

where k is a constant which depends on the measuring device, R is the earth–sun distance. Hence, the relative decrease in intensity per day $-\Delta I/I . \Delta t$ should follow the changes in the electric field. In connexion with magnetic storms and aurorae Alfvén has pointed out that the electric field in the beam is its most important property. This seems to gain support from cosmic ray results as well. If K_p, the planetary index, could be considered as a measure of the electric field, then the relative decrease in intensity per day $-\Delta I/(I.\Delta t)$ should follow the changes in the electric field or the K_p index.

The data that have been chosen for analysis are the Carnegie Institution Ionization Chamber Records[2] for 1945 and 1946. Fig. 2 shows the day-to-day variations in intensity at Huancayo. The former is a fairly undisturbed period while the latter shows heavy decreases. The data for 1946 has been restricted to the period February–October, so that even if we consider 1 month on either side, it is well within the disturbed period.

The superposed epoch method of analysis originally devised by Chree[3] for analysis of geomagnetic activity is used. The procedure is as follows. Five days in each month when the cosmic ray intensity is highest (or lowest) are selected and the intensity on these selected zero days are written down in a column designated 'o' day. The data preceding these selected zero days are written down in columns to the left and are called $-1, -2, -3, \ldots$ days respectively. Similarly the data corresponding to the days following the zero days are written to the right of the 'o' day and are called $+1, +2, +3, \ldots$ days respectively. The average value for each column is determined and a smoothening is carried out by taking the average intensity over 3 days. The values are plotted on a graph against the corresponding day numbers. The same procedure is adopted for the planetary index K_p, the zero days being the same as determined for the cosmic ray data.

Fig. 3 shows the results of the analysis for minimum intensity days for Huancayo for 1945. It is seen that the cosmic ray minimum precedes that of K_p by about 4–5 days. The result agrees with those of Simpson[4] and Kane[5], but differs from that of Van Heerden–Thambyahpillai[6]. The fair correspondence between the relative decrease in cosmic ray intensity per day, $-\Delta I/(I.\Delta t)$, and K_p is seen.

346

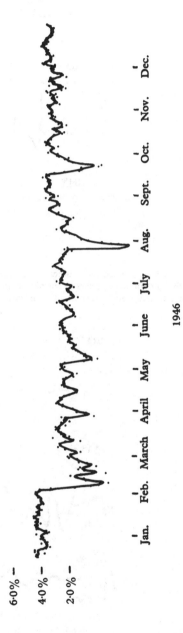

Fig. 2. Daily mean cosmic ray intensities at Huancayo for 1945 and 1946.

347

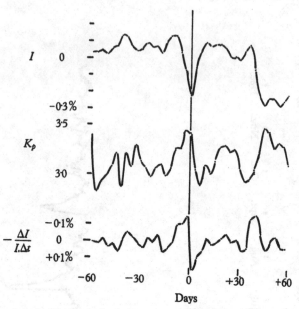

Fig. 3. Chree analysis for Huancayo for 1945 corresponding to five minimum intensity days per month. The top curve refers to cosmic ray intensity, middle one to K_p, and the bottom one to the relative decrease in intensity $-\Delta I/I.\Delta t$ per day, shown reversed.

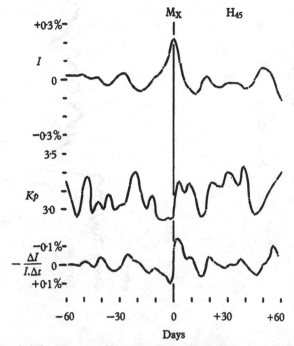

Fig. 4. Chree analysis for Huancayo for 1945 corresponding to five maximum intensity days per month. The top curve refers to cosmic ray intensity, middle one to K_p, and the bottom one to the relative decrease in intensity $-\Delta I/I.\Delta t$ per day, shown reversed.

348

Fig. 4 shows the results of the analysis for Huancayo corresponding to maximum intensity days for the year 1945. The same feature, namely, the cosmic ray maximum preceding that of K_p is observed.

Fig. 5 shows the result for Cheltenham for the same year corresponding to maximum and minimum intensity days. The days of selection are the same as for Huancayo. Table 1 shows the correlation analysis for the various cases. The fair agreement between $-\Delta I/(I.\Delta t)$ and K_p can be seen from the figures and the table.

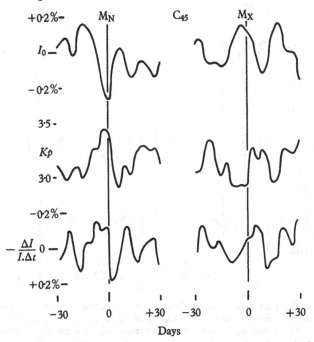

Fig. 5. Chree analysis for Cheltenham for 1945 corresponding to five days of minimum as well as five days of maximum intensity. The top curve refers to cosmic ray intensity, middle one to K_p, and the bottom one to the relative decrease in intensity $-\Delta I/I.\Delta t$ per day, shown reversed.

Table 1. *Correlation coefficient for—1945 between*

No.	Station	I and K_p	$-\dfrac{\Delta I}{I.\Delta t}$ and K_p	I and K_p	$-\dfrac{\Delta I}{I.\Delta t}$ and K_p
		Minimum intensity days		Maximum intensity days	
		For -30 to $+30$ days			
1.	Huancayo	-0.34 ± 0.03	$+0.56\pm0.02$	-0.49 ± 0.02	$+0.65\pm0.02$
2.	Cheltenham	-0.59 ± 0.02	$+0.26\pm0.02$	-0.68 ± 0.02	$+0.48\pm0.02$
		For -15 to $+15$ days			
1.	Huancayo	-0.51 ± 0.04	$+0.59\pm0.08$	-0.36 ± 0.05	$+0.52\pm0.04$
2.	Cheltenham	-0.53 ± 0.04	$+0.33\pm0.05$	-0.59 ± 0.04	$+0.40\pm0.04$

349

Fig. 6 shows the results for Huancayo, Cheltenham and Godhavn for the period February–October 1946. The same features seen in the data for 1945 are observed to an even more pronounced degree. If the analysis is restricted to 14 days on either side, the correspondence between the relative decrease per day, $-\Delta I/(I.\Delta t)$, and K_p is extremely high. This improvement when we restrict the analysis to 14 days instead of 30 days on either side is understandable, because the secondary series usually

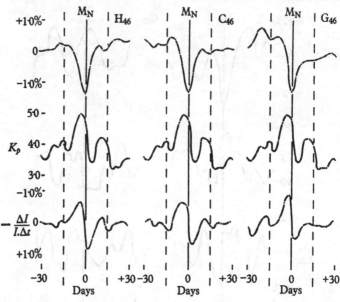

Fig. 6. Chree analysis for Huancayo, Cheltenham and Godhavn for 1946 corresponding to five days of minimum intensity per month. The top curve refers to cosmic ray intensity, the middle one to K_p, and the bottom one to the relative decrease $-\Delta I/I.\Delta t$ per day, shown reversed.

found in the 27-day variations, occurs after a separation of about 14 days, and this is not present to interfere, when we consider only 14 days on either side. Table 2 gives the correlation coefficients in the various cases.

The individual decreases during the period February–October, 1946 are considered for Huancayo, and the correlation between I and K_p, and $-\Delta I/(I.\Delta t)$ and K_p are presented in Table 3. The results are given for 3-day averages as well as for the day-to-day values. In general there is a negative correlation between I and K_p, and a positive correlation between $-\Delta I/(I.\Delta t)$ and K_p, as would be expected from the theory of the beam. This shows that $-\Delta I/(I.\Delta t)$ follows the changes in K_p in a general manner and hence the changes in the electric field.

Table 2. Correlation coefficient for—1946 between

No.	Station	I and K_p	$-\dfrac{\Delta I}{I.\Delta t}$ and K_p	I and K_p	$-\dfrac{\Delta I}{I.\Delta t}$ and K_p
				Minimum intensity days	
		From −30 to +30 days		From −15 to +15 days	
1.	Huancayo	−0·72±0·01	+0·58±0·02	−0·63±0·03	+0·79±0·02
2.	Cheltenham	−0·48±0·02	+0·65±0·02	−0·42±0·05	+0·87±0·01
3.	Godhavn	−0·16±0·02	+0·76±0·01	−0·14±0·05	+0·90±0·01

Table 3. Huancayo—1946. Correlation coefficient between

	Minimum intensity on	3-day averages		Day-to-day values	
		I and K_p	$-\dfrac{\Delta I}{I.\Delta t}$ and K_p	I and K_p	$-\dfrac{\Delta I}{I.\Delta t}$ and K_p
1.	8 February	−0·81±0·02	−0·12±0·06	−0·56±0·04	0±0·05
2.	16 February	−0·30±0·05	+0·67±0·03	−0·35±0·05	+0·76±0·02
3.	21 February	−0·60±0·04	+0·37±0·05	−0·45±0·05	+0·56±0·04
4.	2 March	−0·11±0·06	−0·41±0·05	−0·18±0·38	−0·02±0·06
5.	10 March	−0·70±0·03	+0·03±0·06	−0·69±0·03	−0·22±0·03
6.	29 March	+0·31±0·05	+0·72±0·03	0±0·05	+0·45±0·05
7.	15 April	−0·83±0·02	+0·16±0·06	−0·76±0·02	+0·04±0·06
8.	27 April	−0·23±0·06	+0·65±0·03	0±0·05	+0·40±0·05
9.	8 May	−0·68±0·03	+0·31±0·05	−0·68±0·03	+0·28±0·01
10.	9 June	−0·68±0·03	+0·31±0·05	−0·66±0·03	+0·24±0·02
11.	21 June	+0·76±0·03	+0·25±0·06	+0·65±0·03	−0·07±0·06
12.	29 June	−0·76±0·03	+0·16±0·06	−0·04±0·05	+0·35±0·05
13.	9 July	−0·39±0·05	+0·15±0·06	−0·26±0·02	+0·40±0·05
14.	28 July	−0·69±0·03	+0·50±0·05	−0·42±0·05	+0·60±0·04
15.	15 August	−0·97±0·00	+0·53±0·04	−0·60±0·04	+0·42±0·05
16.	25 August	+0·19±0·06	+0·25±0·06	+0·09±0·06	+0·55±0·04
17.	1 September	+0·68±0·03	+0·43±0·05	+0·18±0·04	+0·83±0·02
18.	23 September	−0·39±0·05	+0·35±0·05	−0·40±0·05	+0·14±0·05

REFERENCES

[1] Alfvén, H. *Cosmical Electrodynamics* (Oxford University Press, 1950), ch. 6; *Tellus*, **6**, 232, 1954; *Tellus*, **7**, 50, 1955.

[2] *Cosmic ray results from Huancayo observatory, Peru*, vol. 16, Carnegie Institution of Washington, 1948.

[3] Chree, C. *Phil. Trans. Roy. Soc.* A, **212**, 76, 1913; Chree and Stagg, *Phil. Trans. Roy. Soc.* A, **227**, 21, 1928.

[4] Simpson, J. A. *Phys. Rev.* **94**, 426, 1954.

[5] Kane, R. P. *Phys. Rev.* **98**, 130, 1955.

[6] Van Heerden and Thambyahpillai, *Phil. Mag.* **46**, 1238, 1955.

Schlüter: Did you measure the time of onset of 23 February 1956 cosmic ray increase at your different stations?

Forbush: Yes. I think from the original records one start of the increase was 3 or 4 min later at Godhavn, Greenland.

Schlüter: We had looked into the question of the 27-day variation of cosmic rays and the geomagnetic data some years ago. A negative correlation was found not only between CR and corpuscular activity (as measured by K_p) but also between CR and the ultra-violet emission of the sun (as measured by the W-numbers) which exceeds that due to the inter-correlation between both kinds of solar activity. I do not see any reasonable theoretical explanation for this effect and I therefore think it worth while to establish or to discard the existence of this effect.

Forbush: If you mean the correlation between cosmic ray intensity and the SW measure of Bartels we have not investigated this. One must take great care of mean correlations.

In connexion with Dr Venkatesan's high correlations referred to in the first part of his talk it should be pointed out that this correlation was between averages. One must remember that means of samples from a population with very low correlations between individuals in a pair, will exhibit a larger correlation which increases with the number of pairs in the mean. Furthermore, statistical tests of the significance of the correlation coefficients must take account of lack of statistical independence between successive samples.

Venkatesan: The necessity of looking into the individual storms has been realized and that is being looked into.

Forbush: An important point is also that successive days are independent.

Ehmert: If I have understood it correctly you find a change in phase between small magnetic perturbations influencing cosmic radiation and great storms on the other side. It seems that the first ones are connected with M-zones and the other ones are flares lying aside.

Biermann: Supplementing Dr Schlüter's remarks I would like to say that the original work on the effect of solar wave-radiation was chiefly done by Dr van Roka at our institute in Göttingen and that has been described in our treatise on *Kosmische Strahlung* (Springer Verlag 1953, ed. by Heisenberg). To this has only to be added that van Roka's theory of that effect advanced at that time now almost certainly appears to be disproved (see e.g. the more recent work by Simpson and his group).

Singer: How often does a cosmic ray decrease not correlate with magnetic storms?

Forbush: They are nearly always correlated.

Singer: Does the cosmic ray decrease arise about one day earlier than magnetic activity? Is that the general rule?

Forbush: No. We have only observed one large decrease starting before a magnetic storm.

Sarabhai: On 23 February 1956 we observed an increase of about 6% in intensity averaged over an hour at stations near the magnetic equator in India,

Unfortunately, we do not know the profile of our increase and are unable to state whether the solar particles travelled in direct trajectories from the sun or were deflected or scattered. A comparison with Huancayo would be interesting.

In relation to Dr Venkatesan's paper I should like to draw attention to the work of the Japanese group where the characteristics of effective and non-effective magnetic storms have been studied. It would be important to understand why, according to the proposed model of Venkatesan, all magnetic storms do not produce a change of cosmic ray intensity.

Could Dr Forbush throw some light on the possible causes which are responsible for the standard deviation, after correcting for the barometric effect and world-wide changes, being more than what one would expect from errors of random sampling?

Forbush: Everything may cause the deviation, except the barometric effect. The data were corrected to constant barometric pressure.

Venkatesan: We have not overlooked possible occurrence of storms without cosmic ray activity. Professor Alfvén has mentioned about the possibility of a capture effect of cosmic ray in an electrically polarized beam. This could explain the presence of a storm which has no associated cosmic ray activity.

Singer: The world-wide character of the Forbush decrease shows that a large region around the earth is affected. Decreases are also observed near the geomagnetic pole and there the cosmic rays would have come from very large distances in a direction perpendicular to the ecliptic plane.

Denisse: The fact that the amplitude of the 27-day variation goes through zero at the moment when the solar activity has minimum intensity suggests that this variation is uniquely correlated with violent magnetic disturbances. However, since the storms do not show up a recurrence of 27 days it is not surprising that the amplitude of the 27-day variation is small.

Ehmert: The famous and expressive picture that Professor Forbush showed of the magnetic storm influence contains the influence of several storms following each other. However, analysis of some great magnetic storms showed that there is an extremely strong correlation between the ring-current field, as measured by the midnight field depression at Huancayo, and the cosmic ray deflection at the same time throughout the individual storm. The coefficient of influence varies from storm to storm but is constant for individual storms. I think that is a criterion on which the theories should be based.

Ferraro: Did you find this correlation with every great storm?

Ehmert: No, I must say there were only 4 or 5 that I was able to analyse.

Lovell: Although ionospheric absorption of the galactic radio emissions has been observed during the main auroral phases following a flare (see, for example, C. G. Little and A. Maxwell: *J. Atmos. Terr. Phys.* **2**, 356, 1952), the observation of a decrease in intensity at the time of the flare reported by Dr Forbush is probably unique. The Jodrell Bank observations of galactic radio emissions during this period are being published in the *Jodrell Bank Annals*. There were marked effects on the scintillation phenomena in the period 30–40 hr following the flare, but no effect whatsoever at the time of the flare. Two equipments were in operation as follows:

1. The 218 ft transit telescope on 90 Mc/s. Beam-width 3° to half power, centred on $13^h 28^m$. R.A.; $+42°\ 20'$ decln., at the time of the flare.

2. An equipment on 80 Mc/s continuously following the Cassiopeia radio source. No significant change in the scintillations or signal strength occurred at the time of the flare.

It is considered that any intensity change exceeding 0·25 dB would have been prominent on these records, and this lack of absorption may be compared with the significant decrease reported by Dr Forbush on the lower frequency. It is hoped that this information will be of interest to Dr Forbush and his colleagues in their consideration of the nature of this effect. In the case mentioned above Little and Maxwell placed the region responsible at a height of well over 100 km and concluded that the effect was due to the reflexion of the incoming radiation near the upper boundary and not absorption during transmission through the region. It is to be hoped that a consideration of all the relevant data will enable the height of the region and the nature of the effect to be assessed in the present instance.

Ehmert: Ionospheric influence by the flare of 23 February 1956 was world-wide and observed also on the night hemisphere as a damping of long waves. From Professor Simpson's ascending values I evaluated the number of free electrons in the height between 40 and 90 km assuming with Budden a recombination coefficient of 10^{-14}. I found 1 to 5 electrons/cm^3. This is the lower limit given by this high coefficient. That is good enough for explaining the anomalies in long-wave propagation. Only in the auroral zones in Canada there seems to have been a further impact of particles.

SOLAR PRODUCTION AND MODULATION OF COSMIC RAYS, AND THEIR PROPAGATION THROUGH INTERPLANETARY SPACE

J. A. SIMPSON

Enrico Fermi Institute for Nuclear Studies and Department of Physics, The University of Chicago, U.S.A.

ABSTRACT

The principal characteristics for changes of cosmic ray intensity as a function of time and primary particle energy are reviewed for those intensity variations which are thought to be of non-terrestrial origin. These variations are either (a) temporary increases of cosmic ray intensity arising from the *de novo* production of cosmic ray particles in the vicinity of the sun in association with some solar flares, or (b) the modulation of extra-solar cosmic radiation within the interplanetary volume by a modulation mechanism related to solar activity.

The study of these variations for low-energy cosmic ray particles is also a unique tool for the investigation of interplanetary magnetic fields and other properties of interplanetary space. As an example, the cosmic ray events associated with the giant solar flare of 23 February 1956 have been studied. The experimental evidence shows that interplanetary magnetic fields must exist for the storage and redistribution of the solar flare cosmic ray particles. A more specific model indicates that disordered magnetic fields lie mainly beyond the orbit of the earth and that diffusion through these irregular magnetic fields is the prominent mechanism for particle storage. In addition, this cosmic ray intensity increase was fortunately superposed in such a way upon a change of intensity arising from a modulation mechanism that it is possible to restrict the kinds of models which account for modulation of cosmic ray intensity within the interplanetary volume.

I. INTRODUCTION

Cosmic ray intensity variations with time have been observed for over 25 years, but only recently has it become clear from experiments that many of these variations are, indeed, changes of primary particle intensity—changes which occur outside the atmosphere and beyond the geomagnetic field. These variations have been correlated with properties of the sun, and it is now obvious that the principal changes observed in the

355

primary cosmic ray spectrum are connected with solar phenomena: the intensity variations have a solar origin [1]. For example, the occasional and transient increases of cosmic ray intensity at the time of large solar flares represent solar cosmic ray production. On the other hand, changes of intensity such as the recurring 27-day variations, or sharp changes in the level of intensity the order of 20 %, as well as gradual changes in the spectrum over the 11-year solar cycle all appear to have their origin in the solar modulation of pre-existent cosmic radiation in the interplanetary medium. Hence, there exist within the solar system both, (a) *de novo* production of cosmic rays at the sun, and (b) the modulation of extra-solar cosmic radiation within the interplanetary volume.

We know that the cosmic ray spectrum undergoes the largest changes with time at low-particle energies, with the magnitude of the variations becoming vanishingly small for energies in excess of 40–50 Bev (Billion electron volts). We conclude that this dependence upon particle energy, along with the change in numbers of particles reaching the earth, requires the presence of magnetic fields of non-terrestrial origin to account for modulation by the sun. The search for a theory of how solar connected phenomena produce cosmic ray modulation effects have led us to re-examine the conditions which prevail in interplanetary space, particularly with regard to these magnetic fields and their description.

Indeed, we use the cosmic ray particles of low magnetic rigidity as a unique 'tool' to study the electrodynamics of the interplanetary medium. Cosmic rays from solar flares are probes for investigating the prevailing conditions at the time of the flare, and place limits upon the magnitudes and configurations of interplanetary magnetic fields and ionized gases. As we shall later show, the solar flare cosmic ray particles will also help us to understand the modulation mechanism.

The application of cosmic rays to problems of the interplanetary medium are not restricted to the study of intensity changes. They have also been used as probes to determine the earth's outer magnetic field distribution. In fact, only recently have we found that the outer geomagnetic field *effective for cosmic rays* is significantly different from earlier predictions—a problem bearing on the question of an inclined and rotating magnetic dipole field interacting with an extensive, ionized gas. However, our discussion in this paper shall be directed to the specific case of cosmic radiation from the solar flare of 23 February 1956, and its implications for a description of the interplanetary medium and cosmic ray modulating mechanism. For this study I wish to acknowledge the collaboration of P. Meyer and E. N. Parker of our laboratory [2].

2. THE SOLAR FLARE OF 23 FEBRUARY 1956

The fifth large increase of cosmic ray intensity known to occur in association with a solar flare took place on 23 February 1956. This was the largest of all the intensity increases since they were first observed in 1942 [3, 4], and it undoubtedly will be the most studied. From these earlier events it was evident that the particles producing the intensity increase occurred predominantly in the low energy portion of the cosmic ray

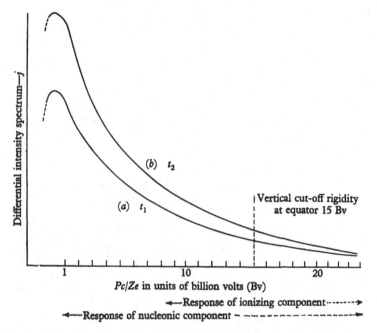

Fig. 1. The range of detector responses for the secondary particles arising from the primary cosmic radiation.

spectrum. Therefore, detectors which respond to the secondary radiation from the low energy portion of the primary cosmic ray spectrum are of special interest in studying the energy spectrum of flare particles. Over the past 8 years a neutron intensity monitor has been developed and used for investigations of this kind at low energies, since more than 0·75 of all cosmic ray particles to which it is sensitive fall within a magnetic rigidity range where we may use the geomagnetic field as a particle rigidity analyser.

To measure changes in primary cosmic ray intensity during the flare we record the intensity of the secondary, nucleonic component generated by the primary radiation. The nucleonic component intensity is indirectly measured by the amount of local neutron production within a pile structure of lead and paraffin. Since we know the response of the neutron detector

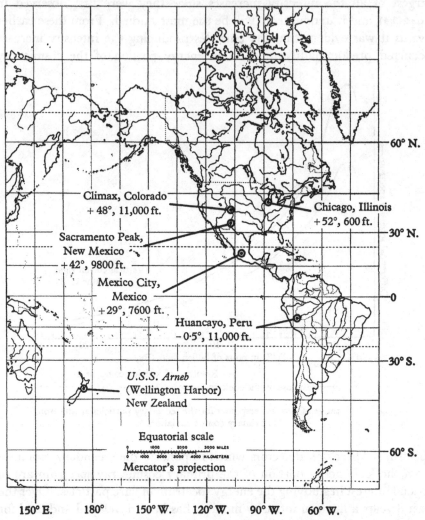

Fig. 2. Locations of neutron intensity monitors established by the University of Chicago.

for the normal cosmic ray spectrum as a function of geomagnetic latitude and atmospheric depth, we may extrapolate any changes of observed intensity to the top of the atmosphere. In this way we deduce the changes in primary intensity. Fig. 1 shows the differential cosmic ray spectrum—

number of incident particles as a function of magnetic rigidity—and illustrates how the nucleonic component extends observations over the range of low magnetic rigidity particles inaccessible to ion chambers or counter telescopes. We have established a network of neutron pile monitors to exploit these principles. Locations of the continuous observing stations are shown on the map, Fig. 2. At the time of the flare the sixth neutron monitor, identical with the units at Chicago and Climax, was returning with the U.S. Antarctic Expedition and was operating in the harbor of Wellington, New Zealand. In addition, neutron detection apparatus was

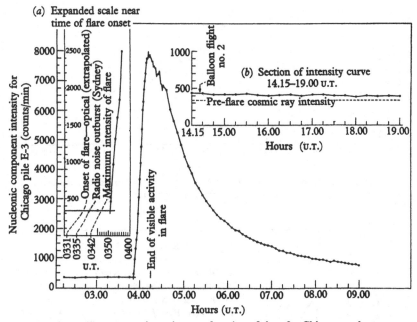

Fig. 3. The neutron intensity as a function of time for Chicago at the time of the 23 February 1956 flare.

carried by a balloon over Chicago during the cosmic ray increase. Thus, the flare of 23 February is of interest since there exists for the first time the means for studying the flare particle intensity as a function of both time and particle rigidity.

We shall report here on the analysis of our experiments and its bearing on the flare particle spectrum, the propagation of the high-energy flare particles in the interplanetary medium, and the relationship of this cosmic ray event to solar phenomena.

The neutron intensity as a function of time outside of impact zones is shown in Fig. 3 for Chicago, and in Fig. 4 for Wellington Harbor. These two widely separated stations yield precise information on the time of onset, the rate of rise, the maximum intensity and the rate of decline of the temporary increase. In Fig. 5 we display the intensity at all six stations as a function of time.

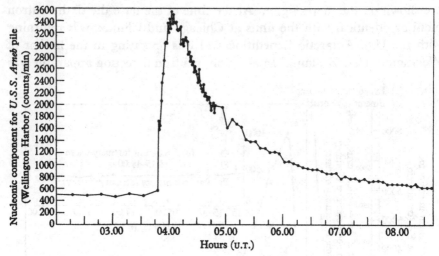

Fig. 4. The neutron intensity as a function of time for Wellington at the time of the 23 February 1956 flare.

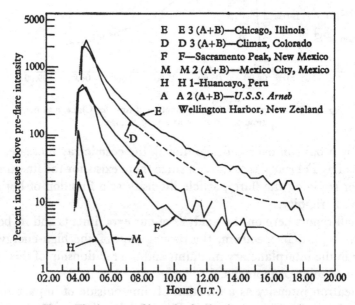

E   E 3 (A+B)—Chicago, Illinois
D   D 3 (A+B)—Climax, Colorado
F   F—Sacramento Peak, New Mexico
M   M 2 (A+B)—Mexico City, Mexico
H   H 1–Huancayo, Peru
A   A 2 (A+B)—*U.S.S. Arneb*
E   Wellington Harbor, New Zealand

Fig. 5. The increase of intensity for the six neutron monitors.

Before undertaking an analysis of the flare particle spectrum, we wish to point out several conclusions which may be drawn directly from the experimental data in Figs. 3–5. They are:

(1) The temporary increase of cosmic ray intensity represents the acceleration of particles to cosmic ray energies in the vicinity of the sun. No hypothesis of focusing, particle storage at the sun or terrestrial phenomena can account for this enormous increase of cosmic ray intensity.

(2) The incident cosmic radiation which produces the 'tail' of the intensity curve at low energies most probably represents particles scattered back to the earth from many directions in the solar system. This conclusion is supported by the evidence that the radiation continues to arrive for more than 15 hr after all indications of activity in the solar region have disappeared. The lack of intensity increases superposed on the flare intensity curves for Chicago, Climax and Sacramento Peak, near 0400 and 0900 local time impact zones, is further strong evidence that the particles at those times were not coming directly from a 'point' source in the direction of the sun.

(3) To preserve the relatively sharp increases of cosmic ray intensity after onset as shown in Figs. 3 and 4, the particles must have traversed relatively uncomplicated orbits—hence it is unlikely that the scattering regions we have invoked to account for the 'tail' could lie inside the orbit of the earth. If the scattering region had been distributed both inside and outside the orbit of the earth the time for rise to maximum intensity would have been the same order of magnitude as the time for decline to background intensity.

On the other hand, since at onset both Chicago and Wellington were in non-impact zones,* the anomalously large scale of the effect measured there could reasonably be accounted for by requiring that the particles arrive more or less isotropically. Hence, although the orbits are not complicated they must involve at least some scattering, with the scatterer located outside the orbit of the earth. This requires that no appreciable magnetic fields lie between the sun and the earth at the time of the flare.

(4) Particles with energies in excess of 15 Bev, and probably 20–30 Bev, were produced at the time of the flare. We derive this result from the relatively large increase of intensity at the geomagnetic equator (Huancayo, Peru) where the minimum energy for arrival of protons from the vertical is ~ 15 Bev.

From the experimental data we have also obtained the flare particle spectrum. There are two assumptions underlying our analysis. First, we

* Note added after the conference: Reference [5] shows that impact zones of high order do exist for Chicago and Wellington.

assume that the incoming radiation is isotropic, and hence we restrict our analysis to the period following the intensity maximum where the incoming radiation becomes isotropic. From this assumption it then follows that we may use Störmer theory in determining geomagnetic cut-off as a function of latitude for the flare particles. Secondly, we assume that the composition of the incoming radiation is not significantly different from the composition of the normal cosmic radiation. Primary neutrons as the

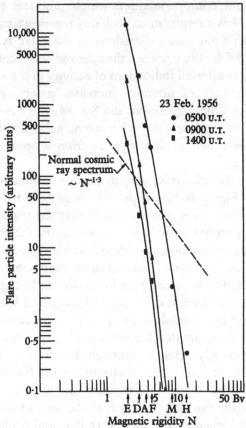

Fig. 6. The primary spectrum for flare particles.

main component are excluded since the flare increase was observed at full intensity on the night side of the earth. The problem then reduces to whether electrons, protons or alpha particles constitute the principal flare radiation. Though electrons may possibly be abundant in the primary flare spectrum, their contributions to the secondary nucleonic component can be estimated to be less than 1 %, assuming that they are just as abundant in the primary radiation as the protons.

The contribution of alpha particles or heavier nuclei to the primary intensity is a more difficult problem, but our analysis of the magnitude of the flare increases at Sacramento Peak, Climax and Chicago proves that the greatest intensity increase at low magnetic rigidity arises almost entirely from protons.

With the above assumptions we have constructed the primary spectrum for flare particles after the first hour of the flare. The results are given in Fig. 6. The spectra follow approximately the power law N^{-7}, although there is evidence that the high rigidity particle intensity tends to decrease more rapidly with time than the intensity for low rigidities.

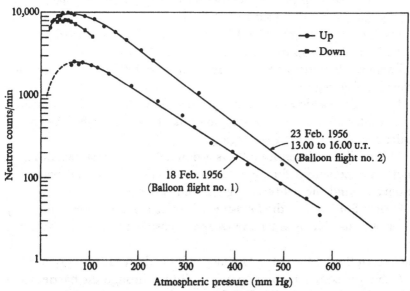

Fig. 7. Balloon flights using neutron detectors. Flight no. 1 was prior to the flare. Flight no. 2 followed the flare.

In Fig. 7 we show the results of two balloon flights: number 1 prior to the flare and number 2 obtained during the progress of the cosmic ray intensity increase. Our analysis shows that the primary spectrum was strongly peaked near magnetic rigidities of 2–4 Bev at 14.30 hour U.T. This is in agreement with the independent observations from the neutron monitor network.

The large difference in onset times of > 5 min between detectors located in impact zones (or at the equator) and detectors outside of impact zones can reasonably be explained by assuming shorter path lengths for the particles arriving in impact zones over those arriving outside of impact zones.

Sittkus *et al.* [6] have suggested that this difference in onset time arises from some reflecting region beyond the orbit of the earth. In this way the particles arriving at regions outside of impact zones will have undergone scattering. This time difference leads to a radius of closest approach for the scattering region to the sun with the value 1·4 to 1·7 a.u.*

3. MODEL FOR THE INTERPLANETARY VOLUME

We now direct our attention to the explanation of the cosmic ray flare effect. It is clear from the experiments that there are several physical conditions within the solar system which must be satisfied in order to develop an explanation for the observations on 23 February 1956. These conditions may be summarized as follows:

(1) There exists a magnetic field-free region extending outward past the orbit of the earth to approximately $r \sim 1\cdot5$ a.u.

(2) There is a boundary region which scatters cosmic ray particles back into the field-free region.

(3) Since the particle intensity declines only slowly after reaching maximum intensity the boundary region must be a barrier for the escape of particles from the 'field-free' region.

(4) The decline of intensity follows a power law $t^{-3/2}$ as shown in Fig. 8 and not an exponential function of time. Consequently, the barrier is continuous around the field-free region, and is not thin.

(5) It then follows that the field-free region is a volume surrounded by a barrier of finite thickness for the escape of cosmic ray particles into the galaxy.

This description of the interplanetary volume derived from experiment suggests that the cosmic ray flare particles diffuse through the barrier from the field-free region. If we let $J(E)\,dE$ represent the *density* of cosmic ray particles with energies in the range $(E, E+dE)$, then we shall assume that $J(E)$ varies according to the diffusion equation

$$\frac{\partial J(E)}{\partial t} = K(E)\nabla^2 J(E), \tag{1}$$

where $K(E)$ is the diffusion coefficient for particles of energy E. We find that there is a solution to the diffusion equation which may have the same $t^{-3/2}$ dependence on time which we found from experiment, Fig. 8; namely,

$$J(E) = \frac{c}{(\pi K t)^{3/2}} e^{-r^2/\pi K t}. \tag{2}$$

* Note added after the conference: It is shown in reference [5] that the differences in onset times are not due to reflexions. The onset time is a smooth function of particle energy over a spread of >9 minutes in onset time.

Eq. (2) is the special solution of Eq. (1) where a burst of radiation is instantaneously released at $t=0$ and at the origin ($r=0$) of an infinitely extensive diffusing medium. This is a problem well known in the theory of heat conduction. If we observe the change in particle density near the source (small values of r) then $J(E) \propto t^{-3/2}$ for all energies.

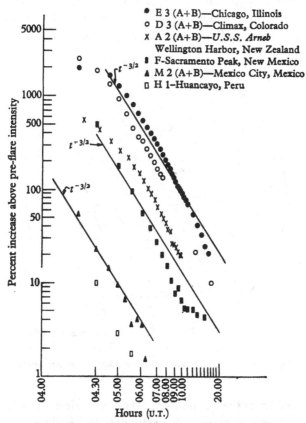

Fig. 8. The decline of cosmic ray intensity follows a power law approximately $t^{-3/2}$ except for late times where the decline approaches an exponential function.

Obviously, this is an over-simplification of the physical conditions if for no other reason than that we know the diffusing region has an inner boundary at $r \approx 1.5$ a.u. and does not exist for $0 \leqslant r \leqslant 1.5$ a.u. Therefore, for simplicity in developing a model we shall assume that the 'field-free' region is a spherical cavity of radius $a = 1.5$ a.u. From Fig. 8 we also know that the function $t^{-3/2}$ does not hold at high energies and, for later times, even at low-particle energies. This indicates, as we shall later show, that the diffusing medium is not infinite in extent. For a barrier of finite thickness we shall assume the outer boundary is spherical and at radius b in order to preserve a simple model. These simplifying assumptions lead us to the picture shown in Fig. 9 for the cross-section of the inner solar system.

Although the model proposed here rests upon the experimental data from the February 1956 flare, it is important to note that the requirement for a field-free region $r > 1 \cdot 0$ a.u. in extent, a scattering region and a delay in the eventual escape of the flare particles from the solar system were

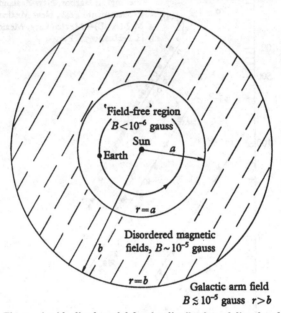

Fig. 9. An idealized model for the distribution of disordered magnetic fields in the solar system.

already deduced from the earlier flare events.[1]. Hence, we hope that the hypothesis proposed here may be extended to all flare particle observations.

We estimate b from the' observed deviation of the intensity from $t^{-3/2}$ for large values of t: $b \approx 5$ a.u. We have also obtained two independent estimates of $K(E) = c/3 \, L(E)$ where

$$L(E) \approx \frac{E_0 \left[\dfrac{E}{E_0} \left(\dfrac{E}{E_0} + 2 \right) \right]^{\frac{1}{2}}}{Z e B}.$$

We note that $K(E)$ is approximately proportional to E for particle energies of several Bev or more.

If we assume that L is the same order of magnitude as the radius of curvature of the particles in the disordered barrier field then the r.m.s. field intensity $B \simeq 2 \times 10^{-5}$ gauss.

Highly ionized field-free clouds of gas may sweep back any ordered solar or galactic magnetic fields which would otherwise pervade the entire interplanetary space [7, 8]. We have assumed that the barrier is represented by these tangled fields piled-up over a region forming a shell with inner radius 1–2 a.u. We do not know the degree of stability of such a barrier against penetration by ordered fields, but we do know from simple hydromagnetic concepts that the time constants are the order of months, or more. For our purposes it is sufficient to point out that the flare event was preceded by months of intense solar activity which should be capable of forming the special conditions required for the barrier region.

4. THE SUPERPOSITION OF FLARE PARTICLES ON A COSMIC RAY MODULATION EFFECT

In view of the severe restrictions we have placed upon magnetic fields and scattering within the solar cavity at the time of the flare it is especially important to understand the origin of the isotropic and rapid decrease of

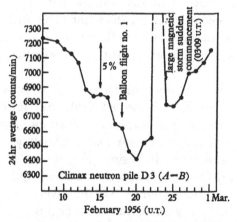

Fig. 10. The decrease of cosmic ray intensity which began before the flare event.

cosmic ray intensity (Forbush-type decrease) which began ~ 10 days prior to the flare event and continued beyond the period of the flare (see Fig. 10). In recent years experiments have shown that this isotropic decrease is not of terrestrial origin, and hence, the mechanism producing it must lie outside the geomagnetic field. Experiments have also shown that the magnitude of this phenomenon is a function of particle rigidity and is a modulation of the pre-existent cosmic radiation by a solar-controlled mechanism [9].

We have discussed elsewhere the possibility for distinguishing among the several hypotheses of cosmic ray modulation by studying the superposition of the flare particle spectrum on the modulated pre-existent cosmic rays [10]; we treat the flare particles as probes for studying electromagnetic conditions in interplanetary space.

The experimental evidence cited above supports modulation by magnetized and ionized clouds. Also the escape of the flare particles from the solar system is strong evidence for the presence of diffusion. The question then arises: How is it possible for these cloud-like regions to expand outward from the sun carrying their tangled magnetic fields without introducing such serious scattering within ~ 1 a.u. that the observed features of the cosmic ray flare event would be destroyed? The magnetic field intensity and scale length of the model clouds proposed by Morrison [11] to account for a rapid intensity decrease and continuing low intensity level ($\sim 10\%$ decrease at climax) are more than an order of magnitude too large to permit observation of the flare event. An alternative explanation [12], wherein the magnetized cloud is captured by the earth and is supported outside the geomagnetic field, namely, a geocentric cloud, does not meet with these objections. For the geocentric model we find that with scattering and diffusion limited to regions near the earth, there is negligible effect on the flare particle orbits.

Let us consider how diffusion of cosmic ray particles through ionized clouds containing twisted and knotted fields may introduce changes of total cosmic ray intensity. For purposes of illustration, we assume that the diffusing region is of infinite extent and thickness y, and that the full galactic intensity is observed in region (1), Fig. 11 (a). Then a detector placed in region (2) after the system has reached equilibrium will detect the full galactic radiation intensity. If, however, region (2) is initially free of radiation, and the galactic radiation begins to diffuse through the scattering barrier, the detector will observe that:

(1) The intensity rises asymptotically to the galactic intensity.

(2) The rigidity dependence of the spectrum is changing and approaches the spectrum of the galaxy.

The flux of particles $F(x, y, z; E)$ in region (2) arising from a gradient in the cosmic ray particle density $J(x, y, z; E)$ for the mean free path

$L(x, y, z; E)$ of magnetic scattering centres $(L \ll y)$ in the twisted fields region may be written:

$$F(x, y, z; E) = -\frac{L}{3} \nabla[VJ]$$

for particles of velocity V.

This is the mechanism proposed by Morrison to account for the intensity variations. A magnetized cloud of ionized matter is ejected from the sun, expanding to an enormous volume which initially contains no cosmic radiation in its inner regions, Fig. 11 (b). If, then, the earth sweeps into

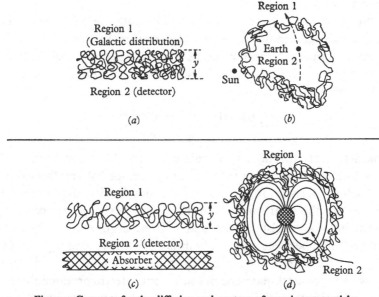

Fig. 11. Concepts for the diffusion and capture of cosmic ray particles.

this volume, a detector will measure a sharply reduced intensity. The outward moving cloud also 'sweeps out' preferentially the low-energy particles due to its velocity V'. The time to reduce the total intensity for reasonable field intensities is the order of a day or more with the time for recovery to normal intensity the order of days.

On the other hand, if we introduce an *absorber* in region (2) which is capable of removing particles as fast as they diffuse through the barrier, as in Fig. 11 (c), it is clear that:

(1) A much less efficient barrier-diffusing region is needed to produce reductions of total intensity.

(2) The intensity will remain low so long as the barrier is present.

Now for a diffusing barrier having the scale size of the solar system, there are no large absorbers capable of producing this effect. But if the diffusing barrier is placed around the earth overlying the earth's magnetic field, then the earth becomes the absorber, Fig. 11 (*d*). This latter proposal forms the basis for the theory developed recently by Parker. It is capable of explaining the sharp Forbush-type decreases, the low rigidity cut-off and its variation with the solar cycle, the 11-year cycle of intensity, all with fields the order of 10^{-5} gauss in the barrier region. The model then becomes a local, geocentric, model. This is in contrast with a barrier centered about the sun and extending throughout the solar system in all directions, i.e. a heliocentric model.

For this geocentric model the earth and its outer field is surrounded by an ionized gaseous nebula of low density retained by gravitational forces but buoyed at an equilibrium distance above the geomagnetic field by the pressures between the geomagnetic field, B(terr.), and the fields in the ionized clouds, B(cloud), i.e.

$$\frac{B^2 \text{ (terrestrial)}}{8\pi} \approx \frac{B^2 \text{ (cloud)}}{8\pi} .$$

This model meets the objections raised by the Morrison model which requires both fields 10^{-1}–10^{-3} gauss and places special conditions on the times at which particles from a solar flare may be detected at the earth.

A theory for the solar modulation of cosmic ray intensity, based upon these principles, is consistent with and a strong argument for the indirect association found between cosmic ray intensity variations and geomagnetic storms, since the interactions of highly ionized gas clouds of solar origin, with or without twisted magnetic fields, are expected to produce observable and transient effects on the geomagnetic field when captured by the earth. These interactions are only now beginning to be studied.

5. CONCLUSION

By extending the study of cosmic ray intensity variations to the low-energy portion of the primary cosmic ray spectrum it becomes clear that the dominant mechanism for both (*a*) modulation of extra-solar radiation, and (*b*) the temporary storage of solar cosmic rays is cosmic ray diffusion through disordered magnetic fields in interplanetary space. The distribution of these disordered fields throughout the interplanetary volume is not known, except that tangled fields surrounding the earth satisfy the conditions for modulating cosmic ray intensity, and disordered magnetic fields beyond the earth's orbit enclosing a field-free region between the sun and the earth are required to account for the diffusion of solar flare-produced cosmic rays away from the vicinity of the sun and earth. The r.m.s. magnetic field intensities required to account for the observed changes in the primary cosmic ray spectrum are no greater than the order of 10^{-5} gauss in the vicinity of the orbit of the earth. The strong correlations between solar activity and the observed changes in the cosmic ray spectrum both for short and long-time scale phenomena leave no doubt as to the solar origin of these magnetized clouds; however, their production and life-history is an unsettled problem.

It is not difficult to imagine that the major geomagnetic storm effects may also find their explanation in the collision of magnetized clouds with the permanent geomagnetic field, and their occasional capture by the field. All of our evidence relating cosmic ray intensity variations to major geomagnetic disturbances support the view that cosmic ray and geomagnetic field variations are linked by a solar-produced mechanism, presumably the production of magnetized clouds in special regions of the sun.

REFERENCES

[1] Simpson, J. A. *Proc. Nat. Acad. Sci.*, **43**, 42, no. 1, 1956.
[2] Meyer, P., Parker, E. N. and Simpson, J. A. *Phys. Rev.* **104**, 768, 1956.
[3] Ehmert, A. *Z. Naturf.* **3** a, 264, 1948.
[4] Forbush, S. E. *Phys. Rev.* **70**, 771, 1946.
[5] Lüst, R. and Simpson, J. A., *Phys. Rev.* **108**, 1563, 1957.
[6] Sittkus, A., Kühn, W. and Andrich, E. *Z. Naturf.* **11** a, 325, 1956; *Z. Naturf.* 1956 (in the press).
[7] Chandrasekhar, S. and Fermi, E. *Astrophys. J.* **118**, 113 and 116, 1953.
[8] Davis, L. *Phys. Rev.* **100**, 1440, 1955.
[9] Simpson, J. A. *Phys. Rev.* **94**, 426, 1954.
[10] Simpson, J. A. *Ann. Géophys.* **2**, 305, 1955.
[11] Morrison, P. *Phys. Rev.* **101**, 1397, 1956.
[12] Parker, E. N. *Phys. Rev.* **103**, 1518, 1956.

Discussion

Eckhartt: I would like to make some comments concerning the question of onset times. As you pointed out there should be a difference in onset times between stations lying within the classical impact zones and those stations lying outside. We have chosen a number of stations where the times of onset were rather sharply defined. For each of these stations we determined the geomagnetic deflexion of cosmic ray particles arriving from zenith in the momentum range up to 10 GeV/c, using Malmfors's curves. Thus we got the initial directions far away from the earth of these flare cosmic ray particles which finally arrive at the chosen stations from zenith directions. These initial or asymptotic directions will be represented by a vector. The end-point of this vector is thought to rest in the center of the earth, whereas the arrow point of the vector is determined by a latitude angle and by a longitude angle in the geographic system of the earth. This is the same manner in which Brunberg demonstrates the asymptotic directions on his globes. In the two-dimensional representation of Fig. 12 we look upon the northern hemisphere. The geographic latitude is

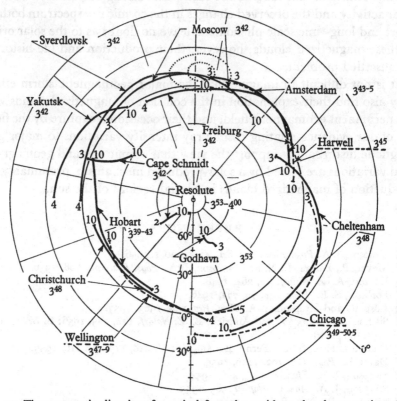

Fig. 12. The asymptotic directions for arrival from the zenith at the chosen stations. The directions towards the sun at the beginning of the solar flare are indicated by the dotted lines at the top of the figure. The time of onset of the cosmic ray increase is written next to the name of each station.

represented by the length of the radius, the geographic longitude by the azimuth angle. For example, take a proton of energy 3 GeV which arrives at Chicago from the zenith. This particle has an asymptotic direction far away from the earth defined by the direction from the earth's center to the point 40° E of Greenwich and 5° S which is shown on Fig. 12. The small figures at the ends of each curve denote the value of the momentum at the particular point. Other values and points have been computed. Thus, the curves represent a smooth interpolation of the asymptotic directions for arrival from the zenith at the chosen stations for particles in the range of momenta denoted by the small figures. But for the low energies in question and for the chosen stations these curves may be representative even for arrival from other small zenith angles. Most of the stations shown here are equipped with ionization chambers. Only the dashed lines refer to stations with neutron monitors or a shower detector at Harwell. At the very moment of the beginning of the flare the sun, as seen from the earth in an angular diameter of 15 and 30° respectively, is given by the dotted lines at the top of the figure. When a curve intersects this area the corresponding station lies in one of the classical impact zones. The time of onset of the cosmic ray increase is written next to the name of each station. You see that Hobart, lying far outside the impact zones, had such an early onset as 3^{39}–3^{43}. The same is valid for the Russian stations Yakutsk and Sverdlovsk.* In general, the stations whose asymptotic directions are between $\pm 90°$ with respect to the direction towards the sun seem to have started earlier, than those stations whose asymptotic directions point away from the sun.

This was my first remark. Let me now present some of the results measured by our own GM directional telescopes. These are declined 30° to the north and to the south. During the whole flare increase the north pointing telescope recorded more particles than the south pointing one. Since particles measured by the two inclined telescopes are almost equally affected by the atmosphere the ratio between the measured pulse rates becomes a measure for the ratio of primary flux above the atmosphere. On Fig. 13 we show the ratio between the relative increases of the north and south pointing telescopes at Stockholm and at Rome. We see that the north pointing telescope measured about 10 % more particles than the south pointing telescope did. This seems to be in conflict with your assumption of isotropy a certain time after the onset of the flare. We interpret our results in the following way. The asymptotic directions—these are the directions of the particles far away from the earth—lie very close to each other for momenta below 10–12 GeV/c. This means that particles below this energy come from nearly the same directions outside the earth's magnetic field. Therefore, there must have been particles with momenta higher than 10 or 12 GeV/c. For the second, the number of particles with higher momenta must have been different for the different asymptotic directions. Isotropic distribution among all asymptotic directions could not cause any difference in the counting rates. Different onset times for the two telescopes would certainly only have affected the first hour's result. If one looks now upon the difference in angles between the direction towards the sun and the asymptotic directions from which the particles have come in order to be measured by our north pointing telescope, one arrives at the values 50 and 70° for momenta 20 and 10 GeV/c respectively. A value of about 6×10^{-6} gauss is found for a mean magnetic field in the interplanetary space which would have caused this deflexion, assuming that the particles are coming from the vicinity of the sun. This value is somewhat higher than the upper limit you have chosen.

Alfvén: I think that if one takes into account only the stations which Professor Simpson showed in his diagrams, then the results could be interpreted with his

* Note added after the conference: According to a recent publication in *Nuclear Physics*, **1**, 585, 1956, the times of onset at the Russian stations should be altered to: Moscow 3^{38}, 3^{40}, Sverdlovsk 3^{40}.

model. On the other hand, if one includes all the stations which are available, for example also the Soviet and Japanese stations, the whole picture changes altogether. As Eckhartt showed in his diagrams we do not get radiation only from the direction of the sun. We also get radiation from other directions. It is very important that the first moment when the increase in cosmic ray intensity is observed on the earth is, as far as could be judged, simultaneous for beams coming from all directions. This is very clearly seen from the recordings of the Soviet stations.

I think the general picture we get here when we include all the stations shows how important is international collaboration and I am very glad of the initiative taken by Gold and Elliot to collect all the data.

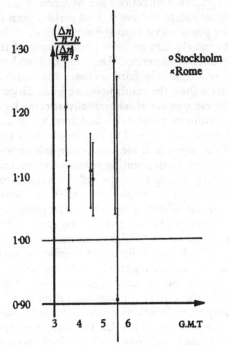

Fig. 13. The ratio between the relative increases in pulse rates of the north and south pointing telescopes at Stockholm and at Rome.

Simpson: Just one remark concerning the whole question of energies. At the high energies the asymmetries certainly seem to exist at the beginning of the flare and are very important. However, after the first hour, most of the particles in the field are coming in below 5 BeV and the arguments given here are based on these very low energies. See also reference [12] of the preceding paper.

Alfvén: You cannot possibly conclude that there is a field-free region between the sun and the earth.

Simpson: I have defined the region containing an average field of 10^{-6} gauss or less as a 'field-free' region, and have included the earth's orbit in this region. The boundary region would begin beyond the earth's orbit.

374

Singer: The solar flare increase on 23 February 1956, was a special case in that it occurred during a Forbush decrease. Statements about the interplanetary field must take account of this. My own view on the Forbush decrease is that it is produced by a turbulent magnetic field which is expanding and therefore decelerating the cosmic radiation. I think that the results that Eckhartt presented which I have not seen before support this point of view.

Elliot: Simpson suggests that the anisotropy might be due to difference in energy between primary particles recorded by telescopes and neutron monitors. I do not agree with this view because I think they very nearly measure the same thing. In the neutron monitor roughly half of the particles are protons and these same protons must be recorded by the counter telescope of the type which Eckhartt has used. I do not see that one can contribute any difference in primary energy to these two types of recording.

Simpson: That is correct. I just want to say that the onset time differences and the asymmetry constitute very crucial points. Most of the particles we have studied were primaries in the region 2-4 Bev, near geomagnetic cut-off energies, and the time for first arrival, i.e. onset time, appears to be a function of particle energy. See also reference [12].

Parker: Professor Alfvén's objection that the observed directions of approach of the primary particles from the solar flare cannot be accounted for by the field-free ($B \leqslant 10^{-6}$ gauss) space is easily avoided by tangled magnetic fields localized about the earth.

Alfvén: Can you construct such a field and what is the order of magnitude of that?

Parker: A rough value would be 10^{-2} gauss in regions at a distance of several earth's radii. Such a possibility is not ruled out even though it conflicts with your theories.

Alfvén: But observations are well interpreted by the electric field model which I proposed yesterday.

Parker: I do not think that observations could pick out a unique theory.

Ferraro: If an interplanetary field is responsible for the influx of cosmic ray particles on the anti-meridian and post-meridian sides of the earth by magnetic deflexion, as Professor Alfvén reports, would we not expect a difference between the time arrivals of particles at the earth in different localities? Is this the case?

Simpson: Also there is the problem of the long storage times observed. How can your model store particles long enough?

Alfvén: In the magnetic field you will have a diffusion outwards.

Parker: We have calculated the diffusion rate of a field of 10^{-5} gauss and it comes out too large by a factor of 10.

Alfvén: I do not believe this. It depends on the shape of the field.

Sarabhai: Dr Simpson has listed a number of variations under modulations of cosmic ray intensity. He has further stated that in these modulations, low energy primaries are more affected than high energy primaries. I would like to point out that this type of energy dependence is by no means always present. As shown by Neher, intensity increases have sometimes been observed for intermediate energies without equally large changes at low energies. Furthermore, the energy dependence in the different types of variations is not the same.

Some years ago, the Chicago group reported the small flare effect in the cosmic ray nucleonic component. Little has since then been heard of this important effect. Has this been confirmed in later measurements? How does the small flare effect fit into the model proposed by Dr Simpson?

Simpson: With respect to the effect of small flares I will say that the apparatus was put into operation just as we came to the declining period of the solar cycle. We have about 15 months period to look for flares. There are about 66 flares available to work with; a statistical treatment is necessary in order to look for a small pulse. The results were just exactly as reported. There seems to be about 1% pulse in the impact zones and nothing evident outside the impact zones. A statistical study, however, is weak because one wants observations for at least one whole solar cycle and we certainly propose to follow this up. But this now requires waiting well into the maximum of the present solar cycle in order to get enough new data. There are two points that one has to consider. First, one may ask if all flares of the same character produce cosmic ray particles arriving here. Secondly, there is the question whether they strictly follow the simplest paths of impact zones worked out by Schlüter and Lüst, and Firor.

Biermann: Would it not seem from the collective evidence we have heard that only a small fraction of the flare radiation everywhere observed on the earth comes directly from the sun? In that case the original outburst must have been of rather short duration (almost a few minutes) and that the 'cavity' was filled quite rapidly. The observed time differences of the onset at various stations would then mainly reflect the irregularities and general shape of the reflecting and diffusing boundary of the cavity assumed by Dr Simpson.

THE ANISOTROPY OF PRIMARY COSMIC RADIATION AND THE ELECTROMAGNETIC STATE IN INTERPLANETARY SPACE

V. SARABHAI, N. W. NERURKAR, S. P. DUGGAL
AND T. S. G. SASTRY

Physical Research Laboratory, Ahmedabad, India

ABSTRACT

Study of the anisotropy of cosmic rays from the measurement of the daily variation of meson intensity has demonstrated that there are significant day-to-day changes in the anisotropy of the radiation. New experimental data pertaining to these changes and their solar and terrestrial relationships are discussed.

An interpretation of these changes of anisotropy in terms of the modulation of cosmic rays by streams of matter emitted by the sun is given. In particular, an explanation for the existence of the recently discovered types of daily variations exhibiting day and night maxima respectively, can be found by an extension of some ideas of Alfvén, Nagashima, and Davies. An integrated attempt is made to interpret the known features of the variation of cosmic ray intensity in conformity with ideas developed above.

The study of the daily variation of cosmic ray intensity provides a unique tool for the evaluation of the anisotropy of the primary cosmic radiation, and changes occurring in it. Since the anisotropy is related to theories of the origin of cosmic radiation and to the electromagnetic state in interplanetary space, it is of importance to summarize the current status of our knowledge derived from measurements of the daily variation of cosmic ray intensity.

1. The daily variation of cosmic rays and the anisotropy of the primary radiation is of a highly variable character. This is seen in large long-term changes of the 12-monthly mean daily variation of meson intensity (Sarabhai and Kane[1], Sarabhai, Desai and Venkatesan[2], Thambyahpillai and Elliott[3], Steinmaurer and Gheri[4]), the day-to-day changes correlated with magnetic character figure and the occurrence of large amplitudes of the daily variation of meson intensity on groups of days

(Sittkus [5], Remy and Sittkus [6]). It is seen from Fig. 1 that on these same days the daily variation at Amsterdam (H. F. Jongen, private communication) at $\lambda = 54°$ N, at Ahmedabad (U. D. Desai) $\lambda = 13°$ N show similar features. However, Kodaikanal (D. Venkatesan) on the magnetic equator

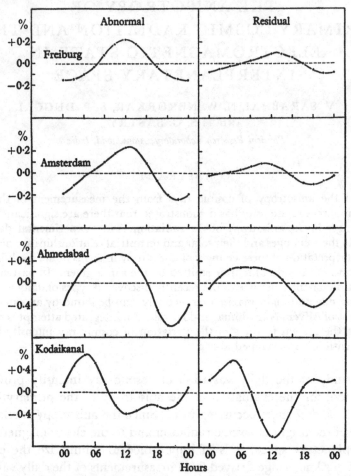

Fig. 1. Comparison of the daily variation at Freiburg (Sittkus), Amsterdam (Jongen), Ahmedabad (Desai) and Kodaikanal (Ventakesan) on days of abnormal amplitudes at Freiburg with the daily variation on residual days.

does not exhibit any marked difference on days on which Sittkus gets abnormal amplitudes.

2. During several years the daily variation, particularly at low latitudes, exhibits two maxima instead of one. This is seen not only in data from Huancayo during 1937-52 (Sarabhai, Desai and Venkatesan [2]) but also

in data at Ahmedabad, Kodaikanal (Venkatesan and Sastry) and Trivandrum (Duggal) during 1950–5 as shown in Fig. 2. The changes of the daily variation indicate that they are primarily due to two types of anisotropies

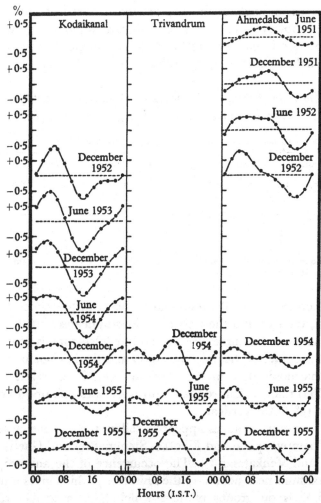

Fig. 2. Twelve-monthly mean daily variation of meson intensity at Ahmedabad, Kodaikanal and Trivandrum centred at six-monthly intervals 1950–5.

which correspond to maxima at 03.00 and 11.00 hours local time respectively (Sarabhai, Desai and Venkatesan[2]).

3. The daily variation of radiation incident in directions very close to the vertical is characterized by large amplitudes and undergoes large changes (Ehmert[7], Sarabhai and Nerurkar[8]). For stations in low

latitudes it exhibits maxima centred at 11.00 hours or 03.00 hours on a majority of days. On some days, which appear to be associated with low values of K_p, the daily variation has two maxima instead of one. The cone within which the radiation with large amplitude appears to be incident is restricted to a semi-angle of about 5° around the vertical. The semi-angle of the cone is however variable, so that during different periods the ratio of the amplitude in narrow angle and in wide angle telescopes varies, as shown in Table 1 (Ehmert[7], Sarabhai and Nerurkar[8]).

Table 1. *The ratio of the diurnal amplitudes measured with telescopes of different semi-angles on identical days during different periods*

| | 1954 | | | | | 1955 | | | | | | | | Refe |
	Aug.	Sept.	Oct.	Nov.	Dec.	Jan.	Feb.	Mar.	Apr.	May	June	July	Total	enc
$\dfrac{CT\,(\pm 5°)}{CT\,(\pm 15°)}$	1·52	1·79	1·72	1·74	2·17	2·40	1·83	1·65	1·68	0·79	1·18	1·04	1·60	1
$\dfrac{CT\,(\pm 45°)}{IC}$	—	—	—	—	—	—	—	—	—	—	—	—	1·5 : 1 to 2·4 : 1	7

The recorders are designated by IC for ionization chamber and CT ($\pm x°$) for counter telescope of semi angle $x°$ in the east–west plane.

4. The comparison of the daily variations with telescopes pointing to the vertical, the east and the west directions indicates that the spread of times of maxima in the three directions is much less than is expected. This is seen in the curves of the daily variation and the harmonic components shown on harmonic dials in Fig. 3 (a) and (b). These relate to a study made by Nerurkar at Ahmedabad with telescopes pointing to directions inclined at 45° to the vertical.

5. The long-term changes appear to follow the 22-year cycle of solar activity (Thambyahpillai and Elliot[3], Steinmaurer and Gheri[4], Sarabhai, Desai and Venkatesan[2]). The occurrence of the anisotropy which gives a maximum near noon and the anisotropy which gives a maximum at night-time is on groups of days which have a 27-day recurrence tendency.

If we consider the above facts, the most important conclusion is that at the present moment our knowledge of the characteristics of a permanent anisotropy, if indeed such exists, is very meagre. Initially, it is appropriate to consider for interpretation only the variable anisotropy about which we have now a number of well-established experimental facts and solar and terrestrial relationships. Theories which have been advanced in the past to explain only the average characteristics of the diurnal component of the daily variation, and those which do not take into consideration the existence of the two types of anisotropies which produce daily variations with maxima separated by approximately 8–10 hr are clearly inadequate. A theory which explains the variable anisotropy by a modulation of the primary intensity seems most promising in this context.

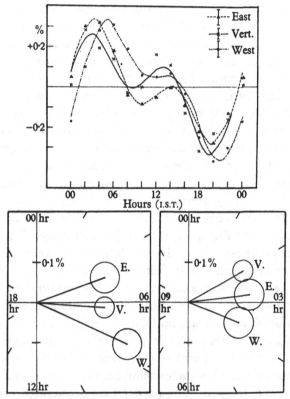

Fig. 3 (a, upper). The daily variation at Ahmedabad in the vertical, the east and the west pointing telescopes. (b, lower), harmonic dials showing the diurnal and the semi-diurnal components of the daily variation in the three telescopes at Ahmedabad.

Nagashima[9] has proposed a theory to explain the magnetic-storm type anisotropy which is directed towards the 12-hr direction. His theory postulates neutral beams of ionized particles ejected from the sun. These carry with them a trapped magnetic field derived from the solar dipole field. The charge separation which occurs in the beams as viewed from the earth, creates an electric field across the beams so that cosmic rays which traverse the beams and come to the earth suffer an acceleration or a deceleration depending on the orientation of the beam and the earth. Nerurkar[10] has studied the implications of an extension of Nagashima's ideas. If the magnetic field trapped within a beam is derived from the high local magnetic field in the neighbourhood of the active region from which the beam is ejected, it would be possible to expect a magnetic field frozen within the beam but having no preferred orientation in relation to the solar dipole field. For the purpose of the theory, the author has assumed that the solar beam would have an outward radial velocity of from 500 to 2000 km/sec, a width of about 5×10^{12} cm and a magnetic field of from 10^{-5} to 5×10^{-6} gauss at the distance of the earth from the sun. These values are consistent with theories of geomagnetic disturbances involving neutral but ionized streams of matter ejected from the sun.

With the daily variation of cosmic ray intensity which arises as an observational effect due to the spinning of the earth, it is only possible to study the anisotropy in the east–west plane. We have therefore to consider the electric field produced in the beam in the east–west plane due to the component of the magnetic field in the direction perpendicular to this plane. We have also to consider the situation that arises because the beam has an angular velocity derived from the spinning of the sun and it approaches the earth, envelops it and then recedes from it. These three cases are illustrated in Fig. 4 (a), (b) and (c) respectively. The top and the bottom series of diagrams of the figure relate to the reversal in direction of the component of the magnetic field perpendicular to the ecliptic. The diagrams illustrate the directions in space, as viewed from the earth, along which cosmic rays suffer acceleration or deceleration after traversing the beam. They also show in a schematic manner the increase or decrease according to local time of mean intensity of a definite primary region due to the spinning of the earth. Some of the consequences of the theory are as follows:

(1) While in the case of particles being accelerated we get an increase of intensity, there is a decrease when particles are decelerated. However, the time of occurrence of the increase and the decrease is separated by 12 hr. Therefore, the maximum of the daily variation occurs at approximately the same time irrespective of the relative position of the beam with respect to the earth.

(2) Depending on the negative or positive sign of the component of the trapped magnetic field, an anisotropy is produced which would give a maximum in the daily variation either at about 08.00 or at about 16.00 hours before deflexion in the geomagnetic field. The measured time of maximum depends on the correction to be applied for bending of the trajectories in the geomagnetic field.

(3) The magnitude of the anisotropy would vary in relation to the magnitude of the component of the trapped magnetic field in the direction perpendicular to the east–west plane. For different orientation of the trapped field of varying magnitude, we can expect the average anisotropy to correspond to a daily variation at low latitude having a most probable time of maximum 03.00 hours or 11.00 hours respectively, even though the

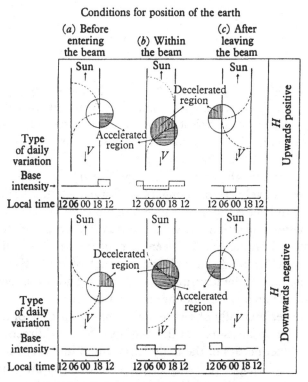

Fig. 4. Accelerated and decelerated regions and the type of daily variation for positive primaries of energy $= 2E_{min}$ when the earth is (a) on the right side, (b) within, and (c) on the left side of the beam. The component of the trapped magnetic field perpendicular to the plane of the paper is positive and towards the reader in upper figures and negative and away from the reader in lower figures.

time of maximum as well as the amplitude on each particular occasion would be different. Table 2 indicates the expected amplitude and time of maximum of the diurnal component of the daily variation at low latitudes.

(4) The anisotropy is only produced for primary cosmic rays of energy above a certain minimum value E_{min}. E_{min} depends on the width of the beam, the magnitude of the trapped magnetic field and its orientation in

Table 2. *The percent amplitude and the time of maximum of the diurnal component of the daily variation of meson intensity in equatorial latitudes.*

Width of the beam $= 5 \times 10^{12}$ cm				When H is positive		When H is negative	
H (gauss)	Velocity (cm/sec)	Change in energy (eV)	E_{min} (Bev)	Amplitude (%)	Time of maximum (hr)	Amplitude (%)	Time of maximum (hr)
10^{-5}	2×10^8	$1 \cdot 3 \times 10^8$	10	0·65	1030	0·65	0300
10^{-5}	10^8	$0 \cdot 65 \times 10^8$	10	0·33	1030	0·33	0300
10^{-5}	5×10^7	$0 \cdot 33 \times 10^8$	10	0·16	1030	0·16	0300
5×10^{-6}	2×10^8	$0 \cdot 65 \times 10^8$	5	0·44	1130	0·39	0130
5×10^{-6}	10^8	$0 \cdot 33 \times 10^8$	5	0·22	1130	0·20	0130
5×10^{-6}	5×10^7	$0 \cdot 17 \times 10^8$	5	0·11	1130	0·10	0130

respect to the east–west plane. No appreciable anisotropy is produced for cosmic ray particles of energy less than E_{min}, while for increasing energy above E_{min}, the per cent anisotropy goes on diminishing. This explains why averaged over long periods the amplitude of the daily variation measured with a neutron monitor is about the same at Huancayo as it is at Climax, even though the mean energy of primary radiation is 19 Bev and 7 Bev respectively (Firor, Fonger and Simpson[11]). Similarly there is almost no latitude effect observable in the amplitude of the daily variation measured with ionization chambers at Huancayo, Cheltenham and Christchurch. Comparison between a neutron monitor and a meson detector cannot be made directly on account of the differences in the response functions of the instruments with respect to directions of arrival of particles. The energy dependence of the anisotropy also reconciles with theory the experimentally observed small change in time of maximum of the daily variation measured in the east and the west directions.

The special properties of the daily variation of intensity measured with narrow angle telescopes pointing towards vertical require much further consideration for experimental study and interpretation.

Our grateful thanks are due to the other research workers of the Physical Research Laboratory whose efforts have contributed to continuous cosmic ray recordings at Ahmedabad, Kodaikanal and Trivandrum and to the Atomic Energy Commission of India for financial assistance.

REFERENCES

[1] Sarabhai, V. and Kane, R. P. *Phys. Rev.* **90**, 204, 1953.
[2] Sarabhai, V., Desai, U. D. and Venkatesan, D. *Phys. Rev.* **99**, 1490, 1955.
[3] Thambyahpillai, T. and Elliot, H. *Nature, Lond.* **171**, 918, 1953.

[4] Steinmaurer, R. and Gheri, H. *Naturwissenschaften,* **42,** 294, 1955.
[5] Sittkus, A. *J. Atmos. Terr. Phys.* **7,** 80, 1955.
[6] Remy, E. and Sittkus, A. *Z. Naturf.* **10***a*, 172, 1955.
[7] Ehmert, A. *Z. Naturf.* **6***a*, 622, 1951.
[8] Sarabhai, V. and Nerurkar, N. W. Proceedings of IUPAP Cosmic Ray Congress, Mexico (in the press).
[9] Nagashima, K. *J. Geomagn. Geoelect., Kyoto,* **7,** 51, 1955.
[10] Nerurkar, N. W. *Proc. Indian Acad. Sci.* **45**A, 341, 1957.
[11] Firor, J. W., Fonger, W. H. and Simpson, J. A. *Phys. Rev.* **94,** 1031, 1954.

SEPARATION OF EXTRA TERRESTRIAL VARIATIONS IN COSMIC RAY INTENSITY AND ATMOSPHERIC EFFECTS BY DIFFERENTIAL MEASUREMENTS WITH G–M TELESCOPES

E. A. BRUNBERG

The Royal Institute of Technology, Stockholm, Sweden

ABSTRACT

The daily variation of cosmic ray intensity can arise partly from atmospheric and partly from non-atmospheric effects. There is at present a difference of opinion whether this latter effect is completely due to extra terrestrial causes or not.

The purpose of the present paper is to suggest a method by which the atmospheric effects could be separated from the other variations without any assumptions about the mechanism of the atmospheric influence.

Assuming that the primary particle intensity around the earth is not quite isotropic it can be shown (see Fig. 1) that, when measuring with a G–M telescope situated at P, the daily intensity variation is written:

$$I = \sum_{\nu} {}^{t}I_{\nu} . \cos \Theta . \cos \Phi . \cos (\omega t + \nu . \Psi - \nu . \Psi'') + \sum_{\nu} {}^{-}C_{\nu} . f(\nu . \omega t), \qquad (1)$$

where the first sum represents variations due to the anisotropy and the second sum is the atmospheric component, due to the fluctuations of the atmosphere. In this expression the angles Θ and Ψ'' are the positional angles of the extra terrestrial cosmic ray source, which represents the anisotropy. Ψ and Φ are the angles through which the particles have been deflected in the geomagnetic field before they are recorded by the telescope, i.e. Ψ and Φ determine the measuring direction or the asymptotic direction of a telescope, when corrected for the geomagnetic deflexion. ${}^{t}I_{\nu} . \cos \Theta . \cos \Phi$ finally, is the amplitude of the νth harmonic of the induced true cosmic ray intensity variation. The number of harmonics and their amplitudes depend on the extension of the cosmic ray source in space, but the inaccuracy of cosmic ray measurements limits our discussion to the first and second harmonics.

Suppose the daily variation in cosmic ray intensity has been measured during a long period by two telescopes, identical as far as regards geometrical dimensions. One telescope, however, is measuring in the north direction with a certain zenith angle, the other telescope in the south direction with the same zenith angle. The averaged variation for the period is then, in accordance with Eq. (1):

$$I_N = \sum_{\nu=1}^{2} {}'I_\nu . \cos \Theta . \cos \Phi_N . \cos (\nu . \omega t + \nu \Psi_N - \nu . \Psi') + \sum_{\nu=1}^{2} C_\nu . f(\nu . \omega t),$$

$$I_S = \sum_{\nu=1}^{2} {}'I_\nu . \cos \Theta . \cos \Phi_S . \cos (\nu . \omega t + \nu . \Psi_S - \nu . \Psi') + \sum_{\nu=1}^{2} C_\nu . f(\nu . \omega t), \quad (2)$$

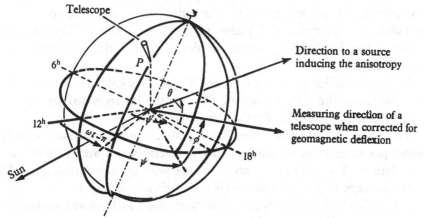

Fig. 1. Relations between the measuring direction of a telescope, direction to a source inducing an anisotropy and the position of the telescope.

where Ψ_N and Φ_N (which represent the geomagnetic deflexion for particles recorded by the *north* pointing telescope) are not equal to Ψ_S and Φ_S depending on the difference in azimuth angles for the two recording telescopes. As the telescopes measure particles, which have passed or come from atmospheric layers not more than about 30 km apart, it seems legitimate to assume, that the atmospheric components are the same for the two telescopes. Thus taking the difference between I_N and I_S the atmospheric components cancel and we get a difference function, which represents only the extra terrestrial variations:

$$I_N - I_S = \sum_{\nu=1}^{2} {}'I_\nu . \cos \Theta$$

$$\times [\cos \Phi_N . \cos (\nu . \omega t + \nu . \Psi_N - \nu \Psi') - \cos \Phi_S . \cos (\omega t + \Psi_S - \Psi')]. \quad (3)$$

25-2

This fact has been pointed out by several authors. Elliot has also pointed out, that the difference function could be used for calculating the true intensity variations.

When comparing this theoretically derived function with experimental results, given in the form of a Fourier series,

$$I_N - I_S = \sum_{\nu=1}^{2} (a_\nu . \cos \nu . \omega t + b_\nu . \sin \nu . \omega t), \qquad (4)$$

it can be shown that
$$a_\nu = {}'I_\nu . f_a(\Psi', p); \qquad (5)$$
$$b_\nu = {}'I_\nu . f_b(\Psi', p),$$

where ${}'I_\nu$ means the amplitude of the true variation, when the measuring direction of a telescope outside the geomagnetic field is parallel to the earth's equatorial plane. Consequently ${}'I_\nu$ is a measure of the strength of the cosmic ray source, inducing the anisotropy and gives its position, measured in the equatorial plane, as mentioned before. The functions f_a and f_b are due to the geomagnetic deflexion, p is the primary particle momentum.

It is now possible to calculate ${}'I_\nu$ and Ψ' for different momenta and then also calculate the true variation, that would be recorded by the two telescopes after correction for the atmospheric effects.

The valuable data from Manchester, where north and south pointing telescopes have been measuring during several years are used for these calculations. Fig. 2 gives the result for the year 1948 assuming 20 GeV/c as the average momentum of the effective primary spectrum.

I_N and I_S are the actually measured variations without any corrections. $(I_N - I_S)$ is the difference, which represents a variation that is not caused by atmospheric fluctuation. ${}'I_N$ and ${}'I_S$ are the calculated true variations, corrected for the atmospheric component and C is the atmospheric component. First and second harmonics are given.

It is of interest to compare different years and to find the dependence of the momentum. This is done in Fig. 3, where the phase and amplitude of the true intensity variation as well as the atmospheric component have been plotted for momenta between 14–32 GeV/c.

The following points may be noted:

(1) The direction to the source depends very little on the chosen momentum while the intensity shows a strong dependence. The phase changes from year to year, which has been pointed out by Elliot and Thambyahpillai [1] and Sarabhai and Kane [2]. The phase of the second harmonic of the true variation is approximately constant during the period considered.

388

(2) The second harmonic of the atmospheric component derived in this way has a constant phase during all the years as should be expected. The first harmonic has approximately the same phase during all the years, if we choose 20 GeV/c as the average momentum (except perhaps for the year 1951).

(3) A value around 20 GeV/c seems to be adequate because for this value the second harmonic of the atmospheric component is opposite in phase with the pressure. Assuming as a first approximation that the second harmonic of the atmospheric component is only pressure dependent, which

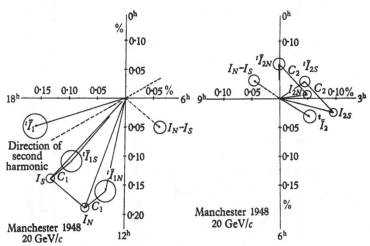

Fig. 2. First and second harmonics of measured and calculated intensity variations. I_N, I_S = Measured intensity variations. ${}^tI$ = The true intensity variation, corrected for geomagnetic deflexion. C = The atmospheric component.

is generally accepted, we can even calculate the pressure coefficient. Thus, as an average value for the first three years we get:

$$- (4 \pm 1)\%/\text{cm Hg}.$$

We must remember that the average value of the momentum, inducing the variation, is not necessarily the same when measuring with different azimuth angles, even if the zenith angles are the same. Also the geomagnetic latitude is important. This depends on the geomagnetic deflexion, but for the north and south pointing telescopes at high latitudes it has been found, that as a first approach the same momentum can be used in the two directions.

The atmospheric effect should be expected to be constant over all the years. The result of the analysis shows that this is the case. The residual

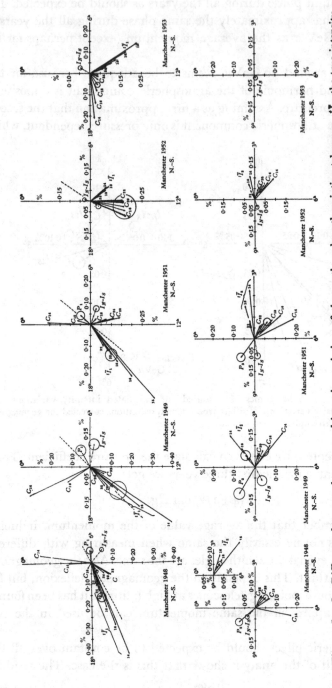

Fig. 3. $I_N - I_S =$ Difference between the measured intensity variations of the north and south pointing telescopes. $†I =$ the true intensity variation, corrected for geomagnetic deflexion. $C_{14}-C_{32} =$ The atmospheric components assuming different average momenta from 14–32 GeV/c. $P =$ Pressure. Dashed line in the 24-hr dial is the direction of the second harmonic of the true intensity variation corrected for geomagnetic deflexion.

390

variation, which changes from year to year is completely free of atmospheric effects. This residual variation is most likely interpreted as being due to an extra terrestrial anisotropy, but the possibility that the variation is caused by some local changes within the geomagnetic field is not ruled out by this differential method.

REFERENCES

[1] Elliot, H. and Thambyahpillai, T. *Nature, Lond.* **171**, 918, 1953.
[2] Sarabhai, V. and Kane, R. P. *Phys. Rev.* **90**, 204, 1953.

PAPER 41

THE SOLAR DAILY VARIATION OF THE COSMIC RAY INTENSITY

H. ELLIOT AND P. ROTHWELL

Imperial College of Science and Technology, London, England

ABSTRACT

Some recent measurements of the solar daily variation for cosmic rays incident from the east and west directions at 45° to the vertical in London are described. The results do not agree with those to be expected if the variation was due to a non-isotropic flux of primary particles entering the earth's magnetic field. This result is discussed in relation to other evidence and it is concluded that the daily variation is probably due to a modulation of the primary cosmic ray intensity in the earth's magnetic field.

I. INTRODUCTION

It is generally believed that the solar daily variation of the cosmic ray intensity is due to a variation of the primary radiation incident on the earth's atmosphere, this intensity variation being produced in some way which is at present not understood. Observations to date have been made at sea-level using ionization chambers, counter telescopes and neutron monitors. Counter telescopes have the advantage that they make it possible to measure the variation for different directions of incidence at the earth's surface whereas ionization chambers and neutron monitors accept radiation within a solid angle which is limited only by atmospheric absorption. Since counter telescopes record primarily either the μ-meson flux or the combined μ-meson and electron components, the intensity observed at sea-level is dependent on atmospheric temperature and pressure. In relating the intensity changes observed at sea-level to changes in the primary intensity, it is therefore necessary to correct for these meteorological variables. In investigations of the solar daily variation it is possible to make an adequate correction for the variation in barometric pressure but in order to correct for temperature it is necessary to know the daily variation in temperature throughout the atmosphere. At present the daily variation in atmospheric temperature is uncertain because of the limitation, in particular the susceptibility to radiation errors, of the instruments used for routine measurements.

In the absence of accurate information about the daily variation in atmospheric temperature, attempts have been made to separate the part of the cosmic ray variation due to atmospheric temperature from that due to variations of the primary intensity by using directional telescopes. These telescopes have been so arranged that they record radiation arriving from quite different parts of the sky but respond in an identical manner to variations in intensity due to atmospheric temperature and pressure changes (Elliot and Dolbear[1], Malmfors[2]). Measurements of this kind together with observations on the nucleonic component, which is not temperature sensitive, have established beyond doubt that the daily variation is largely due to a variation in primary intensity incident on the atmosphere. This variation has been generally attributed to an anisotropic primary intensity entering the earth's magnetic field.

In order to determine the true direction of anisotropy from the observed daily variation, it is necessary to know the deflexion experienced by the primary particles in passing through the earth's magnetic field. The trajectories of cosmic ray particles in the earth's field have been investigated by Brunberg and Dattner[3] by means of scale model experiments. Using the data on the trajectories obtained in this way, Brunberg and Dattner[4] have shown that it is possible to account for the daily variation observed with counter telescopes pointing in the north and south directions at 30° to the vertical if it is assumed that the mean energy of the primary radiation responsible for the variation lies in the region 2 to 4×10^{10} eV. With this assumption, an anisotropy of the primary radiation with a direction lying near the plane of the ecliptic would produce a daily variation of nearly the same amplitude for the north and south directions but with a phase difference of about 2 hr as was indeed observed in 1948 and 1949 (Malmfors[2], Elliot and Dolbear[1]).

Brunberg and Dattner's data on trajectories show that at latitude 50° primary particles of energy 3×10^{10} eV, which have initial directions nearly parallel to the earth's magnetic axis, are deflected in the earth's field so as to arrive from the west at 45° to the vertical. Those with the same energy but with initial directions in the geomagnetic plane, arrive from the east at 45° to the vertical. Consequently, if we point a counter telescope in the east direction at 45° to the vertical it should record the daily variation due to anisotropy of the primaries plus any variation of atmospheric origin since, as the earth rotates, this telescope will scan a strip round the celestial sphere. A telescope pointing at 45° to the west, however, collects radiation from very nearly the same direction through the day and should therefore show only the atmospheric part of the variation.

Observations have been made over a period of one year in London using two counter telescopes arranged in this way and the results are described below.

2. EXPERIMENTAL ARRANGEMENT

Each counter telescope consisted of three trays of counters 60 × 60 cm in coincidence, the extreme trays being separated by 140 cm. The trays were mounted in metal frameworks so that the axes of the telescopes pointed east and west at 45° to the vertical. No absorber was used and the counting rate of each counter set was ~ 15,000 per hr. The apparatus was in operation

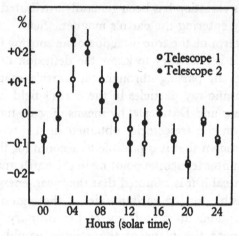

Fig. 1. Bi-hourly departures from the mean for the two telescopes averaged over the period of the measurements.

from May 1954 to April 1955 and during this period the two telescopes were interchanged from time to time in order to eliminate any systematic instrumental difference which might have influenced the daily variation measured by the two telescopes.

As a check on the performance of the equipment, the daily variation data have been added together for each of the two telescopes over the period during which the observations were made. Each telescope having spent the same length of time looking east and west, any instrumental difference would be revealed as a difference between the average daily variation, measured by the two counter sets. Fig. 1 shows the bi-hourly departures from the mean for each of the two telescopes. It can be seen

394

that there is no obvious systematic difference and this is confirmed by Fig. 2 in which the first harmonics for the two sets of data are plotted on a harmonic dial. The harmonic coefficients agree to within the statistical error and we therefore conclude that any systematic difference due to instrumental defects is so small that it can be neglected.

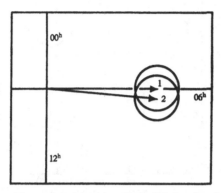

Fig. 2. Data from Fig. 1 plotted on a harmonic dial showing the absence of any systematic difference between the two telescopes.

3. THE DAILY VARIATION FOR THE EAST AND WEST DIRECTIONS

Fig. 3 shows the mean daily variation for the east and west directions for the period April 1954 to April 1955. The data have been corrected for the variation in barometric pressure using a coefficient of 2·7 % per cm Hg. This coefficient was deduced from the day-to-day changes in the rates of the two telescopes due to variations in pressure. The daily variation in barometric pressure in these latitudes is small and the correction does not greatly change the appearance of the curves. Fig. 4 shows the first and second harmonics of these curves plotted on harmonic dials which show that the amplitude of the 12-hr waves are not statistically significant. The amplitude of the 24-hr wave in the west direction is seen to be about three times as great as that for the east direction.

4. DISCUSSION

It is extremely difficult to reconcile this result with the view that the daily variation is produced by an anisotropy of the primary radiation existing at large distances from the earth since, as pointed out in section 2, such an

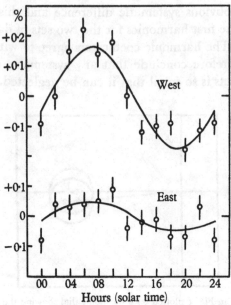

Fig. 3. The mean solar daily variations for the east and west directions after correction for barometric pressure, April 1954 to April 1955 inclusive.

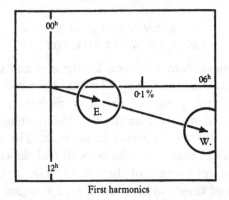

First harmonics

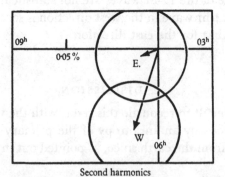

Second harmonics

Fig. 4. Harmonic dials showing the mean 24-hr and 12-hour waves for the east and west directions after correction for barometric pressure.

anisotropy would lead to a larger variation in the east direction than in the west. The basic assumptions involved in this argument are:

(a) that the average energy of the primaries responsible for the variation is in the range 2×10^{10} eV to 4×10^{10} eV, as deduced by Brunberg and Dattner[4] from the observed variation in the north and south directions in 1948, and

(b) that the direction of greatest anisotropy lies in or near the plane of the ecliptic.

The measurements in the north and south directions were made during 1948 and it is possible that the mean energy of the primary radiation responsible for the daily variation has changed since that time. It is certainly true that the amplitude of the variation has been decreased and that the phase has also changed (Thambyahpillai and Elliot[5], Sarabhai and Kane[6]). This could be interpreted as a decrease in average energy of the primaries producing the variation and if one supposes that the energy has decreased to a value of $1 \cdot 5 \times 10^{10}$ eV or less, the trajectories for primaries incident on the earth from either east or west have initial directions which lie in or near the equatorial plane. Under these circumstances both telescopes would be exploring the same strip of sky and should therefore show the same daily variation. It does not seem possible, however, even on this basis, to account for a larger variation from the west than from the east unless one supposes the mean energy to have decreased to a value well below 10^{10} eV when such particles are unable to reach the earth from an easterly direction because of the earth's shadow cone. It then becomes impossible to account for the existence of a daily variation in the equatorial region since the primaries responsible would be unable to reach the earth's equator from any direction.

Turning now to assumption (b), it is possible to envisage some direction of anisotropy which, lying at an angle of 70 or 80° to the plane of the ecliptic, might produce a larger variation on the west pointing telescope than the east. This again leads to difficulty, however, in accounting for the existence of an appreciable daily variation at the equator since the amplitude of the observed variation would be smallest at the equator and increase with increasing latitude. In fact the reverse applies (Elliot[7]).

In summarizing, we may conclude that it is extremely difficult to envisage a state of affairs which enables us to account for the observed variation in the east and west directions in terms of an anisotropy which exists at such a distance from the earth that the asymptotic directions of the primary particles are relevant.

Apart from these results for the east–west directions, there are other characteristics of the variation which are equally difficult to understand on this interpretation and these will now be briefly discussed under (a) and (b).

(a) It is known from comparison of the latitude variations (Fonger[8]) that the nucleonic component at sea-level arises from primaries of lower average energy than those which produce the bulk of the μ-mesons and electrons at sea-level. Because of this difference in primary energy, the deflexion in azimuth of the primaries which produce the nucleonic component must be greater than that for the primaries of the ionizing component. The sense of this deflexion is such that a given direction of anisotropy would produce a daily variation in the nucleon flux with an earlier phase than that for the ionizing component. Simultaneous measurements of the daily variation for the ionizing component and for the nucleonic component were made in Manchester during the period June 1952 to May 1954. During the two periods June 1952 to May 1953 and June 1953 to May 1954, the times of maximum for the nucleon variations were 1330^h and 1300^h respectively, compared with 1040^h and 0840^h for the ionizing component. During both these periods the phase of the daily variation for the ionizing component was in advance of that for the nucleons which is the contrary of what would be expected from consideration of the primary energies involved.

(b) During the period of the present measurements in the east and west directions, simultaneous measurements in the vertical direction in London revealed some remarkable changes in phase of the daily variation for vertical particles (Possener and Van Heerden[9]). During the period June to November 1954 the time of maximum intensity was 0300^h whereas from December 1954 to March 1955 it was 1000^h. No comparable change in phase was observed in either east or west directions and if this phase change represented a genuine change in direction of the anisotropy at this time, it is hardly conceivable that it should not, at the same time, have appeared in the east–west data.

The discussion above leads us to the conclusion that the interpretation of the cosmic ray daily variation as the result of a non-isotropic primary flux entering the earth's magnetic field may well be incorrect. Directional telescope measurements, however, show that the amplitude and phase of the daily variation depend on the direction of observation, so the variation cannot originate in the atmosphere. If these two statements are to be reconciled, it seems that the intensity modulation, which we observe as a solar daily variation, must take place in the earth's magnetic field.

5. CONCLUSION

The results of measurements in the east and west directions together with other known characteristics of the daily variation lead to the conclusion that the daily variation is not due to an anisotropic primary flux entering the earth's magnetic field but is most probably produced by modulation of the primary intensity within the region occupied by the field.

REFERENCES

[1] Elliot, H. and Dolbear, D. W. N. *J. Atmos. Terr. Phys.* **1**, 205, 1951.
[2] Malmfors, K. G. *Tellus*, **1**, 2, 1949.
[3] Brunberg, E. Å. and Dattner, A. *Tellus*, **5**, nos. 2 and 3, 1953.
[4] Brunberg, E. Å. and Dattner, A. *Tellus*, **6**, no. 1, 1954.
[5] Thambyapillai, T. and Elliot, H. *Nature, Lond.* **171**, 918, 1953.
[6] Sarabhai, V. and Kane, R. P. *Phys. Rev.* **90**, 204, 1953.
[7] Elliot, H. *Progress in Cosmic Ray Physics* (North Holland Publishing Company, 1952), p. 468.
[8] Fonger, W. H. *Phys. Rev.* **91**, 351, 1953.
[9] Possener, M. N. A. and Van Heerden, I. J. (in the press).

Discussion on Papers 39, 40 and 41

Singer: At the Mexico conference I suggested that the earth's magnetic field itself might be responsible for the anisotropy. It seems to me that when the earth moves in the interplanetary gas the magnetic field will be deformed by the streaming gas and therefore the field should become anisotropic in longitude. This could produce some anisotropy in cosmic radiation.

Further, as Sarabhai pointed out, the diurnal variation at Kodaikanal is different from other places such as Freiburg and Ahmedabad. Could this be accounted for by a different low energy cut-off? How does it fit in with Elliot's results?

Sarabhai: I do not think that the mean primary energy for the intensity measured at Kodaikanal differs adequately from the mean energy for the intensity at Trivandrum or Ahmedabad to explain the absence of the effect of Kodaikanal. We do not at the present moment see why Kodaikanal behaves differently from the other stations.

Alfvén: In reply to Elliot I should like to say that I cannot see why two different energies should have the same diurnal effect. The anisotropy of cosmic radiation may be a product of the influence of the interplanetary electric and magnetic fields on cosmic radiation but I think it is not in order to assume that low-energy particles and high-energy particles should react in the same way to this field. It depends very much on the radius of curvature, etc. Further, suppose that you have an interplanetary magnetic field somewhat like that given in Fig. 5a. In one case you point the telescope in the direction A and in another case you point it in the direction *B* and it is not necessary at all to believe that the amplitudes measured in these different directions should differ very much. I do not think that at present this is an argument against the assumption of an interplanetary field.

Dr Sarabhai has pointed out that there are maxima occurring at two different times and the conclusion from that was that it is not worth while to take the first harmonic of the variation. I think that although this is a very interesting point of view it does not at all reduce the importance of a first harmonic. If we go from the problem of interplanetary winds to that of terrestrial winds, for example, we may have today here in Sweden a wind from the north or the east.

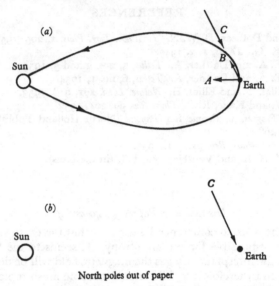

Fig. 5 (*a*). A model of the interplanetary magnetic field in a plane perpendicular to the equatorial plane. (*b*) The direction of the general dift of 10^{10} eV protons in the equatorial plane.

But if you take the average of the whole year you get a south-west wind and if you make the same analysis for Greenland you get a north-east wind. In the same way, even if you have changing directions of the anisotropy, the average means something and there could be drawn important conclusions from it.

Elliot: I want to make two objections to what Alfvén said. First, the picture (Fig. 5*a*) is slightly misleading because in the picture the lines of force are crossing the frame of the ecliptic at right-angles. This means that your lines of force are pointing to a fixed position on the celestial sphere. If one has, e.g. a flow of particles down the lines of force, then one could not see the solar daily variation but the sidereal daily variation.

The second point is that the arrow which points out into the plane of the ecliptic (*A* in Fig. 5*a*) would then indicate that Sarabhai near the equator would not see any daily variation at all.

Sarabhai: In my mind there is an important difference between the analogy of winds and cosmic radiation. I was trying to suggest that, in the case of anisotropy, there are perhaps two different preferred directions and there is no physical reality to an intermediate anisotropy. If you make a harmonic analysis, you will get all the intermediate values of the first harmonic depending on the frequency of occurrence of anisotropy in each of the two preferred directions. Therefore the physical reality to be attached to the movement of the first harmonic is not clear. I cannot say that the first harmonic has no physical reality, but you have also to take account of the second harmonic.

As to the work of Dr Elliot I think a rather crucial point is raised. However, one difficulty in experiments at northern latitudes is, as Brunberg and Dattner have shown, that the corresponding orbits are somewhat complicated. We cannot reconcile the results of east and west observations at the equator, unless we have an anisotropy which becomes negligible below a certain minimum energy.

Regarding the neutron observation that Dr Elliot reported I would like to have from him the following information. What is the relative amplitude in these two cases measured in London, and what is the aperture of the telescope?

Elliot: The relative amplitude was about two. The amplitude of nucleonic variations was about twice that of the ionizing component. A fairly wide aperture was used in the telescope.

Sarabhai: Then I do not think that the two experiments are quite comparable. As I pointed out before there exist components which are fairly well collimated.

Elliot: I do not think that it matters if you choose a wide or a narrow angle. If you choose a wide one the amplitude will be reduced but you will still see the same maximum in intensity. Brunberg and Dattner's curves are quite regular.

Sarabhai: I am not sure that you are right. Taking, e.g. the measuring telescopes pointing east and west, these will not point in the same direction in the sky and you may miss some of the collimated components by averaging over wide angles.

Gold: The enhanced diurnal variation periods earlier this year would seem to argue against any local modulation mechanism. There were groups of days with clear 27-day recurrence tendency, but no detectable relation with magnetic disturbances. For these days the variation appears to be actually due to an addition to the number of particles, not a symmetrical modulation.

Simpson: I wish to ask Dr Elliot a question for information on his excellent results. The 24-hour variations during 1954 show dramatic shifts in time of maximum from one month to the next. How was this taken into account in your analysis?

Elliot: It was taken into account. The point is that precisely the same days were taken for both types of observations, so even if one has shifts in these the relative shift would still be observed and so there should be no bias introduced.

Ehmert: When we first found these abnormal diurnal variations in the years 1950 and 1951 amplitudes of 3 % were observed. Often some consecutive days showed the effect with decreasing amplitude and a slight shifting of the phase from day to day. Often these days also showed a small general decrease of intensity. For large amplitudes the curves sometimes exhibited kinks that occurred in Germany as well as 8–10 hr earlier in Japan. This seems to point to an external influence.

Block: Dr Elliot said that it would be difficult to explain the times of maximum intensity of the cosmic ray components by an anisotropy of the primary cosmic radiation. Also if the magnetic field lines of Alfvén's interplanetary field make a right-angle with the equatorial plane no daily variation of the neutron component would result at the equator.

This is not so because the earth is not generally situated exactly in the equatorial plane as in Fig. 5a. I have recently made some calculations of cosmic ray orbits in this field. They show that 10^9 eV protons (giving the neutron component) move very nearly along the field lines, so there should be a maximum

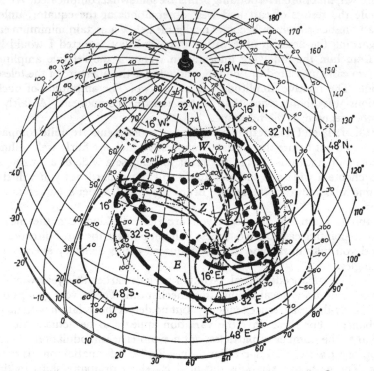

Fig. 6. The three surfaces laid out on the globe for Kiruna by Brunberg (*Tellus*, **8**, 224, 1956) refer to counter telescopes mounted in the directions 30° E., Z, and 30° W. with angular openings of approximately 22° in the east–west direction and 90° in the north–south direction.

at about noon also in the equatorial regions of the earth in fairly good agreement with the time given by Dr Elliot. On the other hand, 10^{10} eV protons show up a general drift in the equatorial plane in the direction indicated by the arrow C in Fig. 5b. This agrees also very well with the time of maximum for this component given by Dr Elliot.

Sandström: I should like to make some remarks on the direction of the primary particles registered by telescopes at a certain zenith angle. Preparations for an experiment of a similar nature to that of Dr Elliot's has induced me to look into this question. There is a marked difference between the influence of a change of energy of the primary particles on the registrations by a telescope directed to a certain zenith angle as compared to one directed to zenith. This is apparent from Brunberg's globes as shown by Fig. 6 which refers to a telescope arrangement in Kiruna. The central surface (dotted boundary) shows the directions in which primary particles arrive when the telescope is directed to the zenith. The surface bounded by a dashed line and indicated by E corresponds to a telescope directed to the east with zenith angle 30°, and the surface indicated by W to another telescope directed to the west with the same zenith angle. If the energy of the primary changes the zenith surface only widens and the average direction will remain the same. The two surfaces representing the directions of the primaries entering the telescopes making zenith angles of 30° will move outwards and sideways thus creating a difference in time correction (which can be mistaken for a component of daily variation of intensity). For the telescopes in the north–south directions this dependence on the energy of the primaries is small as compared to the east–west directions, as long as we keep to high and medium latitudes. Thus, measurements in the east–west directions are complicated. To take care of that difficulty measurements ought to be made at several latitudes with one direction of measurement common.

Incidentally a set of telescopes in Kiruna is now in the east–west direction for a short time. There was no difference at all between the amplitudes in the two directions (averaged over the year 1954).

Elliot: The proposal of Dr Sandström is certainly an excellent one.

SOLAR COSMIC RADIATION AND THE INTERSTELLAR MAGNETIC FIELD*

A. EHMERT

Max Planck Institute for the Physics of the Stratosphere, Weissenau, Germany

ABSTRACT

The increase of cosmic radiation on 23 February 1956 by solar radiation exhibited in the first minutes a high peak at European stations that were lying in direct impact zones for particles coming from a narrow angle near the sun, whilst other stations received no radiation for a further time of 10 minutes and more. An hour later all stations in intermediate and high latitudes recorded solar radiation in a distribution as would be expected if this radiation fell into the geomagnetic field in a fairly isotropic distribution. The intensity of the solar component decreased at this time at all stations according to the same hyperbolic law ($\sim t^{-2}$).

It is shown, that this decreasing law, as well as the increase of the impact zones on the earth, can be understood as the consequence of an interstellar magnetic field in which the particles were running and bent after their ejection from the sun.

Considering the bending in the earth's magnetic field, one can estimate the direction of this field from the times of the very beginning of the increase in Japan and at high latitudes. The lines of magnetic force come to the earth from a point with astronomical co-ordinates near 12·00, 30° N. This implies that within the low accuracy they have the direction of the galactic spiral arm in which we live. The field strength comes out to be about $0·7 \times 10^{-6}$ gauss. There is a close agreement with the field, that Fermi and Chandrasekhar have derived from Hiltner's measurements of the polarization of starlight and the strength of which they had estimated to the same order of magnitude.

I. OBSERVATIONS

A comparison of a number of neutron recordings on 23 February 1956 is given in Fig 1 with data by A. E. Sandström[1] for Stockholm, R. B. Brode and A. Goodwin[2] for Berkeley, P. Meyer and J. A. Simpson[3] for Chicago, C. D. Rose and J. Katzmann[4] for Ottawa, P. L. Marsden, J. W. Berry, P. Fieldhouse and J. G. Wilson[5] for Leeds, B. Meyer[6] for Göttingen,

---

* A paper with more details will be published in the *Zeitschrift für Naturforschung*.

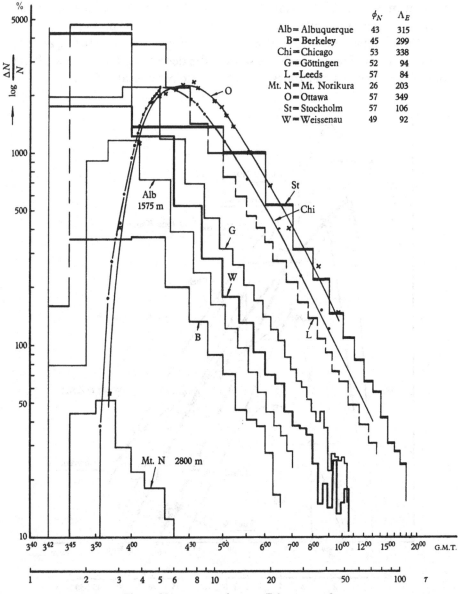

%

$\log \dfrac{\Delta N}{N}$

	ϕ_N	Λ_E
Alb = Albuquerque	43	315
B = Berkeley	45	299
Chi = Chicago	53	338
G = Göttingen	52	94
L = Leeds	57	84
Mt. N = Mt. Norikura	26	203
O = Ottawa	57	349
St = Stockholm	57	106
W = Weissenau	49	92

Fig. 1. Neutron records on 23 February 1956.

A. Ehmert and G. Pfotzer [7] for Weissenau, R. Brown [8] for Albuquerque,
Y. Sekido and C. Ishii [9] for Mount Norijura. A logarithmic scale is used
for the relative increase $\Delta N/N$ of neutron intensity as a function of the time
t which is counted from 3.32 G.M.T. The values of $\tau = t/8\cdot3$ min count this

405

time in manifolds of the time that light needs to travel from the sun to the earth. At later times all curves decrease very nearly in proportion to t^{-2}. The European stations with the earliest beginning of the increase (3·42 G.M.T. $\tau = 1·2$) are the first to reach the state of the decreasing law. Others, lying out of the direct impact zones reach this state considerably later.

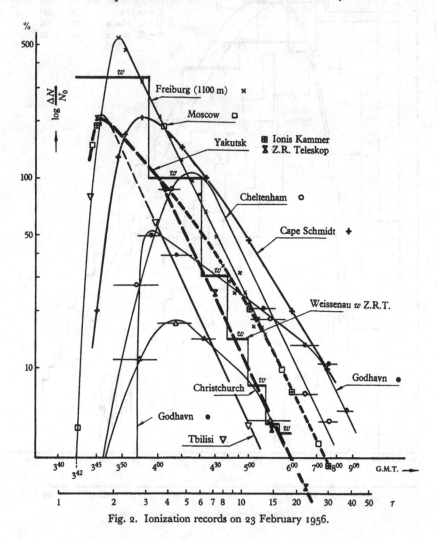

Fig. 2. Ionization records on 23 February 1956.

The same is demonstrated in Fig. 2 by the comparison of some records by ionization chambers with data given by A. Sittkus[10] for Freiburg and S. E. Forbush[11] for Christchurch, Godhavn and Cheltenham and by Vernov, Kopilov, Dorman and Shafer[12] for Moscow, Yakutsk and Tbilisi.

The same character of the curves holds also for earlier cases as is shown by Fig. 3 giving an adequate representation of Forbush's data [13].

We intend to verify, that the characteristic features of this material, i.e. the differences in the times of onset of the increases at different stations

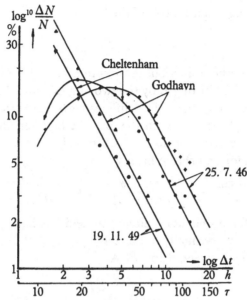

Fig. 3. Ionization records on 19 November 1949.

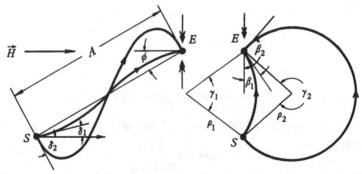

Fig. 4. The orbit conditions for particles moving from the sun S to the earth E in a magnetic field H forming an angle ϕ with SE.

and the general decreasing law, can be explained by the action of an interstellar field bending of the orbits of the solar particles.

Suppose, that at the time $t=0$ there are N_0 particles starting from the sun (denoted by S in Fig. 4) which are distributed isotropically in all directions and spiral along and around the magnetic lines of force. They all move in the relativistic region of energy with nearly the same velocity.

Without the magnetic field a receiver at the point E would in a short time record $N_0/4\pi A^2$ particles/cm^2 at the time $t_0 = A/\beta_1 c$.

With the field present at later times $t > t_0$ particles can arrive at the earth E having run along a spiral orbit of the length

$$L = \beta_1 ct; \quad \beta_1 = \frac{v}{c}, \tag{1}$$

They arrive from other directions than from that of the sun. We define the direction of the orbits hitting E by the two angles β and δ of the projection of the orbit as defined by Fig. 4. Two different orbits are drawn in this figure. We can also use β together with the angle α between the orbit and the magnetic field. α is constant throughout the orbit.

The condition for an orbit to hit E is

$$\frac{A \cos \phi}{\beta_1 c . \cos \alpha} = \frac{pc \sin \alpha . \delta}{\beta_1 c \sin \alpha . cH} . \arc \sin \frac{A \sin \phi \, eH}{2pc \sin \alpha}, \tag{2}$$

where the time for propagation along the field has been put equal to the time for propagation normally to the field.

A is the distance between S and E. As

$$\frac{A \cos \phi}{\beta_1 c \cos \alpha} = t \tag{3}$$

and using

$$\tau = \frac{t}{t_0} \tag{4}$$

we get

$$\tau = \frac{\cos \phi}{\cos \alpha}; \quad \sin \alpha = \left[1 - \left(\frac{\cos \phi}{\tau} \right)^2 \right]^{\frac{1}{2}}. \tag{5}$$

Furthermore, we write

$$\epsilon = \frac{pc}{p_1 c} \quad \text{with} \quad p_1 c = eHA, \tag{6}$$

where for high energies ϵ is a measure of the particle energy. Eq. (2) now gets the form

$$\frac{\tau}{2\epsilon} = \arc \sin \frac{\sin \phi}{2\epsilon \sin \alpha} \tag{7}$$

and

$$\sin \frac{\tau}{2\epsilon} = \left\{ \frac{\sin \phi}{2\epsilon \sqrt{\left[1 - \left(\frac{\cos \phi}{\tau} \right)^2 \right]}} \right\}_{\tau \gg \cos \phi} \longrightarrow \frac{\sin \phi}{2\epsilon}. \tag{8}$$

This is the impact condition connecting τ, ϵ and ϕ. Fig. 5 illustrates Eq. (8) for $\phi = 30°$. A part of the various types of orbits are drawn and numbered as well as the corresponding branches of $\epsilon(\tau)$.

From Fig. 4 it is immediately seen that

$$\beta = \tfrac{1}{2}\delta = \tfrac{1}{2} \text{ arc sin } \frac{A \sin \phi \, eH}{2pc \sin \alpha},$$ (9)

and with (6) and (7) $\qquad\qquad \beta = \tau/2\epsilon.$ (10)

Finally, we have

$$tg\delta = \cos \beta \, tg\alpha = \cos \left(\frac{\tau}{2\epsilon}\right) \cdot \left[\left(\frac{\tau}{\cos \phi}\right)^2 - 1\right]^{\frac{1}{2}},$$ (11)

and for $\tau \gg \cos \phi$ $\qquad\qquad tg\delta \to \tau \cdot \dfrac{\cos (\tau/2\epsilon)}{\cos \phi}.$ (12)

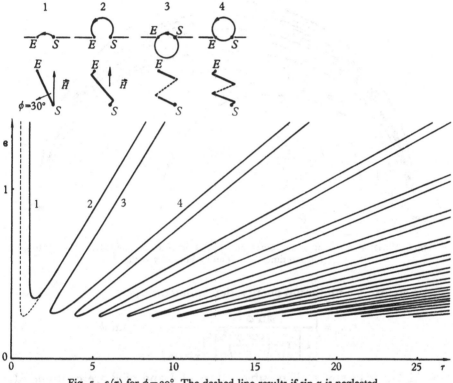

Fig. 5. $\epsilon(\tau)$ for $\phi = 30°$. The dashed line results if sin α is neglected.

Fig. 6 gives the apparent source in the sky for $\phi = 30°$. H denotes the direction of the field (the lines coming to the earth) and S the direction to the sun. The angular distance from the direction of H at the sky equals α in Eq. (2) and is connected with τ by Eq. (5). The later branches are not fully drawn in. Setting the beginning of ejection at 3.32 G.M.T., we have written at the first branch the times in G.M.T. as calculated from the values of τ.

For $(2n-1)\,\pi \leqslant \beta \leqslant 2n\pi$ (with $n=1, 2, 3, \ldots$) the values of β calculated from Eq. (10) must be subtracted by π to give the correct value. In this equation the curves are to be read in the reverse direction in these cases. Besides $\alpha(\tau)$ also $\beta(\tau)$ and $\delta(\tau)$ are given in Fig. 7, all for the case $\phi = 30°$. S denotes the direction to the sun. Particles with very high magnetic rigidity can arrive from here when $\tau = 1$. But in a short time complete deviation dominates and it is very striking, that particles with higher energy are restricted to the neighbourhood of $\delta = \pm 90°$, $\beta = 0°$ or $180°$. These directions normal to the field lines are marked in Fig. 4 by double-arrows.

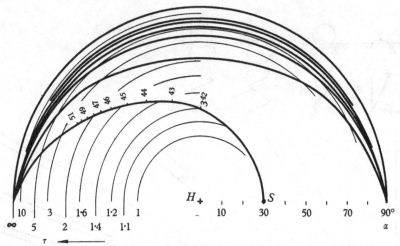

Fig. 6. Variation of the apparent source in the sky from S to ∞. The marks from 3^{42} to 3^{51} refer to time instants on 23 February 1956.

There exists a minimum energy, depending on ϕ;

$$\epsilon_{\min} = \left\{ \frac{\sin \phi}{2 \left[1 - \left(\frac{\cos \phi}{\tau} \right)^2 \right]^{\frac{1}{2}}} \right\}_{\tau \gg \cos \phi} \rightarrow \tfrac{1}{2} \sin \phi. \tag{13}$$

At higher values of τ all possible values of δ and of β are continuously allowed, especially if we assume that the emission lasts over perhaps 10 min. This broadens the lines in Fig. 7 in the direction of τ to overlapping bands. With growing τ the angle α approaches $90°$; this implies that particles finally come in the plane normal to $\mathbf{H}$ within a range of $180°$. This is sufficient to ensure an impact zone for all stations if a suitable direction of the field is assumed. For $\phi = 0$, that means that the field has the direction from the sun to the earth, and at high τ all directions in the plane normal to the field are allowed.

2. COMPARISONS

We do not know the direction of the field and try to find it from the following observations:

(1) Godhavn and Resolute have observed an impact. This is only possible if the direction of **H** is to the north of the sun.

(2) The first beginning of the increase varied with geomagnetic coordinates of the station. We use the stations with continuous reading. They are listed in Table 1.

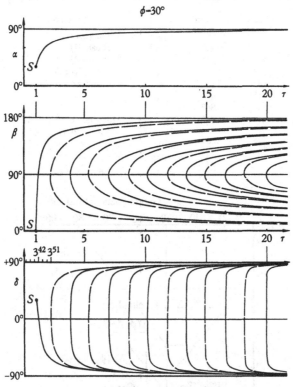

Fig. 7. $\alpha(\tau)$, $\beta(\tau)$ and $\delta(\tau)$ for $\phi = 30°$.

In Fig. 8 these times are plotted by full circles against the geomagnetic latitude. The dispersion of the points indicates an influence of the longitude. Assuming the same coefficient

$$K_{\tau,\phi} = \frac{\Delta t}{\Delta \phi} = \frac{1 \text{ min}}{4°} \tag{14}$$

for all geomagnetic longitudes as indicated by the lines in Fig. 8 we get the distances of these lines in good proportion to the differences of the

411

longitude of the respective stations with exception of Norikura, Yakutsk, and Swerdlowsk and perhaps C. Schmidt.

These stations had, according to this system, to begin before 3.42 G.M.T. They all began at the same time, when the first particles reached the earth. Only Norikura had special conditions and therefore had a later beginning.

The times of the other stations' beginning, reduced for 50° northern geomagnetic latitude with the coefficient of Eq. (14), fit well in a line of dependence on the geomagnetic longitude of these stations. With another

Table 1. *Geomagnetic stations with continuous reading*

	ϕ geom.	Λ geom.	G.M.T.	Observer	Method
Freiburg	49°	90°	3·42	Sittkus[10]	Ionization
Cheltenham	50°	350°	3·48	Forbush[11]	Ionization
Godhavn	79°	32°	3·53	Forbush[11]	Ionization
Moscow	52°	123°	3·42	Vernov, Kopilov	Ionization
Swerdlowsk	48°	141°	3·42	Dorman and Shafer[12]	Ionization
Yakutsk	51°	195°	3·42		Ionization
C. Schmidt	63°	180°	3·42		Ionization
Leeds	57°	84°	3·43–3·45	Marsden, Berry, Field-house and Wilson[5]	Ionization
Chicago	53°	338°	3·50	Meyer and Simpson[3]	Neutrons
Mt. Norikura	26°	208°	3·45	Sekido, Ishii and Migazaka[9]	Neutrons

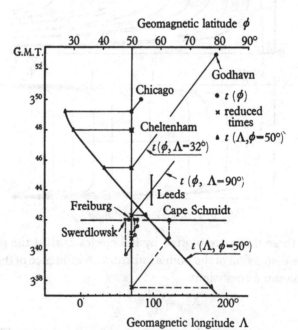

Fig. 8. The time of very first impact as a function of geomagnetic latitude Φ and the dependence of the reduced time on longitude.

coefficient they do not fit such a line. The longitudes of the stations are plotted as triangles against the times reduced for 50° latitude. The coefficient $K_{\tau,\Lambda}$ varies from 20°/min (near Freiburg) to 15°/min (near Cheltenham).

This system is scarcely to be explained by any reflexions of the solar particles. But it is well explained if we assume, that this retarding according to Eq. (14) is connected with the progressing of the first branch in Fig. 6 by interference of the impact zones through the geomagnetic field.

Using Firors representation [14] of Störmer's theory of the orbits we find from his Fig. 4 for 10 GeV/nucleon that a movement of the source by 7° to the north is followed by a movement of the impact zone by 14–30° to the west. The stations fitting the system in Fig. 8 are lying between 100 and 220° western geomagnetic longitude from the sun, C. Schmidt 10° to the west of the sun. This is a confirmation of our analysis in Fig. 8 and its connexion with Fig. 6.

A further help to find out the best fitting value of the direction of **H** is the beginning of the increase in Japan at 3.45 G.M.T. which is only possible by particles with energies > 10 GeV/nucleon and coming to the earth from a source direction with less than 20° northern geomagnetic latitude and a longitude greater than 30° in the east of Japan (Firor's Figs. 3 and 4 [14]). So the best fitting is found with the direction of **H** at this moment 15° to the west and 35° to the north of the sun with $\phi = 40°$. Godhavn is then lying near the upper corner of the bows in Fig. 6 whilst the Russian station C. Schmidt (63° N. and 180°) is in the first impact zone at 3.42 G.M.T.

The direction of **H** may be fixed by this procedure with an accuracy of 30°. The magnetic lines of force come to the earth from the constellation of 'Leo' and thus agree within the accuracy with the direction of the galactic spiral arm we are living in.

Such a magnetic field was assumed by Chandrasekhar and Fermi [15] to explain Hiltner's [16] observations of the polarization of the light from distant stars. They estimated the field strength from the dispersion of the polarization planes to $H = 7 \cdot 2 \times 10^{-6}$ gauss and with quite another method, based on the requirement of equilibrium of the spiral arm with respect to lateral expansion and contraction to $H = 6 \times 10^{-6}$ gauss. The positive or negative direction cannot be distinguished from polarization measurements.

We find the field strength by regarding that, according to our model, the first impact in Japan must be done by particles of the order of 10^{10} eV/nucleon, and that according to Fig. 5 for this time $\epsilon = 0 \cdot 3$. From this we find from Eq. (6)

$$p_1 c \geqslant 3 \cdot 3 \times 10^{10} \text{ eV} \quad \text{and} \quad H = \frac{3 \cdot 3 \times 10^{10}}{300 \times 1 \cdot 5 \times 10^{13}} = 7 \cdot 3 \times 10^{-6} \text{ gauss}, \quad (15)$$

in close agreement with the values by Chandrasekhar and Fermi[15]. The given model explains the retarding effects of the first beginning at all stations. The special curve for Godhavn in Fig. 2 results from the rotating of Godhavn by 3 hr into the best position (as the North American stations do within an hour) and the slow accumulation of the higher energies in this zone according to the branches in Fig. 6. The high spread of the apparent source over a high region of latitude ensures for a long time the staying of all stations in impact zones. Only Mount Norikura leaves it earlier, but at that time the radiation impact in Japan was over.

Furthermore the model explains the high latitude effect of the solar radiation in the end-phase and the narrow energy spectrum, that Pfotzer[17] derived from this latitude effect assuming isotropic radiation outside the earth's field. Fig. 5 demonstrates the suppression of high energies for higher τ in proportion to E^{-2} by the selection in the field.

The hyperbolic decreasing law for the intensities at all stations is another simple consequence of the postulated field.

Particles of homogeneous velocity $(\beta_1 c)$ coming from a narrow source with undeflected propagation reach a surface normal to the direction of propagation at the time t after ejection with a density in proportion to t^{-2}. Particles moving in the direction of a magnetic field are (in large scale) held together. The intensity is in proportion to t°. Particles moving nearly normal to a magnetic field spread only in the direction of the field and not in the direction normal to velocity and field. They arrive after a time t with a density in proportion to t^{-1}. Such particles we have to consider at higher τ, as $\cos\alpha$ then gets very small.

At every time t particles arrive along a path the length of which corresponds to only this time and their front density is

$$N(t) = N_1/t \text{ particles cm}^{-2}.$$

Our curves in Figs. 1, 2 and 3 give the number of particles per second

$$n(t) = \frac{dN(t)}{dt} = \frac{N_1}{2t^2}, \tag{16}$$

whatever the special orbits are.

This law can only be understood, if there is a short time of ejection compared with t itself, if the angular space from which particles fall into the apparatus is constant and if the particle energy is constant. These conditions are fulfilled by our model itself from the moment when the impact reaches the full aperture of the apparatus. And this is depending on the geomagnetic situation.

In Fig. 9 the ratio $$V = N_{measured}/(N_1/t^2)$$

between the measured intensity at time t and the final intensity extrapolated backwards is drawn for some cases given in Figs. 2 and 3. The first beginning of impact is a point on these curves, as far as it is known. The curve for the neutron measurements in Ottawa indicates two components.

Fig. 10 shows on a map in geometrical co-ordinates (and Mercator's projection) the transformed Fig. 6 for the situations on 23 February 1956, 3.42, 19 November 1949, 11.00, 25 July 1956, 16.00 and 28 February 1942, 11.10. These situations are suitable to explain the quite different behaviour of solar cosmic radiation at these occasions, which are also best

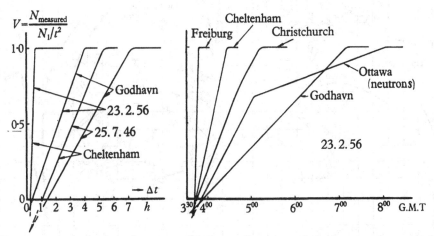

Fig. 9. The ratio between the measured increase, $N_{measured}$, and the increase extrapolated backwards from the decreasing law of the end-phase. The ratio is plotted as a function of time.

seen in Figs. 3 and 9. Especially, the much longer times between the beginning of the flare and the radiation impact on the earth at these occasions proves to be a consequence of the model, caused by the other value and orientation of ϕ. Five hours after the flare Cheltenham was lying in the main impact zone for the branch drawn in Fig. 10 for 16.00 G.M.T., just at the left-hand side, whilst Europe was leaving it. That is in good agreement with observation. For a field in the direction from the sun, America had been in the first impact zone of the eastern branch one hour after the flare. For the galactic field this eastern branch does not exist.

This is a confirmation of the galactic origin of the acting field. Only for 19 November 1949 the impact in Godhavn is no question for a field from the direction of the sun, whilst for the assumed interstellar field there is a difference of the order of 30°.

The decreasing law itself might also be explained by solar fields of the type discussed by Alfvén [18]. They might correspond to our case with $\phi = 0$. In this case, in a figure corresponding to our Fig. 6 the lines, at every time, are full circles round the centre $H \equiv S$; the angular radius being given by $\cos \alpha = 1/\tau$. This model gives a high gain in arriving particles, but it does not explain the long running times of the particles for 25 July 1956. The running of the first impact to the north is somewhat faster than in the case of Fig. 6 but stops before reaching Godhavn. This is because the direction of the sun is now in the centre from which α is to be measured.

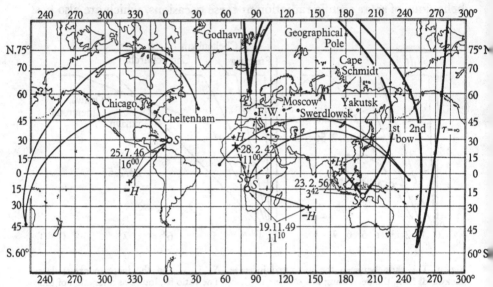

Fig. 10. A transformation of Fig. 6 on a map in geophysical co-ordinates and Mercator's projection. The apparent source outside the geomagnetic field moves along the zenith of the heavy lines. The field estimated from the intensities on 23 February 1956 is assumed to be the same on the other dates. H denotes the direction of the field, S that to the sun at the beginning of the flares. For higher τ-values the lines move according to the earth's rotation.

In this case the running velocity of the first impact towards western longitudes is also slightly greater, but the difference is too small to give a discrimination. But in Japan the first impact ought to be at $\tau = 2 \cdot 0$, i.e. at least 8 min after the first impact in Europe and this rules out this direction of the field.

3. CONCLUSIONS

From the given comparisons of theory and observations we conclude that the interstellar field exists. A severe consequence of this field is that the earth can never be reached by solar particles of energies less than

7×10^9 eV/nucleon (see note added in proof). This value holds for the best positions in spring and autumn. In summer and winter double the energy is needed. But at these times flares on the other side of the sun might also be effective on the earth. The midnight effects which we formerly found[19] fit in this image.

This field is able to make cosmic radiation fully isotropic to very high energies if it exists with a nearly unique direction along the whole width of our spiral arm. The question now arises how do the solar plasma clouds move in this field? Apparently they are not deflected and we think that the very low-energy electrons in these clouds prevent the interstellar field from entering the cloud of a neutral plasma. For this a current sheet must be set up around the border of the cloud in the direction normal to the field. As the field is constant at every place passed by the cloud, there results no further deviation of the velocity vector of the cloud.

For further investigations on this field the dense network of stations and an exact determination of the times of beginning increase is of greatest importance.

The author thanks the *Deutsche Forschungsgemeinschaft* for financial help.

Note added in proof

When this paper was finished for the Stockholm Conference we had no knowledge of the papers of Meyer, Parker and Simpson[20] and of Winckler[21] that give evidence of the impact of low energy primary particles with energies between 1 and 1·5 GeV from balloon measurements in the afternoon[20] and in the evening[21] of 23 February 1956. This is in contradiction to our statement, that energies below 7·5 GeV could not reach the earth from the sun through a magnetic field as evaluated in our paper.

The high-latitude effect between Berkeley and Ottawa or Stockholm indicates furthermore the impact of particles with smaller energy though this might also be understood from the selection of other orbits in higher latitudes. It is essential that the primaries do not arrive in isotropical distribution in our model. This anisotropy was distinctly revealed by measurements with inclined counter-telescopes by Sandström[22] and by Trefall and Trumpy[23]. In connexion with this problem reference should also be given to a recent investigation by Brunberg[24].

Low-energy particles from the sun are possible in our model if ϕ gets small, this implies that the lines of magnetic force are parallel to the line connecting the sun and the earth. We discussed some objections against this case. If an interstellar field has this direction at the end of February and maintains this direction respective to fixed stars, the special curves in Fig. 3

might be understood. The field strength of such a field with $\phi = 0$ comes out to have nearly double the field strength of that discussed in the paper with $\phi = 30°$. Another possibility, we must account for, is that the low-energy particles measured in the stratosphere do not come from the sun but are generated nearer to the earth.

REFERENCES

[1] Sandström, A. E. *Tellus*, **8**, 279, 1956.
[2] Brode, R. B. and Goodwin, A. Jr. *Phys. Rev.* **103**, 377, 1956.
[3] Meyer, P., Parker, E. N. and Simpson, J. A.* *Phys. Rev.* **104**, 768, 1956. *Cosmic radiation studies*, Enrico Fermi Institute for Nuclear Studies (Chicago, 1956).
[4] Rose, C. D. and Katzmann, J.* (in the press).
[5] Marsden, P. L., Berry, J. W., Fieldhouse, P. and Wilson, J. G.* (in the press).
[6] Meyer, B. *Z. Naturf.* **11**a, 326, 1956.
[7] Ehmert, A. and Pfotzer, G. *Z. Naturf.* **11**a, 322, 1956.
[8] Brown, R. *J. Atmos. Terr. Phys.* **8**, 278, 1956.
[9] Sekido, Y., Ishii, C.* and Migazaki, Y.* (in the press).
[10] Sittkus, A., Kühn, W. and Andrich, E.* *Z. Naturf.* **11**a, 325, 1956.
[11] Forbush, S. E.* (in the press).
[12] Vernov, S. N., Kopilov, Yu. M., Dorman, L. I. and Shafer, Yu. G.* (in the press).
[13] Forbush, S. E. *Phys. Rev.* **70**, 771–2, 1946.
[14] Firor, I. *Phys. Rev.* **94**, 1017, 1954.
[15] Chandrasekhar, S. and Fermi, E. *Astrophys. J.* **118**, 113 and 116, 1953.
[16] Hiltner, W. A. *Astrophys. J.* **109**, 471, 1949; **114**, 241, 1951.
[17] Pfotzer, G. *Mitteilungen aus dem Max-Planck-Institut für Physik der Stratosphäre*, no. 7, 1956.
[18] Alfvén, H. *Tellus*, **6**, 232, 1954.
[19] Ehmert, A. *Z. Naturf.* **3**a, 264–85, 1948.
[20] Meyer, P., Parker, E. N. and Simpson, J. A. 'The solar cosmic rays of February 1956 and their propagation through interplanetary space', *Cosmic radiation studies*, Enrico Fermi Institute for Nuclear Studies, Chicago, 1956.
[21] Winckler, J. R. *Phys. Rev.* **104**, 220, 1956.
[22] Sandström, A. E. Collection by Elliot and Gold.
[23] Trefall, H. and Trumpy, B. Collection by Elliot and Gold.
[24] Brunberg, E. Å. This symposium, Paper 40.

Discussion

Eckhartt: As far as I understand the time 3^{42} seems to have a particular meaning in your theory. No onset should be measured before this time at any of the stations. How is this reconcilable with the fact that Hobart measured an earlier onset, 3^{39} as far as I can remember?

Ehmert: The time 3^{42} corresponding to $\tau \approx 1.20$ allows a first impact at $3^{40,3}$ for undeflected particles. Dr Fenton indicates the onset time at Hobart at $3^{41 \pm 2}$ G.M.T.

van de Hulst: Were you able to infer the direction of the magnetic field in the spiral arm from the observations by means of your calculations?

* My thanks are due for a kind communication in exchange of data before publication.

Ehmert: The direction of the field is obtained from fitting the times of the first increase of solar cosmic rays in Japan and in Godhavn. It was not implied.

Alfvén: Would not a solar flare occurring three months before or after an onset exhibit quite different properties?

Ehmert: Yes, if the field is not perpendicular to the ecliptic. With the evaluated direction the particles had to run, at 25 July in 1946, in a direction nearly perpendicular to the field.

Alfvén: Would not the interplanetary magnetic field give about the same result?

Ehmert: Yes, but the difference taken in the direction north–south is not quite the same in our case. The result depends upon the magnetic field direction.

Gold: I like this basic idea very much. But I also think that the spiral arm magnetic field can hardly be expected to preserve its direction in the solar system with the sweeping action of the solar activity which we know to move the intervening gas and hence the field. But that hardly detracts from the attractive theory; it only makes this agreement of direction a little fortuitous.

Ehmert: Yes. It is quite astonishing that a systematic effect due to a rather well-defined magnetic field could last for several days; actually the measurements seem to indicate this.

Gold: Further, impact zones appear to have been absent after about one hour, and one would like to know whether adequate 'washing out' results from the arrival of particles not with isotropy, but from a plane containing the earth. This would be the situation in the presence of a homogenous field and for late particles.

Ehmert: A look on Firor's results giving the connexion between the geomagnetic latitude of the source and the western longitude of the station shows that in the final state there are sufficient impact possibilities. Humps in the intensity-time curves may be possible.

PAPER 43

ON THE VARIATIONS OF THE PRIMARY COSMIC RAY INTENSITY*

E. N. PARKER

Enrico Fermi Institute for Nuclear Studies,
University of Chicago, U.S.A.

ABSTRACT

To construct a model for producing the observed variation in the cosmic ray intensity we consider primarily the Forbush decrease and the general decrease of the cosmic ray intensity during years of solar activity. These are larger variations than the diurnal and 27-day variations and require more drastic assumptions; thus they will better serve to establish a unique model.

It is assumed that the sun does not emit cosmic ray particles except during the time of a solar flare. Thus, decreases in the cosmic ray intensity are to be interpreted as a solar effect which inhibits the arrival of galactic cosmic ray particles at earth. Since the intensity of low rigidity primary cosmic ray particles is observed to vary more than the intensity at higher rigidities, the inhibition has generally been assumed to be caused by magnetic fields.

The necessary depression of the cosmic ray intensity requires both a barrier, to impede their arrival, and a removal mechanism within the barrier, to prevent eventual statistical equilibrium (with uniform particle density). Quantitative development indicates that a heliocentric magnetic dipole, a heliocentric cavity in the galactic field (Davis, *Phys. Rev.* **100**, 1440, 1955), and a heliocentric interplanetary cloud barrier (Morrison, *Phys. Rev.* **101**, 1397, 1956) all encounter serious difficulties in explaining the observed effects, one reason being the ineffective removal that is available.

It is shown that a geocentric magnetic cloud barrier does not encounter these difficulties: it is proposed that during the years of solar activity the terrestrial gravitational field captures magnetic gas of solar origin from interplanetary space, which is then supported by the geomagnetic field; the removal by absorption by the earth is sufficiently effective that only a relatively thin barrier need be maintained; the occasional capture of new magnetic material accounts for the abrupt onset of the Forbush decreases, and the slow decay (0·5 years) of the captured fields for the smooth variation of the mean cosmic ray intensity with the sunspot cycle.

* Assisted in part by the Office of Scientific Research and the Geophysics Research Directorate, Air Force Cambridge Research Center, Air Research and Development Command, U.S. Air Force.

This paper is a summary of the results of a number of formal calculations [1, 2, 3] of the propagation of cosmic ray particles through interplanetary space, and represents a critical analysis of the functioning of the general classes of models that have been proposed to account for the variations in the primary cosmic ray intensity observed at the earth. By focusing our attention on the more extreme of the variations [4], we will be able to eliminate many hypothetical models and arrive within fairly narrow limits at a situation which seems to account in a natural way for the observations.

The largest observed fluctuation in the cosmic ray intensity is the appearance and disappearance of the low energy cut-off with the sunspot cycle. During periods of sunspot activity the energy spectrum of the primary cosmic ray particles drops off rapidly at energies below 1 or 2 GeV/nucleon; to form what is known as the *low energy cut-off*. As solar activity declines during the approach of a sunspot minimum, immense quantities of low energy primary particles gradually appear, to entirely obliterate the cut-off [5, 6] and noticeably increasing the number of particles at all energies up to 30 GeV or more; above 30 GeV the percentage increase is so small as to be unobservable. Isotropy obtains at all times. During the return of solar activity following the minimum the low energy particles disappear bit by bit at irregular intervals of time and after a few years the total number of incoming cosmic ray particles has decreased to the pre-minimum value, exhibiting the low energy cut-off.

The most abrupt fluctuation in the cosmic ray intensity is the Forbush decrease, where the world-wide primary cosmic ray intensity may decrease by as much as 10 % in as little time as 5 or 10 hr and remain low for days or months. Again the variation is largest at low energies and represents a variation in the total number of particles rather than a change in the energy of the individual particles. Only small deviations from isotropy are observed during the onset of the decrease; complete isotropy prevails following the onset.

It is difficult to understand how the above variations can be the result of emission of cosmic ray particles by the sun, and it is generally assumed that they are the result of depression of the general galactic cosmic ray field by processes of solar motivation within the planetary system. The observation that the variations are greatest for particles with low magnetic rigidity and vanishingly small at high rigidities, and the observation that the variations are a result of a change in the number of particles, rather than in particle energies, lead us to the conclusion that the variations result from magnetic deflexion of the particles by interplanetary fields.

The steady form and world-wide character of the depression of the cosmic ray intensity during times of solar activity implies that the deflexion is a statistical process and is not produced by one or two individual regular magnetic fields. Presumably, therefore, the diffusion equation represents a rough approximation to the propagation of the general cosmic ray density through space[7]. We let $j(E, \mathbf{r}, t)\, dE$ represent the number of particles/sec/cm²/steradian with energies in the interval $(E, E+dE)$ at the position $\mathbf{r}$ and time t. We regard the irregular interplanetary magnetic fields as a diffusing medium with coefficient of diffusion $\kappa(E)$ and general velocity field $\mathbf{v}(\mathbf{r})$. Then

$$\frac{\partial j(E, \mathbf{r}, t)}{\partial t} = -\nabla \cdot [j(E, \mathbf{r}, t)\ \mathbf{v}(\mathbf{r})] + \kappa(E)\ \nabla^2 j(E, \mathbf{r}, t). \qquad (1)$$

From elementary kinetic theory the diffusion coefficient κ is equal to $\tfrac{1}{3}wL$ for particles with velocity w and mean free path L. We define the scale $l(\mathbf{r})$ as the mean distance over which the interplanetary magnetic field does not change sign; we let B represent the mean value of the field density over a region of scale $l(\mathbf{r})$. It can be shown[2] that

$$L \cong L_0 \left\{ 1 + \left[\frac{\pi M w c}{2lBq(1 - w^2/c^2)^{\frac{1}{2}}} \right]^2 \right\},$$

where L_0 is the mean free path for passage between regions of field, and M and q are the particle mass and charge.

The formal analysis of the cosmic ray intensity throughout interplanetary space, justifying the exclusive use of (1), has been given elsewhere[1, 2, 3]. The quantitative results may be summarized by the following considerations:

(a) Formal solution of the equations of motion of a charged particle moving in general hydromagnetic fields, varying slowly over space and time as compared to the radius of curvature of the particle trajectory and the Larmor frequency, or abruptly as in a shock wave, show[3] that a particle will experience no increase in its kinetic energy except by the betatron effect[8] and by Fermi's mechanism[9, 10, 11]; both these mechanisms are estimated to be negligible in interplanetary space, in agreement with the observed fact that the cosmic ray intensity variations do not involve changes in the individual particle energy.

(b) The solution of the equations of motion in slowly varying hydromagnetic fields shows that particles can be neither excluded from the solar system nor stored within the solar system by large-scale regular fields, such as a heliocentric magnetic dipole or a heliocentric cavity in the general galactic field [9] unless the large-scale field has very nearly mathematically perfect symmetry and regularity [2]. To significantly trap or exclude particles the field density must not deviate from perfect symmetry by more than one part in 10,000. We believe that the observed solar activity with the associated magnetic and/or material emission from the sun would not allow such regular large-scale fields to occur.

On the basis of (a) and (b) we conclude that in interplanetary space the diffusion equation (1) represents the complete influence of the sun on the cosmic ray particles of galactic origin; the sun is responsible for the production and motion of the interplanetary magnetic fields, represented by $\kappa(E)$ and v.

In order to lower the cosmic ray intensity at earth for the long years of solar activity we must, of course, postulate a tangle of interplanetary fields to impede the arrival of galactic particles. However, unless we can soon remove the particles which manage to diffuse through the tangled interplanetary barrier, then, no matter how dense the barrier, an equilibrium state will soon be achieved and $j(E, \mathbf{r}, t)$ will be uniform throughout interplanetary space with just the cosmic ray intensity found in the interstellar space outside. Therefore, if we wish to depress the cosmic ray intensity for long periods, we must have a removal mechanism inside the interplanetary field barrier to complement the functioning of the barrier; the more effective the removal mechanism, the less dense need be the barrier, etc.

Now the sun is the major absorber of cosmic ray particles in the solar system; the planets and the interplanetary densities of 10^3 atoms/cm^3 are negligible. The sun will absorb a particle confined within the orbit of earth in about 1·5 years. The interplanetary barrier, associated with this solar removal, of sufficient density to produce the observed depression of the intensity would involve closely packed tangled fields of about $0·5 \times 10^{-2}$ gauss surrounding the entire inner solar system. The origin of such a dense interplanetary field is difficult to understand. If it were present between the sun and the earth, we would not expect to see the burst of cosmic ray particles that is observed to accompany some solar flares; we would not expect the almost daily arrival of auroral particles. If the field were present outside the orbit of earth we could not explain the rapid decay of the enhanced cosmic ray intensity following a solar flare; the decay suggests [1] fields of only 10^{-5} gauss. Hence, we do not regard an interplanetary cloud barrier of $0·5 \times 10^{-2}$ gauss as likely.

If we wish to use an interplanetary magnetic barrier more diffuse than $0·5 \times 10^{-2}$ gauss, then we must have a more effective removal mechanism than solar absorption. If we assume that the interplanetary magnetic fields have been ejected from the sun with velocities of the order of 2000 km/sec, then the cosmic ray particles within the orbit of earth will be swept out once each day instead of once each 1·5 years. The field densities need be only 10^{-5} gauss. However, the fields must be ejected more or less isotropically from the sun (even small leaks in the outward rushing cloud barrier nullify the effect); hence we would not expect to be able to see the sharp rise and the terrestrial impact zones [12] of the cosmic ray bursts from solar flares which requires that $B \lesssim 10^{-6}$ gauss inside the orbit of earth [1]. With outward rushing clouds we would expect the general depression of the cosmic ray intensity at earth to depend critically on the day-by-day activity on the observable face of the sun. Hence, we do not believe that there exists such an outward rushing (2000 km/sec) interplanetary cloud barrier of 10^{-5} gauss.

Let us turn our attention from the general depression of the cosmic ray background during years of solar activity to the transient Forbush decrease. The most striking feature of the Forbush decrease is the 5 or 10 % drop in the intensity (as seen in neutron detectors) occurring in as little time as 5 hr. Following such a drop the intensity may level off and remain low for days. Such an abrupt drop implies interplanetary magnetic clouds carrying fields of $0·5 \times 10^{-2}$ gauss, and traveling past earth at 2000 km/sec. Unfortunately we cannot easily reconcile the abrupt drop with the immediate levelling off of the intensity. But even if we overlook these difficulties and use, as originally proposed by Morrison, the somewhat more diffuse field of 10^{-3} gauss, which can produce a decrease only over about 20 hr, we cannot explain how it was possible to observe the abrupt onset of the solar flare of 23 February 1956 while in the minimum of a Forbush decrease: earth was supposedly in the middle of a large magnetic cloud of 10^{-3} gauss; the abrupt onset and terrestrial impact zones required that $B \lesssim 10^{-6}$ gauss.

We wish to suggest on the basis of the above failures of heliocentric models involving tangled interplanetary fields that the observed depressions in the cosmic ray intensity are not heliocentric in origin and do not occur throughout interplanetary space. The most obvious alternative is, of course, that the variations are geocentric in origin and occur only locally about our planet.

We point out that if earth were surrounded by a diffuse cloud of tangled magnetic field ($\sim 3 \times 10^{-2}$ gauss, internal scale of 250 km, and material density 5×10^6 atoms/cm³ or less) then we would observe about a 50 % reduction in the intensity of 2 GeV primaries, with less reduction at higher energies and more at lower energies. The tangled field of such a geocentric cloud should extend out to a distance of several earth's radii. Because the solid bulk of earth absorbs a large fraction of the particles penetrating such a geocentric barrier, the necessary barrier is relatively diffuse and will not obliterate the observed terrestrial impact zones for particles of solar origin. Nor would such a small barrier delay or smooth out the abrupt onset of solar flare particles.

Let us suppose, therefore, that the terrestrial gravitational field occasionally captures passing interplanetary magnetic gas. We suggest that during the gradual onset of solar activity following a sunspot minimum the earth captures and builds a tangled magnetic cloud around itself; since earth is never far from the equatorial plane of the sun, the sun need only eject magnetic matter near its equatorial plane to produce such a cloud. The decay time for the captured magnetic fields is of the order of 0·5 years. Hence, freshly ejected matter need be captured by earth only every month or so to maintain a more or less steady depression of the observed cosmic ray intensity. The geocentric magnetic cloud will gradually disappear when solar activity declines at sunspot minimum. Quantitative calculation [2] shows that the observed depression in the abundance of cosmic ray particles at all energies is easily explained by the accumulation of such a cloud.

If we assume that the capture of passing interplanetary cloud material is not always a continuous process, but that occasionally a relatively large amount may be accumulated by the terrestrial gravitational field all at once, then we can readily account for the Forbush decrease with its abrupt onset and levelling off for long periods following the initial decrease. It is an observed fact that the depression of the cosmic ray intensity during a Forbush decrease initially is not uniform over the earth, but gradually becomes so. The initial non-uniformity is expected from the probable condition that the capture of the new magnetic material is not uniform around the earth; then following capture the material gradually spreads out, arranging itself in a smoother and more or less equilibrium state. Only a local geocentric model can account for the observed non-uniformity.

Now consider the limitations of the calculations on which the geocentric model is based. Given a particular statistical distribution of tangled magnetic fields around earth it is not difficult to calculate the resulting reduction in the cosmic ray intensity; the above description of the expected cosmic ray effects is based on such calculations. However, the dynamics of the capture and formation of a geocentric magnetic cloud form a complex mathematical problem which is beyond our present means to handle in

425

a general way. We can show that the weight of such a cloud is so small that it is easily supported by the geomagnetic field without significant magnetic effects occurring at the surface of earth, but we can do little more that is not merely speculation. Therefore, we would like very much to obtain an indication of the presence and structure of the geocentric magnetic cloud which is independent of the cosmic ray observations. Unfortunately with the complex thermodynamic structure of the solar atmosphere, the immense quantities of interplanetary hydrogen, and the dubious thermodynamic state of the geocentric gas, one is led to the conclusion that even such obvious measures as rocket observations at high resolution of the solar L_α line may not yield unambiguous results.

REFERENCES

[1] Meyer, P., Parker, E. N. and Simpson, J. A. *Phys. Rev.* (in the press).
[2] Parker, E. N. *Phys. Rev.* **103**, 1518, 1956.
[3] Parker, E. N. *Phys. Rev.* (in the press).
[4] Simpson, J. A., Paper 38 of this volume.
[5] Meyer, P. and Simpson, J. A. *Phys. Rev.* **99**, 1517, 1955.
[6] Neher, H. V. *Phys. Rev.* **103**, 228, 1956.
[7] Morrison, P. *Phys. Rev.* **101**, 1397, 1956.
[8] Swann, W. F. G. *Phys. Rev.* **43**, 217, 1933.
[9] Davis, L. *Phys. Rev.* **101**, 351, 1956.
[10] Fan, C. Y. *Phys. Rev.* **82**, 211, 1951; **101**, 314, 1956.
[11] Fermi, E. *Phys. Rev.* **75**, 1169, 1949; *Astrophys. J.* **119**, 1, 1954.
[12] Firor, J. *Phys. Rev.* **94**, 1017, 1954.

Discussion

Singer: I agree that the mechanism that Morrison proposed probably does not work, because it is a transient mechanism and it requires a high field. But I do not think that your mechanism will work either; rather particles have to be decelerated by an electric field. You have stated an objection to the electric field mechanism which would lead to a (non-observed) anisotropy. This applies to the picture of a polarized beam; my own view is that particles are decelerated by expanding turbulence set up by beams or clouds. This cloud, when coming from the sun, must expand and give an inverse Fermi effect (to be published in *Phys. Rev.*). The electric field effects are very efficient due to the Liouville factor: $i \propto D p^2 \beta c$, where i is the directional intensity in flux, D the density in phase space, p the momentum of the particle and βc, which has been put in here for the sake of completeness, is usually equal to one. The reason for putting β in here is that I want to explain the production of a knee. In this deceleration mechanism, which I have in mind, when the energy loss is such as to make a particle non-relativistic, we get $\beta c < 1$. Since in a turbulent gas the gas density and the magnetic field are coupled, the ionization loss now becomes important and this is a removal mechanism which I think is most effective for low-energy particles forming a trap and a knee.

Parker: We have looked into both the inverse betatron effect and ionization loss and concluded that they were not sufficient to produce it.

Singer: Let me make a remark about measurements. According to your view the cosmic ray intensity during a Forbush decrease should be low in the top of the atmosphere and should rise when you get out several earth radii. According to my point of view the cosmic ray intensity would be low until you get out of the solar system. Further: according to your view there should be no shift in the position of low energy cut-off, whereas I should find a northward shift in the knee. Concerning these different points of view I will say that one might at the moment be able to decide this by measurements near the poles.

Can you hold your cloud also near the earth at the magnetic pole so that it completely surrounds the earth?

Parker: It should completely surround the earth and perhaps be slightly thinner at the poles because the earth's field is denser there.

Ferraro: I would like to ask you about the leading ideas of the size of that cloud with a magnetic field of about 10^{-2} gauss.

Parker: This is a tangled field and the scale of the inhomogeneity which we estimated from cosmic ray intensities was about 300 km. The cloud would be of the order of 2–3 earth radii thick.

Ferraro: What is the inner boundary?

Parker: For the inner boundary I can only give you a lower limit of half an earth's radius above the earth's surface. It may be more than that.

Ferraro: But in that case if you get variations in the magnetic field would you not expect to observe this at the earth?

Parker: We have tried to estimate the effects which will be produced by the magnetic field we have assumed and we find that there are two competing effects. The orders of magnitude are difficult to estimate; I am sorry I cannot give you a definite reply to your question.

Schlüter: May I ask whether the fast rising time of a few minutes during a big flare is compatible with this model?

Parker: Yes. The transit time through this cloud around the earth is of the order of a fraction of a second. The effect that it will produce is that it deflects in a random way and impedes the particles coming in. A 1 GeV particle is seriously impeded but particles of 4 or 5 GeV come through with not more than 30° deflexion.

ELECTROMAGNETIC ACCELERATION OF PARTICLES TO COSMIC RAY ENERGIES*

W. F. G. SWANN

*Bartol Research Foundation of the Franklin Institute,
Swarthmore, Pennsylvania, U.S.A.*

ABSTRACT

It has been demonstrated that if a particle is accelerated in an electromagnetic field between two points designated by subscripts (1) and (2), the particle will gain energy in passing from (1) to (2) provided that the quantity J defined by

$$J = s \frac{\partial}{\partial s} (U_{s2} - U_{s1})^2 - 2 \frac{\partial U_{s'}}{\partial t} \int S \frac{\partial U_{s'}}{\partial s} \, dt$$

is positive, where ds is an element of path, U_{s1} and U_{s2} are the initial and final values of the vector potential along the path, and t is the time. Moreover, if the particle is at rest at the point (1), its energy W_2 at the point (2) is such that

$$W_2 > e \, | \, (U_{s2} - U_{s1}) \, |,$$

here e is the charge on the particle.

A study has been made of the problem in which the motivating agency responsible for the electromagnetic field is a toroid with currents circulating in such fashion as they would circulate if the anchor ring of the toroid were wound with a wire in which a current decayed with time. The particular case studied is that where a particle moves along the axes of symmetry, and dimensions are chosen of astronomical size such as to make them apply to such phenomena as are observed in certain nebulae.

The magnitudes chosen are as follows:

$a \equiv$ Radius of cross section of the toroidal winding = 1 light year.

$r_0 \equiv$ Mean radius of toroid = 2000 light years.

$H_0 \equiv$ Initial field in the toroid = 10^{-3} gauss.

$v/\alpha \equiv$ Time for the current to decay to $1/e$ of its initial value = 1000 years.

With the above assumptions, a particle of electronic charge, starting from the center of the toroid and travelling along the axes of symmetry would acquire an energy in excess of 3×10^{14} eV.

* Supported in part by the joint program of the Office of Naval Research and the U.S. Atomic Energy Commission.

The general theory of acceleration of charged particles by magnetic induction invokes the application of Lagrange's equations to a Lagrangian function for a charge in an external electromagnetic field defined by a vector potential $\mathbf{U}$ and a scalar potential ϕ. In many problems ϕ is zero and the Lagrangian function becomes

$$L = -m_0 c^2 (1 - u^2/c^2)^{1/2} + \frac{e}{c} (\mathbf{U}.\mathbf{u}). \tag{1}$$

The magnetic field $\mathbf{H}$, and the electric field $\mathbf{E}$ are given by

$$\mathbf{H} = \operatorname{curl} \mathbf{U}; \quad \mathbf{E} = -\frac{1}{c} \frac{\partial \mathbf{U}}{\partial t}. \tag{2}$$

In many problems of changing magnetic fields it is possible, rather readily, to calculate line integrals of the electric field $\mathbf{E}$ along assigned paths, but such calculations are of no avail for calculating increase of particle energy unless we can show that the particles can describe paths such as to make use of the line integrals to the end of acquiring energy continually, at least over sufficiently long periods of time. The complexity of the particle motions is such that, usually it is not practicable to seek solutions for energy increase by calculating the path and calculating the increase of energy as the particle traverses it. In view of the foregoing considerations it is useful to develop criteria for continual increase of energy, and theorems which give information as to lower limits of energy attained in certain specified cases. A few of these matters are discussed in the following.

I. GENERAL CONSIDERATIONS PERTAINING TO THE CONTINUAL INCREASE OF ENERGY OF A PARTICLE STARTING FROM REST IN AN ELECTROMAGNETIC FIELD

It is clear that the particle, starting from rest, will move initially so as to make an acute angle with the electric field, i.e. with the vector

$$- (1/c) (\partial \mathbf{U}/\partial t).$$

As long as it continues to move so as to make an acute angle with the positive direction of $\mathbf{E}$, the energy will continue to increase. It can only decrease by the motion developing a character in which the particle makes an obtuse angle with $\mathbf{E}$, so that it has a component opposite to $\mathbf{E}$. In order to acquire this condition, however, it would have to pass through a condition, at some point P, in which it moved perpendicular to $\mathbf{E}$ at the point. If there were no magnetic field, it certainly could not pass through this latter condition because, at the point P, the particle would have acting on it a field tending to increase the component velocity parallel to $\mathbf{E}$, and so to bring the particle's path back to the condition in which it made an acute angle with $\mathbf{E}$.

If there is a magnetic field when the particle is at P, with its path perpendicular to $\mathbf{E}$, there will arise from this magnetic field a force $\mathbf{v} \times \mathbf{H}/c$ perpendicular to $\mathbf{v}$ and parallel to $\mathbf{E}$. This force may be in the direction of $\mathbf{E}$ or in the opposite direction, depending on the circumstances. If, however, $|\mathbf{E}| > |\mathbf{H}|$, we shall certainly have $|\mathbf{E}| > |\mathbf{v} \times \mathbf{H}/c|$ so that even if $\mathbf{v} \times \mathbf{H}/c$ is in the opposite direction to $\mathbf{E}$, the resultant force will be in the direction of $\mathbf{E}$ and will bring the particle back to the condition in which its path makes an acute angle with $\mathbf{E}$, and so the particle gains energy at P.

The condition $|\mathbf{E}| > |\mathbf{H}|$ as a criterion for continual gain of energy is thus *sufficient** but not always a necessary condition for continual increase of energy.

There is one exception to the above theorem. It is to be found in the case where the particle passes through a place where $\mathbf{E}$ reverses sign. In this case, the argument fails. An example is to be found in the case of a particle traveling along the general direction of propagation of a plane wave. It will be acted on by the field of the wave, which will oscillate in sign along a direction perpendicular to the general direction of the particle, and indeed, in this case the particle will alternately gain and lose energy. Of course, in the plane wave we have cited we have $\mathbf{H} = \mathbf{E}$ at all times, so that, strictly speaking, the test of our theorem is too severe. However, we are certainly on the safe side if we exclude from the theorem cases where the particle passes through a place of reversal of $\mathbf{E}$.

Concerning the looping of a particle around a line

If $|\mathbf{E}| > |\mathbf{H}|$, or less stringently, if the path of the particle always makes an acute angle with $\mathbf{E}$, the particle can never describe an angle 2π around any line, OP, unless there is a finite line integral of $\mathbf{E}$ (at the particle) taken along the path of the particle projected in a plane perpendicular to OP. The reason is as follows: In the light of the hypothesis, there is a finite component of $\mathbf{E}$ in the direction of the projected path at each point thereof, and therefore, if the projected path curves through an angle 2π, $\mathbf{E}$ (at the particle) projected on that path will have a finite line integral taken over the range 2π.

* The sufficiency of the condition for an axially symmetrical field with the z and r components of U zero, was established in the writer's first paper on this subject[1]. A simplified version of the theory is given by the writer in [2]; also in a later paper, 'The Acquirement of Cosmic Ray energies by Electromagnetic Induction in Galaxies'[3].

The lower limit of the energy gained along a path

If T is the kinetic energy of the particle, and $W \equiv T + mc^2$, it is readily possible, by the application of Lagrange's equations, to show that

$$W_2^2 - W_1^2 = e^2 (U_{s2} - U_{s1})^2 + \int \left[e^2 \dot{s} \frac{\partial}{\partial s} (U_s - U_{s1})^2 - 2e^2 \frac{\partial U_s}{\partial t} \int \dot{s} \frac{\partial U_s}{\partial s} dt \right] dt,$$

(3)

where U_s refers to the vector potential resolved along the direction of the path, at an arbitrary point on the path. $\dot{s}$ is the velocity along the path, subscripts (1) and (2) refer respectively to values at the point occupied at $t = 0$ and the point occupied at some later time. If we define J as

$$J \equiv \dot{s} \frac{\partial}{\partial s} (U_s - U_{s1})^2 - 2 \frac{\partial U_s}{\partial t} \int \dot{s} \frac{\partial U_s}{\partial s} dt$$

(4)

then, in cases where, at all instants, J is positive, we can write

$$W_2^2 - W_1^2 > e^2 (U_{s2} - U_{s1})^2.$$

In cases where the particle starts from rest and where, for large kinetic energies, mc^2 is neglected, this expression assumes the simple form

$$| W_2 | > e | (U_{s2} - U_{s1}) |.$$

(5)

This relation is of value in certain cases.

2. CASE OF AN AXIALLY SYMMETRICAL FIELD, IN WHICH THE VECTOR POTENTIAL U HAS NO COMPONENT ALONG THE r OR z DIRECTIONS

The case cited has many interesting features, some of which have been developed by the writer in a paper [3]. We shall summarize a few of these. It results from Lagrange's equations that

$$\frac{1}{2} \frac{dW^2}{dt} = e^2 \left(U_\theta - \frac{r_0 U_0}{r} \right) \frac{\partial U_\theta}{\partial t},$$

(6)

where $r_0 U_0$ apply at the instant and position when the particle commences to change its kinetic energy.

Criteria for continual increases of energy in the axially symmetrical case

Confining ourselves, without loss of generality, to the case where $\partial U_\theta / \partial t$ is positive, we see that the *necessary* condition for continued increase of energy is

$$U_\theta - \frac{r_0 U_0}{r} > 0$$

(7)

except at $t = 0$.

A *sufficient*, but not always a necessary condition for (7) to hold is that, for all positions of the particle

$$\frac{d}{dt}(rU_\theta - r_0 U_0) > 0$$

or, since $r_0 U_0$ is a constant $\quad \dfrac{d}{dt}(rU_\theta) > 0.$ (8)

It can readily be shown that

$$\frac{d}{dt}(rU_\theta) = [-E_\theta - (\mathbf{v} \times \mathbf{H})_\theta/c],$$ (9)

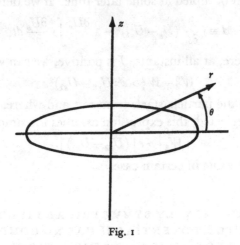

Fig. 1

where $\mathbf{v}$ is the velocity of the particle. Since $-E_\theta \ (=\partial U_\theta/\partial t)$ is positive, a sufficient, but not always necessary, condition for (8) to be true is therefore

$$|E_\theta| > |\mathbf{H}|,$$ (10)

which is a condition already cited for the more general case in which axial symmetry is not demanded. Thus conditions (7), (8) and (10) stand in order of stringency. Condition (10) is sufficient for (7) and (8). Condition (8) is sufficient for (7), while (7) is necessary and sufficient.

Case where, in axial symmetry, the particle starts from rest at the place and instant when $U_\theta = 0$

In this case, (6) becomes

$$\frac{dW^2}{dt} = e^2 \frac{\partial}{\partial t}(U_\theta)^2$$

so that the energy in this case increases continually under all circumstances.

432

Criterion for absence of 'looping', in the case of axial symmetry

We have seen that (7) is the necessary and sufficient condition for continual increase of energy. Now it can readily be shown from Lagrange's equations that

$$r W \dot\theta = -ce(U_\theta - r_0 U_0/r).$$

Hence, if there is continual increase of energy, $\dot\theta$ must always be of the same sign and finite. Thus, in such a case, the particle can never 'loop', except aroung the z-axis; for if the particle should loop in any other manner, there would have to be a place where $\dot\theta$ was zero.

Mechanisms in which there is no magnetic field in the space surrounding the motivating currents when those currents are steady

Tempting problems are presented by the discussion of an infinite solenoid, and by an anchor ring-winding. Here, there is no external magnetic field in the steady state, and even if the currents vary with the time, the magnetic field remains small for slow variations, in spite of the existence of a very definite electromotive force over a path encircling the solenoid outside thereof, or encircling the anchor ring so as to thread it.* If we could neglect the external magnetic field completely on such problems, we should always have $|\mathbf{E}| > |\mathbf{H}|$ and the sufficient criterion for continual gain of energy would be assured.

Although, in cases of the type cited, there is no magnetic field in the space surrounding the motivating currents when these currents are steady, there certainly is some magnetic field when the currents vary with the time. This can most readily be seen from the electromagnetic equations for free space, which demand a finite value for curl $\mathbf{H}$ and so for $\mathbf{H}$ if there is a finite value for $\partial \mathbf{E}/\partial t$. In free space we have in fact, all of the electromagnetic vectors, $\mathbf{E}$, $\mathbf{H}$, $\mathbf{U}$, obeying the wave equation; and, as far as our interests are concerned, we need confine our attentions only to the wave equation for $\mathbf{U}$.

$$\nabla^2 \mathbf{U} - \frac{1}{c^2} \cdot \frac{\partial^2 U}{\partial t^2} = 0.$$

---

* In such problems, the role of the vector potential presents a more realistic picture of the origin of the electric field at a point than does the behavior of the magnetic field. Thus, in the steady state problem for a solenoid, there is no magnetic field at a point P outside the solenoid, but there is a very definite vector potential. It is true that the Faraday law survives to the extent of predicting that around a path encircling the solenoid there is an electromotive force equal to the magnetic flux through the path; but this magnetic flux is confined for the most part to the area inside the solenoid. As a matter of fact, even when, as in the case of a long finite solenoid, there is a small magnetic field outside the solenoid, that field is in a direction opposite to that of the flux in the solenoid.

For the case of axial symmetry exemplified in Fig. 1, this equation assumes the form

$$\frac{\partial^2 U_\theta}{\partial r^2} + \frac{\partial^2 U_\theta}{\partial z^2} + \frac{\partial}{\partial r}\left(\frac{U_\theta}{r}\right) = \frac{1}{c^2}\frac{\partial^2 U}{\partial t^2}. \tag{11}$$

A useful problem which serves as a basis for the discussion of other problems is that of a circular current of small size, flowing around the axis of z at the origin. For the steady state case, this entity acts like a magnet of moment μ given by

$$\mu = \pi a^2 I,$$

where a is the radius and I the current flowing in the positive direction of θ. Thus μ is in the positive direction of the axis of z. It can easily be

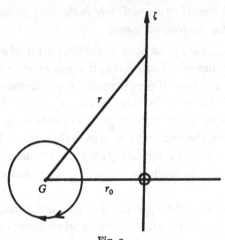

Fig. 2

verified that, for the case where I varies with the time according to the law $I = I_0 f(t)$, where $f(0) = 1$, the appropriate solution of (11) is

$$U_\theta = \frac{\mu_0 r f(t - R/c)}{R^3} + \frac{\mu_0 r f'(t - R/C)}{CR^2}, \tag{12}$$

where $\mu_0 = \pi a^2 I_0$; $R^2 = z^2 + r^2$. The solution (12) can by the combination of a number of such circular current elements, serve as the basis for the solution of a solenoid, or of a toroidal winding in which the current varies with the time.

Case of a toroidal winding

It is convenient to transpose the current ring, for which (12) is the solution, to the position shown in Fig. 2, where it now forms an element of a toroidal winding, the anchor ring of the toroid having its plane of symmetry perpendicular to the axis of ζ.

434

Let the old axis of z corresponding to (12) be in the downward direction through the paper, so that the current is in the direction shown. The plane of the paper shall be the old plane $z=0$, and r shall continue to have its original meaning. The positive direction of θ is in the direction of the arrow and the value of U_θ at the point P is, from (12)

$$U_\theta = \frac{\mu_0}{r^2} f(t-r/c) + \frac{\mu_0 r_0}{cr} f'(t-r/c).$$

The component of $\mathbf{U}$ in the positive direction of ζ is thus

$$U_\zeta = -\frac{\mu_0 r_0}{r^3} f(t-r/c) - \frac{\mu_0 r_0}{cr^2} f'(t-r/c), \tag{13}$$

where r_0 is as indicated.

If now we consider the toroidal winding to be made up of a large number of circular currents like that shown, each of them will make a contribution to U_ζ. If I_0 as defined above, is taken to be the current per unit length measured around the solenoid, and if ϕ is measured around the ζ axis, the moment appropriate to the element of angle $d\phi$ is $\pi a^2 I_0 r_0 d\phi$ which replaces μ_0 in (13) for the contribution of $d\phi$ to the total vector potential A_ζ along the ζ direction. Integrating with respect to ϕ, we find for A_ζ itself the value

$$A_\zeta = -\frac{2\pi^2 a^2 I_0 r_0^2}{r^3}\left[f(t-r/c) + \frac{r}{c} f'(t-r/c)\right]. \tag{14}$$

It is convenient to express I_0 in terms of the magnetic field inside the toroid for the steady case. If H_0 is this field, we have

$$2\pi r_0 H_0 = 4\pi (2\pi r_0 I_0)$$

so that

$$H_0 = 4\pi I_0.$$

Writing

$$\frac{\pi a^2 H_0}{2} \equiv N_0 \tag{15}$$

we have

$$A_\zeta = -\frac{N_0 r_0^2}{r^3}\left[f(t-r/c) + \frac{r}{c} f'(t-r/c)\right]. \tag{16}$$

All components of the vector potential perpendicular to the ζ axis cancel, from symmetry, so that A_ζ, represents the complete vector potential along the ζ axis.

In the steady state, there is no magnetic field anywhere outside the toroid. Such is not the case for the non-steady state, however, for in general there is in such a case, a rate of change of vector potential and so an electric field at all points in the space around the toroid, and there is in general a rate of change of electric field which demands the existence of a magnetic

435

field. However, along the ζ axis there is no magnetic field in a plane perpendicular to that axis; for if there were a field perpendicular to the axis at any part P, there would from symmetry be a magnetic flux towards or away from the axis at that point, and this is impossible. Any magnetic field in the vicinity of the ζ axis must take the form of circular lines around that axis, the situation conforming to

$$\frac{1}{c}\frac{\partial E_\zeta}{\partial t} = (\text{curl } \mathbf{H})_\zeta.$$

Applying this to a small tube of radius σ surrounding the axis we have

$$\frac{\pi\sigma^2}{c}\frac{\partial E_\zeta}{\partial t} = 2\pi\sigma H_\phi,$$

where H_ϕ is the magnetic field in the direction of increase of ϕ, thus,

$$H_\phi = \left(\frac{\sigma}{2c}\right)\frac{\partial E_\zeta}{\partial t}$$

and H_ϕ vanishes with the vanishing of σ for a finite value of $\partial E_\zeta/\partial t$.

In the light of the above we see that the only force on a charged particle on the axis of ζ is a force along that axis, and the force is $-(1/c)\,\partial A_\zeta/\partial t$ per unit charge.

We are now in a position to make use of (16). We shall choose $f(t)$ to qe of the form

$$f(t) = e^{-\alpha t}$$

so that

$$f(t-r/c) = e^{-\alpha(t-r/c)} = e^{-\alpha t}e^{\alpha r/c}$$

and

$$f'(t-r/c) = -\alpha a^{-\alpha(t-r/c)} = -\alpha e^{-\alpha t}e^{\alpha r/c}.$$

Thus

$$A_\zeta = -\frac{N_0 r_0^2}{r^3}\left[1 - \frac{\alpha r}{c}\right]e^{-\alpha t}e^{\alpha r/c}. \tag{17}$$

The electric field E_ζ along the ζ axis is given by

$$E_\zeta = -\frac{1}{c}\frac{\partial A_\zeta}{\partial t} = -\frac{N_0 r_0^2 \alpha}{cr^3}\left[1 - \frac{\alpha r}{c}\right]e^{-\alpha t}e^{\alpha r/c}. \tag{18}$$

The electric field reverses sign at the value of r given by $\alpha r/c = 1$. There are thus two categories of interest corresponding to $\alpha r/c < 1$ and $\alpha r/c > 1$. We shall call them cases A and B, respectively. The second contribution (involving $\alpha r/c$) on the right-hand side of (18) is of course a close mathematical relative of the radiation field from an electric dipole, which field varies less rapidly with the distance—to the extent of one power of r—than does the non-radiation field, whose counterpart, is the first term on the right-hand side of (18). At small distances the non-radiation term dominates, but at great distances the radiation term dominates.

Regarding the role played by the scale of the phenomena

For such phenomena as occur on stars, $(1/\alpha)$ may be expected to be of the order of a few days, as in the case of sunspots, for example. For these cases α may be taken to be of the order 10^{-6} or less, so that the quantity c/α is of the order 3×10^{16} cm. This is much larger than stellar dimensions* so that in general for particles accelerated in stars we shall be concerned with the case where $\alpha r/c \ll 1$.

For phenomena on a galactic scale such as we encounter in the nebulae, α may be expected to be much smaller than 10^{-6}. On the other hand, the dimensions available in a nebula are such that $\alpha r/c$ can approach or even exceed unity, and the scale of the space occupied by the motivating currents can afford to be correspondingly large so as to provide for significant acceleration at the great distances concerned. Thus, both cases A and B are likely to be of interest to us in cosmological speculations.

Case A, where $\alpha r/c < 1$

This corresponds to $(\zeta^2 + r_0^2) < c^2/\alpha^2$, i.e. to the region

$$(c^2/\alpha^2 - r_0^2)^{1/2} > \zeta > -(c^2/\alpha^2 - r_0^2)^{1/2},$$

i.e. to the region
$$\left(\frac{c^2}{\alpha^2 r_0^2} - 1\right)^{1/2} > \zeta/r_0 > -(c^2/\alpha^2 r_0^2 - 1)^{1/2}. \tag{19}$$

For $\alpha = 10^{-6}$, and $r_0 \sim 10^{11}$ for stellar dimensions $c^2/\alpha^2 r_0^2 = 9 \times 10^{10}$. Thus (19) becomes

$$\frac{c}{\alpha r_0} > \frac{\zeta}{r_0} > -\frac{c}{\alpha r_0}$$

or, for the magnitudes cited

$$3 \times 10^5 > \frac{\zeta}{r_0} > -3 \times 10^5.$$

In this region, the field is always negative, and a positive particle starting in the region moves in the negative direction and gains energy continually.

Let us consider a case where $a = 10^9$ cm; $r_0 = 10^{10}$ cm; $\alpha = 10^{-5}$; $H_0 = 10^4$, so that, from (15), $N_0 = 1\cdot5 \times 10^{22}$.

Let us calculate the gain in energy of a proton in traveling from $-\zeta = 0$ to $-\zeta = r_0 = 10^{10}$ cm. The quantity $\alpha r/c$ will, over the whole path,

---

* Of course for special phenomena in which $(1/\alpha)$ might be of the order of 1 sec, $\alpha r/c$ would be comparable with unity even for stellar dimensions.

always be less than 0.5×10^{-5}, so that, replacing the factor $e^{-\alpha t}e^{\alpha r/c}$ by unity, as will subsequently be justified,

$$E_\zeta = -\frac{N_0 r_0^2 \alpha}{cr^3}.$$

The energy gained will be given by

$$W = -\frac{N_0 r_0^2 \alpha e}{c} \int \frac{d\zeta}{r^3}.$$

Writing $-\zeta/r_0 = \tan \lambda$, we have $\pi/4$ for the upper limit of λ, and

$$W = \frac{N_0 \alpha e}{c} \int_0^{\pi/4} \frac{\sec^2 \lambda \, d\lambda}{(1 + \tan^2 r)^{\frac{3}{2}}} = \frac{N_0 \alpha e}{c} \int_0^{\pi/4} \cos \lambda \, d\lambda.$$

Thus

$$W = \frac{N_0 \alpha e}{c} \sin \frac{\pi}{4} = \frac{N_0 \alpha e}{1 \cdot 4 c} = \frac{300 \, N_0 \alpha}{1 \cdot 4 c} \text{ eV.,}$$

$$W = 10^9 \text{ eV.}$$

If E_m is the field at the end of the above path, the time taken to describe the path is less than τ as given by $r_0 = E_m e \tau^2 / 2m$, where m is the relativistic mass at the end of the path. m is not greatly different from the rest mass, 1.6×10^{-24}, so that it results that $r_0 = N_0 \alpha r_0^2 e \tau^2 / 2mr^3 c$ and, on inserting the values with $r/r_0 = 2^{\frac{1}{2}}$, we find τ of the order 1 second. Thus the replacement of $e^{-\alpha t}e^{\alpha r/c}$ by unity as above is justified.

Case B where $\alpha r/c > 1$ over the path

Let us write $\alpha r_0/c \equiv \eta$, and let us consider a case where $a = 10^{18}$ cm ($= 1$ light year); $H_0 = 10^{-3}$ gauss, (so that $N_0 = 1.5 \times 10^{33}$); $1/\alpha = 3 \times 10^{10}$, (corresponding to the decay of the motivating currents to $1/e$ of their initial values in 1000 years); $\eta = 2$, so that $r_0 = 2 \times 10^{21}$ cm ($=$ about 2000 light years).

The exponential factor in the expression for E_ζ is composed of two factors $\exp \alpha r/c$ and $\exp (-\alpha t)$. We shall first examine the consequences of neglecting the second factor, so that

$$E_\zeta = \frac{N_0 r_0^2 \alpha}{cr^3} \left[\frac{\alpha r}{c} - 1 \right] e^{\alpha r/c}.$$

It is easy to show that E_ζ increases continually with r. We have

$$W > \frac{N_0 r_0^2 \alpha e}{c} \int \left[\frac{\alpha}{cr^2} - \frac{1}{r^3} \right] d\zeta$$

438

over the range of integration concerned. Writing $\zeta/r_0 = \tan \lambda$

$$W > \frac{N_0 \alpha e}{c}\left[\frac{\alpha r_0}{c}\int_0^\lambda d\lambda - \int_0^\lambda \cos \lambda\, d\lambda\right],$$

$$W > \frac{300\, N_0 \eta}{r_0}\,[\eta\lambda - \sin \lambda] \quad \text{electron volts.}$$

Now a proton with a velocity 99 per cent of the velocity of light has an energy of 6×10^9 eV. It has a mass m equal to $7\, m_0$. For such a proton, and with $\eta = 2$

$$2\lambda - \sin \lambda < \frac{2 \times 10^7 r_0}{2 \times N_0} < 1\cdot5 \times 10^{-5}$$

so that $\lambda = 1\cdot5 \times 10^{-5}$ and the corresponding value of ζ is $1\cdot5 \times 10^{-5} r_0$. Since the minimum value of E_ζ occurs at $r = r_0$, and in this case is given by $E_{\min.} = (N_0/r_0^2)\,(\alpha r_0/c)\,(\alpha r_0/c - 1)\,\exp\,(\alpha r_0/c)$, we have, for $\eta = 2$, $E_{\min.} = 2(7\cdot3)\, N_0/r_0^2$, and if τ is the time for the particle to reach the point $\zeta = 1\cdot5 \times 10^{-5} r_0$, we have, with $m = 7 \times 1\cdot6 \times 10^{-24}$

$$1\cdot5 \times 10^{-5} r_0 > \frac{14\cdot6\, N_0 e \tau^2}{2 r_0^2 m}.$$

Hence $\tau < 4 \times 10^5$ sec and $\alpha\tau < 1\cdot3 \times 10^{-5}$. Thus, the neglect of the factor $\exp\,(-\alpha t)$ in the expression for E_ζ is valid for the above calculation, in which λ is limited to the small value $1\cdot5 \times 10^{-5}$.

We may now evaluate the total situation as follows: The least value of $\alpha r/c$ is $\alpha r_0/c$. The total time to travel a distance ζ larger than the value $1\cdot5 \times 10^{-5} r_0$ considered above will be less than $\zeta/v + \tau$, where $v = 99c/100$, and $\tau = 4 \times 10^5$ sec. Thus, if ζ is such that $\alpha\zeta/v + \alpha\tau$ is not greater than $\alpha r_0/c$, the exponent $(\alpha r/c - \alpha t)$ in the general expression for E_ζ will always be positive. This gives

$$\zeta \leqslant r_0 v/c - v\tau \leqslant (99/100)\,(r_0 - 4 \times 10^5 c).$$

Since $r_0 = 2 \times 10^{21}$, the quantity $4 \times 10^5 c$ may be neglected and ζ may be permitted a value sensibly as large as r_0, so that the upper limit $\lambda = \pi/4$ may be used in the expression for W. We thus find that, in travelling over the said range, the particle gains energy W such that

$$W > \frac{300\, N_0 \eta}{r_0}\left[\frac{\pi}{4}\eta - \frac{1}{1\cdot4}\right].$$

Putting $\eta = 2$, $N_0 = 1\cdot5 \times 10^{33}$ and $r_0 = 2 \times 10^{21}$

$$W > 3 \times 10^{14} \text{ eV.}$$

439

Of course, the numbers here cited are susceptible of enormous variations without exceeding the realm of reason, and our example is taken simply as an illustration.

REFERENCES

[1] Swann, W. F. G. *Phys. Rev.* **43**, 217, 1933; *J. Franklin Inst.* **215**, 273, 1933.
[2] Swann, W. F. G. *J. Franklin Inst.* **258**, 205, 1954.
[3] Swann, W. F. G. *J. Franklin Inst.* **255**, 383, 1954.

Discussion

Alfvén: My only remark is that I think we all are very glad that the senior of all accelerating processes has worked very well!

Singer: Would your theory be applicable to all values of $(dH/dt)/H$, say, when the line integral cannot be defined?

Swann: Yes, the mechanism works for all values of $(dH/dt)/H$. This indeed is my quantity $-\alpha$.

Singer: Could the process work statistically, i.e. the particle gains and loses energy, but on the average gets accelerated?

Swann: Yes, it could work under suitable circumstances, but I have confined my attention to a problem where the energy increases continually.

Schlüter: Has your theory been worked out solely for the case of vacuum, without electrical conductivity?

Swann: Yes.

Bunemann: What gauge of potentials has one to use in order to make the formula $W > e \mid U_{s2} - U_{s1} \mid$ right? You seem to have used a particular gauge, the retarded potentials.

Swann: The potentials used are the retarded potentials defined by

$$\nabla^2 U - \frac{1}{c^2} \frac{\partial^2 U}{\partial t^2} = -\rho u/c; \quad \Delta^2 \phi - \frac{1}{c^2} \frac{\partial^2 \phi}{\partial t^2} = -\rho.$$

However, any equivalent pair of potentials could give equivalent results but in general with great analytical difficulty.

FURTHER OBSERVATIONS OF THE POINT SOURCE OF COSMIC RAYS*

Y. SEKIDO, S. YOSHIDA AND Y. KAMIYA

Physical Institute, University of Nagoya, Japan

ABSTRACT

A point source of cosmic rays at $\delta = 0°$, $\alpha = 5^h\,30^m$, was reported at the Mexico Meeting of IUPAP in September 1955 (*Nature, Lond.* **177**, 35, 1956). The existence of the point source was verified by further observation. This phenomenon suggests the possibility of a direct method of exploring interplanetary space.

The authors[1, 2] reported the existence of a point source of cosmic rays at declination $\delta = 0°$, right ascension $\alpha = 5^h\,30^m$, before entering into the geomagnetic field, observed with two Geiger–Müller counter telescopes until 18 August 1955. The observation was continued with the two telescopes, and the same point source was observed again as shown in Fig. 1. The count at the point source is $N = 2395$, while the average of the other seven positions is $N_0 = 2196 \cdot 1$. Therefore

$$N - N_0 = +0 \cdot 091 N_0 = +4 \cdot 3 N_0^{\frac{1}{2}}. \quad (1)$$

Since April 1954, observations were done at the zenith distance $Z = 80°$,

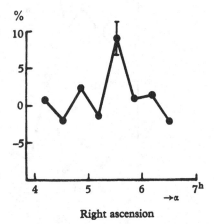

Fig. 1. Further observation of cosmic rays from the declination $\delta = 0°$. Telescope no. 1: 19 August 1955–March 1956. Telescope no. 2: 19 August 1955–June 1956.

twice a day at two azimuths, $A = 85°$ and $A = 255°$, respectively. In Fig. 2, the results of the total observations (1954–6) were divided into four independent observations. Every curve shows the existence of the same point source. That is to say, the point source was observed four times, with two telescopes and at two azimuths, respectively. The total count at the

* Presented by Dr Y. Fujita.

point source is $N = 3688$, while the average of the other seven positions is $N_0 = 3322 \cdot 6 \pm 22$. Therefore

$$N - N_0 = +0 \cdot 110 N_0 = +6 \cdot 3 N_0^{\frac{1}{2}}. \tag{2}$$

In the previous report, the corresponding values were

$$N - N_0 = +0 \cdot 157 N_0 = +5 \cdot 2 N_0^{\frac{1}{2}}. \tag{3}$$

Therefore, the existence of the point source was confirmed with greater accuracy than in the previous report.

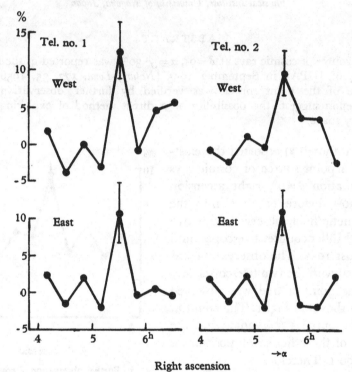

Fig. 2. Four independent observations of cosmic rays from the declination $\delta = 0°$. Telescope no. 1, April 1954–March 1956. Telescope no. 2, July 1955–June 1956. East: Azimuth $A = 85°$. West: $A = 255°$.

At the same declination and in a wider range of right ascension shown in Fig. 3, there are eighteen positions, though the time intervals of the observation of them are not strictly equal to each other. The expectation of the count at the point source, determined from the observation of the other seventeen positions is $\{N\} = 3365 \pm 15$, and

$$N - \{N\} = +0 \cdot 097 \{N\} = +4 \cdot 9\sigma, \tag{4}$$

442

where σ is the standard deviation of the intensities determined from the seventeen positions. σ is 2·0 %, while that expected from simple Poisson distribution is 1·8 %. Therefore, there is not much difference between them. σ includes anisotropies of cosmic rays in the seventeen positions and the effect of geophysical and instrumental instabilities, even if they exist. Still, the above result shows that the existence of the point source is sufficiently significant compared with the standard deviation σ.

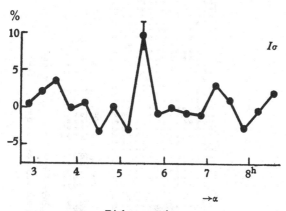

Right ascension

Fig. 3. Cosmic rays from the declination $\delta = 0°$. Telescope no. 1: April 1954–March 1956. Telescope no. 2: July 1955–June 1956. σ: Standard deviation determined from 17 points, excepting the point source of cosmic rays at $\alpha = 5^h\ 30^m$.

Fig. 4 shows the intensity distribution over a little larger part of the celestial sphere, observed with telescope no. 2, which can observe five adjacent declination bands at once. This figure shows that the above described point source is the most significant position in this part of the celestial sphere.

We thank Professor Y. Hagihara and Professor T. Hatanaka for their interest and discussions, and also Dr K. Nagashima and Mr H. Ueno for their help in this work.

REFERENCES

[1] Sekido, Y., Yoshida, S. and Kamiya, Y. *J. Geomagn. Geoelect., Kyoto*, **6**, 22, 1954.
[2] Sekido, Y., Yoshida, S. and Kamiya, Y. *Nature, Lond.* **177**, 35, 1956; Communication to Meeting of IUPAP Mexico, September 1955, p. 45.

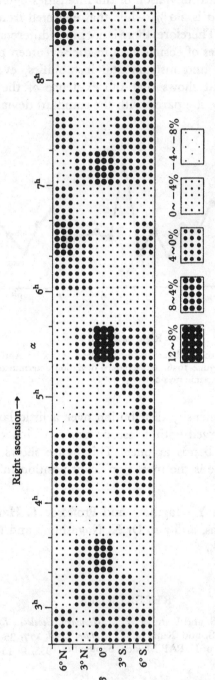

Fig. 4. Distribution of cosmic ray intensities in a part of the celestial sphere. Telescope no. 2, July 1955–June 1956. Graduation: +12%, +8%, +4%, 0%, −4%, −8%. The point source is at $\delta = 0°$, $\alpha = 5^h\ 30^m$.

Discussion

Parker: What are the particle energies in the observations made?

Fujita: Nearly 3×10^{11} eV, I should suppose.

Parker: How are such particles observed?

Ehmert: An energy of 9 GeV is necessary for mesons to penetrate the lead shield between the counters of the telescope. The primary energies are extremely high.

Parker: Could someone give a rough order of magnitude of the energy the primaries might have to produce such particles?

Heidmann: At the Mexico conference the primary particles were assumed to have an energy of 200–300 GeV.

THE STÖRMERTRON*

WILLARD H. BENNETT

U.S. Naval Research Laboratory, Washington, D.C., U.S.A.

A tube has been developed in which the shapes of streams of charged particles moving in the earth's magnetic field can be produced accurately to scale. The tube has been named the Störmertron in honor of Carl Störmer who calculated many such orbits. New developments which have made this tube possible include a method for coating the inside of large glass tubes with a transparent electrically conducting film, and an electron gun producing gas-focused streams in less than $\frac{1}{2}$ micron of mercury vapor, a nearly vapor-free grease joint, and a nearly vapor-free carbon black. The magnetic dipole field of the earth is simulated with an Alnico magnet capped with properly shaped soft iron caps. The stream is deflected using two pairs of yoke coils near the gun.

Traversing the stream approximately parallel with the equatorial plane of the earth produces successive shapes assumed by the stream as its direction of projection is advanced past the earth as the stream would be advanced due to the rotation of the sun. Using electronically timed stop motion cameras, motion pictures have been made of the shapes of the advancing streams. It is observed that particles can approach the earth in free orbits and then leave the earth in free orbits if unscattered by collisions, or if properly scattered by collisions while near the earth, the particles become captured into periodic orbits and constrained to move in a ring until scattered out. These rings, as revealed by the Störmertron, consist of particles in periodic orbits which are smeared out in magnetic longitude but are confined to a volume generated by rotating a crescent about the magnetic axis with horns of the crescent impinging on the earth in the two auroral zones.

A larger Störmertron is being built in which the dipole magnetic field at any desired strength can be produced by passing a current through the coil representing the earth. This coil has been designed to produce the magnetic field of the dipole at the center with an error of less than 3 %.

* Presented at the Symposium by E. O. Hulburt, Washington. D.C.

This tube is to be used to plot out various auroral areas in order to compare with the observed geographical distribution of aurorae to be obtained during the International Geophysical Year.

Discussion

Ferraro: Is it possible to do this experiment with wider beams?

Hulburt: We have not tried that. I think it is difficult to perform it with an electron gun.

Dattner: Was the reason of the discharge inside the forbidden zone the scattering of the electrons by collisions?

Hulburt: Yes.

Block: I wonder if you have calculated the influence of the positive charges which may be produced by ionizing the gas in the vicinity of the beam. This may affect the orbit.

Hulburt: I think in this experiment with this position of the electron gun the electron orbits behave as in field-free space.

Alfvén: We saw very nice ring-currents. Could you say of what type they were?

Hulburt: They are probably due to electrons scattered into ring-shaped curves around the dipole. I think these results would have some bearing on geomagnetic effects such as magnetic storms.

Singer: I would like to point out that the scattering mechanism in these model experiments is different from the scattering mechanism I proposed yesterday. The scattering in the Störmertron is produced by collisions with gas atoms. The scattering process I proposed was due to perturbations of the dipole field several earth radii out. However, the particles that I started with are protons and will of course describe analogous orbits in the dipole field.

Dattner: I would like to ask Dr Hulburt about the mean free path in the chamber.

Hulburt: The pressure is probably about 10^{-4} mm Hg and the mean free path some centimeters.

Block: What was the energy of the particles? 100 volts or so?

Hulburt: Yes, a few hundred volts.

Block: I believe that the mean free path would be about 1 m.

PART VI
HIGH CURRENT DISCHARGES

UNTERSUCHUNGEN ÜBER IMPULSENT-LADUNGEN IM ZUSAMMENHANG MIT DER MÖGLICHKEIT VON KONTROLLIERBAREN THERMONUKLEAREN REAKTIONEN

L. A. ARTSIMOVICH

Academy of Sciences, Moscow, U.S.S.R.

Eine notwendige Voraussetzung für die Entstehung von thermonuklearen Reaktionen ist eine hohe Temperatur des Stoffes. Die ersten Spuren vom Entstehen solcher Reaktionen können wir bei der Erwärmung des Stoffes (Deuterium oder einer Mischung von Deuterium und Tritium) bis zu $T \sim 10^6$ zu entdecken hoffen. Aber nur bei $T \sim 10^8$ können die thermonuklearen Reaktionen als eine neue Energiequelle ein Interesse erwecken.

Die grösste Schwierigkeit, welcher wir beim Versuch, steuerbare thermonukleare Reaktionen hervorzurufen, zu begegnen haben, besteht darin, dass bei einer gewissen Temperatursteigerung die Wärmeverluste sehr stark wachsen. Bei Temperaturen, die der Entstehung von intensiven thermonuklearen Reaktionen entsprechen, muss der Stoff ein Plasma mit einem sehr hohen Ionisationsgrad vorstellen. Die Wärmeverluste im Plasma, die vom gewöhnlichen Mechanismus der Wärmeleitung verursacht werden, steigen proportional zu $T^{7/2}$. Bei einer Erwärmung des Stoffes nur bis zu $T = 10^5$, wenn man nicht spezielle Massnahmen zur Beseitigung dieser Verluste trifft, werden sie so gross, dass eine weitere Erhöhung der Temperatur unmöglich wird. Darum wird bei der Frage über die Hervorrufung steuerbarer thermonuklearer Reaktionen zur Hauptaufgabe die Ausarbeitung von Metoden, die für eine kräftige Herabsetzung der Energieverluste bürgen, die vom Mechanismus der Wärmeverluste verursacht werden (d.h. solcher Verluste, die mit der Wärmeübertragung durch Elektronen und Ionen des Plasmas verbunden sind).

Sollten wir den Mechanismus der Wärmeverluste im Plasma in einer etwas vereinfachten Form betrachten, so könnten wir sagen, dass unsere Aufgabe dahin kommt, solche Bedingungen zu schaffen, bei welchen die schnellen Teilchen genügend lange im Plasma aufgehalten werden, damit für die Ionen eine kennbare Möglichkeit geschaffen wird miteinander durchzureagieren. Bei einer solchen vereinfachten (und nicht strengen) Aufgabestellung, wird diejenige Grösse, die den energetischen Wirkungsgrad einer sich gedachten thermonuklearen Anordnung kennzeichnet, folgendermassen ausgedrückt:

$$\tau n . \overline{v\sigma} . \frac{\overline{W}}{kT}$$

Hier ist τ die mittlere Lebensdauer eines schnellen Iones im System, n—die Ionenkonzentration, $\overline{v\sigma}$—der Mittelwert des Produktes aus der Multiplikation von mittlerer Wärmegeschwindigkeit mit dem effektiven Querschnitt der Kernreaktion und W—die Energie, die beim Elementarakt der Reaktion frei wird. Es ist zu bemerken, dass bei einem vorgegebenen Wert der Temperatur, der Wärmewirkungsgrad durch die Multiplikationsgrösse τn angegeben wird. Das Produkt aus der Multiplikation der übrigen Faktoren, die im Ausdruck des Koeffizienten enthalten sind, ist nur von T abhängig und erreicht bei einem bestimmten Wert das Maximum—für die DD-Reaktion bei $T \sim 5 \times 10^8$. Bei dieser vorausgesetzt 'optimalen' Temperatur wird der energetische Wirkungsgrad für Deuterium in der Grössenordnung von $\sim 10^{-15} \tau n$. Damit der energetische Wärmewirkungsgrad einer thermonuklearen Anordnung genügend hoch wird, muss man bestrebt sein, die Lebensdauer des schnellen Iones im System zu verlängern. Es ist nur dann zu erreichen, wenn es gelingt, die Bewegung der Teilchen im Plasma in irgendwelcher Art und Weise zu begrenzen.

Der erste Gedanke dieser Art, der uns bekannt ist, wurde von A. D. Sakharov und I. E. Tamm im Jahre 1950 ausgesprochen. Er besteht darin, ein starkes Magnetfeld zur Begrenzung der Teilchenbewegung im Plasma zu benutzen. In einem starken Magnetfeld können die Elektronen und Ionen nur längs der Kraftlinien auf grössere Entfernungen sich frei bewegen. In der Querrichtung zu den Kraftlinien können die Teilchen nur durch Zusammenstösse in Bewegung und zu einer Energieübergabe kommen. Die Theorie der Vorgänge in einem vollständig ionisierten Plasma zeigt, dass die Strömung der Wärmeenergie in einer zum Vektor **H** senkrechten Richtung, bei hohen Werten von H und T, um viele Grössenordnungen kleiner wird, im Vergleich zum Wert, der in Abwesenheit des Feldes vorhanden wäre. Die Auswertung dieser generellen Idee kann verschiedene konkrete Formen geben. Man kann sich Methoden der Hervorrufung von thermonuklearen Reaktionen vorstellen, die auf *dauernde* Plasmaerwärmungsverfahren gegründet sind. Es sind auch andere Verfahren möglich, bei welchen eine *momentane* Temperatursteigerung des

Stoffes mit Hilfe eines kräftigen Impulsverfahrens von kurzer Dauer erreicht wird. Aber in allen Fällen ist der Wirkungsgrad proportional zu H^2. Der Arbeitszyklus, von welchem ich hier im kurzen berichten werde (ihm werden noch zwei Vorträge eines mehr speziellen Charakters gewidmet werden) ist mit der Untersuchung der Möglichkeit von Plasmaerwärmung durch kräftige elektrische Entladungen verbunden.

Es ist klar, dass man ein starkes Magnetfeld im Plasma beim Durchlassen eines genügend starken Stromes erzeugen kann. Dabei entsteht auch eine Nebenerscheinung von Thermoisolierung. Bei der Zusammenwirkung mit seinem eigenen Magnetfeld, zieht sich der durch das Plasma fliessende Strom zusammen, rückt im Kompressionsverlauf das Plasma mit und isoliert es von den Gefässwandungen, in welchen sich dieses befindet. Gleichzeitig übt der Strom noch eine notwendige Funktion aus—er erwärmt das Plasma. Die Plasmaerwärmung muss auf Kosten der Kompressionskräfte, sowie auch des Jouleeffektes geschehen. Die Theorie der Erscheinungen, die in Anwesenheit eines Stromes im Plasma auftreten, wurde zuerst auf Grund von Voraussetzungen über einen quasistationären Charakter des Prozesses gebaut. Unter dieser Voraussetzung rechnen wir damit, dass in jedem Zeitmoment die elektrodynamischen Kräfte der Zusammenziehung mit dem Gasdruck des Plasmas im Gleichgewicht stehen. Die grundsätzlichen Folgerungen aus dieser Anfangstheorie waren die folgenden:

1. Die Temperatur des Plasmas ist zum Quadrat der Stromstärke proportional und in jedem Augenblick durch die folgende Formel bestimmt:

$$T = \frac{I^2}{4kN}.$$

Hier ist I die Stromstärke, k die Boltzmannkonstante und N—die Anzahl der Teilchen mit gleichem Vorzeichen, auf eine Längeneinheit der Plasmasäule bezogen. Es ist zu bemerken, dass diese Formel eigentlich von Prof. Alfvén stammt (weil sein ausgezeichnetes Buch *Kosmische Elektrodynamik* sie in einer allgemeineren Form enthält). Etwas anders ausgedrückt, kommt sie auch in einer Arbeit von Dr. Schlüter vor. Die angeführte Formel für eine quasistationäre Erwärmung des Plasmas, die Wasserstoff oder Deuterium enthält, ist gültig für den Fall, dass $N \gg 10^{16}$ ist. (Auf einen Abschnitt der Plasmasäule, der zahlenmässig dem klassischen Radius des Protons gleich ist, muss eine Anzahl von Teilchen zukommen, die bedeutend grösser als Eins ist.) Bei dieser Bedingung muss ein Wärmegleichgewicht zwischen Elektronen und Ionen herrschen, d.h.

$$T_i = T_e.$$

453

2. Eine notwendige Bedingung für das Vorhandensein einer von den Wänden isolierten Plasmasäule ist ein stetiges Wachsen des im Plasma fliessenden Stromes. Wenn der Strom zu steigen aufhört, bricht die Plasmaschnur zusammen und berührt die Wände. Die Veränderung des Radius der Plasmaschnur mit der Zeit ist vom Gesetz der Stromsteigerung abhängig.

Obwohl, wie wir weiter sehen werden, unter den Umständen unserer Versuche, die Anfangstheorie, die auf die Voraussetzung von Gleichheit der Kräfte zwischen dem magnetischen Druck und dem Gasdruck baut, sich als falsch erwiesen hat, hat sie doch eine bestimmte Bedeutung, soweit es nicht ausgeschlossen ist, dass bei einer gewissen Veränderung der Versuchsbedingungen ein Gleichgewicht der Drucke zu erreichen ist. Es ist noch zu bemerken, dass diese Theorie ihren heuristischen Wert für die Analyse verschiedener Arten von elektrodynamischer Labilität der stromführenden Plasmasäule aufrecht erhält. In einer von unseren mehr speziellen Mitteilungen wird über die Anwendung dieser Theorie zur Untersuchung der Frage über die Einwirkung des äusseren magnetischen Feldes auf die Stabilität der stromführenden Plasmaschnur vorgetragen. Die experimentelle Untersuchung von Impulsentladungen von kurzer Dauer, aber grosser Stromstärke (bis zu 2 Millionen Ampere) führte zu einer grundsätzlichen Veränderung unserer Vorstellungen über den Vorgangsmechanismus, der im Plasma auftritt.

Die Analyse eines umfangreichen experimentellen Materials, das von Messungen auf oszillographischem Wege von magnetischen und elektrischen Feldern, Druckimpulsen, Intensität und Breite der Spektrallinien, Intensität der Röntgen- und Neutronenausstrahlungen des Plasmas eingesammelt wurde, gestattete einige Grundzüge aufzudecken, die die Vorgänge der Impulsentladung kennzeichnen.

Im Anfangsstadium der Entladung entsteht immer in der Nähe der Wände des Entladungsrohres eine dünne Skin-Schicht. Innerhalb dieser Schicht ist Gas vorhanden, das von der Einwirkung der Entladung praktisch unberührt ist. Die Stromdichte innerhalb dieses inneren Bereiches ist gleich Null. Nachher, unter der Einwirkung der elektrodynamischen Kräfte, beginnt die zylindrische Plasmaschicht sich mit einer wachsenden Geschwindigkeit in die Richtung der Achse des Entladungsrohres zusammenzuziehen. Die Geschwindigkeit des Zusammenziehens ist von der Anfangsdichte des Gases, dem Radius des Rohrs und vom Anfangswert des dI/dt abhängig. Für Gase mit kleinem Atomgewicht und kleinem Anfangsdruck hat diese Geschwindigkeit in unseren Versuchen bis zu $1 \cdot 5 \times 10^7$ cm/sec erreicht. Im Moment der maximalen Zusammendrückung, wenn die innere Grenze der zylindrischen Plasmaschicht die Achse erreicht, ist die Stromdichte im zentralen Teil der Entladung, die einen sehr kleinen Teil des ganzen Rohrquerschnittes einnimmt, mehrere Mal die zehnfache der mittleren Stromdichte im Rohr. Aber die Stromverteilungskurve hat keine scharfe Grenze, und auf die zentrale Zone (mit einem Radius von 0·1 des Rohrradius) entfällt weniger als die Hälfte des ganzen Stromes. Nachdem die maximale Zusammendrückung stattgefunden hat, beginnt eine schnelle Ausbreitung der Plasmaschnur und nachher wird die zweite Zusammenziehungsphase beobachtet. Bei niedrigen Werten der Anfangsdichte gelingt es drei nacheinanderfolgende Zusammenziehungs- und Ausbreitungsphasen der Plasmaschnur zu beobachten. Bei radialen Pulsierungen der Plasmaschnur, die dem ersten maximalen Zusammenziehungsmoment folgen, hält sich die Stromdichte in der zentralen Zone der Entladung die ganze Zeit sehr hoch, obwohl sie bedeutende Schwankungen erfährt. Als eine Illustration wird in der Abb. 1 die Verteilung des Stromes über den Rohrquerschnitt bei einer Entladung im Deuterium bei einem Anfangsdruck von 0·05 mm Hg widergegeben. Sie entspricht der zweiten maximalen Zusammenziehungsphase. Die Stromstärke war in diesem Augenblick ca 400 kA. Als eine charakteristische Eigenart dieser Verteilung ist anzugeben, dass in einer gewissen Entladungzone der Strom die Richtung ändert (wegen des Skin-Effektes). Die Leitfähigkeit des Plasmas, nach der Dicke der Skin-Schicht geschätzt, ist von der Grössenordnung 10^{14} CGS El-stat. Einheiten (in der ersten Zusammenziehungsphase).

Messungen von Druckimpulsen im Plasma mit Hilfe von piezoelektrischen Elementen, die von A. M. Andrianov und H. V. Filippov durchgeführt wurden, zeigen, dass in der ersten Zusammenschnürungsphase die Druckwelle zusammen mit der inneren Grenze des Stromes in die Richtung der Achse wandelt. Bis zum Moment der maximalen Zusammenschnürung ist der Druck in der Achsennähe sehr gering. Im selben Moment, wenn der Strom die Achse erreicht, steigt der Druck in der zentralen Zone bis zu 25–50 at (bei einem anfänglichen Gasdruck im Rohr von der Grössenordnung von 0·1 mm Hg).

Im Anfangsstadium der Entladung ist der Ionisationsgrad des Gases nicht hoch. Spektrometrische Messungen, die von C. J. Lukjanov und V. I. Sinizin ausgeführt wurden, zeigen, dass am Schlussmoment der ersten Zusammenschnürungsphase aus der ganzen Anzahl der Gasatome nicht mehr als 5–10 % ionisiert waren (für Entladungen im Wasserstoff und Deuterium mit einer Maximalstromstärke von ca 300 kA, bei einem Anfangsdruck von 0·01–0·1 mm Hg). Hieraus können wir folgern, dass in diesem Entladungsstadium die mittlere Elektronenenergie nicht gross ist (also einige Elektronenvolt nicht übertrifft). Nach der ersten Zusammenschnürung kann der Ionisationsgrad in der zentralen Zone der Entladung offenbar sehr hoch sein. Spektrometrische Untersuchungen zeigen noch, dass bei einer hohen Stromstärke im Entladungsraum nach der zweiten Zusammenschnürung eine merkbare Quantität von Beimischungen auf-

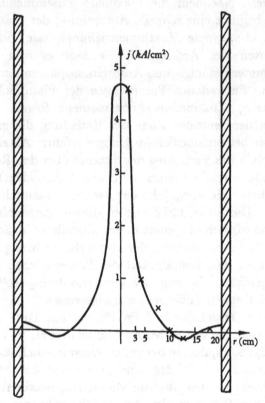

Abb. 1. Verteilung der Stromdichte über dem Entladungsquerschnitt im Deuterium bei $U_0 = 40$ kV und $P_0 = 0.05$ mm Hg für die zweite Kompressionsphase.

tritt, die von der Wechselwirkung zwischen dem Plasma und den Wänden entsteht. Darum erweckt die Untersuchung der späteren Stadien der Entladung kein grösseres Interesse.

Die vorgetragenen experimentellen Tatsachen führen zu folgenden Vorstellungen über den Mechanismus der grundsätzlichen Vorgänge bei einer Impulsentladung. Bei schnellem Anwachsen des Stromes in der Anfangsphase der Entladung können die elektrodynamischen Kräfte, die proportional zu I^2 anwachsen, nicht durch den inneren Druck des ioni-

456

sierten Gases kompensiert werden. Eine solche Kompensation ist deswegen
unmöglich, weil im ersten Stadium des Vorgangs der Strom nur in der
Nähe der Rohrwände fliesst und innerhalb der zylindrischen Plasmaschicht sich Gas befindet, das seinen ursprünglichen kleinen Druck aufrechterhalten hat. Unter der Einwirkung der elektrodynamischen Kräfte,
erfährt darum die zylindrische Plasmaschicht, die anfänglich an den
Rohrwänden lag, eine Beschleunigung die nach der Achse des Rohres
gerichtet ist.

Ein bedeutender Teil der Arbeit, die von den elektrodynamischen
Kräften in diesem Stadium ausgeführt wird, geht in die kinetische
Energie der gerichteten Bewegung derjenigen Teilchen über, die zur
konvergierenden Plasmaschicht gehören. Da aufgeladene Teilchen von
verschiedenen Vorzeichen sich mit derselben Geschwindigkeit bewegen,
erhalten die Ionen eine grosse kinetische Energie, während die Energie
der Elektronen, ihrer kleinen Masse wegen, sich kaum ändert. Den Zu-
sammenschnürungsvorgang kann man auch als das Entstehen einer
Stosswelle betrachten, die zur Achse konvergiert. Vor der inneren Front
dieser Welle befindet sich ursprünglich neutrales Gas. Bei der Bewegung
der Plasmaschicht wird das Gas teilweise von den aufgeladenen Teilchen
des Plasmas mitgenommen (wegen des grossen effektiven Querschnitts der
Ionenumladung) und ionisiert. Darum wächst die Stoffmasse, die in
Bewegung gerät, stetig an und die Gesamtzahl an Elektronen und Ionen
im Plasma steigt. Die quantitative Theorie der Plasmazusammenschnürung wurde von M. A. Leontovitch und C. M. Ossovez entwickelt.
Diese Theorie ruht auf der Voraussetzung, dass im Anfangsstadium der
Kompression der Gasdruck in Verhältnis zur Grösse $(d/dt)\,Mv$, die die
Änderung der Bewegungsgrösse des Plasmas angibt, klein ist. Mit Hilfe
einiger anderer Annahmen, die die Berechnungen vereinfachen (insbesondere wird angenommen, dass der Strom von der Zeit linear abhängig
ist), liefert diese Theorie die folgende Formel für die Bestimmung der
Dauer des Kompressionsvorgangs:

$$\tau_c = 1{,}4\, M^{\frac{1}{4}} a_0^{\frac{1}{2}} \left(\frac{dI}{dt}\right)^{-\frac{1}{2}}.$$

In dieser Formel ist M—die Gasmasse auf einer Längeneinheit der Plasma-
säule, a_0—Rohrradius, I—Stromstärke in CGS El-magn. Einheiten. Der
Vergleich zwischen gemessenen und berechneten Werten für τ_c ist in der
Abb. 2 angeführt. (Die theoretischen Werte sind mit geraden Linien dar-
gestellt.) Im weiten Umfang der Änderung der Grösse M ist eine gute
Übereinstimmung zwischen Theorie und Experiment festzustellen.

Das Stadium der magnetischen Beschleunigung des Plasmas schliesst sich in dem Moment ab, wenn die innere Grenze der Plasmaschnur die Achse erreicht. In diesem letzteren und sehr kurzen Kompressionsstadium geht ein bedeutender Teil der Energie der gerichteten Ionenbewegung in Wärme über, was zu einer kräftigen Druck- und Temperatur-steigerung führt. Nach einer groben Schätzung erreichte die Temperatur des Plasmas in den Phasen maximaler Kompression in unseren Versuchen die Grössen-ordnung von 10^6 Grad (für Wasserstoff und Deuterium bei einem Anfangsdruck von der Grössenordnung von 10^{-2} mm Hg). Diese Schät-

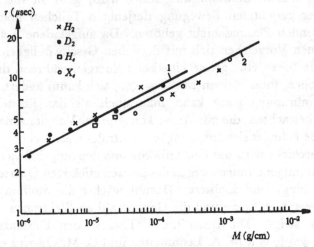

Abb. 2. τ—Abhängigkeit von M bei $U_0 = 30$ kV, $a_0 = 20$ cm. (1) $dI/dt = 6 \times 10^{10}$ A/sek (für leichtes Gas); (2) $dI/dt = 7 \cdot 5 \times 10^{10}$ A/sek (für schweres Gas).

zung kann, erstens auf Grund von Messungen an Grössen, die die energetische Bilanz der Entladung bestimmen, zweitens—auf Grund von Messungen der Druckimpulse mit Hilfe von piezoelektrischen Elementen, und, drittens—auf dem Wege einer Bestimmung der Arbeit der elektro-dynamischen Kompressionskräfte, gemacht werden. Um Missverständnisse zu vermeiden, ist es notwendig zu bemerken, dass wir bei der Rede über die Temperatur des Plasmas die Temperatur der schweren Teilchen—Ionen und neutraler Atome—ins Auge fassen. Wir können sagen, dass infolge des grossen Wertes des effektiven Querschnittes der Umladung, die Tem-peratur der Ionen und Atome in der sich zusammenziehenden Plasma-schnur beinahe gleich ist.

Wie schon oben angeführt wurde, wird beim Kompressionsvorgang ein bedeutender Teil der Arbeit, die von den elektrodynamischen Kräften geleistet wird, auf die kinetische Energie der gerichteten Bewegung

schwerer Teilchen aufgewandt. Gleichzeitig ist zu bemerken, dass bei
der Konvergenz der zylindrischen Plasmaschicht in die Richtung der
Achse, die Masse, die in Bewegung gerät, stetig anwächst, daher muss ein
Teil der Arbeit auf unelastische Prozesse, d.h. auf die Erhöhung der
Plasmatemperatur, verwendet werden. Die Leistung, die auf unelastische
Kräfte verbraucht wird, ist $\sim \frac{v^2}{2} \frac{dM}{dt}$ gleich (v—die Geschwindigkeit der
Plasmazusammenschnürung). Die Daten über die energetische Bilanz der
Impulsentladung machen die Annahme ganz wahrscheinlich, dass noch
vor dem Momente der maximalen Kompression die Temperatur der
Ionen und Atome einige hunderttausend Grade erreicht (was der mitt-
leren Energie einer chaotischen Bewegung von der Grössenordnung von
einigen zehn Elektronvolt entspricht). Im Gegenteil verbleibt die
Temperatur des Elektronenkomponenten des Plasmas im Stadium der
ersten Zusammenschnürung sehr niedrig und entspricht der mittleren
Elektronenenergie von einigen Elektronenvolt. Dies folgt aus den Ergebnissen der spektrometrischen Messungen, die auf einen nicht hohen
Ionisationsgrad des Plasmas weisen.

Falls die mittlere Elektronenenergie einige zehn Elektronvolt erreichen
würde, so wäre das Gas schon vor dem maximalen Zusammenschnürungs-
moment des Stromes vollständig ionisiert.

Der Charakter der Vorgänge, die nach dem Moment der maximalen
Kompression auftreten, ist noch nicht klar. Aber es ist offenbar, dass nach
der maximalen Kompression eine auseinandergehende Stosswelle entstehen muss, die das Plasma gegen die Wände mitnimmt. Die auseinandergehende Welle muss durch die Einwirkung der dynamischen Kräfte,
die den Strom zusammenschnüren, schnell gebremst werden, weswegen
wieder eine Kompressionsphase eintritt, auf die wieder die zweite
Schnurverbreitung folgt. Die Frequenz der radialen Pulsierungen des
Plasmas ist von der Grössenordnung $\frac{\overline{H}}{M}$, wo $\overline{H}$ einen gewissen Mittelwert des Magnetfeldes der Schnur darstellt. In diesem Stadium des
Prozesses müssen offenbar verschiedene Arten von Labilität auftreten, die
einer Plasmaschnur bei grosser Stromstärke eigen sind, wodurch die
Schnurenform sich stark ändern kann. Die charakteristische Zeit, die die
Geschwindigkeit bestimmt, mit welcher die Schnur von der zylindrischen
Form abweicht, ist der Grösse nach mit der Periode der radialen Pul-
sierungen vergleichbar.

Der Verlust an Stabilität bei der Plasmaschnur und die starke Wechselwirkung zwischen dem Plasma und den Wänden führen zu einer

wesentlichen Veränderung des Charakters der Prozesse. In späteren Stadien der Entladung erscheint im Volumen eine grosse Quantität von Nebengasen und die Plasmatemperatur sinkt ganz erheblich.

Die theoretische Analyse verschiedener Arten elektrodynamischer Labilität, die einer Plasmaschnur bei starkem Strom eigen sind, zeigt, dass ein starkes äusseres Magnetfeld, das längs der Schnurachse gerichtet wird, bei gewissen Verhältnissen die Rolle eines effektiven stabilisierenden Faktors spielen kann, mit dessen Hilfe gewisse Arten von Labilität vollständig beseitigt und andere kräftig herabgesetzt werden können. Daher schien es interessant, die Einwirkung eines äusseren Magnetfeldes auf eine Impulsladung von hoher Stromstärke zu untersuchen.

A priori konnte man vermuten, dass im Anfangsstadium des Zusammenziehens der Plasmaschnur in Anwesenheit eines kräftigen äusseren Feldes, die Geschwindigkeit der Plasmabewegung gegen die Achse kleiner werden wird. Solche Verminderung muss dadurch bedingt sein, dass die sich zusammenziehende Schnur bei grosser Leitfähigkeit des Plasmas die Kraftlinien des äusseren Feldes mitnehmen wird, weswegen die Feldstärke des Längsfeldes innerhalb der Schnur steigen und ein Überschuss an magnetischem Druck zustandekommen wird, der den elektrodynamischen Stromkräften, die das Plasma zusammenziehen, entgegen gerichtet ist. Man kann sagen, dass im Prozess der Zusammendrückung die Plasmaschnur sich als ein Stoff mit paramagnetischen Eigenschaften verhalten muss.

Die experimentellen Untersuchungen des Einflusses vom äusseren Magnetfeld auf die Impulsentladung, die von A. L. Besbatschenko, I. N. Golovin, D. P. Ivanov, V. D. Kirillov und H. A. Javlinsky ausgeführt wurden, bestätigten die Voraussetzung, dass das Plasma der Impulsentladung paramagnetische Eigenschaften hat. Gleichzeitig haben sie zu einem unerwarteten Ergebnis geführt—es stellete sich heraus, dass das Plasma die paramagnetischen Eigenschaften nicht nur im Kompressionsstadium behält. Daraus folgt offenbar, dass der Paramagnetismus des Plasmas nicht nur durch den klassischen Mechanismus vom Mitnehmen der Kraftlinien durch den sich zusammenziehenden Leiter, sondern auch durch andere Effekte (besonders durch Anisotropie der Leitfähigkeit im magnetischen Felde) verursacht werden kann. Die Ergebnisse dieser experimentellen Untersuchungen werden zum Gegenstand einer Mitteilung von I. N. Golovin gemacht werden.

Eine interessante Eigenschaft der kräftigen Impulsentladungen ist die, dass sie bei gewissen Verhältnissen zu einer Quelle von Neutronen- und harten Röntgenstrahlen werden können. Diese Erscheinung, von uns im Jahre 1952 entdeckt, hat noch keine entgültige Erklärung erhalten. Bei einer Entladung im Deuterium und einem Anfangsdruck von einigen Tausendstel mm bis zu 0,5 mm Hg werden sehr kurze Impulse von Neutronen- und harter Röntgenstrahlung beobachtet, die zeitmässig mit der Phase der zweiten (und manchmal der dritten) maximalen Kompression zusammenfallen. In Rohren mit metallischen Seitenwänden kommen die Neutronen und harten Röntgenstrahlen auch bei bedeutend höheren Anfangsdrucken—bis zu 10 mm—auf. (Es ist zu bemerken, dass in diesem Falle die Entladung einen mehr komplizierten Charakter erhält.) Impulse von harter Röntgenstrahlung werden auch bei Entladungen im Wasserstoff beobachtet.

In dem Falle, wenn die Entladung im Deuterium stattfindet, können die von der Röntgenstrahlung und Neutronen erzeugten Impulse oszillographenmässig auf Phase eingestellt werden. Dabei stellt es sich heraus, das ihre Erscheinungsmomente genau zusammenfallen. Es spricht ohne Zweifel dafür, dass die beiden Arten von Strahlungen eine gemeinsame Ursache haben. Da das Erscheinen von harten Röntgenstrahlen darauf hinweist, dass im Plasma in einem gewissen Zeitmoment sehr schnelle Elektronen mit einer Energie von einigen hundert keV auftreten, so ist die Ursache dieses Effektes in einem gewissen Beschleunigungsmechanismus zu suchen.

Wenn wir diesen Mechanismus zu erraten suchen, müssen wir die spezifischen Verhältnisse bei der Impulsentladung mit kräftiger Stromstärke in Rechnung ziehen. Bei einer solchen Entladung ist im Plasma ein starkes Magnetfeld vorhanden. Falls der Strom im Plasma über eine zylindrische Symmetrie verfügt, so bilden die Kraftlinien des Feldes konzentrische Kreise, dessen Zentra auf der Achse des Entladungsrohrs liegen. Elektrische Felder im Plasma, beim Vorhandensein einer Stromsymmetrie, können Komponenten nur in der zur Achse des Rohres parallelen oder in radialer Richtung haben. Daher wird in jedem Punkt des Plasmas der Feldstärkevektor des elektrischen Feldes zum Vektor des magnetischen Feldes senkrecht stehen. Bei diesen Verhältnissen kann die Beschleunigung der Teilchen nur in dem Teil des Raumes zustandekommen, wo die Feldstärke des Magnetfeldes nahe an Null liegt, d.h. nur in der Nähe der Entladungsachse. Die zweite notwendige Bedingung für die Beschleunigung von geladenen Teilchen im Plasma ist die, dass die Dichte des Stoffes im Raume, wo die Beschleunigung stattfindet, sehr klein sein soll.

Auf Grund des oben Gesagten können wir eine der möglichen Ursachen vom Auftreten der schnellen Teilchen im Plasma aufweisen. Die Teilchen können in der Nähe der Entladungsachse im elektrischen Längsfelde induktiven Ursprungs beschleunigt werden. Eine solche Hypothese kann das Auftrittmoment der schnellen geladenen Teilchen erklären. Wie die Messungen zeigen, hat das elektrische Längsfeld im Plasma im Moment der zweiten maximalen Konvergenz einen sehr hohen Wert. Die Spannung dieses Feldes kann vielfach diejenige Grösse überschreiten, die nur durch die an das Rohr in diesem Moment angelegte äussere Spannung verursacht wird. Gleichzeitig wird der Stoff unter der zweiten Kompressionsphase in der Nähe der Achse des Entladungsrohrs praktisch vollständig ionisiert, und seine Dichte ist in der zentralen Zone im Anfangsstadium der zweiten Konvergenzphase gering. Daher sind unter der zweiten Kompressionsphase offenbar die notwendigen Bedingungen zur Beschleunigung für eine gewisse Gruppe von Teilchen, die in der Nähe der Entladungsachse liegen, vorhanden. Aber wir können nicht zur Zeit behaupten, dass gerade dieser Mechanismus die Ursache der Erscheinung von schnellen Teilchen bei der Entladung ist. Es ist nicht ausgeschlossen, dass verschiedene Arten von Labilität, die der Plasmaschnur bei hoher Stromstärke eigen sind, im Beschleunigungsprozess der Teilchen eine bedeutende oder sogar eine bestimmende Rolle spielen. Wir können hier auf zwei Arten solcher Labilität hinweisen, die von der Theorie vorausgesehen und beim Versuch wirklich beobachtet werden. Eine von ihnen besteht darin, dass die Plasmaschnur sich in der Längsrichtung ungleichmässig zusammenzieht. Es hat zur Folge, dass in den Phasen der maximalen Zusammenziehung auf einzelnen Stellen der Plasmaschnur sehr schmale Einschnürungen entstehen können, die dann wieder in schnelle Verbreitungen übergehen. Im Moment einer solchen Verbreitung der Einschnürung können sehr grosse lokale Überspannungen auftreten. Die zweite Art von Labilität, die beim Versuch beobachtet wird, besteht im willkürlichen Entstehen eines magnetischen Längsfeldes im Plasma, das durch die wirbelartige Verdrillung der Plasmaschnur verursacht wird. Das Entstehen des magnetischen Längsfeldes führt zu einer Formveränderung der magnetischen Kraftlinien, weswegen eine Beschleunigung von Teilchen auch ausserhalb der Grenzen der zentralen Zone der Entladung möglich wird.

Es ist auch möglich, das im Prozesse der Entstehung von elektrischen Feldern, die die Teilchen im Plasma beschleunigen, die Wechselwirkung zwischen dem Plasma und den Wänden der Entladungskammer eine gewisse Rolle spielt.

Jedenfalls kann man als festgestellt ansehen, dass die von uns beobachteten harten Strahlungen vielleicht kein direktes Verhältnis zu den thermonuklearen Vorgängen haben. Bei den Temperaturen, welche bei Impulsentladungen von einer Stromstärke in der Höhe von einigen hundert Kiloampere bis zu einer Million Ampere erreicht werden, müssen die thermonuklearen Reaktionen eine ganz geringe Intensität haben (bedeutend kleiner als die Reaktionsintensität, die bei unseren Versuchen beobachtet wurde). Ausserdem kann man das Auftreten der harten Röntgenstrahlen in der Impulsentladung nicht mit Hilfe des thermonuklearen Mechanismus erklären.

Wir haben einige Ergebnisse der Untersuchung von kräftigen Impulsentladungen geprüft. Das wichtigste von diesen Ergebnissen ist die Feststellung der experimentellen Möglichkeit, sehr hohe Temperaturen zu erreichen—von der Grössenordnung von einer Million Grad. Eine weitere Erhöhung der Temperatur ist nur durch einen Übergang zu einer noch höheren Steigerungsgeschwindigkeit des Stromes bei der Entladung möglich (weil die von Ionen im Kompressionsprozess angeeignete Energie proportional zu der Steigerungsgeschwindigkeit des Stromes ist). Bei einer genügend grossen Steigerungsgeschwindigkeit des Stromes kann man mit einem Auftritt von intensiven thermonuklearen Reaktionen im Moment der ersten Kompression rechnen. Die praktischen Aussichten für eine weitere Arbeit in dieser Richtung ist im grossen ganzen davon abhängig, ob es gelingt solche Voraussetzungen zu schaffen, bei welchen die Plasmaschnur beim Anwachsen des Stromes vielfache Schwingungen aussteht, ohne dass sie zerfällt oder die Wände berüht. Die der Schnur eigene Labilität kann zur Ursache werden, dass man solche Vorbedingungen nicht schaffen kann. In diesem Falle wird die Grösse τ (d.h. die mittlere Lebensdauer der schnellen Ionen im Plasma) zu klein, daher wird auch der energetische Wirkungsgrad, der zu τn proportional ist, gering. Auf Kosten der Steigerung von n kann man hier nicht weit kommen, weil die absolute Energiegrösse, die zur Erzeugung einer Impulsentladung notwendig ist, bei der gegebenen mittleren kinetischen Energie des Ions proportional zu n wächst. Bei der Betrachtung des Problemes im ganzen ist zu beachten dass der Weg über die kurzzeitige Temperatursteigerung des Plasmas bei einer kurzen Impulsentladung nur eine von sehr vielen Richtungen darstellt, die man zur Lösung der Aufgabe über die Erregung von kontrollierbaren thermonuklearen Reaktionen einschlagen kann.

UNTERSUCHUNG EINER STARKSTROMGASENTLADUNG IM MAGNETISCHEN LÄNGSFELDE*

A. L. BESBATSCHENKO, I. N. GOLOVIN, D. P. IVANOV,
V. D. KIRILLOV UND N. A. JAVLINSKIJ

Academy of Sciences, Moscow, U.S.S.R.

ZUSAMMENFASSUNG

Die Gasentladung im Deuterium bei Stromstärken bis zu 700 tausend Ampere im magnetischen Längsfelde bis zu 12000 Oersted wurde untersucht. Die Einwirkung des Feldes auf den Verlauf der Entladung wurde festgestellt und eine Zunahme des Magnetfeldes im Innern der Entladungssäule entdeckt. Eine Bewertung der Plasmaleitfähigkeit und der Ionisierungszahl wird gegeben.

I. EINLEITUNG

Die Starkstromgasentladung hat in der letzten Zeit die Aufmerksamkeit der Forscher im Zusammenhang mit dem Suchen nach einer Möglichkeit der Erregung von steuerbaren thermonuklearen Reaktionen auf sich gezogen. In den Arbeiten [1] und [2] wurden die Erwärmungsverfahren des Plasmas auf dem Wege einer direkten Beschleunigung der Ionen durch das Magnetfeld des durch das Gas fliessenden Stromes untersucht. Wie aber in diesen Arbeiten gezeigt wurde, kann das heisse Plasma nicht lange aufrecht-erhalten werden, weil die Entladungssäule schnell zusammenbricht, das Plasma die Wände berührt und sich abkühlt.

Die Vermutung schien also natürlich, dass ein magnetisches Längsfeld die Labilität der Plasmasäule aufheben oder jedenfalls den Vorgang des Zusammenbruches und der darauffolgenden Abkühlung der Plasmasäule verzögern könnte.

In diesem Artikel ist die Beschreibung von Untersuchungen der Gasentladung im Deuterium bei einem Druck von 0·05 bis zu 0·4 mm Hg enthalten. Die verschiedenen Stadien der Entladung beim Anwachsen des Stromes von Null bis zum Maximum wurden studiert. Die Strom-

---

* Presented at the Symposium by I. N. Golovin.

stärke erreichte 700 tausend Ampere, die Feldstärke des Längsfeldes 12000 Oersted. Die Untersuchungen wurden bei einer mit der Feldstärke des Entladungsstromes vergleichbaren Längsfeldstärke ausgeführt.

2. DIE BESCHREIBUNG DER ANLAGE

Die Untersuchung der Gasentladung wurde mit einer Impulsanlage durchgeführt, deren Aufbau in der Abb. 1 dargestellt ist. Eine 65 bis

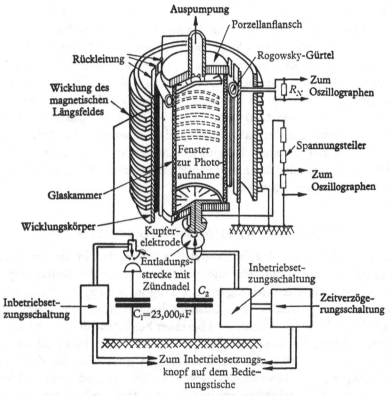

Abb. 1. Aufbau der experimentellen Anlage.

70 cm lange Entladungskammer aus Glas oder in einigen Fällen aus Porzellan von 18–20 cm Durchmesser mit flachen Kupferelektroden an den Enden, ist innerhalb einer Spule von 36 cm Diameter untergebracht. Die Öffnungen zwischen den Wicklungswindungen ermöglichen es, das Leuchten der Entladung zu beobachten.

Eine Kondensatorenbatterie $C_1 = 23000\,\mu F$, die über eine Kugelfunken-strecke in die Spule entladen wird, erregt im Kreise gedämpfte elektrische

Schwingungen, die eine Frequenz von 73 Hz haben. Das magnetische Längsfeld ist in dem von der Entladungskammer eingenommenen Volumen genügend homogen und erreicht in der ersten Halbperiode eine Feldstärke von 12000 Oersted für eine Aufladung der Kondensatorenbatterie bis zu einer Potentialdifferenz von 2 kV.

In dem Augenblick, wenn das Magnetfeld der Spule durch sein Maximum geht, wird von einem steuerbaren Entlader die Entladung der Kondensatorenbatterie C_2 durch die Gasentladungskammer zustande gebracht. Die Frequenz des Schwingungskreises, der aus C_2, den stromzuführenden Schienen und der Entladungskammer besteht, ist hoch und im Verhältnis zu ihr kann man das magnetische Längsfeld als konstant betrachten. Für die bei den Versuchen gebrauchten Kapazitäten von C_2, die 30 μF und 180 μF ausmachten, waren die entsprechenden Frequenzen 43 bzw. 17 kHz, und die maximale Stromstärke durch die Gasentladungskammer 330 kA und 700 kA bei einer Aufladung der Kondensatoren bis zu einem Potentialunterschied von 40 kV. Die Kugelfunkenstrecken im Kreise der Kondensatorenbatterien C_1 und C_2 haben eine Zündvorrichtung, die von einer Thyratronenschaltung gesteuert wird. Zum Einschalten der Batterie C_2 im erforderlichen Augenblick dient eine Röhrenschaltung mit einstellbarer Zeitverzögerung.

3. DIE MESSMETHODEN UND ERGEBNISSE

In den Arbeiten [1] und [2] ist festgestellt worden, dass in der Abwesenheit des magnetischen Längsfeldes die Bestehungsdauer der Entladungssäule klein und von der Geschwindigkeit der Trägheitsbewegung des ionisierten Gases von den Wänden der Gasentladungskammer zur deren Mitte abhängig ist. Wir entdeckten, dass bei einer Feldstärke des Längsfeldes von einigen tausend Oersted der Zusammenbruch der Plasmasäule später eintritt. Es ist wichtig, dass ein merkbarer Unterschied im Zusammenbruchsvorgang in Abwesenheit des äusseren magnetischen Feldes zu beobachten ist. Tatsächlich kommt in der Abwesenheit eines äusseren magnetischen Feldes der Zusammenbruch gleich nach der ersten Zusammenziehung zustande. Falls ein magnetisches Feld z. B. von einer Stärke von 0·7 oder 2 tausend Oersted vorhanden ist (siehe Abb. 3), führt die Schnur eine volle Schwingung aus, ohne zusammenzubrechen.

Bei allen von uns untersuchten Arbeitsverhältnissen, wenn der Strom im Gas sein Maximum erreicht, treten im Leuchtspektrum der Entladung Siliziumlinien auf, die auf eine Verdampfung der von der Entladung abgebrannten Kammerwände hinweisen.

Bei den Versuchen wurden die Stromstärke der Entladung, die Spannung zwischen den Elektroden, der Radius der Entladungssäule und der Mittelwert des magnetischen Längsfeldes in derselben gleichzeitig gemessen.

Die Stromstärke der Entladung wurde mit einem Rogovsky-Gürtel gemessen, der mit einem Nebenwiderstand R_N überbrückt war, wobei $R_N \ll \omega L$, wo ω die Kreisfrequenz des Entladungskreises und L die Selbstinduktion des Gürtels bezeichnen. Die Spannung auf dem Widerstande R_N, die dem vom Gürtel umfassten Strom proportional ist, wurde mit einem Doppelstrahl-Impulselektronenstrahloszillographen OK-17 registriert.

Die Spannung zwischen den Elektroden wurde über einen niederohmigen Spannungsteiler, der in der Abb. 1 dargestellt ist, durch den anderen Strahl desselben Oszillographen registriert. Einige Beispiele der Strom- und Spannungsoszillogramme werden in der Abb. 2 widergegeben.

Eine regelmässige Sinusform des Stromes wird vom Kreise bestimmt, dessen Wellenwiderstand grösser als der Widerstand der Gasentladung ist. Die Spitze im Spannungsdiagramm bei $H = 0$ ist dadurch zu erklären, dass beim schnellen Zusammenziehen der Entladungssäule eine zusätzliche induktive Komponente der Elektrodenspannung $2V = \dfrac{\dot{a}}{a} lI$ auftritt, wo a den Säulenradius, l den Abstand zwischen den Elektroden und I die Stromstärke der Entladung bedeuten. Mit dem Anwachsen des magnetischen Feldes wird die Zusammenziehung der Säule verzögert und die zusätzliche induktive Komponente kleiner, die Spannung nimmt einen gleichmässigen cosinusförmigen Verlauf.

Eine Vergleichung der Oszillogramme 2 a und 2 b zeigt, dass auch eine Erhöhung des Anfangdruckes des Gases die Spannungkurve glättet.

Der Radius der Entladungssäule wurde nach Entladungsaufnahmen auf photographischem Wege gemessen, die mit einem Schnellphotoregistriergerät mit einer Aufnahmegeschwindigkeit von 2×10^6 Aufnahmen/Sek. gemacht wurden. Einige Beispiele solcher Aufnahmen sind in der Abb. 3 angegeben. Zur Bestimmung derjenigen Aufnahme, die dem Anfang des Vorgangs entspricht, wurde gleichzeitig mit der Entladungssäule auch der Funke in der Kugelentladungsstrecke photographiert. Der Radius der Entladungssäule, der durch die Breitenmessung des sichtbaren Schwärzungsgebietes des Films bestimmt wurde, ist als eine Zeitfunktion bei verschiedenen Entladungsverhältnissen in der Abb. 4 aufgetragen. Dabei wurden die Aufnahmen für das Zeitmoment ausgenützt, bevor die Entladungssäule die zylindrische Symmetrie verliert.

Eine Bestimmung des Radius von den Aufnahmen ist nur nach den

30-2

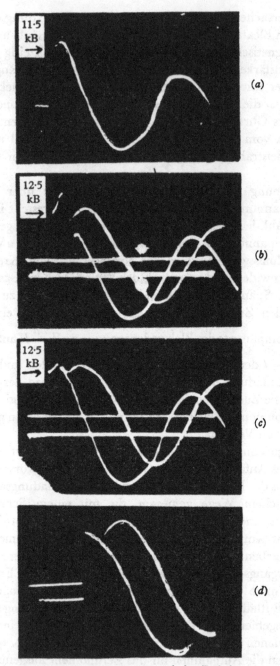

Abb. 2. Oszillogramme der Elektrodenspannung und des Entladungsstromes. $f=43$ kHz, $I_{max}=250$ kA. (a) $H_0=0$, $p=0.3$ mm Hg, (b) $H_0=0$, $p=0.05$ mm Hg, (c) $H_0=1500$ Oersted, $b=0.05$ mm Hg, (d) $H_0=8000$ Oersted, $p=0.05$ mm Hg.

468

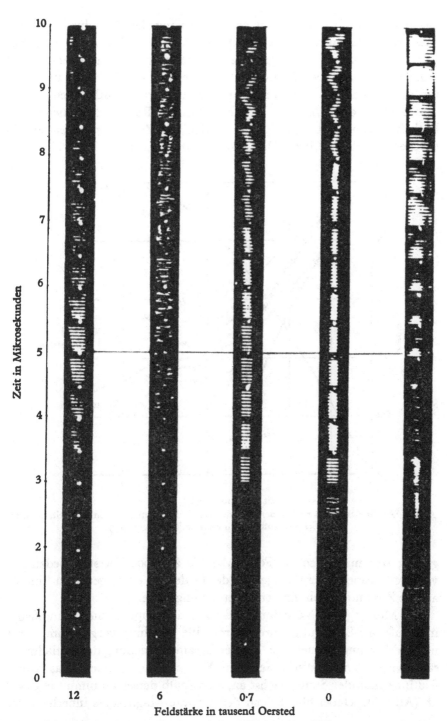

Abb. 3. Schnellphotoaufnahmen der Gasentladung. Die dunklen Linien quer über der Entladung sind Schatten von Spulenwindungen für magnetische Längsfelderregung.

469

ersten 1·5–2 Mikrosekunden nach dem Anfang der Entladung möglich, weil im Anfangsmoment die Leuchthelligkeit zu klein ist. Eine Extrapolierung der angeführten Kurven zu Null hin zeigt eine zufriedenstellende Übereinstimmung zwischen dem Anfangsradius der Säule a_0 und dem inneren Radius der Entladungskammer. Weiterhin wird angenommen, dass der auf diese Art und Weise bestimmte Radius der leuchtenden Entladungszone mit dem Radius derjenigen Zone zusammenfällt, in welcher der Strom fliesst. Die Zusammenziehung wird durch eine Stei-

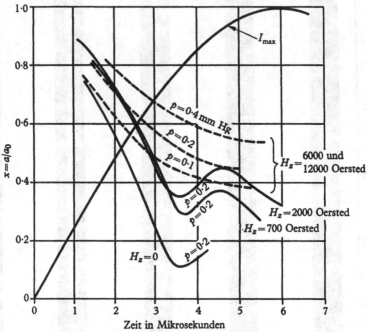

Abb. 4. Abhängigkeit des Radius der Entladungssäule im Deuterium von der Zeit, erhalten von Schnellphotoaufnahmen vom Typ der Abb. 3.

gerung der magnetischen Feldstärke bis zu 6000 Oersted bedeutend verzögert. Eine weitere Steigerung des Feldes hat einen geringen Einfluss auf die Zusammenziehung der Gasentladungssäule.

Die Messung der Verteilungsänderung des magnetischen Längsfeldes im Laufe der Entladung wurde durch eine Windung vorgenommen, die nach Abb. 5 angeordnet war. Bei der Zusammenziehung der Entladungssäule wird das magnetische Feld vom Plasma mitgenommen. Das Längsfeld innerhalb der Säule wächst an, ausserhalb derselben nimmt es etwas ab (Abb. 6). Dabei bleibt doch der gesamte Magnetfluss innerhalb der

470

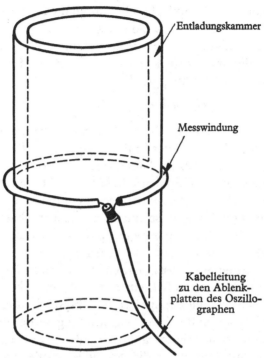

Abb. 5. Anordnung der Windung, zur Messung der Änderung des Längsfeldes bei der Entladung.

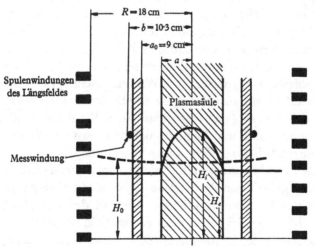

Abb. 6. Änderung des magnetischen Längsfeldes während der Entladung.

Spule konstant, weil für die Frequenz des Entladungskreises die Spule, die das magnetische Längsfeld bildet, durch die sie speisende Kondensatorenbatterie praktisch kurzgeschlossen ist.

Das Anwachsen des Magnetflusses innerhalb der Messwindung, der auf ihr eine elektromotorische Kraft E erregt, ist der Abnahme des Flusses ausserhalb der Windung gleich:

$$\pi a^2 \Delta \bar{H}_i - \pi (b^2 - a^2) \, \Delta H_e = \int_0^t E\,dt = \pi (R^2 - b^2) \, \Delta H_e. \qquad (3\cdot1)$$

Es bedeuten: $\Delta \bar{H} = \bar{H}_i - H_e$ mittlere Zunahme der Feldstärke dem Querschnitt nach innerhalb der Entladungssäule; $\Delta H_e = H_0 - H_e$ Abnahme der Feldstärke ausserhalb der Säule; a Radius der Entladungssäule; b Radius der Messwindung; R Spulenradius; t die Zeit vom Anfang der Entladung gerechnet; H_0 Längsfeldstärke vor der Entladung; H_i und H_e die Grösse des Feldes beziehungsweise innerhalb und ausserhalb der Entladungssäule.

Die an der Windung erregte Spannung wurde durch ein angepasstes Koaxialkabel direkt oder über einen Spannungsteiler aus Widerständen auf die Platten eines Oszillographen übertragen. Gleichzeitig wurde der andere Strahl zum Aufschreiben der Stromstärke benutzt. Die Oszillogramme wurden photographiert und graphisch integriert. Einige Beispiele der Oszillogramme der Spannungswindung werden in der Abb. 7 angeführt.

Aus den Schnellphotographien (Abb. 3) ist ersichtlich, dass im Laufe von 5–6 Mikrosekunden die Entladungssäule eine regelmässige zylindrische Form beibehält. Die Oszillogramme (Abb. 7) zeigen bei $H \neq 0$ in dieser Zeit eine gleichmässig verlaufende Kurve, die ein Anwachsen des Längsfeldes innerhalb der Entladungssäule aufweist, aber im Falle $H = 0$ ist in dieser Zeit die elektromotorische Kraft an der Windung gleich Null.

Nach 6 Mikrosekunden kommen auf den Aufnahmen Ausbiegungen der Säule zum Vorschein, und auf den Windungsoszillogrammen treten mit diesen Ausbiegungen verknüpfte unregelmässige Schwingungen auf.

Eine graphische Integrierung der Windungsoszillogramme wurde nur im ersten Zeitabschnitt ausgeführt, wo der Säulenradius a bekannt ist.

Wie aus der Formel $(3\cdot1)$ folgt, kann man für diesen Fall nicht nur ΔH_e, sondern auch $\Delta \bar{H}_i$ und folglich auch den Fluss im Inneren der Säule $\phi_i = \pi a^2 \bar{H}_i$ für ein beliebiges Zeitmoment bestimmen.

In der Abb. 8 geben die punktierten Kurven das gemessene Verhältnis ϕ_i/ϕ_0 in Abhängigkeit von der Zeit an $(\phi_0 = \pi a^2 H_0)$. Die ganzen Kurven wurden aus der Berechnung erhalten, die mit der Aufnahme gemacht wurde, dass $\bar{H}_i^2 - H_e^2 = H_\phi^2$, wo H_ϕ das eigene Feld des Entladungsstromes auf der Oberfläche der Entladungssäule bedeutet. Das Zusammenfallen der Kurven führt zur Folgerung, dass der Gasdruck im Vergleich zum

472

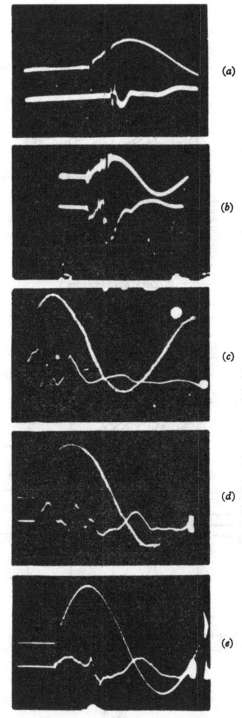

Abb. 7. Oszillogramme des Entladungsstromes und der Spannung an der Windung, die die Änderung des magnetischen Längsfeldes während der Entladung misst. $f = 43$ kHz, $p = 0.1$ mm Hg, $I_{max} = 300$ kA. (a) $H_0 = 0$, (b) $H_0 = 170$, (c) $H_0 = 2 \times 10^3$, (d) $H_0 = 6 \times 10^3$ und (e) $H_0 = 12 \times 10^3$ Oersted.

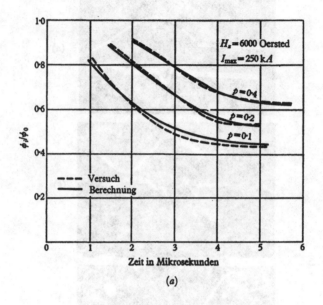

(a)

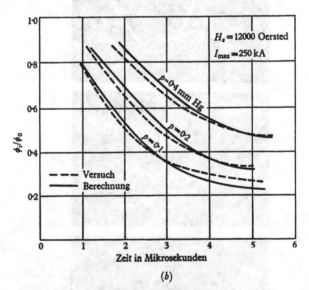

(b)

Abb. 8. Änderung des Verhältnisses zwischen dem in der Säule zurückgebliebenen magnetischen Fluss ϕ_t und dem anfänglichen Fluss $\phi_0 = \pi a_0^2 H_0$.

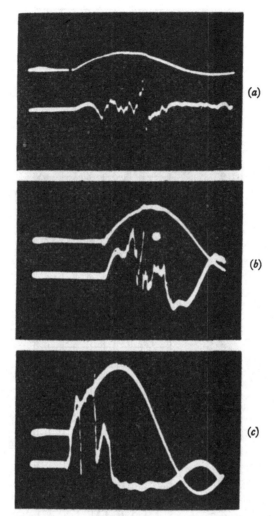

Abb. 9. Spannungsoszillogramme an der Windung für verschiedene Werte des maximalen Stromes. $f = 17$ kHz, $p = 0 \cdot 1$ mm Hg und $H_0 = 4000$ Oersted. (a) $I_{max} = 160$, (b) $I_{max} = 360$ und (c) $I_{max} = 700$ kA.

magnetischen Druck klein ist. Dafür spricht auch die Bearbeitung der Oszillogramme der Windungsspannung E_0 für verschiedene Werte des Entladungsstromes I (Abb. 9), aus welcher zu ersehen ist, dass im jeden Zeitmoment $\int E_b\, dt$ näherungsweise dem Quadrat von I proportional ist.

Tatsächlich ist einerseits $\int E_b\, dt \sim I^2$, anderseits $\int E_b\, dt \sim (H_1^2 - H_e^2)\, a^2$. Die Abnahme von ϕ_i / ϕ_0 mit der Abnahme des Radius zeigt eine verhältnis-

475

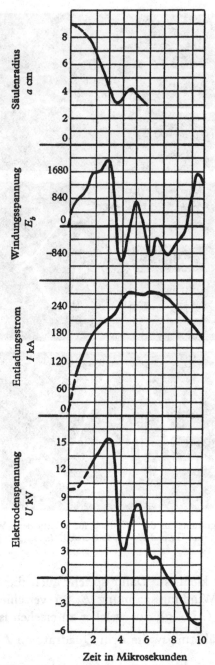

Zeit in Mikrosekunden

Abb. 10. Abhängigkeit von der Zeit des Entladungssäulenradius, der Windungsspannung, des Entladungsstromes und der Elektrodenspannung bei einem Längsfelde von 2000 Oersted.

476

mässig geringe Leitfähigkeit des Plasmas unter den ersten drei Mikro-sekunden an, so dass ein bedeutender Teil des Flusses ϕ_0 (unter gewissen Verhältnissen grösser als die Hälfte) im Kompressionsvorgang die Säule verlässt. Während der fünften und sechsten Mikrosekunde, wie aus der Abb. 8 vorgeht, hört das Herauskriechen des Längsfeldes aus der Entladungssäule praktisch auf.

Als eine Illustration dazu, wie weit die Änderungen aller unserer zu messenden Grössen mit der Zeit von der Kompression abhängig sind, die auf den Schnellphotoaufnahmen beobachtet wird, sind in der Abb. 10 der Radius a der Entladungssäule, die Wicklungsspannung E, die Ent-ladungsstromstärke I und die Spannung zwischen den Elektroden V in Abhängigkeit von der Zeit aufgetragen. Alle Kurven sind bei einem Druck von $p = 0\cdot 2$ mm Hg und $H = 2000$ Oersted aufgenommen. Dem Moment der ersten maximalen Zusammenziehung der Säule entspricht ein Durchgang der Windungsspannung durch Null. Das Anwachsen des Entladungs-stromes wird in diesem Moment kleiner, aber die Spannung zwischen den Elektroden erreicht ihr Maximum. Bei der Ausbreitung der Säule ändert die Windungsspannung ihr Zeichen, der Entladungsstrom wächst ver-hältnismässig rasch an, die Spannung zwischen den Elektroden sinkt.

4. RADIALE BEWEGUNGEN IN DER ENTLADUNGSSÄULE

Eine vollständige theoretische Untersuchung des Plasmazustandes wäh-rend der Entladung würde sehr kompliziert werden. Darum muss man die Gasdynamik der Säulenkompression und die dabei auftretenden kine-tischen Vorgänge (Ionisation, Streuung, Umladung u.a.) einzeln studieren. Wir wollen nun die Frage über die Kompression der Plasmasäule be-trachten.

Wie aus den Aufnahmen hervorgeht, die in der Abb. 3 wiedergegeben werden, zieht sich die Entladungssäule bei allen gegebenen Werten des Längsfeldes in den 5–6 Mikrosekunden zusammen und bleibt dabei der Länge nach homogen. Die Entladungszonen, die in der Nähe von Elek-troden liegen, erscheinen auf den Aufnahmen nicht, und wir werden ihnen in diesem Aufsatz kein Interesse schenken. Wir nehmen zur Beschreibung des Anfangstadiums der Entladung also an, dass alle die die Entladung kennzeichnenden Grössen nur Funktionen des Abstandes von der Ent-ladungsachse sind. Wir vernachlässigen die innere Friktion, d.h. wir nehmen an, dass die kinetische Energie der radialen Bewegung nicht über andere Freiheitsgrade verfügt.

Unter gemachten Voraussetzungen kann die Kompression der Plasma-säule durch die folgende Gleichung beschrieben werden:

$$\frac{d}{dt}\left(Mn\alpha\,\frac{dr}{dt}\right) = -\frac{d}{dr}\,(nT) + \frac{1}{c}\,[\mathbf{j}\,\mathbf{H}]. \tag{4·1}$$

Es bedeuten: $\mathbf{j}$ die Stromdichte in der Entladungssäule; M die Ionenmasse, gleich der Atommasse; n die Summe der Atom- und Ionenzahlen in 1 cm³ des Plasmas; T die Plasmatemperatur im ganzen. Die Grösse α kenn-zeichnet das Mitnehmen des neutralen Gases durch das sich zusammen-ziehende Plasma.

Die Gleichung (4·1) beschreibt die Bewegung eines beliebigen Plasma-punktes innerhalb der Entladungssäule. Die Schnellphotoaufnahmen (Abb. 3) zeigen, wie der Radius a der Oberfläche der Entladungssäule mit der Zeit sich ändert. Um eine Differentialgleichung aus (4·1) erhalten zu können, die die Abhängigkeit des Radius a von der Zeit bei einem gegebenen Änderungsgesetz des vollen Stromes und bei einem gegebenen Längsfeld bestimmt, wollen wir die folgenden vereinfachenden Annahmen machen:

1. Wir betrachten die Säulenmasse, die an der Bewegung teilnimmt, als gleichmässig über den Querschnitt verteilt und konstant in der Zeit.

2. Wir nehmen an, dass die Geschwindigkeit und damit auch die Beschleunigung der Punkte innerhalb der Säule proportional zum Abstand von der Entladungsachse sind.

3. Wir vernachlässigen den Druck nT im Vergleich zu den elektro-dynamischen Kräften.

Unter Voraussetzung dieser Annahmen und nach einer Umformung, die im Anhang angegeben worden ist, und unter Einführung von dimen-sionslosen Variablen, erhalten wir eine Gleichung, die den Radius der Entladungssäule mit der Zeit in folgender Art und Weise verbindet:

$$x\ddot{x} - \left(\frac{t_{0c}}{t_M}\right)^2 \left[\frac{1}{x^2} - x^2 \left(\frac{b_1^2 - 1}{b_1^2 - x^2}\right)^2\right] + \frac{\sin^2 \omega_0 \tau}{\omega_0^2} = 0. \tag{4·10}$$

Diese Gleichung kann für gegebene Parameter t_{0c}, t_M, b und ω_0 zahlen-mässig integriert werden. Die Ergebnisse einer Integration für zwei kennzeichnende Fälle sind in den Abb. 11 und 12 mit ganzen Linien eingetragen worden.* In denselben Abbildungen sind die Ergebnisse der Messung des Säulenradius nach Schnellphotographien eingezeichnet. Wir sehen, dass für ein Feld $H = 2000$ Oersted (Abb. 11) die berechnete Kurve

---

* Die Integration wurde von G. A. Michailev auf einer Elektronenrechenmaschine CEM-1 durchgeführt.

den experimentellen Punkten ganz nahe kommt, während für ein Feld von 6000 Oersted (Abb. 12) die berechnete Kurve bedeutend höher über den experimentellen Punkten liegt.

Wie die Messungen mit Hilfe der Windung (siehe[3]) gezeigt haben, verbleibt das Längsfeld nicht in der Säule, sondern kriecht aus ihr heraus. Eine genaue Lösung der Aufgabe unter Berücksichtigung des Herauskriechens des Feldes ausserhalb der Säule ist kompliziert, und bei

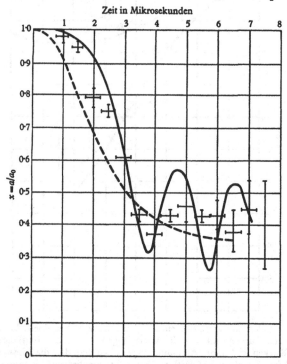

Abb. 11. Änderung des Entladungssäulenradius mit der Zeit bei einem Längsfeld von 2600 Oersted. Die ganzen Kurven sind nach den Gleichungen (4·10) und (4·13) berechnet worden. Die punktierten Kurven sind in der Annahme erhalten worden, dass die Masse der Entladungssäule gleich Null ist, d.h. dass ein Gleichgewicht der magnetischen Spannungen $H_0^2 - H_i^2 = H^2$ auf der Oberfläche der Entladungssäule herrscht. $H_0 = 2000$ Oersted, $I_0 = 250$ kA, $\omega = 2·5 \times 10^6$ sek^{-1}, $p = 0·2$ mm Hg.

unserer nicht besonders grossen Präzision der Messungen und zahlreichen Vereinfachungen, die bei der Aufstellung der Gleichung (4·10) gemacht wurden, hat sie keinen Sinn. Daher werden wir bei der Berechnung der radialen Schwingungen das Herauskriechen nur dadurch in Rechnung ziehen, dass wir einen solchen Koeffizienten $\beta^2(x) < 1$ einführen, dass

$$H_1^2 = H_0^2 \frac{\beta^2}{x^4} \text{ wird.}$$

Bis zur 6ten Mikrosekunde bei Feldern von 6 und 12 tausend Oersted, wie es die Kurven der Abb. 8 zeigen, ist der Fluss in der Säule mit einer genügenden Genauigkeit zu ihrem Radius proportional, d.h.

$$\beta^2(x) = x^2. \tag{4.12}$$

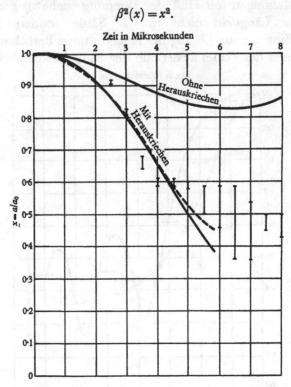

Abb. 12. Änderung des Entladungssäulenradius mit der Zeit bei einem Längsfeld von 6000 Oersted. Die ganzen Kurven sind nach den Gleichungen (4.10) und (4.13) berechnet worden. Die punktierten Kurven sind in der Annahme erhalten worden, dass die Masse der Entladungssäule gleich Null ist, d.h. dass ein Gleichgewicht der magnetischen Spannungen $H_i^2 - H_e^2 = H^2$ auf der Oberfläche der Entladungssäule herrscht. $H_0 = 6000$ Oersted, $I_0 = 250$ kA, $\omega = 25 \times 10^5$ sek^{-1}, $p = 0.2$ mm Hg.

Wenn wir (4.11), (4.12) und (4.9) in (4.4) einsetzen (siehe Anhang), so erhalten wir die Gleichung der radialen Säulenbewegung mit Rücksicht auf das Herauskriechen des Längsfeldes:

$$x\ddot{x} - \left(\frac{t_{0c}}{t_M}\right)^2 \left[\frac{\beta^2}{x^2} - x^2\left(\frac{b_1^2 - \beta}{b_1^2 - x^2}\right)^2\right] + \frac{\sin^2 \omega_0 \tau}{\omega_0^2} = 0. \tag{4.13}$$

Das Ergebnis aus der Integration der Gleichung (4.13) ist in der Abb. 12 wiedergegeben.

Obwohl in den Gleichungen (4.10) und (4.13) der Gasdruck vernachlässigt wurde, ist es zu ersehen, dass für die ersten vier Mikrosekunden die experimentellen Punkte zufriedenstellend auf die berechnete Kurve fallen. Es ist ein Beweis dafür, dass die grundlegenden Faktoren, die den Kompressionsvorgang bestimmen, richtig berücksichtigt wurden. Die Erhitzung des Plasmas ist offenbar so gering, dass sie keine bedeutende Abweichung zwischen dem Versuch und der Berechnung hervorruft. Aus der Bewertung folgt, dass die Mitteltemperatur des Plasmas 15–20 eV nicht überschreiten kann.

5. ÜBER DIE LEITFÄHIGKEIT DES PLASMAS

Das Anwachsen des magnetischen Längsfeldes innerhalb der Entladungssäule kann nicht nur durch das Mitnehmen des Feldes durch das sich zusammenziehende Plasma verursacht werden. Die theoretischen Untersuchungen zeigen, dass, falls die freie Laufzeit eines Elektrons von der Geschwindigkeit seiner Bewegung abhängt, so ist die stationäre Leitfähigkeit des Plasmas längs und quer des magnetischen Feldes verschieden. Berechnungen, die in Rücksicht auf die Coulomb'sche Dispersion durchgeführt wurden[4], ergeben, dass die Leitfähigkeit in der Querrichtung des Feldes zweimal kleiner ist als in der Längsrichtung. In der Entladungssäule ergibt das eigene magnetische Feld in Zusammensetzung mit dem äusseren Längsfeld spiralförmig gewundene Kraftlinien. Infolge der Anisotropie der Leitfähigkeit ist der Strom zur Achse nicht parallel gerichtet, sondern wird krumm, wie es die Abb. 13 zeigt, und gibt eine Längskomponente des Magnetfeldes, die immer das anfängliche Feld steigert. Die Grösse dieses Effektes ist vom Verhältnis H_ϕ/H_z abhängig. Bei kleinen Längsfeldern $H_z < H_\phi$ kann der Effekt sehr bedeutend ausfallen, was auch beim Versuch beobachtet wird.

Falls die Säule, die sich bis auf einen kleinen Radius zusammenzieht, das ganze Längsfeld mitnimmt, das bis zum Entladungsbeginn das Entladungsrohr füllt, besetzt nun, infolge der Einschliessung des ganzen Flusses des Längsfeldes in der Säule, der ausserhalb des Entladungsrohres gewesene Fluss beinahe die ganze Spulenfläche. Die Abnahme des äusseren Feldes wird dabei zu

$$\Delta H_e = H_0 \frac{a_0^2 - a^2}{R^2 - a^2} \approx H_0 \frac{a_0^2}{R^2} \quad \text{bei} \quad a \ll a_0.$$

Aus diesem Grunde kann das Integral der elektromotorischen Kraft nach der Zeit $\int_0^t E \, dt$, das von der Windung aufgenommen wird, nicht grösser sein als

$$\pi(R^2 - b^2) \, \Delta H_e = \pi H_0 (R^2 - b^2) \frac{a_0^2}{R^2}.$$

Bei einem nicht vollständigen Mitnehmen wird das Anwachsen des Flusses innerhalb der Windung und dementsprechend $\int E\,dt$ kleiner.

Bei Versuchen mit einem Längsfeld von 100 Oersted gibt jedoch die Anzeige der Windung ein Integral, das mehr als zweimal die angegebene Grösse übersteigt. Es bedeutet, dass der Fluss nicht nur die Säule nicht verlässt, sondern sogar in ihr anwächst. Der beobachtete Effekt liegt weit

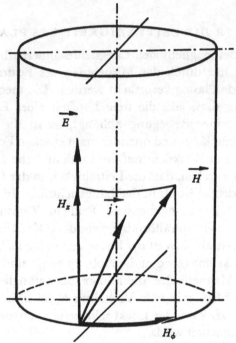

Abb. 13. Die Stromrichtung in der Entladungssäule bei anisotropischer Leitfähigkeit.

ausserhalb der Messfehlergrenzen. Er könnte durch die Verdrehung der Säule in eine zylindrische Spirale erklärt werden. Wir haben auf keiner der Aufnahmen eine Verdrehung der Säule in den ersten drei Mikrosekunden beobachtet. Unterdessen ist das Integral der Windungsspannung auf die Rechnung des Mitnehmens des Feldes grösser, als es sein sollte, und wird schon in diesem Anfangsmoment beobachtet. Daher ist es meist wahrscheinlich, dass das beobachtete Anwachsen des Flusses des Längsfeldes in der Entladungssäule bei kleinen anfänglichen Feldstärken durch die Anisotropie der Plasmaleitfähigkeit im magnetischen Felde verursacht wird.

Wir wollen nun zur Bewertung der Grösse der Leitfähigkeit übergehen. Nachdem wir durch Versuch das Herauskriechen des magnetischen Feldes aus der konvergierenden Säule bestimmt haben, können wir die Plasma-leitfähigkeit bewerten. Um die Berechnung nicht zu komplizieren, werden wir die Schätzung der Leitfähigkeit des Entladungsplasmas auf ein starkes magnetisches Feld ($H_0 > 3000$ Oersted) begrenzen, wobei wir die Anisotropie der Leitfähigkeit vernachlässigen. In diesem Falle ist die Änderung des magnetischen Längsflusses innerhalb der Säule $d\phi_i/dt$ mit der Leitfähigkeit σ durch die folgende Beziehung verknüpft:

$$\frac{d\phi_i}{dt} = \frac{ac^2}{2\sigma}\left(\frac{\partial Z_z}{\partial r}\right)_{r=a} \tag{5.1}$$

die aus Maxwells Gleichungen abgeleitet worden ist.

Für die Berechnung der Leitfähigkeit nach der Formel (5.1) ist es notwendig, nicht nur die Geschwindigkeit zu wissen, mit welcher das Feld aus der Säule 'herauskriecht', sondern auch die Verteilung des Längsfeldes über seinen Querschnitt.*

Wenn wir die Aufgabe annähernd lösen wollen, kann man bei einer geringen Leitfähigkeit als eine erste Annäherung eine parabolische Verteilung des Feldes innerhalb der Säule annehmen, zum Beispiel:

$$H_i = H_e\left[1 + A\left(1 - \frac{r^2}{a^2}\right)\right]. \tag{5.2}$$

Es ist nicht schwer, den Parameter A zu bestimmen, wenn wir die mittlere Feldstärke in der Säule und die Feldstärke ausserhalb dieser kennen, weil

$$\bar{H}_i = H_e\left(1 + \frac{A}{2}\right).$$

Für den Fall der Beständigkeit des magnetischen Flusses in der Spule vom Radius R ist

$$A = \frac{2R^2}{a^2} \cdot \frac{H_0 - H_e}{H_e}. \tag{5.3}$$

Bei solcher Feldverteilung ist

$$\left(\frac{\partial H_z}{\partial r}\right)_{r=a} = -\frac{2AH_e}{a}. \tag{5.4}$$

Falls wir (5.3) und (5.4) in (5.1) einsetzen, erhalten wir:

$$\sigma = 2c^2\frac{R^2}{a^2} \cdot \frac{H_0 - H_e}{d\phi_i/dt}.$$

* V. D. Shafranov schlug eine Methode zur Berechnung der Verteilung des Längsfeldes über den Querschnitt vor und nahm auch an der Besprechung über die Ergebnisse teil.

$H_0 - H_e$ wird durch die Windung gemessen, a von der Photoaufnahme, $d\phi_i/dt$ kann man aus den Kurven der Abb. 8 erhalten. Eine auf diese Art und Weise berechnete Leitfähigkeit bei verschiedenen Arbeitsverhältnissen wird in der Abb. 14 angeführt. Für die erhaltenen Werte für σ muss die Eindringungstiefe des Stromes j_ϕ, der das magnetische Längsfeld aufrechterhält, wegen des Skin-Effektes sehr klein sein. Die Verteilung

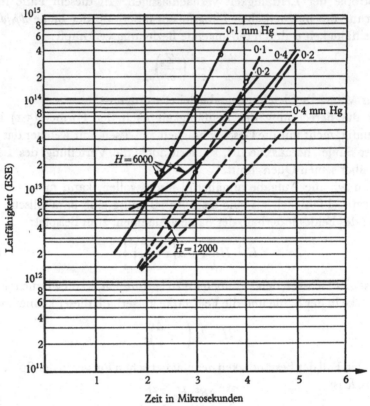

Abb. 14. Leitfähigkeit des Plasmas, berechnet nach der gemessenen Geschwindigkeit des Herauskriechens des Längsfeldes aus der Entladungssäule. Innerhalb der Säule ist eine parabolische Verteilung des Längsfeldes angenommen worden. Formel (5·2).

des Längsfeldes wird innerhalb der Schnur mehr gleichmässig und an deren Grenzen mehr steil, als dies von der Formel (5·2) angegeben wird. Dann wird $(\partial H_z/\partial r)_{r=a}$ und folglich werden auch die Werte von σ grösser sein als die in der Abb. 14 angeführten. Jedoch unterscheiden sich die Werte für σ, die mit Rücksicht auf den Skin-Effekt berechnet werden, von den angeführten nicht mehr als zweimal, und auch dieses geschieht in der vierten bis fünften Mikrosekunde, wenn die Bestimmung von σ aus der Geschwindigkeit des Herauskriechens wegen der Abnahme der Konvergenzgeschwindigkeit der Säule sehr ungenau wird.

Falls man andererseits auf Anisotropie der Leitfähigkeit Rücksicht nimmt, so werden die σ-Werte beispielsweise 1·5–2mal kleiner. Jedoch erklären diese Fehler nicht die Tatsache, dass, wenn man auch die Stromdichte j_z und die Leitfähigkeit als konstant über dem Säulenquerschnitt annimmt, der Spannungsfall im Längswiderstand der Säule für die angeführten Werte in der zweiten und dritten Mikrosekunde die ganze gemessene Spannung bedeutend übersteigt, während aus der Phasenverschiebung zwischen Strom und Spannung ersichtlich ist, dass die Wirkspannungskomponente die volle Entladungsspannung bedeutend unterschreitet. Darum sind die wirklichen Leitfähigkeitswerte grösser als die in der Abb. 14 angeführten.

6. ÜBER DIE IONISIERUNGSZAHL

Wenn wir die Plasmaleitfähigkeit kennen, können wir die Ionisierungszahl bestimmen. Wir können die bekannte Formel für Leitfähigkeit benutzen:

$$\sigma = \frac{e^2 n_e \tau_e}{m}. \tag{6·1}$$

Es bedeuten: n_e die Zahl von Elektronen in 1 cm³ und τ_e die freie Elektronenlaufzeit. Die freie Elektronenlaufzeit zwischen zwei Zusammenstössen mit Wasserstoff-Molekülen und Atomen wird aus der folgenden Formel bestimmt, die für Elektronentemperaturen zwei bis mehrere Zehner Elektronvolt gibt:

$$\tau_{ea} = \frac{2·2 . 10^{-10}}{p(1-\eta)}. \tag{6·2}$$

Hier bedeuten: p den Gasdruck in mm Hg und $\eta = n_e/n_0$ die Ionisierungszahl (n_0 ist die Summe der Atom- und Ionenzahl in 1 cm³ des Plasmas).

Für die Formel (6·2) nehmen wir den Koeffizienten 2·2 . 10⁻¹⁰ an, der am besten mit den Angaben von Hayley und Reed[5] und mit dem vollen Wechselwirkungsquerschnitt des Elektrons mit dem Molekül H_2 nach den Messungen von Townsend und Ramsauer[6] übereinstimmt.

Die freie Elektronenlaufzeit zwischen aufeinanderfolgenden Coulomb-Streuungen auf Ionen ist verschieden, in Abhängigkeit davon, ob die Elektronen eine regulierende Komponente längs oder quer zu den magnetischen Kraftlinien haben. Für die Bewegung längs der magnetischen Kraftlinien gibt die folgende Formel:

$$\frac{1}{\tau_{eu}} = \frac{3\pi^2}{32} \left(\frac{e^2}{kT}\right)^2 n_e V \ln\left|\frac{1}{(2\pi n_e)^{1/3}} \frac{kT}{e^2}\right|. \tag{6·3}$$

485

Für die Bewegung quer zu den magnetischen Kraftlinien ist τ_{eu} zweimal kleiner. In der Formel (6·3) wurden die Coulombkräfte an dem Debye-Radius abgeschnitten für ein Plasma mit einer Ionentemperatur gleich derjenigen von Elektronen. Wenn $T_e \gg T_u$, so wird der Logarithmus in der Formel (6·3) für $T_e = 10$ eV und $T_u = \frac{1}{40}$ eV 1·5mal kleiner als für $T_e = T_u = 10$ eV. Soweit wir die Ionentemperatur nicht kennen, können wir ganz willkürlich für alle weiteren Bewertungen den aus der Formel (6·3) entnommenen Wert für τ_{eu} annehmen. Für $n_e = 10^5$, $T = 10$ eV erhalten wir:

$$\tau_{eu} = 1\cdot3 \cdot 10^5 \frac{T_{eV}^{3/2}}{n_e} = 1\cdot8 \cdot 10^{-12} \frac{T_{eV}^{3/2}}{\eta p}, \qquad (6\cdot4)$$

wo T_{eV} die Temperatur in Elektronenvolt bedeutet. Die freie Elektronenlaufzeit in der Formel (6·1) ist mit (6·2) und (6·4) durch die folgende Beziehung verbunden:

$$\frac{1}{\tau_e} = \frac{1}{\tau_{ea}} + \frac{1}{\tau_{eu}}. \qquad (6\cdot5)$$

Wenn wir (6·5), (6·4) und (6·2) in (6·1) einsetzen, erhalten wir:

$$\sigma = \frac{4 \times 10^{15}}{1/\eta - 1 + \dfrac{122}{T_{eV}^{3/2}}}. \qquad (6\cdot6)$$

Durch die Messung der Leitfähigkeit erhalten wir also nur eine Beziehung zwischen möglichen Werten der Ionisierungszahl η und der Temperatur T_{eV}.

In der Abb. 15 sind Kurven $\eta(T)$ für verschiedene Werte von σ wiedergegeben, die nach der Formel (6·6) berechnet worden sind. Der höchste Wert des Leitvermögens in der vierten-fünften Mikrosekunde (Abb. 14) ist gleich 4×10^4 (ESE). Wir sehen von den Kurven der Abb. 15, dass bei der Annahme einer sogar hohen Elektronentemperatur diesem Leitvermögen eine Ionisierungszahl grösser als 15 % entspricht, was mit den Angaben in [7] übereinstimmt.

7. SCHLUSSFOLGERUNGEN

Die in dieser Arbeit beschriebenen Untersuchungen gestatten die folgenden Schlüsse zu machen:

1. Das magnetische Längsfeld verzögert das Zusammenziehen der Entladungssäule unter Einwirkung des eigenen Stromfeldes. Der Zusammenbruch der Entladungssäule tritt später ein als bei $H_z = 0$. Nach der ersten Konvergenz bei $H_z \leqslant 2000$ Oersted werden radiale Schwingungen

der Säule beobachtet, worauf die Säule sich ausbiegt und, endlich, deutliche Umrisse verliert. Auf diese Art und Weise unterscheidet sich der Zusammenbruchsprozess in Anwesenheit eines Längsfeldes wesentlich vom Zusammenbruch der Säule in Abwesenheit eines äusseren Längsfeldes.

$$\frac{1}{\eta} = \frac{4 \cdot 0 \cdot 10^{15}}{\sigma} - \frac{122}{T^{3/2}} + 1$$

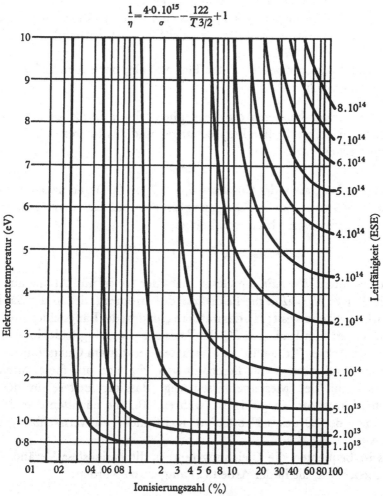

Abb. 15 Zusammenhang zwischen der Ionisierungszahl und der Elektronentemperatur bei einigen Leitfähigkeitswerten des Plasmas.

2. Es ist festgestellt worden, dass das Anwachsen der Feldstärke des magnetischen Längsfeldes innerhalb der Gasentladungssäule nicht durch die Zusammenziehung der Säule allein erklärt werden kann. Offenbar ist der beobachtete Effekt durch die Anisotropie des Plasmas im magnetischen Felde verursacht.

3. Im Rahmen der Fehlergrenzen des Versuches geschieht die Zusammenziehung der Entladungssäule bei einem Druckgleichgewicht des Längsfeldes und des eigenen magnetischen Feldes, was auf eine verhältnismässig niedrige Plasmatemperatur hinweist. Die Bewertung zeigt, dass die Plasmatemperatur 15–20 eV nicht übersteigt.

4. Das Leitvermögen des Plasmas, auf Grund des Mitnehmens des magnetischen Flusses bewertet, erreicht in der vierten-fünften Mikrosekunde 4×10^{14} (ESE).

5. Die Ionisierungszahl in der Plasmasäule, nach dem Leitvermögen bewertet, erreicht 15 %.

Die Verfasser benützen die Gelgenheit, ihren Dank an L A. Artsimovitch für die wertvollen Besprechungen und das Interesse für die Arbeit auszusprechen.

Anhang

Nach dem Weglassen des Gliedes $-d/dr\,(nT)$ in (4·1) und nach der Multiplikation der so erhaltenen Gleichung mit $r^2\,dr$ und nach deren Integration in den Grenzen von o bis a, erhalten wir:

$$MN\alpha a\ddot{a} = -N\overline{T} + \frac{a^2}{2}\,[H_\phi^2 - (\overline{H_{zi}^2} - H_{ze}^2)]. \qquad (4\cdot2)$$

Hier bedeuten: $N = \pi a^2 n$ die Gesamtzahl der Ionen und Atome auf 1 cm Säulenlänge; $H_\phi = 2I/ca$ die Feldstärke des Stromes I, der in der Säule an deren Oberfläche fliesst; $\overline{H_{zi}^2}$ das mittlere Quadrat des Längsfeldes innerhalb des Gebietes vom Radius a; H_{ze} die Feldstärke des Längsfeldes im Abstand a von der Achse; T die mittlere Plasmatemperatur am Querschnitt.

Eine Vergleichung von Berechnungen, die sich auf diese Annahme gründen, mit den Messergebnissen zeigt, dass die letztere für die drei ersten Mikrosekunden gültig ist, während für die nächsten drei Mikrosekunden der Druck nT offenbar nicht mehr zu vernachlässigen ist (siehe 4).

Es ist bequemer, die Gleichung (4·2) in dimensionslose Veränderliche umzuschreiben, die in der Arbeit (2) ausgenutzt wurden:

$$x = \frac{a}{a_0}; \quad \tau = \frac{t}{t_{0c}}. \qquad (4\cdot3)$$

Wollen wir annehmen, dass der Strom sich nach dem Gesetz $I = I_0 \sin \omega t$ ändert. Dabei bekommt die Gleichung (4·2), in dimensionslose Veränderliche umgeschrieben, die folgende Form:

$$x\ddot{x} - \left(\frac{t_{0c}}{t_M}\right)^2 \frac{\overline{H_i^2} - H_e^2}{H_0^2}\,x^2 + \frac{\sin^2 \omega_0 \tau}{\omega_0^2} = 0, \qquad (4\cdot4)$$

488

wo ausser der 'charakteristischen Zeit',

$$t_{0c} = \sqrt{\left[\frac{ca_0}{I_0 \omega} \sqrt{\left(\frac{MN\alpha}{2} \right)} \right]}, \qquad (4 \cdot 5)$$

die laut der Arbeit[2] die Konvergenzzeit der Säule in Abwesenheit des magnetischen Längsfeldes bestimmt, noch eine andere Zeitcharakteristik, die 'magnetische Zeit' erschienen ist:

$$t_M = \frac{\sqrt{(2MN\alpha)}}{H_0}. \qquad (4 \cdot 6)$$

Wie wir weiter sehen werden, wird die Periode der radialen Säulenschwingungen im Felde von der magnetischen Zeit bestimmt. Es ist bequem, den Wert der magnetischen Feldstärke vor der Entladung gleich H_0 zu setzen und a_0 gleich dem anfänglichen Radius der Entladungssäule anzunehmen. Die dimensionslose Frequenz in (4·4) wird aus der folgenden Formel bestimmt:

$$\omega_0 = \omega t_{0c}. \qquad (4 \cdot 7)$$

Zur Integrierung der Gleichung (4·4) muss man die Abhängigkeit des $\overline{H_i^2}$ und H_e vom Säulenradius wissen.

Bei einer grossen Plasmaleitfähigkeit ist das magnetische Feld 'eingefroren', das Längsfeld ist über dem Querschnitt konstant, und der Fluss innerhalb der Entladungssäule ändert sich mit der Zeit nicht. In diesem Falle:

$$\overline{H_i^2} = H_i^2 = \frac{H_0^2}{x^4}. \qquad (4 \cdot 8)$$

Bei allen Versuchen (siehe 3) wird ausserdem der Fluss des Längsfeldes innerhalb der Spule von Radius b aufrechterhalten, die das magnetische Feld erregt, d.h. in jedem Moment ist

$$H_0 b^2 = H_e(b^2 - a^2) + H_i a^2. \qquad (4 \cdot 9)$$

Unter Anwendung von (4·8) und (4·9), sowie der Bezeichnungseinführung $b/a_0 = b_1$, erhalten wir die Bewegungsgleichung (4·10).

Die Gleichung (4·10) wurde von uns für die Anfangsverhältnisse $x = 1$, $\dot{x} = 0$ bei $t = 0$ ausgerechnet. Es sei δ die durch eine Störung hervorgerufene Abweichung von der Lage x_1, die durch die Gleichung (4·10) bei den gegebenen Anfangsverhältnissen bestimmt wird.

Falls wir $x = x_1 + \delta$ in die Gleichung (4·10) einsetzen und die Grösse δ^2 gegenüber 1 vernachlässigen, sowie auch wenn wir für das Gebiet von Schwingungen $\ddot{x}_1 \frac{\delta}{x_1} \ll \delta$ ansehen, erhalten wir:

$$\ddot{\delta} + \left(\frac{t_{0c}}{t_M} \right)^2 R^2 \delta = 0, \qquad (4 \cdot 11)$$

d.h. der Säulenradius schwingt herum um den Gleichgewichtswert x_1. Der Faktor R wird nur durch die Geometrie der Anlage, den Radius x_1 der zusammengedrückten Säule, sowie auch durch die Bedingungen der Aufrechterhaltung von Fluss in der Säule und der Spule, die das Längsfeld erregt, bestimmt:

$$R = 2\left[\frac{\beta^2}{x_1^4} + \left(\frac{b_1^2 - \beta}{b_1 - x_1^2}\right)^2 \left(1 + \frac{2x_1}{b_1^2 - x_1^2}\right)\right]. \tag{4.14}$$

Folglich ist die dimensionale Schwingungsfrequenz der zusammengedrückten Entladungssäule:

$$\Omega = \frac{R}{t_M} = \frac{RH_0}{\sqrt{(2MN\alpha)}}. \tag{4.15}$$

LITERATUR

[1] Andrianov, A. M., Artsimovitch, L. A., Basilevskaja, O. A., Prohorow, I. G. and Filippow, N. W. *Atomnaja Energija*, **1**, no. 3, 76, 1956.
[2] Leontowitsch, M. A. and Osowetz, S. M. *Atomnaja Energija*, **1**, no. 3, 81, 1956.
[3] Ginsburg, V. L. *Theorie der Ausbreitung von Radiowellen in der Ionoshpere.* OGIS, GTI (1949).
[4] Braginskij, S. I. Privatmitteilung.
[5] Hayley, R. H. and Reed, I. W. *The Behaviour of Slow Electrons in Gases* (1941).
[6] Siehe, zum Beispiel, Massey, H. S. W. and Burhop, E. H. *Electronic and Ionic Impact Phenomena* (1952), Seite 206.
[7] Lukjanow, S. J. and Sinizin, W. I. *Atomnaja Energija*, **1**, no. 3, 88, 1956.

ÜBER DIE STABILITÄT EINES ZYLINDRISCHEN GASLEITERS IM MAGNETISCHEN FELDE*

W. D. SCHAFRANOW

Academy of Sciences, Moscow, U.S.S.R.

In der Arbeit [1] wurde die Stabilität eines Zylinders aus vollständig ionisiertem Plasma untersucht, dessen Druck durch die magnetischen Kräfte des längs des Zylinders fliessenden Stromes im Gleichgewicht gehalten wird. Störungen wurden untersucht, bei welchen der Zylinder eine schraubenförmige Gestaltung erhält. Gegenüber diesen Störungen zeigte sich das Gleichgewicht labil. In der vorliegenden Arbeit, gerade wie in [1], wird mit Hilfe der Methode von kleinen Schwingungen und unter der Annahme einer idealen Leitfähigkeit die Stabilität gegenüber beliebigen Störungen in Anwesenheit einer magnetischen Feldkomponente, die längs des Zylinders gerichtet ist ('Längsfeld'), untersucht. Kriterien der Stabilität werden festgestellt.

Das Ausgangssystem der Gleichungen besteht aus Gleichungen der magnetischen Hydrodynamik für ein ideal leitendes Medium:

$$\frac{d\rho}{dt} + \rho \operatorname{div} \vec{\mathbf{v}} = 0, \qquad \frac{\partial \vec{\mathbf{H}}}{\partial t} = \operatorname{rot}\left[\vec{\mathbf{v}}\,\vec{\mathbf{H}}\right], \tag{1}$$

$$p = \text{const.}\ \rho^{\gamma}, \qquad \rho\,\frac{d\vec{v}}{dt} = -\nabla p + \frac{1}{c}\left[\vec{\mathbf{j}}\,\vec{\mathbf{H}}\right].$$

Beim Gleichgewicht ist $v = 0$, $\partial/\partial t = 0$, der Zylinder ist in der Richtung der Achse und des Azimuts homogen $\partial/\partial z = \partial/\partial\phi = 0$. Die Komponenten des Feldes H_z^0 und H_ϕ^0 sind von Null verschieden. Wir nehmen an, dass innerhalb des Zylinders $H_\phi^0 = 0$ und H_z^0 überall homogen ist.

$$\left.\begin{aligned}
H_{\phi i}^0 &= 0, \qquad H_{\phi e}(r) = \frac{2I}{ca}, \\[4pt]
H_{zi}^0 &= h_i H_{\phi e}^0(a) = \text{const.} \qquad H_{ze}^0 = h_e H_{\phi e}^0(a) = \text{const.}
\end{aligned}\right\} \tag{2}$$

---

* Presented at the Symposium by L. A. Artsimovich.

Die Indexe i, e beziehen sich bzw. auf das innere und äussere Feld. In diesem Falle ist die Dichte und der Druck über dem Querschnitt konstant, wobei

$$8\pi p^0 = H_{ze}^{02} + H_{\phi e}^{02}(a) - H_{zi}^{02} = H_{\phi e}^{02}(a)\,(1 + h_e^2 - h_i^2). \tag{3}$$

Es ist bequem, die Störungen in Koordinaten von Lagrange zu untersuchen. Es sollen die Gasteilchen eine Verschiebung $\boldsymbol{\xi}(\mathbf{r})\,e^{i(kz+m\phi+\omega t)}$ erfahren. In einer linearen Annäherung in Beziehung auf die Störungen sind die Korrektionen aller Grössen zur Verschiebung proportional: $\mathbf{H} = \mathbf{H}^0 + \mathbf{H}^{(1)}\,e^{i(kz+m\phi+\omega t)}$ u.d.gl. Aus (1) erhalten wir Gleichungen für diese Korrektionen:

$$\rho^{(1)} = -\rho^2\,\mathrm{div}\,\boldsymbol{\xi}, \quad \mathbf{H}_i^{(1)} = \mathrm{rot}\,[\boldsymbol{\xi}\mathbf{H}_i^0], \tag{4}$$

$$p^{(1)} = -\gamma p^0\,\mathrm{div}\,\boldsymbol{\xi}, \quad \omega^2\boldsymbol{\xi} + c^2\nabla\,\mathrm{div}\,\boldsymbol{\xi} + \frac{c_H^2}{H_{zi}^{02}}\,[\mathrm{rot}\,\mathbf{H}_i \cdot \mathbf{H}_i^0] = 0,$$

$$c^2 = \frac{\gamma p^0}{\rho^0}, \quad c_H^c = \frac{H_i^{02}}{4\pi\rho^0}. \tag{5}$$

Die Gleichung für $\boldsymbol{\xi}$ hat die folgende Lösung:

$$\xi_z = C_m I_m(\alpha r), \quad \xi_\phi = C_m\,\frac{k^2c^2 - \omega^2}{\alpha^2 c^2}\,I_m(\alpha r),$$

$$\xi_r = iC_m\,\frac{k^2c^2 - \omega^2}{k\alpha c^2}\left[\frac{m}{\alpha r}\,I_m(\alpha r) - I_{m-1}(\alpha r)\right], \tag{6}$$

$$\alpha^2 = \frac{\left(k^2 - \dfrac{\omega^2}{c^2}\right)\left(k^2 - \dfrac{\omega^2}{c_H^2}\right)}{k^2 - \omega^2\left(\dfrac{1}{c^2} + \dfrac{1}{c_H^2}\right)}. \tag{7}$$

Als Längenmasstab wird überall der Zylinderradius a angenommen.

Korrektionen zum Felde ausserhalb des Zylinders werden aus den Gleichungen $\mathbf{H} = \nabla\psi$, $\nabla\psi = 0$ erhalten. Unter Annahme von

$$\psi = \psi^0 = \psi^{(1)}\,e^{i(kz+m\phi+\omega t)}$$

erhalten wir für $\psi^{(1)}$ eine Besselgleichung m-er Ordnung vom imaginären Argument, deren Lösung, auf $r \to \infty$ begrenzt, $\psi^{(1)}(r) = \mathrm{const}\,K_m(kr)$ ist. Die Integrationskonstante wird aus der Kontinuitätsbedingung der normalen Feldkomponente zur Zylinderoberfläche bestimmt. Infolge der idealen Leitfähigkeit werden die magnetischen Kraftlinien vom Stoff mitgenommen, wobei sie parallel zur Oberfläche verbleiben. Daher ist die normale Feldkomponente gleich Null und das äussere Feld vom Inneren unabhängig. Wollen wir den Wert der ersteren auf der Oberfläche des erregten Zylinders hinschreiben

$$H_{ze}^{(1)} = i\xi_r(a)\ (mH_{\phi e}^0 + kH_{ze}^0),$$

$$H_{\phi e}^{(1)} = \xi_r(a)\ m(mH_{\phi e}^0 + kH_{ze}^0)\Big/\left[k\,\frac{K_{m-1}(k)}{K_m(k)} + m\right],$$

$$H_{ze}^{(1)} = \xi_r(a)\ k(mH_{\phi e}^0 + kH_{ze}^0)\Big/\left[k\,\frac{K_{m-1}(k)}{K_m(k)} + m\right].$$

(8)

Die erhaltene Lösung für die Korrektionen ist für einen vollständig bestimmten Eigenwert von ω^2 gültig, dessen Vorzeichen sagt, ob das Gleichgewicht gegen die gegebene Störung stabil ($\omega^2 > 0$) oder labil ($\omega^2 < 0$) ist. Dieser Eigenwert wird aus der Grenzbedingung erhalten, welche aus der Bewegungsgleichung hergeleitet wird. Bei unserer Aufgabestellung kommt es auf die Forderung hinaus, dass auf der Zylinderoberfläche die folgende Bedingung erfüllt werden soll:

$$8\pi p = H_{\phi e}^2 + H_{ze}^2 - H_{zi}^2.$$

(9)

In Anbetracht, dass

$$H_\phi^2(a + \xi_r) = H_\phi^{02}(a) + 2H_\phi^0(a)\ H_\phi^{(1)}(a) + \frac{\partial H_\phi^{02}}{\partial r}\ \xi_r(a)$$

u.s.w. ist, und unter Berücksichtigung von (3) und (4) bei $r = a = 1$ erhalten

wir $\quad -\gamma p^0\ \mathrm{div}\,\vec{\xi} = -\frac{\xi_r}{4\pi}\left\{H_{\phi e}^{02} - \frac{(mH_{\phi e}^0 + kH_{ze}^0)^2}{k\,\dfrac{K_{m-1}(k)}{K_m(k)} + m}\right\} - i(\mathrm{div}\,\vec{\xi} - ik\xi_z)\,\frac{H_{zi}^{02}}{4\pi}.$

(10)

Wenn wir die Werte von $\vec{\xi}$ aus (6) einsetzen, können wir die Bedingung folgendermassen niederschreiben:

$$\frac{\gamma}{2}(1 + h_e^2 - h_i^2)\,\frac{\omega^2}{\omega^2 - k^2 c^2} = \left(\frac{I_{m-1}(\alpha)}{\alpha I_m(\alpha)} - \frac{m}{\alpha^2}\right)\left[1 - \frac{(m + k.h_e)^2}{k\,\dfrac{K_{m-1}(k)}{K_m(k)} + m}\right] - h_i^2 \equiv f(\omega^2).$$

(11)

Diese Gleichung hat ausser dem positiven Spektrum von Lösungen $\omega^2 > 0$, die den Schall- und Alfvénwellen eines Gases im gestörten Zylinder entsprechen, noch einen Zweig der Eigenwerte $\omega_m^2(k)$, die in einem gewissen Gebiete ein negatives Vorzeichen besitzen. Falls das Längsfeld gleich Null ist, liegt dieser Zweig bei $m = 0$ und $m = 1$ vollständig im negativen Gebiet, wobei für $k \ll 1$

$$\omega_0^2 = -\frac{k^2 c^2}{\gamma - 1}\quad (\gamma \neq 1), \qquad \omega_0^2 = -2\sqrt{2}\,\frac{kc^2}{a}\quad (\gamma = 1),$$

$$\omega_1^2 = -\frac{2}{\gamma}\,k^2 c^2 \ln\frac{1}{k}.$$

(12)

Für $m \geqslant 2$ ist dieser Zweig im Gebiete von kleinen k positiv; bei einem Werte von $k = K_m$, der von der Gleichung $kK_{m-1}/K_m + m - m^2 = 0$ bestimmt wird, geht er in das negative Gebiet über. Diese Werte von K_m sind die folgenden: $K_2 = 3$, $K_3 = 8$, $K_4 = 14$, ..., $K_m = m^2$ ($m \gg 1$). Im Gebiete von kurzwelligen Störungen haben alle Abzweigungen von $\omega_m^2(k)$ eine Assymptotik

$$k \to \infty \quad \omega_m^2(k) \to -\frac{2}{\gamma}\frac{kc^2}{a}. \tag{13}$$

In Anwesenheit eines Längsfeldes geht dieser Zweig bei grossen k in das positive Gebiet über. Das Stabilitätskriterium erhalten wir aus der Forderung, dass die Dispersionsgleichung (11) keine Lösung $\omega^2 < 0$ haben soll. Bei negativen ω^2 wächst der linke Teil der Gleichung monoton von $\omega^2 = 0$ bis $\omega^2 = -\infty$, während der rechte Teil abnimmt. Eine Lösung $\omega^2 < 0$ fehlt folglich, wenn $f(0) < 0$, oder wenn

$$k^2 h_i^2 + (kh_e + m)^2 \frac{k\dfrac{I_{m-1}(k)}{I_m(k)} - m}{k\dfrac{K_{m-1}(k)}{K_m(k)} + m} > k\frac{I_{m-1}(k)}{I_m(k)} - m \text{ ist.} \tag{14}$$

Hierzu gehörende Kombinationen

$$\phi_1(k_1 m) = k\frac{I_{m-1}(k)}{I_m(k)} - m, \quad \phi_e(k_1 m) = k\frac{K_{m-1}(k)}{K_m(k)} + m \tag{15}$$

sind vom Vorzeichen des m nicht abhängig, während der Faktor $(kh_e + m)^2$ bei der gegebenen Störungsform wesentlich vom Vorzeichen des m/kh_e abhängt. Bei der Bestimmung des Stabilitätskriteriums muss man beiden Vorzeichen Rechnung tragen. Der Bestimmtheit wegen nehmen wir weiter $k > 0$ und $m > 0$ an.

Wollen wir die folgenden Sonderfälle beachten:

(1) $H_\phi^0 = 0$. Das Gleichgewicht ist immer stabil, weil (14) in

$$\phi_2(k_1 m)\, H_{ei}^2 + \phi_1(k_1 m)\, H_{ze}^2 > 0 \quad (\phi_2 > 0,\ \phi_1 > 0)$$

übergeht.

(2) $H_{ze}^0 = h_e = 0$. Das Stabilitätskriterium für $m = 0$ ist $h_1^2 > \dfrac{I_1(k)}{kI_0(k)}$. Für $m = 2, 3, ...$ wird ein kleinerer Wert von h_i notwendig sein. Der Maximalwert des rechten Teiles der letzten Ungleichheit ist gleich mit 0.5. Folglich ist das Gleichgewicht bei $H_{zi}^0 > \dfrac{\sqrt{2}\,I}{ca}$ für Störungen mit $m \neq 1$ stabil.

Für $m = 1$ wächst das für eine Stabilität notwendige Längsfeld mit $k \to 0$ logarithmisch. Inwieweit $h < 1$ (3) ist, liegt das Gebiet der Stabilität bei $k > 0.46$.

(3) $H_{zi}^0 = h_i = 0$. Die Stabilitätsbedingung ist

$$|h_e| > 1/k[\sqrt{(kK_{m-1}/K_m+m)} \pm m],$$

das Vorzeichen '−' entspricht dem Falle $m/kh_e > 0$, das Vorzeichen '+' dem entgegengesetzten Falle. Da $|h_e|$ mit Anwachsen von m wächst, so sichert das äussere Längsfeld von sich allein die Stabilität nicht. Störungen von einer Wellenlänge $\Lambda \simeq 2\pi a \dfrac{|h_e|}{m}$ werden vom Felde nicht stabilisiert.

Für einen allgemeinen Fall $h_e \neq 0$, $h_i \neq 0$ und nach der Einführung des Verhältnisses $\epsilon = \left|\dfrac{h_e}{h_i}\right|$, erhalten wir das Stabilitätskriterium $|h_i| > h_0$;

$$h_0 = \frac{\epsilon m \phi_1 + \sqrt{[\phi_2 + \epsilon^2 \phi_1 - m^2)}\, \phi_1 \phi_2]}{k(\phi_2 + \epsilon^2 \phi_1)}. \qquad (16)$$

Die Grenzfälle sind:

$$\begin{matrix} (k \ll 1) \\ m \neq 0 \end{matrix} \qquad h_0 = \frac{\epsilon m + \sqrt{[(1+\epsilon^2)\, m - m^2]}}{k(1+\epsilon^2)},$$

$$(k \gg 1) \qquad h_0 = \frac{\epsilon m + \sqrt{[(1+\epsilon^2)\, k - m^2]}}{k(1+\epsilon^2)}. \qquad (17)$$

Das Stabilitätsgebiet von langwelligen Störungen $k < K_m$ für $m \geq 2$ verengt sich bei $1 + \epsilon \geq m$ bis auf Null.

Auf den Abb. 1 und 2 liegt das Stabilitätsgebiet zwischen der Abszissen-achse und der Kurve $h(k)$.

Wollen wir notieren, dass die Stabilitätsbedingung (14) auf die Forderung hinauskommt, dass die von der Zylinderverschiebung entstehende zusätz-liche Kraft F, die seitens des magnetischen Feldes wirkt, gegen die Ver-schiebungsrichtung wirken soll. Tatsächlich, wie aus den Gleichungen (10) und (11) folgt, ist

$$F = \xi_2 \cdot \frac{H_{\phi e}^{02}}{4\pi a} \frac{f(0)}{\phi_1(k)}.$$

Falls sich der Gaszylinder innerhalb des koaxialen leitenden Zylinders vom Radius b befindet, muss man in der Bedingung (16) $\phi_2(k)$ gegen

$$[\phi_2(k) - \chi(kb)\, \phi_1(k)\, I_m(k)/K_m(k)]/[1 + \chi(kb)\, I_m(k)/K_m(k)]$$

auswechseln, wo

$$\chi(kb) = [kbK_{m-1}(kb) + mK_m(kb)]/[kbI_{m-1}(kb) - mI_m(kb)].$$

Wollen wir uns auf die Untersuchung des Einflusses des koaxialen Zy-linders auf die langwelligen Störungen $kb \ll 1$ beim Vorhandensein eines äusseren Längsfeldes $(h_i = 0)$ begrenzen. Für $k = 0$ ist die Kraft, die auf den Zylinder seitens des magnetischen Feldes wirkt, gleich:

$$F = -\frac{I^2 \xi_r}{c^2 a} \frac{b^{2m}(m-1)+m+1}{b^{2m}-m}, \quad m \neq 0,$$

$$F = \frac{I^2 \xi_r}{c^2 a}\left(1 - \frac{2h_e^2}{b^2-1}\right), \quad m = 0.$$

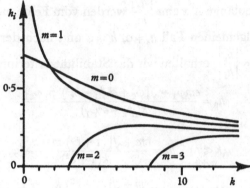

Abb. 1. h_i für verschiedene Werte der Wellenzahl k bei $h_e = 0$. Die Stabilitätsgebiete liegen zwischen der Abszissenachse und den Kurven $h_i(k)$.

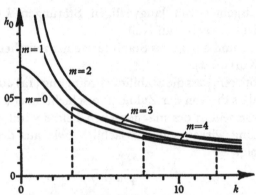

Abb. 2. h_0 für verschiedene Werte von k bei $h_i = h_e = h_0$. Die Stabilitätsgebiete wie in Abb. 1.

Die Stabilitätsbedingung ist $F/\xi_2 < 0$. Es ist ersichtlich, dass das magnetische Feld zwischen den Zylindern die Störungen bei $m = 0 \cdot 1$ dämpft.

Zum Schluss danke ich den Akademikern M. A. Leontowitsch und S. I. Braginsky für wertvolle Anweisungen.

LITERATUR

[1] Kruskal, M. und Schwarzschild, M. *Proc. Roy. Soc.* A, **223**, 348, 1954.

PART VII
ADDITIONAL CONTRIBUTIONS

MAGNETO-HYDROSTATIC EQUILIBRIUM IN AN EXTERNAL MAGNETIC FIELD

P. A. SWEET

University of London Observatory, London, England

ABSTRACT

The expression $\iiint_{\text{all space}} \Delta H^2\, dv$ for a change in magnetic energy is shown to be incorrect when applied to a body carrying an electric current and situated in an external magnetic field. A modified expression is derived.

Chandrasekhar's form of the virial theorem in a magnetic field is extended to the case where an external magnetic field is present.

I. INTRODUCTION

Gjellestad [1] has determined the elongation of a gravitating, homogeneous, incompressible, perfectly conducting fluid sphere with no interior magnetic field, but situated in a uniform external magnetic field. Problems of this nature have a bearing on the behaviour of interstellar gas clouds in a galactic magnetic field. In the above work the equilibrium configuration was found by a method, involving the principle of virtual work, along lines developed by Chandrasekhar and Fermi [2]. The expression adopted for the change $\Delta\mathfrak{M}$ in the magnetic energy of the body, consequent on a deformation, was given by

$$\Delta\mathfrak{M} = \frac{1}{8\pi} \iiint_{\text{all space}} \Delta H^2\, dv, \qquad (1)$$

where H is the magnetic intensity. This formula is not correct, and, in view of possible future work along these lines, it seems worth while to reinvestigate it *ab initio*.

The stability of an interstellar gas cloud in a galactic magnetic field has been examined in a recent paper by Mestel and Spitzer [3] using a form of the virial theorem extended by Chandrasekhar and Fermi [2] to include forces of magnetic origin. The expression derived by Chandrasekhar and Fermi applies, however, only to a body carrying its own magnetic field, and it is therefore of interest to extend their results to the case where a magnetic field of external origin is present.

32-2

2. THE MAGNETIC ENERGY OF A CURRENT
SYSTEM IN AN EXTERNAL FIELD

The rate of working of an electromagnetic field on a system of electric currents and charges is $\mathbf{E}.\mathbf{j}$ per unit volume, where $\mathbf{E}$ is the electric intensity and $\mathbf{j}$ is the electric current density. Gaussian units are used, and unit magnetic permeability and dielectric constants assumed. This expression is derived by considering a charge q moving with velocity $\mathbf{v}$; the Lorentz force on the charge is $q\mathbf{E} + q\mathbf{v} \times \mathbf{H}/c$, where c is the velocity of light. The rate of working on the charge is thus $q\mathbf{E}.\mathbf{v}$, and the above result follows on summing over all charges in a unit volume. The rate of working dW/dt on the whole body is therefore given by

$$dW/dt = \iiint_D \mathbf{E}.\mathbf{j}\, dv, \tag{2}$$

where D is any region containing the body. D, however, must not contain any part of the charge and current system responsible for the external field. This is because, when the body is deformed, the disturbances produced in the field may cause the external system to extract or supply electrostatic and magnetic energy, the amount depending on the mechanism maintaining the system.

It will be convenient to assume that the external field is produced by a fixed charge and current system, although the type of external system does not influence the final result. Now

$$\mathbf{H} = \mathbf{H}^{ext} + \mathbf{H}_1, \tag{3}$$

where $\mathbf{H}^{ext}$ is the magnetic field produced by the external current system, and $\mathbf{H}_1$ is the field arising from the current system in the body. In the region D curl $\mathbf{H}^{ext} = 0$ since D contains only the currents in the body. Further, $\mathbf{H}^{ext}$ is constant in time, hence Maxwell's equations for this region can be written

$$\text{curl } \mathbf{H}_1 = \frac{4\pi\mathbf{j}}{c} + \frac{1}{c}\frac{\partial \mathbf{E}}{\partial t}, \quad \text{curl } \mathbf{E} = -\frac{1}{c}\frac{\partial \mathbf{H}_1}{\partial t}. \tag{4}$$

The right-hand side of (2) can now be transformed as in Poynting's theorem, and dW/dt thereby expressed in the form

$$\frac{dW}{dt} = -\frac{1}{8\pi} \iiint_D \frac{\partial}{\partial t} (\mathbf{E}^2 + \mathbf{H}_1^2)\, dv - \frac{c}{4\pi} \iint_S (\mathbf{E} \times \mathbf{H}_1).d\mathbf{S}, \tag{5}$$

where S is the surface which bounds D.

500

The circumstances of a body in a uniform external field are realized when the typical distance L of variation of the external field is large compared with the linear dimensions of the body. The dimensions of D can then be made much larger than those of the body while remaining small compared with L. If the term 'all space' is used in a restricted sense to denote such a region D, then (5) can be written

$$\frac{dW}{dt} = -\frac{1}{8\pi} \iiint_{\text{all space}} \frac{\partial}{\partial t} (\mathbf{E}^2 + \mathbf{H}_1^2) \, dv. \tag{6}$$

The surface integral in (5) has vanished because $\mathbf{H}_1$ tends to zero at least as fast as r^{-3} at large distances r from the body and $\mathbf{E}$ is bounded. The change $\Delta \mathfrak{M}_1$ in the magnetic energy of the body in a virtual deformation is, by definition, minus the magnetic part of the work done on the body by the field. From (6) it therefore follows that

$$\Delta \mathfrak{M}_1 = \frac{1}{8\pi} \iiint_{\text{all space}} \Delta \mathbf{H}_1^2 \, dv \tag{7}$$

in contrast to the expression in (1). In using the expression $\Delta \mathfrak{M}_1$ in an energy equation it must be remembered that this only represents the total change in field energy provided that there is no charge density in the body, and provided that the deformation is made sufficiently slowly to avoid electromagnetic radiation.

It is of interest to examine $\Delta \mathfrak{M} - \Delta \mathfrak{M}_1$. Since $\Delta \mathbf{H}^{\text{ext}} = 0$ by definition, then

$$\Delta \mathbf{H}^2 - \Delta \mathbf{H}_1^2 = 2 \mathbf{H}^{\text{ext}} . \Delta \mathbf{H}_1. \tag{8}$$

Thus

$$\Delta \mathfrak{M} - \Delta \mathfrak{M}_1 = \frac{1}{4\pi} \mathbf{H}^{\text{ext}} . \Delta \iiint_{\text{all space}} \mathbf{H}_1 \, dv. \tag{9}$$

But

$$\iiint_{\text{all space}} \mathbf{H}_1 \, dv = \tfrac{8}{3}\pi \mathbf{M}, \tag{10}$$

where $\mathbf{M}$ is the magnetic dipole moment of the body. To demonstrate this consider the identity

$$\iiint_D \mathbf{H}_1 \, dv = -\iint_S \mathbf{A}_1 \times d\mathbf{S}, \tag{11}$$

where $\mathbf{A}_1$ is a vector potential for $\mathbf{H}_1$, D is a region containing the body, and S is the surface bounding D. Then

$$\mathbf{A}_1 = \mathbf{M} \times \mathbf{r}/r^3 + \mathrm{o}\left(\frac{1}{r^3}\right) \tag{12}$$

for large values of r where $\mathbf{r}$ is the radius vector from an origin in the neighbourhood of the body. Eq. (10) follows on substituting this expression into the identity and allowing the region D to expand to include 'all space'. Hence

$$\Delta\mathfrak{M}-\Delta\mathfrak{M}_1=\tfrac{2}{3}\mathbf{H}^{\text{ext}}.\Delta\mathbf{M}. \tag{13}$$

The expression $\Delta\mathfrak{M}$ is therefore incorrect whenever the magnetic dipole moment of the body changes.

3. THE VIRIAL THEOREM

The equations of motion of an inviscid fluid in a magnetic field and in its own gravitational field are

$$\rho\frac{\partial\mathbf{v}}{\partial t}+\rho(\mathbf{v}.\operatorname{grad})\,\mathbf{v}=-\operatorname{grad}\,p+\rho\operatorname{grad}\,V+\frac{1}{4\pi}\operatorname{curl}\mathbf{H}\times\mathbf{H}, \tag{14}$$

where p is the hydrostatic pressure, ρ is the density, $\mathbf{v}$ is the fluid velocity and V is the gravitational potential. Electrostatic and electromagnetic forces are ignored in this application.

The virial theorem is derived by taking the scalar product of both sides of (14) with $\mathbf{r}$ the position vector and integrating over all space. The term 'all space' here is used in the restricted sense described in the previous section. After making the usual transformations of the integrals involving the mechanical terms the virial theorem is given by

$$\frac{1}{2}\frac{d^2I}{dt^2}=2T+3(\gamma-1)\,U+\Omega+\frac{1}{4\pi}\iiint_{\text{all space}}\mathbf{r}.(\operatorname{curl}\mathbf{H}\times\mathbf{H})\,dv, \tag{15}$$

where $I=\iiint_{\text{all space}}\rho r^2\,dv$, T is the kinetic energy of mass motion, U is the internal energy and Ω is the gravitational energy of the body. The ratio of specific heats is taken as uniform. By substituting $\mathbf{H}=\mathbf{H}_1+\mathbf{H}^{\text{ext}}$ and noting that $\mathbf{H}^{\text{ext}}$ is constant the magnetic term in (15) can be written

$$\mathscr{I}=\frac{1}{4\pi}\iiint_{\text{all space}}\mathbf{r}.(\operatorname{curl}\mathbf{H}_1\times\mathbf{H}_1)\,dv+\frac{1}{4\pi}\iiint_{\text{all space}}\mathbf{r}.(\operatorname{curl}\mathbf{H}_1\times\mathbf{H}^{\text{ext}})\,dv. \tag{16}$$

Since $\mathbf{H}_1$ is solenoidal the first integrand in (16) may be transformed as follows:

$$\mathbf{r}.(\operatorname{curl}\mathbf{H}_1\times\mathbf{H}_1)=\tfrac{1}{2}\mathbf{H}_1^2+\operatorname{div}\,(\mathbf{r}.\mathbf{H}_1\mathbf{H}_1-\tfrac{1}{2}\mathbf{r}\mathbf{H}_1^2). \tag{17}$$

The second integral in (16) can be expressed in the form

$$\mathbf{H}^{\text{ext}}.\left[\frac{1}{4\pi}\iiint_{\text{all space}}\mathbf{r}\times\operatorname{curl}\mathbf{H}_1\,dv\right].$$

502

By a well-known formula (Stratton[4])

$$\frac{1}{4\pi}\iiint_{\text{all space}} \mathbf{r} \times \text{curl } \mathbf{H}_1 \, dv = 2\mathbf{M}, \qquad (18)$$

hence, on substituting the expressions given by (17) and (18) into (16), and using Gauss's theorem,

$$\mathscr{I} = \frac{1}{8\pi}\iiint_{\text{all space}} \mathbf{H}_1^2 \, dv + \frac{1}{4\pi}\underset{R=\infty}{\mathscr{L}t}\iint_{\Sigma_R} (\mathbf{r}.\mathbf{H}_1\mathbf{H}_1 - \tfrac{1}{2}\mathbf{H}_1^2\mathbf{r}).\,d\mathbf{S} + 2\mathbf{M}.\mathbf{H}^{\text{ext}},$$

$$(19)$$

where Σ_R is a sphere of radius R centred at the origin. At large distances from the origin $\mathbf{H}_1 = o(r^{-3})$. The second integral on the right-hand side of (19) therefore vanishes.

The generalized virial theorem is therefore

$$\frac{1}{2}\frac{d^2}{dt^2} T = 2T + 3(\gamma - 1)\,U + \Omega + \frac{1}{8\pi}\iiint_{\text{all space}} \mathbf{H}_1^2 \, dv + 2\mathbf{M}.\mathbf{H}^{\text{ext}}. \qquad (20)$$

The first member of the magnetic part agrees with the expression derived by Chandrasekhar[2] in the absence of an external magnetic field. The second term gives the contribution when an external magnetic field is present.

REFERENCES

[1] Gjellestad, G. *Astrophys. J.* **120**, 172, 1954.
[2] Chandrasekhar, S. and Fermi, E. *Astrophys. J.* **118**, 116, 1953.
[3] Mestel, L. and Spitzer, L. *Mon. Not. R. Astr. Soc.* **116**, 503, 1956.
[4] Stratton, J. A. *Electromagnetic Theory* (McGraw Hill, 1941), p. 235.

THEORY OF ISOTROPIC MAGNETIC
TURBULENCE IN GASES

S. A. KAPLAN

Astronomical Observatory, University of L'vov, U.S.S.R.

ABSTRACT

A system of spectral equations of magnetic turbulence in gases differing from that given by Chandrasekhar is suggested. The solution of this system is examined. Correlation and structure functions of the turbulence of interstellar gases, determined according to the data on radial velocities of interstellar clouds from Adams's catalogue, are given. For motions of a scale less than the fundamental one (l less than 80 pc) the spectral function $F(k)$ is about $k^{-1.71}$, which agrees with the theoretical conclusions.

I

1. The theory of isotropic turbulence of gases in the magnetic field (gasomagnetic turbulence) can be developed at present only by means of spectral methods. The correlative method does not permit, in general, to take into account the dissipation of energy in the shock-waves, arising as a result of 'supersonic' turbulence.

The theory of the isotropic turbulence in gases can be applied only in the absence both of the mean directed flow of gases and of the mean directed magnetic field, i.e. only in the case, when all directions of the velocity vectors and of the magnetic field are equally probable. Such is the case, for instance, when an originally weak magnetic field has been increased as a result of an 'entanglement' of the magnetic lines of force caused by turbulent movements of ionized gases. The properties of motions in the interstellar space and in the nebulae can evidently be explained in the same mannner.

2. The author offered in 1953–4 the following system of spectral equations [1], [2] of magnetic turbulence in gases:

$$\epsilon_k = 2\left(\nu + \kappa_f \int_k^\infty \sqrt{\frac{F(k)}{k^3}} \, dk\right) \cdot \int_0^k F(k) \, k^2 \, dk$$

$$+ 2 \int_0^k \sqrt{[F(k) \, k^3]} \, [\zeta_f(k) \, F(k) + \mu(k) \, G(k)] \, dk, \qquad (1)$$

504

$$\epsilon_m = 2\left(\lambda + \kappa_g \int_k^\infty \sqrt{\frac{F(k)}{k^3}}\, dk\right) \int_0^k G(k)\, k^2\, dk$$

$$-2 \int_0^k \sqrt{[F(k)\, k^3]}\, [\zeta_g(k) + \mu(k)]\, G(k)\, dk, \qquad (2)$$

where ν is the viscosity; $\lambda = c^2/4\pi\sigma$ ($\sigma = $ conductivity); $k = 2\pi/r$ (r is the characteristic scale of motion); $F(k)$ is the spectral density of kinetic energy, $G(k)$ is the spectral density of magnetic energy (referred to the unity of mass); κ_f and κ_g are some dimensionless quantities of the order of unity, $\zeta_f(k)$, $\zeta_g(k)$ and $\mu(k)$ are dimensionless, slowly varying functions;

$$0 \lesssim \zeta_g(k) \lesssim \zeta_f(k) \lesssim 1, \quad -1 \lesssim \mu(k) \lesssim +1.$$

The first two members of the right part of Eq. (1) have the same meaning as in Heisenberg's theory[3] of turbulence of incompressible fluids. The first two members of the right part of Eq. (2) are analogous. In detail, the member with κ_g describes the transmission of magnetic energy from big vortices to the lesser ones, occurring simultaneously with the decay of kinetic energy (member with κ_f) of big vortices. The member

$$2 \int_0^k \mu(k)\, \sqrt{[F(k)\, k^3]}\, G(k)\, dk,$$

positive in (1) and negative in (2), describes an increase or a decrease of magnetic energy as a result of 'entanglements' or respectively 'disentanglements' of the lines of force. The function $\mu(k)$ depends, actually, upon the relation F/G, being positive at $F \gtrsim G$ and negative at $F \gtrsim G$ and $\mu = 0$ at $F \approx G$.

The member with $\zeta_f(k)$ describes the dissipation of the kinetic energy in shock-waves and the member with $\zeta_g(k)$ describes the corresponding increase of the magnetic energy. Functions ζ_f and ζ_g depend upon the relation of gas velocities to the velocity of sound and tend to acquire constant and positive values with the increase of these relations.

Finally, the members ϵ_k and ϵ_m respectively describe the total dissipations of the kinetic and magnetic energies in the vortices with wave numbers in the intervals between 0 and κ. In steady states $\epsilon_k = $ constant and $\epsilon_m = $ constant. In the case of a decay of turbulence:

$$\epsilon_k = -\frac{\partial}{\partial t} \int_0^k F(k, t)\, dk, \quad \epsilon_m = -\frac{\partial}{\partial t} \int_0^k G(k, t)\, dk. \qquad (3)$$

It is necessary to note that, though the system of spectral equations (1) and (2) is postulated arbitrarily, the choice of members with magnetic energy is substantially limited by the linear character of these equations in respect to $G(k)$ inasmuch as the corresponding equations of magnetic gas dynamics are also linear. The choice of the two members in (2), describing the changes of magnetic energy (with κ_g and μ) is due to two members from the corresponding correlative equations (11). There are several other physical considerations in favour of this choice of systems (1) and (2). The values κ_f and κ_g, as well as the functions $\zeta_f(k)$ and $\zeta_g(k)$ should be known, the function $\mu(k)$ is determined from the conditions of compatibility of these equations.

3. Systems (1) and (2) have two solutions for the spectral region of small wave numbers (analogous to the spectral region by Kolmogoroff):

(A)
$$F(k) = G(k) = F_0(k_0/k)^{[5/3+(32/27)(\zeta_f-\zeta_g)/(\kappa_f+\kappa_g)+\dots]},$$
$$\mu = -(\kappa_g\zeta_f + \kappa_f\zeta_g)/(\kappa_g + \kappa_f),$$
(4)

F_0 and k_0 are arbitrary constants for the case of steady state. This solution corresponds to the case, when the magnetic and kinetic energies are in equilibrium.

(B)
$$F(k) = F_0(k_0/k)^{[5/3+(32/27)\zeta_f/\kappa_f+\dots]},$$
$$G(k) = \frac{2}{3}\frac{3\kappa_f\lambda - 4\kappa_g\nu}{\kappa_g^2}\sqrt{(F_0 k_0)}\left(\frac{k_0}{k}\right)^{[1/3+(16/27)\zeta_f/\kappa_f+\dots]},$$
$$\mu = \tfrac{5}{3}\kappa_g - \frac{5}{12}\frac{\kappa_g}{\kappa_f}\zeta_f - \zeta_g - \dots$$
(5)

The second solution corresponds to the case when the magnetic energy is mainly concentrated in vortices of the inner scale of turbulence (i.e. the case investigated by Batchelor in 1950[4]). When analyzing the solution (B) we suggested $\kappa_g \approx \kappa_f \lambda/\nu$, if $\lambda < \nu$ and $\kappa_g \approx \kappa_f$, if $\lambda \gtrsim \nu$. In this case $\epsilon_k/\epsilon_m \approx \sqrt{(F_0/k_0)}/\nu \approx Re$. Solution (B) may be conventionally called a quasi-stationary one.

If the dissipation of energy in shock waves can be disregarded (hydromagnetic turbulence) we shall have in equations (1), (2), (4) and (5) $\zeta_f = \zeta_g = 0$.

The author investigated also the structure of spectra at different values of wave numbers. Unsteady magnetic turbulence in gases was also studied.

4. It was often supposed that the equilibrium of kinetic and magnetic energies (solution (A)) cannot occur in the presence of magnetic turbulence in gases, because the magnetic field of a large scale suppresses gas motions of lesser scales. This makes the movements more regular, which contradicts the statistical character of turbulence. Such regulation of the movements does not occur in the case of solution (B). We may suppose that solution (A) can also be realized, but in this case vortices of different scales

should be more isolated than in the case of absence of a magnetic field (or solution (B)). Here the space fluctuation in the density of magnetic energy and consequently of the kinetic energy must be far greater.

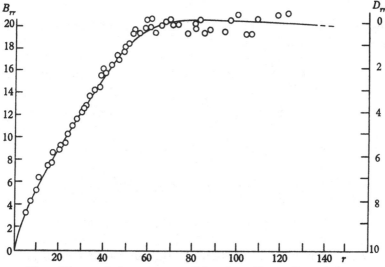

Fig. 1. The correlative and structural functions of interstellar turbulence.

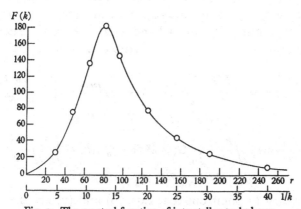

Fig. 2. The spectral function of interstellar turbulence.

5. The theory of magnetic turbulence in gases describes satisfactorily the properties of chaotic motions in interstellar gases and nebulae. Fig. 1 shows the correlation (B_{rr}) and structural (D_{rr}) functions of the turbulence of interstellar gases, found by the author [5] on the basis of Adams's catalogue of radial velocities of interstellar clouds. Fig. 2 shows spectral function

507

$F(k)$ calculated according to the data of Fig. 1. In the region of motions of lesser scale than the principal one $(r < 90 \text{ pc})$ we have $F(k) \sim k^{-1.71}$, which is in good agreement with (4) and (5).

2

In 1955 Chandrasekhar proposed another system of spectral equations of hydromagnetic turbulence[6]

$$\frac{1}{2} \frac{\partial F(k)}{\partial t} = \kappa \sqrt{\frac{F(k)}{k^3}} \int_0^k F(k) \, k^2 \, dk + \kappa \sqrt{\frac{G(k)}{k^3}} \int_0^k G(k) \, k^2 \, dk$$

$$- \kappa F(k) \, k^2 \int_k^\infty \sqrt{\frac{F(k)}{k^3}} \, dk - \kappa G(k) \, k^2 \int_k^\infty \sqrt{\frac{F(k)}{k^3}} \, dk$$

$$- \kappa F(k) \, k^2 \int_k^\infty \sqrt{\frac{G(k)}{k^3}} \, dk - \nu F(k) \, k^2, \qquad (6)$$

$$\frac{1}{2} \frac{\partial G(k)}{\partial t} = \kappa \sqrt{\frac{F(k)}{k^3}} \int_0^k G(k) \, k^2 \, dk + \kappa \sqrt{\frac{G(k)}{k^3}} \int_0^k F(k) \, k^2 \, dk$$

$$- \kappa G(k) \, k^2 \int_k^\infty \sqrt{\frac{G(k)}{k^3}} \, dk - \lambda G(k) \, k^2. \qquad (7)$$

The symbols used in these equations are the same as defined in § 1, κ is the numerical constant equal to all members.

In our theory is also taken into account the dissipation of energy in shock-waves. But here, as we are interested in making a comparison of the system of Eqs. (1) and (2) with Chandrasekhar's we have excluded these members. Our system in this case, written in a differential form is as follows:

$$\frac{1}{2} \frac{\partial F(k)}{\partial t} = \kappa_f \sqrt{\frac{F(k)}{k^3}} \int_0^k F(k) \, k^2 \, dk - \kappa_f F(k) \, k^2 \int_k^\infty \sqrt{\frac{F(k)}{k^3}} \, dk$$

$$- \mu(k) \, G(k) \, \sqrt{[F(k) \, k^3]} - \nu F(k) \, k^2, \qquad (8)$$

$$\frac{1}{2} \frac{\partial G(k)}{\partial t} = \kappa_g \sqrt{\frac{F(k)}{k^3}} \int_0^k G(k) \, k^2 \, dk - \kappa_g G(k) \, k^2 \int_k^\infty \sqrt{\frac{F(k)}{k^3}} \, dk$$

$$+ \mu(k) \, G(k) \, \sqrt{[F(k) \, k^3]} - \lambda G(k) \, k^2. \qquad (9)$$

It is necessary to point out that Eqs. (6) and (7) are the reduction to spectral language of the system of correlative equations of isotropic hydromagnetic turbulence, found by Chandrasekhar[6]:

$$\left. \begin{aligned} \frac{\partial}{\partial r} \left(\frac{\partial^2}{\partial t^2} - \nu^2 D_5^2 \right) Q &= -2Q \frac{\partial}{\partial r} D_5 Q - 2H \frac{\partial}{\partial r} D_5 H, \\ \left(\frac{\partial^2}{\partial t^2} - \lambda^2 D_5^2 \right) H &= -2QD_5 H - 2HD_5 Q - 2 \frac{\partial Q}{\partial r} \frac{\partial H}{\partial r}, \end{aligned} \right\} \qquad (10)$$

508

where $Q(r, t)$ and $H(r, t)$ are correlative scalars, determining the correlation of second-order tensors with the components of velocity and components of strength of the magnetic field in two points of the fluid M' and M'', respectively, so that $r = |r'' - r'|$ and $t = t'' - t'$. D_5 is a differential operator.

The system (8) and (9) is a reduction to spectral language of another system of correlative equations, also found by Chandrasekhar[7].

$$\frac{\partial Q}{\partial t} - 2\nu D_5 Q = 2\left(r\frac{\partial}{\partial r} + 5\right)(T - S),$$

$$\frac{\partial H}{\partial t} - 2\lambda D_5 H = 2P. \qquad \left.\right\} \qquad (11)$$

Here the correlative scalars were taken for the same moments, i.e. $t'' = t'$ and T, S, P are correlative scalars, determining the correlative third-order tensors. System (11) is derived from equations of magnetic hydrodynamics, supposing the turbulence to be of homogeneous and isotropic nature, while for the derivation of system (10) Chandrasekhar used the hypothesis by Millionschchikov[8] about the relation of the fourth correlative moments to the second correlative moment. The correctness of this hypothesis in the complicated case of hydromagnetic turbulence was not clear. Moreover, it is necessary to show that this supposition (being not very correct even in the case of hydrodynamic turbulence) does not lead to wrong results. An introduction of this hypothesis may be justified, if the derivation of system (10) is the final aim, because system (10) is total and may be solved, contrary to Eq. (11). However, the use of this hypothesis was not suitable as the heuristic ground of the spectral theory, in the case when some other physical hypotheses must be made. System (11) is better for this purpose, because at its derivation no arbitrary mathematical hypothesis has been made.

System (10) did not suit as a heuristic ground for the derivation of a spectral theory also for the reason shown below. It is evident that the left part of correlative equations was directly reduced to the spectral theory. For the reduction to the spectral theory of the right part of this equation, i.e. namely of the non-linear term, the hypotheses must be introduced. Comparing the left part of Eqs. (8) and (9) with (11) we see that Eqs. (8), (9)–(11) pass directly from one to the other. In (8) and (9) and (11) the members of the left part are also the first derivative of the energy with time. But the left part of Eq. (10) is not the first derivatives of energy with time. Therefore the right part of Eq. (10) cannot be directly related to the right part of Eqs. (6) and (7).

From these considerations one may say that system (11) is much more suitable than system (10), as the heuristic ground for the postulation of the spectral theory.

But Chandrasekhar's system of spectral equations proceeds, in the main, from system (10). For instance, the term $\sqrt{[G(k)/k^3]}$ describes 'turbulent resistance', in the same way as the term $\sqrt{[F(k)/k^3]}$ describes 'turbulent viscosity'. This is explained only by the symmetry of the right part of (10) in respect to Q and H. The same explanation was given by him in respect to the equality of the values of κ in Eqs. (6) and (7). However, the appearance of the term $\sqrt{[G(k)/k^3]}$ is difficult to interpret from a physical point of view. Indeed, there is dissipation of the magnetic energy, the same as of the kinetic energy (i.e. a transfer from big scale motions to small scale motions is taking place). But it must be kept in view, that the transfer of magnetic energy between the vortices of different scales is not determined by the magnetic field (because the Maxwell equations are linear), but are connected with the motions of fluids, transferring the magnetic energy according to the principle of 'frozen' lines of magnetic forces. Thus, the 'turbulent conductivity' does not depend upon the parameters $\sqrt{[G(k)/k^3]}$ but depends upon the parameters $\sqrt{[F(k)/k^3]}$. There was an analogy with molecular viscosity and conductivity, which depend also upon the velocity and the length of the free path and does not depend upon the magnetic field (in isotropic conductor). It is also clear that the 'turbulent viscosity' $\kappa_f\sqrt{[F(k)/k^3]}$ and the 'turbulent resistance' $\kappa_g\sqrt{[F(k)/k^3]}$ could not be quite equal and we choose therefore different values for the constants κ_f and κ_g.

We want particularly to point out that, owing to the term $\sqrt{[G(k)/k^3]}$ Eqs. (6) and (7) of Chandrasekhar are not linear in respect to the magnetic energy, while the fundamental hydromagnetic equation and Maxwell's equation, as well as the correlative equations, are linear in respect to magnetic energy. The non-linearity of (10) was made artificially by introducing tensors of higher orders. The linearity of the systems of spectral equations in respect to $G(k)$ was one of the chief requirements followed in the derivation of Eqs. (8) and (9).

The member $\mu(k)\, G(k)\, \sqrt{[F(k)\, k^3]}$ in Eqs. (8) and (9) describes the transfer of kinetic energy into the magnetic energy, owing to 'entanglements' or 'disentanglements' of magnetic lines of force. Indeed, it is known that the density of magnetic energy increases owing to an 'entanglement' of the magnetic lines of force, proportionally to $\left(\dfrac{\partial v}{\partial x}\right)_H \cdot \dfrac{H^2}{8\pi}$, where $\left(\dfrac{\partial v}{\partial x}\right)_H$ is the gradient of velocity in the direction of the vectors of the

510

magnetic field. In spectral terms this expression was written in the form given above. The presence of dimensionless functions $\mu(k)$ is explained by the necessity to describe the direction of the transfer of energy. This function may be defined by the conditions of compatibility of Eqs. (8) and (9). We may also suppose that an exchange of kinetic and magnetic energies in a definite scale of motions does not depend on the motion in other scales. Therefore, the member $\mu(k) \, G(k) \, \sqrt{[F(k) \, k^3]}$ was written in an integral form. It is possible, however, that this simple form would be insufficient in the future, we shall then easily write it in an integral form. This is not necessary at present. There are no members describing this process clearly in Chandrasekhar's system.

As a summary of all that has been said above we arrive at the conclusion that the system of spectral equations of hydromagnetic turbulence proposed by Chandrasekhar does not reflect the physical process taking place in this case. This system does not satisfy the requirements of linearity in respect to the magnetic energy. According to our opinion, the system (8) and (9) is better suited to the physical picture of phenomena of hydromagnetic turbulence.

We point out in conclusion that the solution of Eqs. (6) and (7) found by Chandrasekhar is also difficult to explain from the physical point of view. Indeed, Chandrasekhar [6] found two solutions in both of which there is the equipartition of the kinetic and magnetic energies in vortices of the biggest scales $(k \rightarrow 0)$, i.e. in such scales of motions, in which systems (6) and (7) are not quite correct. The relation $G(k)/F(k)$ tends to zero in the first Chandrasekhar's solution and to 2·6 in the second solution with increasing k. In the second case the magnetic energy is always greater, than the kinetic one.

Meanwhile, we can think that on the whole the inequality $G(k) \leqslant F(k)$ must be fulfilled, because in the reverse case the magnetic field suppresses the motion of fluids. Furthermore, as the external energy is transferred into turbulence in the shape of big vortices we can think that it must be $G(k)/F(k) \rightarrow 0$ since $k \rightarrow 0$. We can suppose, at last, that among the possible solutions of spectral equations, there must be an equipartitional solution for sufficiently large intervals of wave numbers. Solutions of systems (6) and (7) do not satisfy any of these requirements.

The solutions of our system of spectral equations, given in paragraph 1, satisfy these requirements, because they are more probable as compared with Chandrasekhar's solution.

REFERENCES

[1] Kaplan, S. A. *Circ. Lvov astron. obs.* no. 25, 1953.
[2] Kaplan, S. A. *C.R. Acad. Sci. U.R.S.S.* **94**, 33, 1954; *J. Exp. Theor. Phys.* **27**, 699, 1954.
[3] Heisenberg, W. *Z. Phys.* **124**, 628, 1948.
[4] Batchelor, G. K. *Proc. Roy. Soc.* A, **201**, 405, 1950.
[5] Kaplan, S. A. *A.J. U.S.S.R.* **32**, 255, 1955.
[6] Chandrasekhar, S. *Proc. Roy. Soc.* A, **233**, 322 and 390, 1955.
[7] Chandrasekhar, S. *Proc. Roy. Soc.* A, **204**, 435, 1951.
[8] Millionschchikov, M. D. *C.R. Acad. Sci. U.R.S.S.* **32**, 611 and 615, 1941.

AN APPROXIMATIVE CALCULATION OF ELECTRIC CONDUCTIVITY IN THE LOWER LAYERS OF THE SOLAR ATMOSPHERE

M. KOPECKÝ

Astronomical Institute of the Czechoslovak Academy, Ondřejov, Czechoslovakia

ABSTRACT

Recently Nagasawa determined a method of calculation of the electric conductivity in a lowly ionized gas, which leads to more precise results than the formula given by Alfvén. The numerical calculation, however, is much more complicated for the Nagasawa method. This paper simplifies the calculation giving a relation, by means of which the Alfvén values of Nagasawa are obtained in a rather simple way.

For an approximative calculation of the electric conductivity in a lowly ionized gas we may use the relation [1]

$$\sigma = \frac{e_0^2}{m_e^{1/2}} \cdot \frac{1}{S(3kT)^{1/2}} \cdot \frac{n_e}{n_n},$$ (1)

where e_0, and m_e denote the electric charge and the mass of an electron, n_e the number of free electrons, n_n the number of neutral atoms, k the Boltzmann constant, T the temperature, and S the effective cross-section of the particles in question.

A much more precise method of calculation of the electric conductivity was recently developed by Nagasawa [2]. His method leads to a system of equations, the approximative solution of which gives the following relation for σ_0

$$\sigma_0 = \tfrac{3}{16} e_0^2 \frac{n_e(B_{13} + B_{23})}{n_e B_{12}(B_{13} + B_{23}) + n_n B_{13} B_{23}},$$ (2)

where

$$B_{12} = \frac{\sqrt{\pi}}{2} \left(\frac{m_e m_i}{m_e + m_i} \right)^{1/2} A e_0^4 (2kT)^{-3/2},$$ (3)

$$B_{13} = \frac{\sqrt{\pi}}{2} \left(\frac{m_e m_n}{m_e + m_n} \right)^{1/2} S(2kT)^{1/2}$$ (4)

$$B_{23} = \frac{\sqrt{\pi}}{2} \left(\frac{m_i m_n}{m_i + m_n} \right)^{1/2} S(2kT)^{1/2}, \qquad (5)$$

$$A = \ln \left(\frac{4kT}{e_0^2 n_e^{1/3}} \right)^2. \qquad (6)$$

In these equations m_i and m_n denote the mass of ions and neutral atoms. The given solution assumes that atoms are only singly ionized.

The electric conductivity σ_0 determined from (2) is closer to the real conditions than the value of σ, given by Eq. (1). The numerical calculation of σ_0, however, is much more complicated than the procedure connected with the calculation of σ. To simplify this calculation, we will show that there exists a simple relation between σ_0 and σ, which may be used for the numerical calculation of the electric conductivity in some cases, especially in the lower layers of the solar atmosphere.

For these layers we may put

$$\frac{m_e m_i}{m_e + m_i} = m_e, \qquad (7)$$

$$\frac{m_e \cdot m_n}{m_e + m_n} = m_e, \qquad (8)$$

$$\frac{m_i m_n}{m_i + m_n} = 1 \cdot 5 m_H, \qquad (9)$$

where m_H denotes the mass of a hydrogen atom. Then the relation (2) can be written in the form

$$\sigma_0 = \frac{3}{8 \sqrt{\pi}} \cdot \frac{e_0^2 n_e}{m_e^{1/2} S n_n (2kT)^{1/2}} \cdot \frac{m_e^{1/2} + 1 \cdot 22 m_H^{1/2}}{\frac{n_e}{n_n} \cdot \frac{A}{S} \, e_0^4 (2kT)^{-2} (m_e^{1/2} + 1 \cdot 22 m_H^{1/2}) + 1 \cdot 22 m_H^{1/2}}. \qquad (10)$$

The effective cross-section, S, may be assumed according to Alfvén[1] equal to $10^{-15}\,\mathrm{cm^2}$. The last measurements show, however, that the hydrogen effective cross-section should be taken lower, about $10^{-14}\,\mathrm{cm^2}$. For these values of the cross-section and in the case of low ionization ($n_e/n_n < 10^{-2}$) we may put in the first approximation

$$\frac{m_e^{1/2} + 1 \cdot 22 m_H^{1/2}}{\frac{n_e}{n_n} \cdot \frac{A}{S} \, e_0^4 (2kT)^{-2} (m_e^{1/2} + 1 \cdot 22 m_H^{1/2}) + 1 \cdot 22 m_H^{1/2}}. \qquad (11)$$

Then we get, with regard to the relation (1), that σ_0 is proportional to σ. Therefore, as far as the assumed conditions are fulfilled, we may determine more precise values of the electric conductivity σ_0 by means of the values of σ, the calculation of which is substantially easier. We denote the electric conductivity deduced in this way by $\bar{\sigma}$. We get

$$\bar{\sigma} = 0\cdot26\sigma. \tag{12}$$

An error which appears by using the approximation (11) is the same or less than the difference between various models of the solar photosphere. This fact is well demonstrated in Fig. 1, which contains the following

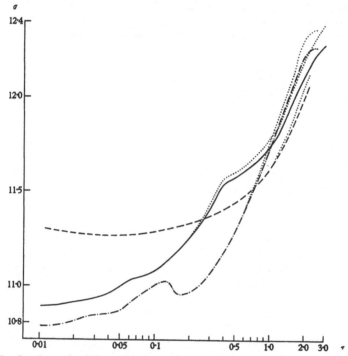

Fig. 1. The electric conductivity σ_0 calculated according to Eq. (2) for the photospheric models of Minnaert (dashed curve), de Jager (full curve), Berdičevskaja (dash and dot curve) and the deviation from these results calculated from Eq. (12) (dotted curve).

curves: the course of the electric conductivity σ_0, calculated according to (2) for the photospheric model of Minnaert[3] (dashed curve), of de Jager[4] (full curve) and Berdičevskaja[5] (dash and dot curve); and the course of the electric conductivity, calculated according to (12), as far as it differs for the various models from the course of σ_0 (dotted curve). All throughout these calculations we used the less favourable value of the effective cross-section S for the approximation (11), i.e. $S = 10^{-15}\,\mathrm{cm^{-2}}$.

33·2

Fig. 1 demonstrates that the calculation of the electric conductivity by means of the relation (12) is fully satisfactory.

A more detailed paper on this theme has been published in [6].

REFERENCES

[1] Alfvén, H. *Cosmical Electrodynamics* (Oxford University Press, 1950).
[2] Nagasawa, S. *Publ. Astr. Soc. Japan*, **7**, 9, 1955.
[3] *The Sun*, ed. by Kuiper, G. P. (Chicago, 1953).
[4] de Jager, C. *Recherches astr. de l'Obs. d. 'Utrecht*, **13**, no. 1, 1952.
[5] *Teoretičeskaja Astrofisika*, ed. by Ambarzumjan, B. A. (Moskva, 1952).
[6] Kopecký, M. *Bulletin of the Astronomical Institutes of Czechoslovakia*, **8**, 71, 1957.

ON THE NATURE OF THE EMISSION
FROM THE GALAXY NGC 4486

I. S. SHKLOVSKY

Sternberg Astronomical Institute, Moscow, U.S.S.R.

As is known, a striking peculiarity of the radio galaxy NGC 4486 is the presence of a small and very bright 'jet' in its central part. As it seems to us, the key to the understanding of the nature of the radio galaxy NGC 4486 is the purely continuous spectrum of the above jet, where not even a slightest trace or emission or absorption lines is present.

In the case of the Crab nebula we also meet intense continuous optical emission of some specific nature accompanied by powerful radio emission. We gave some time ago a new interpretation of the continuous optical emission from the Crab nebula. This emission is caused by the same mechanism of non-thermal character, which causes radio emission, namely by relativistic electrons in magnetic fields.

There are reasons to suggest that the nature of the continuous optical emission of the 'jet' of NGC 4486 is similar. From our calculations[1] it follows, that the intensity of the magnetic field in the region of the 'jet' is of the order of 10^{-4} gauss, namely approximately the same as in the 'amorphous' part of the Crab nebula. It follows as a result that the energy of the relativistic electrons, responsible for the optical emission of the 'jet' must be of the order of 10^{11}–10^{12} eV, and their concentration about 5×10^{-9} cm^{-3}. A much larger amount of electrons with energies equalling 10^9–10^{10} eV should be present in this case. The region of the 'jet' is, consequently, a powerful generator of relativistic particles. Unusual conditions prevailing there, are apparently favourable for the acceleration of particles. Obviously, the acceleration is caused by a Fermi statistical mechanism.

Relativistic electrons formed in the central part of NGC 4486 will diffuse into the surrounding space and in the course of 1–2 millions of years (the time during which the 'jet' is getting formed) they are filling a considerable part of the volume of NGC 4486. Wandering through faint interstellar magnetic fields of that galaxy, the relativistic electrons will radiate in the range of radio waves, which is the cause of an anomalously high radio emission capacity of NGC 4486.

From the quantitative theory developed by us it follows that the mean concentration of relativistic electrons with energies E greater than 5×10^{-8} eV is in NGC 4486 of the order of 5×10^{-8} cm^{-3}, and their differential energy spectrum $dn(E) = K/E^3$. The total energy of the relativistic electrons is about 5×10^{56} ergs over the whole radio galaxy NGC 4486. The number of particularly energetic 'luminous' relativistic electrons with E about 10^{11}–10^{12} eV, located in the region of the 'jet', is millions of times less than the number of less energetic relativistic electrons filling a considerable part of the volume of NGC 4486 and causing its radio emission.

Supposing that the optical emission of the 'jet' continues to exist for 10^6 years with the observed intensity, we shall find that during this interval of time the radiated relativistic electrons loose about 2×10^{55} ergs of their energy. This new estimation of the process, which is the initial cause of the anomalous phenomena going on in NGC 4486, is one and a half order lower than the above estimate (total energy of relativistic electrons in NGC 4486). It may be considered as the lower boundary of the energy emitted for the formation of the 'jet'.

Let us point out that the radio emission of NGC 4486 for 10^6 years, integrated over the spectrum, will be about 10^{54} ergs, i.e. considerably less than the amount of energy lost by the electrons of the 'jet' in the course of the same interval of time.

This is explained by the fact that the relativistic electrons in the interstellar medium of NGC 4486 are loosing their energy extremely slowly. They will cause radio emission for at least 5×10^8 years. The 'jet' will get dispersed in the course of that time and its visible traces will disappear.

An extremely interesting and important conclusion follows from this suggestion: there may be observed radio galaxies with absolute and respective radio emission similar to NGC 4486, but without any peculiarities in their optical rays!

It may be considered that all relativistic electrons that fill the volume of NGC 4486 were found in result of *single* outbursts. It is most natural, however, to suggest that the 'jets' in NGC 4486 are a recurrent phenomenon and that relativistic electrons that have filled NGC 4486 were formed as a result of 10–20 outbursts.

What is the nature of 'a jet' (or of 'jets')? Two hypotheses may be suggested.

(a) The 'jet' has originated as the result of a certain enormous explosion in the central part of NGC 4486. However, it should be accepted in this case that the energy emitted during such an explosion is extraordinarily enormous—it is hundreds of millions times greater than during an outburst of a super-nova. Similar phenomena are unknown in modern physics or astrophysics. Difficulties that arise when such a hypothesis is subsequently developed are extremely numerous.

(*b*) The anomalous conditions in the central part of NGC 4486 are caused by collisions of massive aggregates (of the type of large globular clusters) containing interstellar gas. These aggregates, extremely numerous in the spheroidal galaxy NGC 4486, must have velocities of motion about 500–800 km/sec and 'frontal' collisions between them may, similarly as in the case of the colliding galaxy Cygnus A, be the cause of a generation of considerable amounts of relativistic particles.

REFERENCE

[1] Shklovsky, I. S. *A.J. U.S.S.R.* **33**, no. 3, 1955.

OPTICAL EMISSION FROM THE CRAB NEBULA IN THE CONTINUOUS SPECTRUM*

I. S. SHKLOVSKY

Sternberg Astronomical Institute, Moscow, U.S.S.R.

Extensive progress of radio astronomical methods of investigation during recent years has totally changed our ideas about the nature of the enevelopes ejected during the outburst of super-novae and, consequently, also about the phenomenon of the outburst of the super-nova[1]. The only envelope of a super-nova, investigated in sufficient details by means of optical astronomy, is the Crab nebula (see W. Baade[2], Minkowski[3], Greenstein and Minkowski[4], Barbier[5]. All available explanations concerning the nature of the Crab nebula are based upon these investigations. They were also applied in general to envelopes of other super-novae. Considering the Crab nebula as a typical super-nova remnant let us shortly discuss these statements.

According to[2] the Crab nebula consists of two mutually penetrating parts: a system of comparatively thin filaments, located on the periphery of a nebula, expanding with the velocity of 1000–1300 km/sec and an amorphous mass filling the inner part of the nebula. The expanding 'network' of the filaments gives the line emission, the amorphous mass a strictly continuous emission spectrum. According to [3] the radiation from the Crab nebula in the emission lines constitutes only several per cent of the total

---

* The new interpretation of the emission from the Crab nebula generally accepted at present was published by me in 1953 (see [1]).

During Professor Oort's visit of the U.S.S.R. in the summer of 1954 in connexion with the opening of the restored Pulkovo Observatory, I informed him about this new interpretation and excited his most keen interest to this problem. Professor Oort together with Dr Walraven developed and extended our investigations (J. Oort and T. W. Walraven, *B.A.N.* 12, no. 462, 285, 1956). In particular, the extremely interesting observational data obtained by the U.S.A. investigators was used by Professor Oort.

The new important results on the polarization of the Crab nebula obtained by the Dutch, American and Soviet astronomers were not taken into account in the present paper, since it was prepared earlier (in the spring of 1955—to be presented at the Dublin Assembly of the I.A.U., where it could not be published for some technical reasons).

Some new important results confirming our interpretation of the nature of the Crab nebula were obtained recently in the U.S.S.R.

luminosity of the amorphous mass in the continuous spectrum. From colour indices given in [3] (interstellar absorption taken into account) it follows that the intensity of the radiation in the continuous optical spectrum per unit interval of frequencies, decreases with the growth of frequency approximately as $\sim \nu^{-1}$.

All statements concerning the physical conditions in the Crab nebula were until recently based upon the interpretations of its continuous spectrum, Baade [2] and Minkowski [3] do not admit the possibility of some mechanism of radiation in the continuous spectrum, except the free-free and free-bound transitions in the strongly ionized gas matter of the nebula. This radiation is, according to [3] excited by the exclusively hot star, a former super-nova. Such a mechanism of radiation is natural at first sight. No processes of scattering could, naturally, explain the observed radiation of the Crab nebula, the magnitude of the nebula equalling 9^m, and the magnitude of the two stars in the centre of the nebula about 16^m. The radiation of that nebula might only be its proper radiation.

Other mechanisms in a diffuse medium creating a continuous spectrum were unknown.

Later on in order to explain the faint continuous spectrum of the planetary nebulae A. J. Kipper introduced successfully the mechanism of 'splitting of L alpha quanta' [6]. If, however, this mechanism should be responsible for the emission from the amorphous part of the Crab nebula intense H-lines should be produced, which is not confirmed by observations. The distribution of energy in the continuous spectrum would be different from what is observed as well. The modification of Kipper's mechanism— 'splitting' of the quanta of the helium resonance line—cannot explain the continuous spectrum of the Crab nebula owing to the same cause.

Supposing, however, that the continuous spectrum of the Crab nebula is caused by free-free transitions we meet with extreme difficulties and contradictions. Many of these difficulties were known earlier, but no attempts were made to analyse them critically, because there were no doubts that the Baade-Minkowski's mechanism of continuous emission may be erroneous.

Let us shortly discuss these difficulties.

An inevitable consequence of the accepted mechanisms of emission is the conclusion that (a) the kinetic temperature of the Crab nebula is extremely high—of the order of hundreds of thousands degrees, or even higher (because the discontinuity of the Balmer series and He^+ is observed), (b) the concentration of electrons in the Crab nebula is of the order of $10^3 cm^{-3}$, (c) the mass of the nebula is about $20 M_\odot$.

If one considers that the radiation emitted by the nebula is caused by the central hot star—a former super-nova—the temperature of the surface of that star must be excessively high, higher than 500,000°, its radius being extremely small, less than $0 \cdot 02 R_\odot$ [3]. Professor Oort, retaining the mechanism of emission from the amorphous mass of the Crab nebula, believes in consequence of the evidently fantastic characteristics of the hypothetical central star, that the hot amorphous mass is not excited at all [7]. According to Oort's opinion, the hot nebula has retained the temperature of the inner part of the almost totally exploded central star. He suggests that the process of cooling of such an extremely hot extended mass of gas is going on sufficiently slowly. It is, however, difficult to admit the hypothesis that during an outburst of a super-nova it is getting destroyed and scattered. Spectral and photometric observations of the outburst of super-novae in other galaxies and energetic considerations contradict this hypothesis.

Quite recently Ramsey advanced a hypothesis that the high temperature of the amorphous mass of the Crab nebula may be maintained by the processes of radio-active decay of some non-stable isotopes, formed in the process of the explosion of a super-nova [8]. However, as it may be shown, this hypothesis is beneath criticism.

Thus, the suggestion that the amorphous mass represents a totally exploded star and that no external agent (like the ultra-violet emission of the central star, for example) is required to maintain the extremely high kinetic temperature, is deprived of any serious reasons. There are still less reasons to believe that the central star—a former super-nova—possesses the same characteristics, which result from Baade and Minkowski's interpretation of the continuous spectrum of the Crab nebula.

The morphological peculiarities of the Crab nebula seem altogether incomprehensible if such interpretation is admitted. The kinetic temperature of the filaments is rather low—about 10,000°. Their density cannot, therefore, be very high. The most intense lines in the spectrum of the filaments are $\lambda 3727$ (O II) and $\lambda 6548$–6584 (N II). The transition probability is extremely small for such lines. They are, therefore, getting intensified in the case of nebulae with low densities. The concentration of the particles in the filaments hardly exceeds 300–400 cm^{-3}. The fact of a long-lasting co-existence of comparatively cold and sufficiently diffuse filaments and an extremely hot diffuse mass with a not lesser density seems improbable. How are these filaments moving through the amorphous mass for 900 years?

The low state of excitation and ionization is, further, also quite incomprehensible. The rapid and 'energetic' electrons must inevitably enter into the filaments from the amorphous mass and cause ionization of atoms. It is also unclear, why the powerful ultra-violet emission of the central star (or of the diffuse hot mass) does not cause strong ionization in the filaments.

Let us point out that we observe co-existence of an extremely hot coronal matter and comparatively cold protuberances. However, in this case the picture is altogether different: the density of the 'cold' protuberanes is 2–3 orders higher, than that of the 'hot' corona, while in the Crab nebula the density of the filaments and of the amorphous mass is similar.

Finally, it is unclear why the nebulae—remnants of super-novae outbursts of 1572 and 1604—are so weak as compared with the Crab nebula. If the outburst of a super-nova signifies a complete destruction and scattering of the star, why do we not observe, in the places where super-novae have flared up, bright nebulae with continuous spectrum, remnants of the outbursts of such super-novae?

We underline that all these difficulties are the consequence of the interpretation of the continuous spectrum of the Crab nebula according to Baade and Minkowski.

In so far as these difficulties are insurpassable, according to our opinion, some other explanation of the continuous spectrum of the Crab nebula should be searched for.

New and important facts, which may throw light upon the nature of the Crab nebulae were revealed after the earlier studies by Baade and Minkowski had been published. The discovery of the radio emission may be given as an example. The spectrum of that emission is much more slow than for other sources. In the enormous spectral interval from $\lambda = 750$ cm to $\lambda = 9\cdot4$ cm, embracing about seven octaves, the flux F_ν of radio emission decreases for only $2\cdot5$ times. The Soviet radio astronomers discovered recently the radio emission of the Crab nebula on the $3\cdot2$ cm wave. But the value of the flux in this wave-length is somewhat less than that on the $9\cdot4$ cm waves [9]. It may, thus, be stated, that in the interval of eight octaves the flux of radio emission from the Crab nebula decreases 3–$3\cdot5$ times. The law of the variation of the flux with the growth of frequency in the range of decimetre and centimetre waves may be written as $F_\nu \sim \nu^{-0\cdot2}$.

It is evident that the radio emission of the Crab nebula cannot stop abruptly at $\lambda_1 = 3\cdot2$ cm, being zero for $\lambda < \lambda_1$. It is beyond doubt that the flux of radio emission exists also for $\lambda \ll \lambda_1$, but the modern radio astronomical technique does not make possible the discovery of such emission.

It is to be questioned quite naturally whether the optical emission of the

Crab nebula with continuous spectrum does not form a continuation of its radio emission. In other words, cannot the radio and the optical emission of that nebula be caused by the same, but undoubtedly non-thermal, mechanism? It was shown in [10] and [4] that the radio emission of the Crab nebula cannot be considered as the prolongation of its optical emission in the continuous spectrum, assuming that the latter is of thermal origin, caused by free-free transitions. The problem that is advanced now is altogether different: it is not the radio emission that should be explained by the optical thermal emission, but vice versa—the optical emission must necessarily be explained by the non-thermal radio emission [1]. Thus, the mechanism of the optical emission of the Crab nebula with continuous spectrum must, according to this conception, be an extraordinary one, altogether different as compared with all thermal mechanisms of emission, which were known in astrophysics.

As it was found in [10] the flux of emission in the continuous optical spectrum per unit interval of frequency is in the case of the Crab nebula a thousand times less than in the range of metre waves. If in the interval of eight octaves of the studied range of radio emission the flux decreases three times, then it seems quite natural that it may become decreased for 300 times more in the interval of fifteen octaves that remain up to the optical range ($\lambda \sim 8000$ Å). The dependence of the intensity of the frequency in this range of spectrum may be approximately represented as $F_\nu \propto \nu^{-0.5}$, it is to be much more 'steep' than in the range of decimetre and centimetre waves. In the range of the optical frequencies the spectrum becomes still more steep, $F_\nu \propto \nu^{-1}$, which is seen from the colour temperature of the optical continuous spectrum of the Crab nebula.

The only acceptable mechanism of radio emission of the Crab nebula may be the 'synchrotron'-emission of the relativistic electrons in magnetic fields [10].

Let us show the main equations describing this process.

The energy emitted by a relativistic electron, moving in a magnetic field will equal:

$$P(\nu, E)\, d\nu = 16 \cdot \frac{e^3 H}{mc^2}\, \bar{\bar{P}}\left(\frac{\nu}{\nu_m}\right) d\nu, \tag{1}$$

where E is the energy of the electron, H the component of the magnetic field, perpendicular to the direction of the velocity. The function $\bar{\bar{P}}(\nu/\nu_m)$ reaches maximum for $\nu/\nu_m = 1$. In this case $\bar{\bar{P}} = 0.1$. Further,

$$\nu_m = \frac{eH}{2\pi mc} \cdot \left(\frac{E}{mc^2}\right)^2. \tag{2}$$

524

Let the differential energetic spectrum of relativistic electrons be

$$N(E)\,dE = \frac{K}{E^\gamma}\,dE.$$

The intensity of emission is

$$I_\nu = \frac{1}{4\pi}\iint P(\nu, E)\,N(E)\,dE\,dR = (2\pi)^{\frac{1}{2}(1-\gamma)}\cdot\frac{e^3 H}{mc^2}\left(\frac{2eH}{m^3 e^5}\right)^{\frac{1}{2}(\gamma-1)}\cdot U(\gamma)\,K\,\nu^{\frac{1}{2}(1-\gamma)}$$

$$= 1\cdot6\times 10^{-21}\cdot(2\cdot8\times 10^8)^{\frac{1}{2}(\gamma-1)}\cdot U(\gamma)\,K\cdot H^{\frac{1}{2}(\gamma-1)}\lambda^{\frac{1}{2}(\gamma-1)}$$

$$\times R \text{ erg.cm}^{-2}.\text{cycles}^{-1}.\text{steradian}^{-1}.$$

where R is the length of the emitting region, $U(\gamma)$ for $\gamma = 1\cdot2$ and 3 equals $0\cdot37, 0\cdot125$ and $0\cdot087$, respectively [11].

The flux of emission $F_\nu = \int I_\nu\,d\Omega = I_\nu.\bar\Omega$.

The solid angle of the Crab nebula is $\bar\Omega = 2\times 10^{-6}$. The length of the nebula is $R \approx 1$ pc $= 3\times 10^{18}$ cm.

It may be expected that in the Crab nebula $H \sim 10^{-3}$ gauss [12]. In the radio interval of the spectrum $F_\nu = 1\cdot8\times 10^{-23}$ watts/m^2 and changes as $\lambda^{0\cdot2}$. Here $\gamma = 1\cdot5$ and according to [3] $K \sim 3\times 10^{-8}$.

The concentration of relativistic electrons in the Crab nebula responsible for its radio emission will then be:

$$N = \int_{E_1}^{E_2} N(E)\,dE = K\int_{E_1}^{E_2}\frac{dE}{E^\gamma}.$$

For the metre waves of radio emission $E_1 \approx 3\times 10^7$, $E_2 = 3\times 10^9$ eV, from which $N \approx 10^{-5}$ cm^{-3}. The total energy of these electrons equals

$$E = V\int_{E_1}^{E_2} N(E)\,E.dE \approx 4\times 10^{47} \text{ ergs},$$

where the volume of the Crab nebula is $V \approx 10^{56}$ cm^3. The energy emitted by the super-nova during its outburst may reach 10^{49}–10^{50} ergs.

We shall assume that the optical emission with continuous spectrum is caused in the main by relativistic electrons [1]. However, if the electrons with energies 10^7–10^9 eV are responsible for radio emission, the optical emission will be caused in the main by the electrons with energies about 5×10^{11}–10^{12} eV (in so far as $\nu_m \propto E^2$; see [2]). Let us, in the same way as above, estimate the concentration of such electrons.

For $\lambda = 5\times 10^{-5}$ cm $(\nu = 6\times 10^{14}$ sec$^{-1})$, $F_\nu = 1\cdot5\times 10^{-23}$ ergs/cm^2/sec cycle/sec [10], the exponent of the energy spectrum of the quick electrons $\gamma = 3$, $U(\gamma) = 0\cdot087$. According to [8], $K = 3\times 10^{-9}$ the concentration of electrons with energies $E > E_0 = 5\times 10^{11}$ eV and

$$K.\int_{E_0}^{\infty}\frac{dE}{E^\gamma} = \frac{K}{2E_0^2} \approx 2\times 10^{-9} \text{ cm}^{-3}.$$

The energy density of the electrons with energies $E > E_0$ will equal

$$K \cdot 1/E_0 \approx 4 \times 10^{-9} \, \text{ergs/cm}^3.$$

It will namely be of the same order as the energy density of softer relativistic electrons, responsible for its radio emission.

Thus, the altogether insignificant amount of relativistic electrons is the cause of a comparatively powerful optical emission. An extremely important conclusion may be made from it: the mass of the 'amorphous' part of the Crab nebula cannot be very great. If in the internal part of the nebula only relativistic particles would be present, its mass should be of the order of $10^{-6} M_\odot$.

It cannot be assumed that extremely interlaced magnetic fields can be present in vacuum. A sufficiently rarefied gas, which does not show itself optically, owing to its rarefication, must be present in the inner part of the nebula. The origin of this gas may, possibly, be the ejection of matter from the super-nova after the maximum. From an analysis of the diffusion velocity of relativistic particles in the Crab nebula and the dimensions of the turbulent elements contained in it it appears that $l = 3 \times 10^{16}$ cm. However, it must be, at least, several times greater than the mean free path. An estimation of the lower limit boundary of the mass of the amorphous part of the Crab nebula may be established from it, which equals $M_1 = 10^{32}$ g. The real value of the mass of the amorphous part of the Crab nebula must be close to M_1. This follows from energetic considerations. The density of the kinetic energy must be close to the density of the magnetic energy $H^2/8\pi$. H cannot exceed appreciably 3×10^{-4} gauss. This means that $H^2/8\pi \leqslant 4 \times 10^{-9}$ erg/cm^3. Consequently, for $V \sim 3 \times 10^7$ cm/sec the density $\rho \lesssim 10^{-24}$ g/cm^3. It follows from it that the mass of the amorphous part of the Crab nebula is of the order of 10^{32} g, i.e. $0 \cdot 05_\odot$.

The mass of the filament system is also hardly exceeding several hundredths of the solar mass. This results from the estimation of the volume occupied by the filaments and from the density of the filaments $\rho < 7 \times 10^{-22}$ g/cm^3.

Thus, the mass of gases ejected during the outburst of the super-novae 1054 does not exceed, apparently, $0 \cdot 1 M_\odot$, it is, namely, one hundred times less than the value assumed formerly.*

* Pikelner has recently explained that the well-known mysterious acceleration of the Crab nebula is caused by the pressure of the magnetic field in the nebula. Independent considerations permitted him to estimate the mean strength of the magnetic field of this nebula: $H \approx 3 \times 10^{-4}$. Hence, owing to the acceleration of the system of filaments in the Crab nebula he determined its mass, established by him to be $0 \cdot 1 M_\odot$, which coincides satisfactorily with our estimates (A.J. U.S.S.R. **33**, no. 6, 1956).

Such a comparatively small value of the mass of envelopes, ejected during the outbursts of super-novae, is of essential importance for the whole problem. It signifies that the outburst of the super-nova does by no means signify a disruption and scattering of stars. The process of an outburst of a super-nova does not differ much from the process of a nova outburst. The difference lies only in the scale of the phenomenon. There are of course qualitative differences between super-novae and novae outbursts, too. For instance, the brightest (according to their absolute magnitude in maximum) novae—are the rapid novae (see[13]), whereas in the light curves of the super-novae an enormous luminosity in maximum co-exist with a rather gradual decrease of light with time.

It may be understood now why no bright nebulae with continuous spectrum similar to that of the amorphous part of the Crab nebula are observed in the places of other galactic super-novae. It is certain that a sufficiently large number of relativistic electrons with $E > 5 \times 10^{11}$ eV is not originating in all outbursts of super-novae. Therefore, only a small number of radio nebulae should have a sufficiently strong optical spectrum. Special conditions are also required in order that relativistic electrons of high energies, originated at a definite stage of the development of a nebula, should not loose a considerable part of their energy during several centuries.

The apparent stellar magnitude of the systems of filaments of the Crab nebula will be about 12^m–13; only 3^m brighter than the magnitude of the nebula remnant of the nova 1604. This super-nova was 4^m fainter in maximum, than the super-nova of 1054.

If the optical emission of the Crab nebula with continuous spectrum is caused by relativistic electrons a polarization of this emission should be expected[14].

We paid attention to the fact that the expected polarization must have a small-cell character[13]. The light polarized in a given direction must arrive from a region, where the magnetic field is almost homogeneous. As we have seen above, the dimensions of such regions, $l \sim 3 \times 10^{16}$ cm, constitute approximately $1/50$ of the dimension of the nebula, or 2–3″. An averaging of the polarization along the line of sight should take place, but statistically one must expect a 'non-compensated' polarization. The polarization can even reach 5–10 %.

The polarization of the Crab nebula, which has been predicted theoretically, was recently observed by Dombrovsky[15]. It was found that the polarization is of a rather regular nature. The main direction of the polarization is oriented along the axis of the Crab nebula. Such a character of the polarization may, possibly, be caused by the superposition of homogeneous interstellar magnetic field, which existed in the region of the space, where the super-novae of 1054 had burst. Thus, a randomly oriented magnetic field in the Crab nebula must have a component (equalling about 10 %) of a regular nature. The presence of such a component will not affect essentially the diffusion velocity of the relativistic particles, but will assist its 'spreading' in this direction. The elongated form of the amorphous mass of the Crab nebula, may, possibly, be explained by it. Let us mention in this connexion that G. A. Shajn paid attention to the existence of preferential direction in IC 443 and the Crab nebula—doubtless remnants of old outbursts of super-novae [16]. He connected this fact with the existence of a general interstellar field in the place where the super-nova had outbursted.

Further detailed study of the polarization of the optical continuous emission from the Crab nebula is needed.*

REFERENCES

[1] Shklovsky, I. S. *C.R. Acad. Sci. U.R.S.S.*, **90**, 983, 1953.
[2] Baade, W. *Astrophys. J.* **96**, 188, 1942.
[3] Minkowski, R. *Astrophys. J.* **96**, 199, 1942.
[4] Greenstein, J. and Minkowski, R. *Astrophys. J.* **118**, 1, 1953.
[5] Barbier, D. *Ann. Astrophys.* **8**, nos. 1–2, 35, 1945.
[6] Kipper, A. J. *Progress of Soviet Science in the Estonian S. S.R. since 1940 to 1950* (Tallin, 1950).
[7] Oort, J. *Problems of Cosmical Aerodynamics* (Dayton, Ohio, 1951).
[8] Cowling, T. G. *Les Processes Nucléaires dans les Astres* (Liège, 1953).
[9] Kardashev, N. A., Kajdanovsky, N. L., Shklovsky, I. S. *C.R. Acad. Sci. U.R.S.S.*, 1955 (in the press).
[10] Shklovsky, I. S. *A.J. U.S.S.R.* **90**, 15, 1953.
[11] Ginsburg, V. L. *Progress of Phys. Sci.* **51**, 343, 1953.
[12] Shklovsky, I. S. *C.R. Acad. Sci. U.R.S.S.*, **91**, 475, 1953.
[13] Kopylov, I. M. *Publ. Crim. Astrophy. Obs.* **10**, 200, 1953.
[14] Shklovsky, I. S. *Publ. of 3rd Conference on Cosmogony, devoted to the origin of cosmic rays* (Moscow, 1954).
[15] Dombrovsky, V. A. *C.R. Acad. Sci. U.R.S.S.*, **94**, 21, 1954.
[16] Shajn, G. A. and Hase, V. F. *C.R. Acad. Sci. U.R.S.S.*, **95**, 713, 1954.

* After this paper had been written (in the spring of 1955) important investigations of the Crab nebula have appeared. These are namely the following: G. A. Shajn, S. B. Pikelner, R. N. Ikhsanov, *A.J. U.S.S.R.* **32**, 395, 1953; E. K. Khatchikian, *C.R. Acad. Sci. Arménie*, **21**, 63, 1955; J. Oort and T. Walraven, *B.A.N.* **12**, no. 462, 285, 1956; W. Baade, *B.A.N.* **21**, no. 462, 312, 1956.

MAGNETIC FIELDS IN RADIO SOURCES

G. R. BURBIDGE

Mount Wilson and Palomar Observatories, Pasadena, California, U.S.A.

ABSTRACT

Recent work has suggested very strongly that most non-thermal radio sources emit by the synchrotron mechanism—the radiation of relativistic electrons and positrons in magnetic fields. In this paper a summary of calculations of the total energy in particles and magnetic field in a number of radio sources has been given. Magnetic fields estimated in this way for the Crab, Cassiopeia A, our Galaxy, M 87, NGC 5128, NGC 1316 and Cygnus A are tabulated.

Apart from the method of detection of stellar magnetic fields by measuring the Zeeman effect on stellar spectrum lines, which has been carried out extensively and exclusively by H. W. Babcock, the only other method of measuring cosmical magnetic fields so far devised has been very indirect. In particular, measurements of the polarization of starlight in our own Galaxy have been made by Hiltner[1], Hall (Hall and Mikesell[2]) and Mrs Smith[3]. Polarization measures in extra-galactic nebulae have been made for NGC 5055 and NGC 7331 by Mrs Elvius[4,5]. If the polarization is attributed to scattering by interstellar grains which have been aligned by an interstellar magnetic field, some idea of the direction of the field and its strength can be obtained from these results if a theory of grain alignment is used. The most plausible theory is that of Davis and Greenstein[6] which suggests that in spiral systems the gross structure of the magnetic field is such that the lines of force lie along the spiral arms, and that the mean field strength is near 10^{-5} gauss.

Two other methods of estimating the strengths of cosmical magnetic fields have now become available. The first of these which we shall briefly mention has recently been suggested by Bolton and Wild[7]. They have proposed that it may be possible to measure the Zeeman splitting of the 21-cm line radiation emitted by neutral hydrogen in the interstellar gas. They estimate that, using present techniques, and a radio telescope with an aperture of 150 ft (several instruments as large as this are under construction) it may be possible to detect magnetic fields as weak as 3×10^{-6} gauss. This method clearly has great potentialities.

The second method which we wish to discuss in the remainder of this paper is that of estimating magnetic field strengths in radio sources. This again is an indirect method, but it does afford some possibility of obtaining information about magnetic fields in very distant extra-galactic nebulae.

Recent work has strongly supported the original suggestion of Alfvén and Herlofson[8] that the mechanism of radio emission, in most strong discrete sources (with the possible exception of the sun), is the synchrotron mechanism in which electrons (and positrons) emit acceleration radiation while spiralling in magnetic fields. The strongest confirmation of this theory has come following the work of Shklovsky[9, 10] who suggested that the high degree of polarization associated with acceleration radiation might be detectable in the Crab Nebula, and in the jet in M 87 (NGC 4486) in the strong optical continua which both of these radio sources emit. The attempts to detect this polarization in the Crab by Vashakidze[11], Dombrovsky[12], Oort and Walraven[13] and Baade[14] and in M 87 by Baade[15] proved entirely successful, thus providing very strong confirmation of the theory.

The theory underlying this type of radiation is well known (Schott[16], Schwinger[17]). If the spectrum of the radiation and the total power emitted have both been measured, it is a fairly straightforward matter to compute the total energy which must be currently present, both in the electron-positron flux and in the magnetic field, as a function of the mean magnetic field strength H. If the frequency spectrum of the radiation is determined sufficiently accurately, a value for the index of the assumed particle energy spectrum $(N(E) \propto E^{-n})$ can be deduced. However, the value of the total particle energy does not depend very sensitively on n. These calculations have been done in detail for the Crab (Oort and Walraven[13],) for M 87 (Burbidge[18]), and for NGC 5128 and NGC 1316 (Burbidge and Burbidge[19]). In all of these cases a series of magnetic field strengths have been assumed and the corresponding total energies have been calculated. To give some idea of the ranges of energies involved we reproduce in Table 1 a portion of Table 4 given in the paper on M 87 (Burbidge[18]) for the radio emission (here a value of n has been deduced from the observed radio frequency spectrum). To obtain the most probable value of the mean magnetic field strength a further postulate has to be made. The most reasonable further condition which may be imposed is to demand that the total energy (particle energy + magnetic energy) is a minimum.

530

Table 1

H (gauss)	E (electron-positron energy) (ergs)	$\mathfrak{M}$ (magnetic energy) (ergs)
10^{-2}	$5 \cdot 6 \times 10^{61}$	$4 \cdot 7 \times 10^{61}$
10^{-3}	$1 \cdot 8 \times 10^{53}$	$4 \cdot 7 \times 10^{59}$
10^{-4}	$5 \cdot 6 \times 10^{54}$	$4 \cdot 7 \times 10^{57}$
10^{-5}	$1 \cdot 8 \times 10^{56}$	$4 \cdot 7 \times 10^{55}$
10^{-6}	$5 \cdot 6 \times 10^{57}$	$4 \cdot 7 \times 10^{53}$

The question now arises as to whether the total particle energy involved is simply the electron-positron energy, or whether a contribution is also to be expected from a proton flux which may be associated with the electrons. This question can only be settled if the mechanism by which the flux of particles has gained its energy is understood. There appear to be three possibilities.

(1) The electrons have been accelerated after being produced at very low energies. In this case a corresponding number of protons will have been accelerated with them, and if any induction-type mechanism of the Fermi type is responsible, the protons will gain kinetic energies M/m times those of the electrons. Since at very low energies (below $\sim$ 100 MeV) the energy losses of the electrons by atomic processes are very large under most astrophysical circumstances, and in most cases those energy losses will overcome the energy gain by any type of Fermi mechanism, however efficient it may be, it appears that this mode of electron production is unlikely.

(2) The electrons and positrons have been produced following nuclear collisions between the quiescent interstellar gas atoms and a flux of high-energy protons. In this case the electrons and positrons are already produced at high energies (10^8–10^9 eV are entirely possible for protons with high enough energy), so that the difficulties inherent in process (1) are avoided. Theoretical work on the radio emission from our own Galaxy (Burbidge [20, 21, 22]) suggests, for example, that the total power emitted may be accounted for by the flux of electrons and positrons produced by the known cosmic ray flux interacting with the interstellar gas.

(3) The electrons and positrons may be produced following the annihilation of protons and anti-protons in the sources. This possibility has been explored elsewhere (Burbidge [18], Burbidge and Hoyle [23]). The advantage obtained by postulating that some anti-matter is present in the sources, is that it provides a very large energy supply with electrons and positrons already having energies of the order of 10^8 eV, and in this case the total energy is just the electron-positron energy without any proton flux.

531

34-2

Table 2. *Estimates of total energies and magnetic field strengths in radio sources*

	Total energy (I)* (magnetic+ particles) (ergs)	Total energy (II)† (magnetic+ particles) (ergs)	$\bar{H}$ (I) (gauss)	$\bar{H}$ (II) (gauss)
Crab	$1 \cdot 5 \times 10^{48}$	6×10^{49}	10^{-3}	10^{-3}
Cassiopeia A	6×10^{47}	$1 \cdot 7 \times 10^{49}$	3×10^{-4}	2×10^{-3}
Galaxy (disk)		$\sim 10^{55}$		10^{-5}
(halo)	$\sim 10^{55}$			$1-2 \times 10^{-6}$
M 87 (NGC 4486) (optical jet)	2×10^{55}	4×10^{56}	10^{-3}	10^{-2}
(radio source)	5×10^{55}	10^{57}	10^{-4}	10^{-3}
NGC 5128 (central region)	10^{55}	10^{57}	10^{-6}	10^{-5}
(halo)	10^{57}	10^{59}	10^{-5}	2×10^{-4}
NGC 1316 (central region)	10^{55}	10^{57}	10^{-6}	10^{-5}
(halo)	10^{57}	10^{59}	10^{-5}	2×10^{-4}
Cygnus A	$10^{58}-10^{59}$	10^{60}	$5 \times 10^{-5}-$ 5×10^{-6}	5×10^{-4}

* Assuming that only electrons and positrons are present.

† Assuming that a primary proton flux produces electrons and positrons in nuclear collisions.

The results which are given in Table 2 have been computed by supposing that either (2) or (3) is operative. For (2) it is found, in general, that the energy in the total proton flux is about 10^2–10^3 times greater than that in the electron-positron flux. Thus the magnetic fields may vary between the two assumptions in some cases by factors ~ 10. Though our final values of H are somewhat uncertain they do show that it is probable that magnetic fields ranging from 10^{-2}–10^{-3} gauss in the Crab Nebula and in M 87 to 10^{-5}–10^{-6} gauss in the halo regions of NGC 5128 and in our own Galaxy, are present. Details and descriptions of most of the sources listed in Table 2 have been given by Baade and Minkowski[24, 25] and Pawsey[26]. The dimensions and hence the volumes of the extra-galactic sources have been estimated by using a value of the Hubble constant = 180 km/sec/megaparsec. Estimates for the Crab have been taken with some modifications from the paper of Oort and Walraven[13]. The others have been taken from work of the author (Burbidge[18, 20, 27] and Burbidge and Burbidge[19]). When more radio astronomical data become available, estimates of fields in a large number of radio sources may be made.

REFERENCES

[1] Hiltner, W. A. *Astrophys. J.* **114**, 241, 1951.
[2] Hall, J. S. and Mikesell, A. H. *Publ. U.S. Nav. Obs.* **17**, Part 1, 1950.
[3] Smith, E. van P. *Astrophys. J.* **124**, 43, 1956.
[4] Elvius, A. *Stockholms Observatoriums Annaler*, **17**, no. 4, 1951.
[5] Elvius, A. *Stockholms Observatoriums Annaler*, **19**, no. 1, 1956.
[6] Davis, L. and Greenstein, J. L. *Astrophys. J.* **114**, 206, 1951.
[7] Bolton, J. G. and Wild, J. P. *Astrophys. J.* **125**, 296, 1956.
[8] Alfvén, H. and Herlofson, N. *Phys. Rev.* **78**, 616, 1950.
[9] Shklovsky, I. S. *Dokl. Akad. Nauk S.S.S.R.* **90**, 983, 1953.
[10] Shklovsky, I. S. *Astro. J., Moscow*, **32**, 215, 1955.
[11] Vashakidze, *Russian Astr. Circ.* no. 147, 1954.
[12] Dombrovsky, *Dokl. Akad. Nauk. S.S.S.R.* **94**, 1021, 1954.
[13] Oort, J. H. and Walraven, T. *B.A.N.* **12**, 285, 1956.
[14] Baade, W. *B.A.N.* **12**, 312, 1956.
[15] Baade, W. *Astrophys. J.* **123**, 550, 1956.
[16] Schott, G. A. *Electromagnetic Radiation* (Cambridge University Press, 1912), p. 109.
[17] Schwinger, J. *Phys. Rev.* **75**, 1912, 1949.
[18] Burbidge, G. R. *Astrophys. J.* **124**, 416, 1956.
[19] Burbidge, G. R. and Burbidge, E. M. *Astrophys. J.* **125**, 1, 1957.
[20] Burbidge, G. R. *Astrophys. J.* **123**, 178, 1956.
[21] Burbidge, G. R. *Phys. Rev.* **101**, 906, 1956.
[22] Burbidge, G. R. *Phys. Rev.* **103**, 264, 1956.
[23] Burbidge, G. R. and Hoyle, F. *Nuovo Cim.* **4**, 558, 1956.
[24] Baade, W. and Minkowski, R. *Astrophys. J.* **119**, 206, 1953.
[25] Baade, W. and Minkowski, R. *Astrophys. J.* **119**, 215, 1953.
[26] Pawsey J. L. *Astrophys. J.* **121**, 1, 1955.
[27] Burbidge, G. R. Unpublished (1956).

SOME EFFECTS WHICH CAN
ACCOMPANY MAGNETIC STORMS

S. B. PIKELNER

Astrophysical Observatory, Crimea, U.S.S.R.

ABSTRACT

Some effects are examined—the probability of penetration of a part of the flow of particles from the sun to the earth's atmosphere, as a consequence of the compression of the plasma—and formation of an electrical current along the motion of the plasma.

Apparently, the main results of observations cannot be explained by these effects. But the latter may probably cause some secondary phenomena which could perhaps be discovered by means of special observations.

Modern theories of magnetic storms and aurorae proceed from the suggestions by S. Chapman and V. Ferraro[1] of a quasi neutral stream of corpuscules from the sun, flowing around the earth at a distance of several radii. These theories explain satisfactorily the first stage of the magnetic storm. As a consequence of these theories it was found by D. Martyn that the mean concentration of the stream is about 20 cm^{-3}, its distance from the earth constituting 5·5 radii. According to D. Martyn the polarization of the stream creates an electric field, which causes an acceleration up to large energies of the particles moving towards the earth.

Self-induction is not taken into account in these considerations. This self-induction reduces strongly the energy of the particles because the time of relaxation of the current through the earth's atmosphere is several years. H. Alfvén's theory[2] introduces also currents flowing from the internal surface of the stream towards the earth. Storm-time variations represent, according to the above theories, the magnetic field of these currents. Observations show, however, that magnetic disturbances consist of short fluctuations, the amplitudes of which usually exceed the mean variations. At the same time currents of large scales are changing more slowly. In addition, it was shown by A. Nickolski that the magnetic field of the earth during quiet hours of stormy days does not differ from the usual undisturbed field.

534

This was the reason for the author to investigate the principal arguments which lead to the conclusion of the impossibility for the stream to reach the earth. This impossibility is the sequence of diamagnetic properties of the plasma. The gradient of the magnetic pressure retards the motion of the plasma in places where the density of the magnetic energy is about equal to the initial density of the kinetic energy of the flow. However, the compressibility of the plasma is not taken into account in this suggestion. In the course of the first phase of the magnetic storm, when the stream compresses the magnetic field of the earth, the state of the stream is non-stable. Individual elements and jets are formed, which penetrate into the field of the earth, as the denser parts of the stream are less retarded by the magnetic field than the surrounding parts.

The field contracts these elements, their density will be increased and they will be able to penetrate further, than in the case of constant density.

We suppose for quantitative calculation the internal pressure of elements to be equal to the external magnetic pressure. The pressure of the earth's atmosphere is not taken into account. The internal magnetic pressure for isothropic compression is proportional to $\rho^{4/3}$, and the gas pressure is proportional to $\rho^{5/3}$, this being an adiabatic compression. When the initial field of the stream is about 10^{-6}–10^{-5} [3] the magnetic pressure inside the elements will be less than the gas pressure and may be omitted. The system of equations is:

$$P = \frac{1}{8\pi} H^2,$$

$$\frac{P}{\rho^{5/3}} = \frac{P_0}{\rho_0^{5/3}} = \frac{kT_0}{n_0^{2/3} m_H^{5/3}},$$

$$\tfrac{1}{2}\rho V^2 = \tfrac{3}{2} P_1 + \frac{1}{8\pi} H_1^2 = \frac{5}{2} \frac{1}{8\pi} H_1^2,$$

where the index 0 or 1 means that the value is taken previous to the compression, or at the moment when the element is stopped.

The solution of the system is the following:

$$H_1 = \frac{\sqrt{(8\pi)}}{5^{5/4}} \frac{n_0^{1/2} m_H^{5/4}}{(kT_0)^{3/4}} V^{5/2} \approx 2,$$

if $n_0 \approx 20$, T_0 about $5000°$, V about 10^8 cm/sec. The stream may penetrate into the upper layers of the atmosphere ($H_1 \approx 0.2$) at 'direct' collisions, if $V \approx 4 \cdot 10^7$ cm/sec.

The concentration and the temperature are determined during compression by the expressions:

$$n = \left(\frac{1}{8\pi} H^2\right)^{3/5} \frac{n_0^{2/5}}{(kT_0)^{3/5}}; \quad T = \left(\frac{n}{n_0}\right)^{2/3} T_0.$$

If $H = 0 \cdot 2$, the concentration $n \approx 10^6$, $T \approx 6 \times 10^6$ degrees.

The linear dimensions are reduced considerably and probably constitute some tens of kilometers.

Hot dense clouds may be expected to penetrate during magnetic storms into the upper layers of the atmosphere. They are optically unobservable, but may perhaps be discovered by means of radio methods.

Only a part of the retardation forces is taken into account by the diamagnetic effect. If the cloud propagates across the lines of force, both the condensation of these lines in front of the cloud and their elongation must be taken into account.

The condensation of the lines of force may be estimated by means of the theory of magnetic-hydrodynamical shock waves by F. Hoffman and E. Teller[4] and by L. Helfer[5]. The degree of compression of the matter is about $1 \cdot 2$ to $1 \cdot 5$ in a comparatively strong magnetic field.

The increase of H is close to this value. The retardation is estimated similarly as before, but a value of H^2 two times greater than before is accepted. The motion across the magnetic field will be retarded, if the transversal component of the velocity will be less than 5×10^7 cm/sec.

The upper limit of the effect of elongation of the lines of force may be obtained on the assumption that a plane layer is moving inside of which the outer field has already penetrated. The time of retardation of this layer may be approximately taken as the time of crossing of the layer by the magnetic-hydrodynamical waves. The computations including the inhomogeneity of the magnetic field of the earth show that the movement normal to the field is retarded most strongly in the vicinity of the earth. Actually the field penetrates into the element comparatively slowly and not very deep, especially on great distances from the earth, where the element itself is of much greater dimensions. The element may, probably, move across the lines of force, when the distances from the earth are considerable but the transverse motion close to the earth is rapidly retarded. A more definite calculation is impossible at present.

Besides the above mechanisms there is another effect, which may influence essentially the motion of an element. If the plasma moves through some other plasma, retardation of the electrons is much more strong than the retardation of protons. This was indicated by W. Bennet and E. Hulburt[6], who suggested that the stream must consist of high-velocity protons and slow electrons. However, such combination of different velocities represents an electrical current, the increase of which from zero being restricted by self-induction. In reality the value of the current must be deduced from Maxwell's equations.

Let the space-charge always be zero (this being not quite correct, but causing no large errors). Then the following equations are valid:

$$\Delta \mathbf{A} - \frac{4\pi\sigma}{c^2}\frac{\partial \mathbf{A}}{\partial t} = -\frac{4\pi}{c}\mathbf{j}^{(o)}; \quad \mathbf{j} = \mathbf{j}^{(o)} - \frac{\sigma}{c}\frac{\partial \mathbf{A}}{\partial t},$$

where $\mathbf{j}^{(o)} = ne\mathbf{V}$. In the co-ordinate system $u = Vt - z$ the value of $j^{(e)}$ does not depend on time.

It may be accepted that $j^{(e)} = j_0 \exp(-r^2/r_0^2)$ for $u > 0$. Boundary conditions are: for $u = 0$, $j = 0$; for $u \to \infty$ j is restricted; for $r \to \infty$, $j \to 0$ and $\partial j/\partial r \to 0$.

Let us pass to the system u and solve the equations by means of successive approximations. Then for $u \ll r_0^2$

$$j = \frac{c^2}{\pi\sigma V}\left(1 - \frac{r^2}{r_0^2}\right)j^{(e)}\frac{u}{r_0^2} \ll j^{(e)};$$

the electrons and ions are moving with about the same velocity. The current increases linearly with the distance from the front of the stream. The radial current and space-charge must, consequently, be present. The magnetic field of the current is insignificant.

The current in the magnetic field of the earth must be influenced by the force

$$\mathbf{f} = \frac{1}{c}\mathbf{j} \times \mathbf{H}.$$

This force differs as compared with Lorentz forces by the scalar coefficient only, as the direction of $\mathbf{j}$ is the same as that of $\mathbf{v}$. The equation of the motion of the element of plasma is identical with the equation of the motion of a charged particle in the field of the earth. This equation was investigated by C. Störmer. The plasma must move according to Störmer's trajectory calculated for particles of a mass

$$M = \frac{\pi\sigma V}{c^2}\frac{r_0^2}{u}m_H.$$

The value of M will be increased, if we take into account that the influence of the field exists only in the layer, into which the field has penetrated. The effect of the force $\mathbf{f}$ being therefore essential only close to the earth where the dimensions of the element is of the order of several tens of kilometers. Therefore we may expect that the above compressed gaseous clouds will enter the atmosphere in places located on a spiral, close to the one by Störmer, but sufficiently far from the poles, since $M \gg m_H$.

This may possibly be connected with the regularity recently discovered by A. P. Nickolski [7], according to which the magnetic disturbances in polar regions commence in places, distributed along Störmer's spiral. These disturbances are the greatest in places where this spiral intersects the auroral zone. It cannot be supposed that the magnetic disturbances may be explained by the moving clouds, as the magnetic field of an element is

$$H = \text{curl}_\alpha \ A = \frac{2c}{\sigma V} \frac{ur}{r_0^2} j_0 e^{-(r/r_0)^2}.$$

This field is not small, but it decreases rapidly with distance.

A shock-wave arises when the moving clouds penetrate into the earth's atmosphere. This wave is becoming rapidly damped as it propagates in a medium of increasing density. The wave may ionize atoms and cause excitation of the upper layers of the atmosphere, which will be accompanied by a faint luminosity (time of recombination from a day to a year). The aurorae cannot be, obviously, explained by the above mechanism.

The giant local pulsations of the earths magnetic field and the reflexion of the radio waves from the heights up to 1500 km observed by Harang [8] may be associated with the penetration of such clouds into the upper layers of the atmosphere.

REFERENCES

[1] Chapman, S. and Ferraro, V. C. A. *Terr. Mag.* **36**, 77 and 171, 1931; **37**, 147 and 421, 1932; **38**, 79, 1933; **45**, 245, 1940.
[2] Alfvén, H. *Cosmical Electrodynamics* (Oxford University Press, 1950).
[3] Dorman, L. I. *Bull. Acad. Sci. U.R.S.S.* **20**, no. 1, 1956.
[4] Hoffman, F. and Teller, E. *Phys. Rev.* **80**, 692, 1950.
[5] Helfer, L. *Astrophys. J.* **117**, 177, 1953.
[6] Bennet, W. and Hulburt, E. O. *Phys. Rev.* **95**, 315, 1954; *J. Atmos. Terr. Phys.* **5**, 211, 1954.
[7] Nickolski, A. P. *Bull. Acad. Sci. U.R.S.S.*, Série géographique, no. 5, 1954.
[8] Harang, L. *Geof. Publ.* **13**, no. 3, 3, 1941.

NAME INDEX

SUBJECT INDEX

Printed in the United States
By Bookmasters